Student Solutions Manual

Laurel Technical Services

CALCULUS

Preliminary Edition

ROBERT DECKER

DALE VARBERG

PRENTICE HALL Upper Saddle River, NJ 07458

Acquisitions Editor: ***George Lobell***
Supplement Acquisitions Editor: ***Audra Walsh***
Production Editor: ***Barbara Kraemer***
Production Supervisor: ***Joan Eurell***
Production Coordinator: ***Alan Fischer***

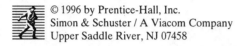

Simon & Schuster / A Viacom Company
Upper Saddle River, NJ 07458

Printed in the United States of America

10 9 8 7 6 5 4 3 2 1

ISBN 0-13-504481-2

Prentice-Hall International (UK) Limited, *London*
Prentice-Hall of Australia Pty. Limited, *Sydney*
Prentice-Hall Canada, Inc., *Toronto*
Prentice-Hall Hispanoamericana, S.A., *Mexico*
Prentice-Hall of India Private Limited, *New Delhi*
Prentice-Hall of Japan, Inc., *Tokyo*
Simon & Schuster Asia Pte. Ltd., *Singapore*
Editora Prentice-Hall do Brasil, Ltda., *Rio de Janeiro*

Table of Contents

Chapter 0

Section 0.1

Concepts Review

1. $y = 8$; $x = -2, 1, 4$

3. Circle; $y = \sqrt{4 - x^2}$, $y = -\sqrt{4 - x^2}$

5. The actual curve may not behave as expected between the plotted points; likewise, it may have characteristic features beyond the range of the plotted points. In either case, the result may miss important features of the curve unless the points are chosen carefully.

Problem Set 0.1

1. $y = -x^2 + 4$;
   y-intercept: $y = 4$;
   Solve $y = (2 - x)(2 + x) = 0$.
   x-intercepts: $x = -2, 2$

x	y
–3	–5
–2	0
–1	3
0	4
1	3
2	0
3	5

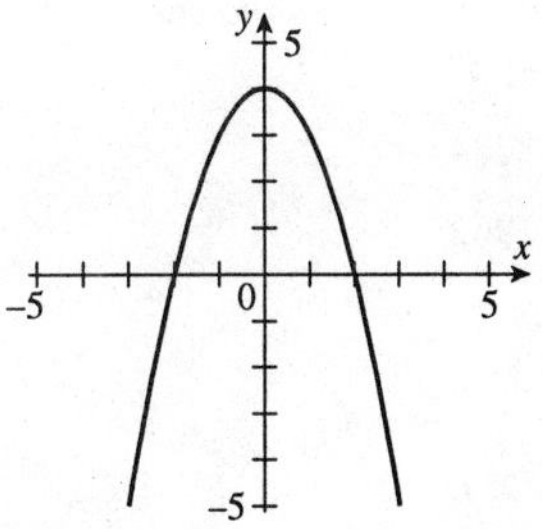

3. $3x^2 + 4y = 0$; $y = -\frac{3}{4}x^2$;
   y-intercept: $y = 0$
   x-intercept: $x = 0$

x	y
–2	–3
–1	$-\frac{3}{4}$
0	0
1	$\frac{3}{4}$
2	3

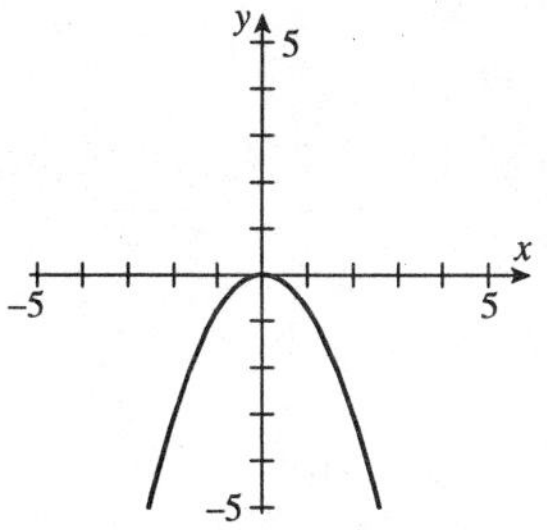

5. $x^2 + y^2 = 36$;
   $y = \pm\sqrt{36 - x^2}$
   y-intercepts: $y = -6, 6$
   x-intercepts: $x = -6, 6$

x	y
–6	0
–4	±4.4721
–2	±5.6569
0	±6
2	±5.6569
4	±4.4721
6	0

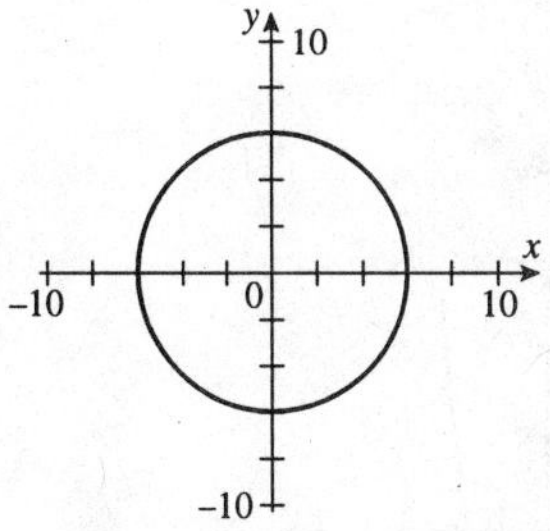

7. $4x^2 + 9y^2 = 36;$

$y = \pm\frac{1}{3}\sqrt{36 - 4x^2}$

Solve $9y^2 = 36$.

y-intercepts: $y = -2, 2$

Solve $4x^2 = 36$.

x-intercepts: $x = -3, 3$

x	y
–3	0
–2	±1.4907
–1	±1.8856
0	±2
1	±1.8856
2	±1.4907
3	0

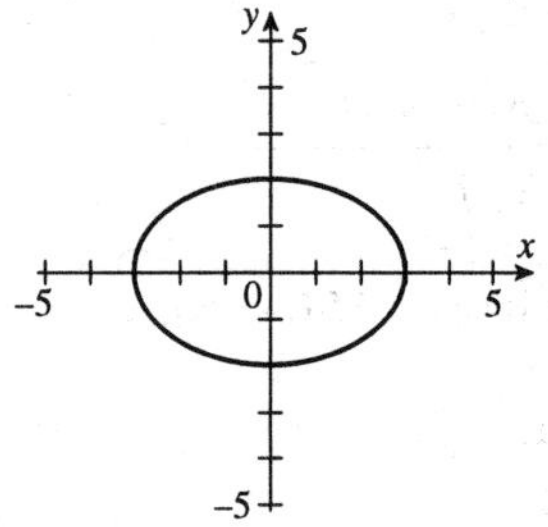

9. $y = x^3 - 3x$

y-intercept: $y = 0$

Solve $y = x(x^2 - 3) = 0$.

x-intercepts: $x = -\sqrt{3},\ 0,\ \sqrt{3}$

x	y
–2	–2
–1	2
0	0
1	–2
2	2

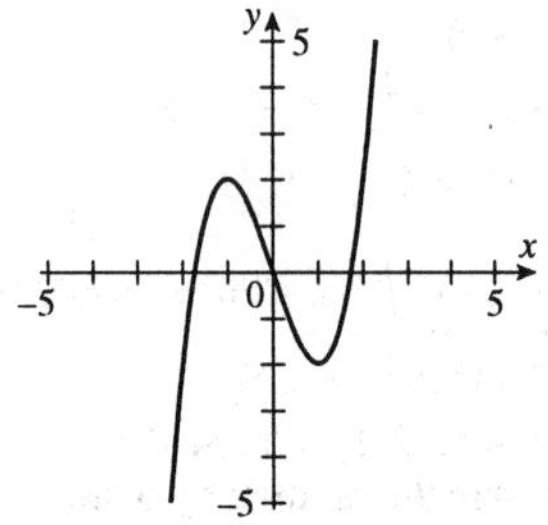

11. $y = \frac{1}{x^2 + 1}$

y-intercept: $y = 1$

No x-intercepts

x	y
–4	0.0588
–2	0.2
–1	0.5
0	1
1	0.5
2	0.2
4	0.0588

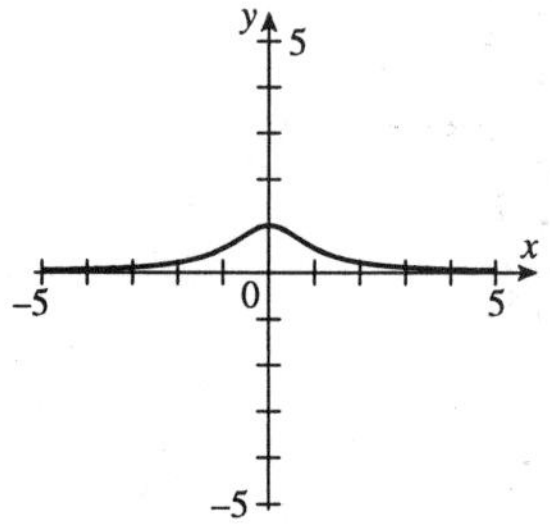

13. $x^3 - y^2 = 0$

$y = \pm\sqrt{x^3}$

y-intercept: $y = 0$

x-intercept: $x = 0$

x	y
0	0
1	±1
2	±2.8284
3	±5.1962

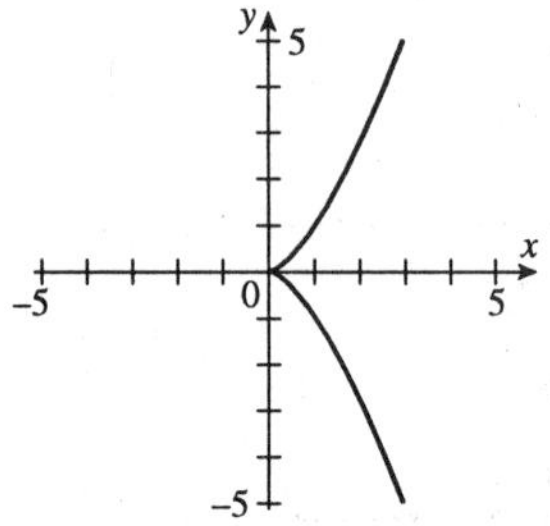

15. $y = (x - 2)(x + 1)(x + 3)$

y-intercept: $y = -6$

x-intercepts: $x = -3, -1, 2$

x	y
–4	–18
–3	0
–2	4
–1	0
0	–6
1	–8
2	0
3	24

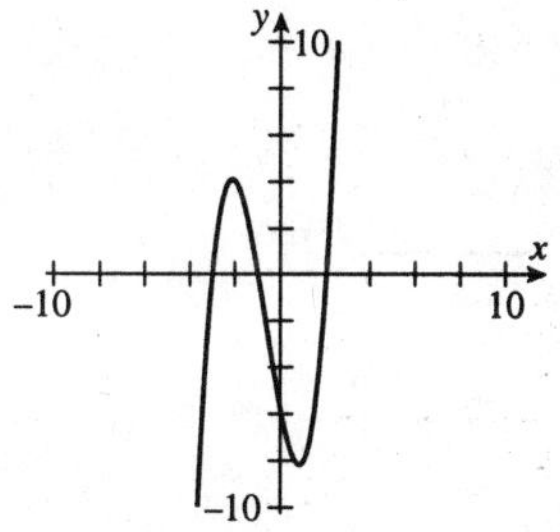

17. $-x+1=x^2+2x+1$

$x^2+3x=0$

$x(x+3)=0$

$x=-3, 0$

Intersection points: (–3, 4), (0, 1)

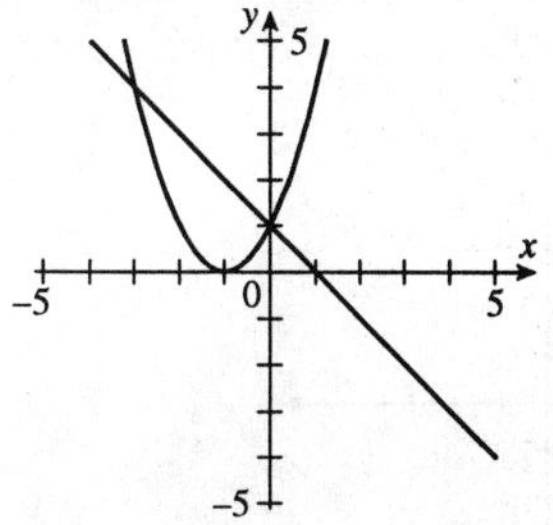

19. $-2x+1=-x^2-x+3$

$x^2-x-2=0$

$(x-2)(x+1)=0$

$x=-1, 2$

Intersection points: (–1, 3), (2, –3)

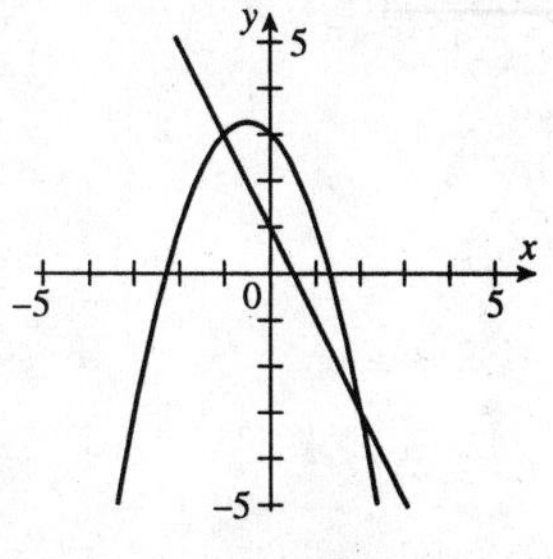

21. $1.5x+3.2=x^2-2.9x$

$x^2-4.4x-3.2=0$

$$x=\frac{4.4\pm\sqrt{(-4.4)^2+4(3.2)}}{2}$$

$x\approx -0.635,\ 5.035$

Intersection points:
(–0.635, 2.247), (5.035, 10.753)

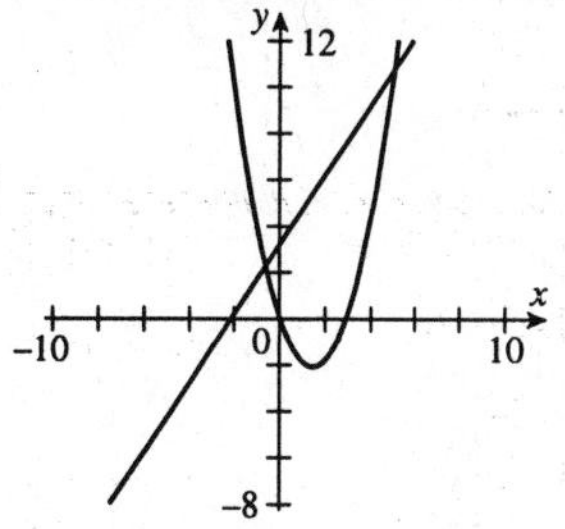

23. $x^2+(4x+3)^2=4$

$x^2+16x^2+24x+9=4$

$17x^2+24x+5=0$

$$x=\frac{-24\pm\sqrt{(24)^2-4(17)(5)}}{2(17)}$$

$x\approx -1.158,\ -0.254$

Intersection points: (–1.158, –1.631), (–0.254, 1.984)

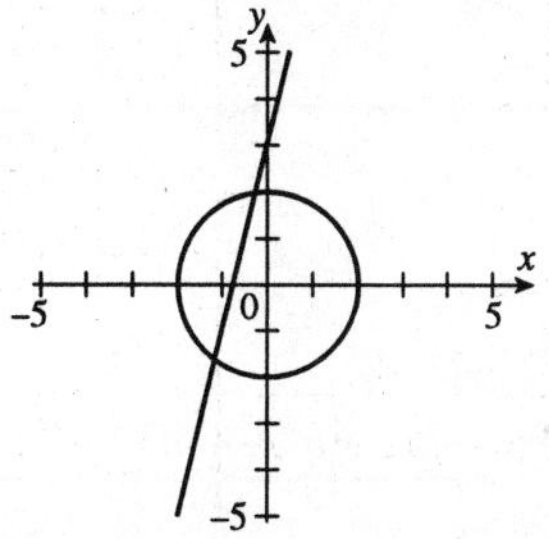

25. Center: $\left(\frac{2+6}{2}, \frac{3-1}{2}\right)=(4, 1)$

Inscribed circle:

Midpoint of one side $=\left(\frac{2+6}{2}, \frac{3+3}{2}\right)=(4, 3)$

$r^2=(4-4)^2+(1-3)^2=4$

$(x-4)^2+(y-1)^2=4$

Circumscribed circle:

$r^2=(4-2)^2+(1-3)^2=4+4=8$

$(x-4)^2+(y-1)^2=8$

27. Cost by truck = 3.71(214 + 179) = \$1458.03

$AC=\sqrt{(214)^2+(179)^2}\approx 279$

Cost by plane ≈ 4.82(279) = \$1344.78

It is cheaper to ship the product by plane.

29. $y = 3x^4 - 2x + 1$

At $x = -1$, $y = 3(-1)^4 - 2(-1) + 1 = 6$.

At $x = 1$, $y = 3(1)^4 - 2(1) + 1 = 2$.

$d = \sqrt{(-1-1)^2 + (6-2)^2} = \sqrt{20} = 2\sqrt{5}$

31.

x	y
0.25	4.5
0.5	2.7071
0.75	2.1994
1	2
2	1.9142
4	2.25
8	2.9534
12	3.5474
16	4.0625

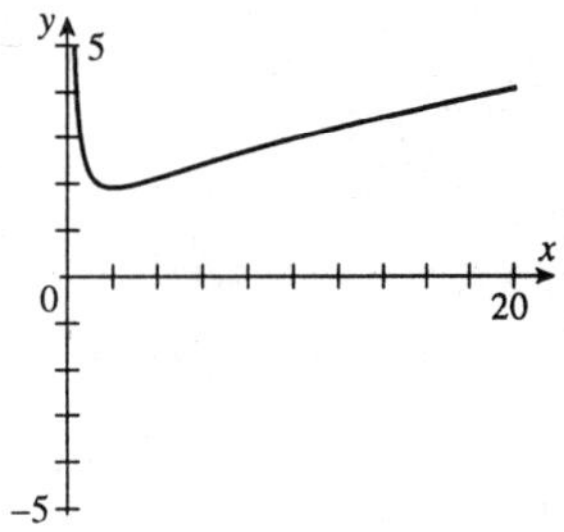

33. $y = ax^2 + bx + c$

$d = b^2 - 4ac$

Setting $y = 0$, $x = \dfrac{-b \pm \sqrt{b^2 - 4ac}}{2a} = \dfrac{-b \pm \sqrt{d}}{2a}$.

When $d > 0$, there are two x-intercepts. When $d = 0$, there is only one x-intercept. When $d < 0$, the graph does not touch the x-axis.

35.

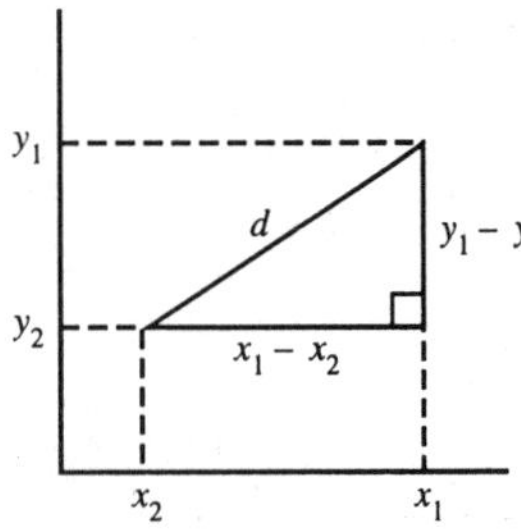

The distance between (x_1, y_1) and (x_2, y_2) is the hypotenuse of the right triangle in the figure. The legs of the triangle have length $x_1 - x_2$ and $y_1 - y_2$, so the distance d is

$d = \sqrt{(x_1 - x_2)^2 + (y_1 - y_2)^2}$.

Section 0.2

Concepts Review

1. Domain; range

3. $f(2) = 12$; $f(a) = 3a^2$

5. A mathematical model is a mathematical description of a real-life situation, sometimes using equations or graphs. It makes it possible to answer questions about the situation being modeled, by manipulating the equations, or by observing the behavior of the graphs.

Problem Set 0.2

1. $f(x) = -4$

x	y
–3	–4
–2	–4
–1	–4
0	–4
1	–4
2	–4
3	–4

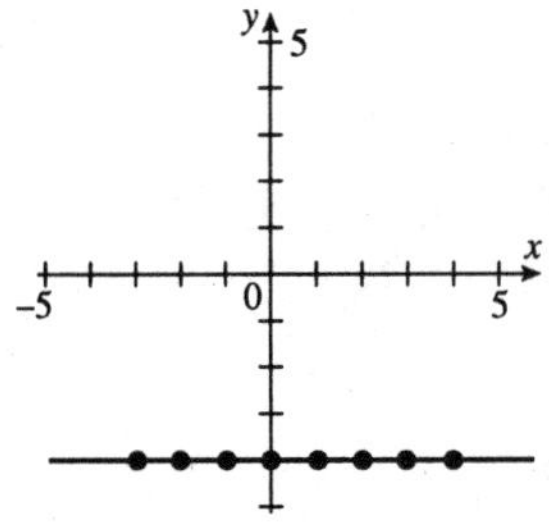

3. $F(x) = 2x + 1$

x	y
–2	–3
–1	–1
0	1
1	3

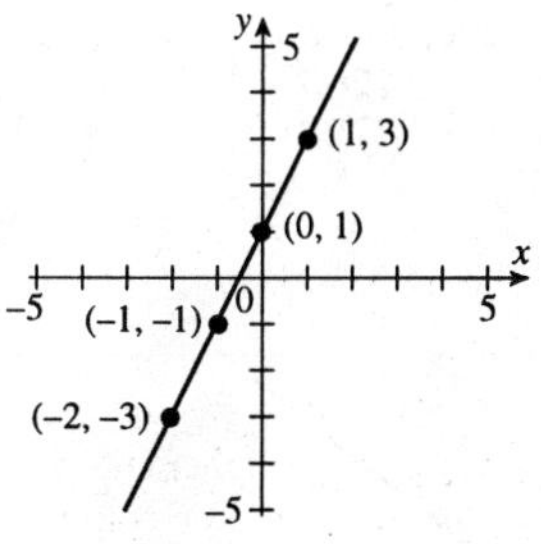

5. $g(x) = 3x^2 + 2x - 1$

x	y
−1	0
0	−1
1	4

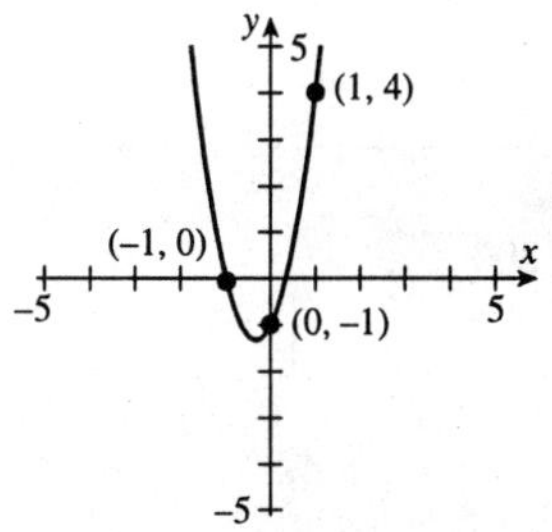

7. $g(x) = \dfrac{x}{x^2 - 1}$

x	y
−3	$-\frac{3}{8}$
−2	$-\frac{2}{3}$
0	0
2	$\frac{2}{3}$
3	$\frac{3}{8}$

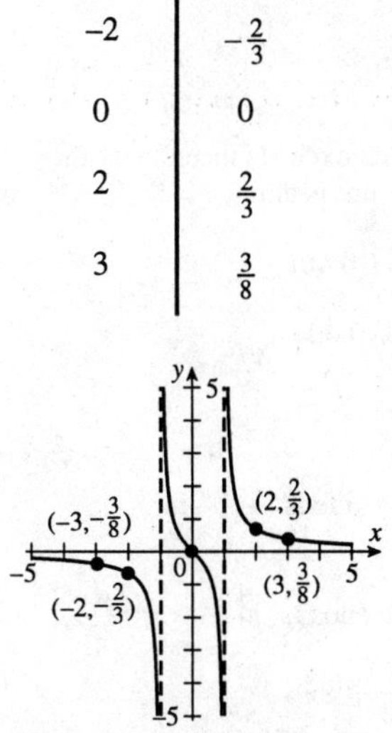

9. $f(w) = \sqrt{w - 1}$

w	y
1	0
2	1
5	2

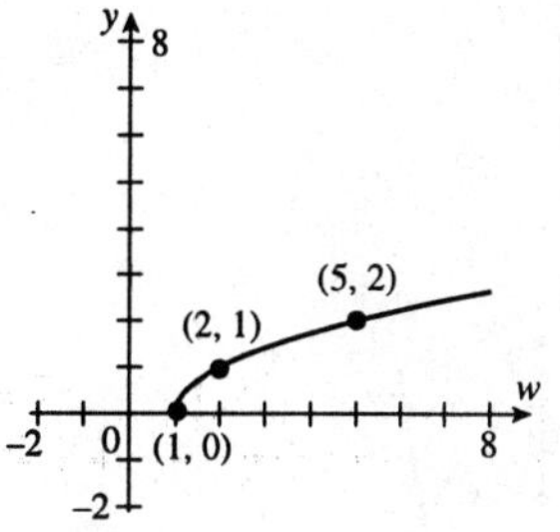

11. $f(x) = |2x|$

x	y
−3	6
−2	4
−1	2
0	0
1	2
2	4
3	6

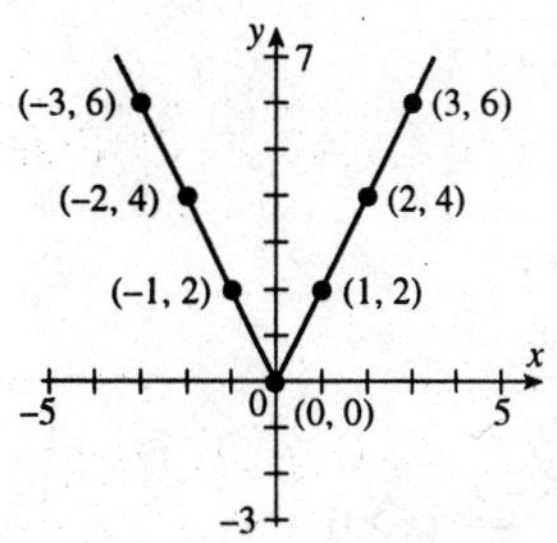

13. $g(t)=\begin{cases}1 & \text{if } t\le 0\\ t+1 & \text{if } 0<t<2\\ t^2-1 & \text{if } t\ge 2\end{cases}$

t	y
–2	1
0	1
1	2
2	3
3	8

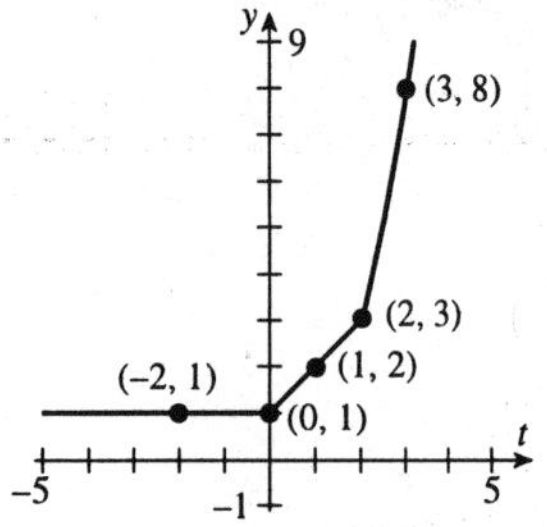

15. a. $x^2+y^2=4$

$y^2=4-x^2$

$y=\pm\sqrt{4-x^2}$

Not a function

b. $xy+y+3x=4$

$y(x+1)=4-3x$

$y=\dfrac{4-3x}{x+1}$

$f(x)=\dfrac{4-3x}{x+1},\ x\ne -1$

c. $x=\sqrt{3y+1}$

$x^2=3y+1$

$y=\dfrac{x^2-1}{3}$

$f(x)=\dfrac{x^2-1}{3},\ x\ge 0$

d. $3x=\dfrac{y}{y+1}$

$3xy+3x-y=0$

$y(1-3x)=3x$

$y=\dfrac{3x}{1-3x}$

$f(x)=\dfrac{3x}{1-3x},\ x\ne\dfrac{1}{3}$

17. $T(x)=151x+2200$, domain: $0\le x\le 100$

$u(x)=\dfrac{T(x)}{x}=151+\dfrac{2200}{x}$, domain: $0<x\le 100$

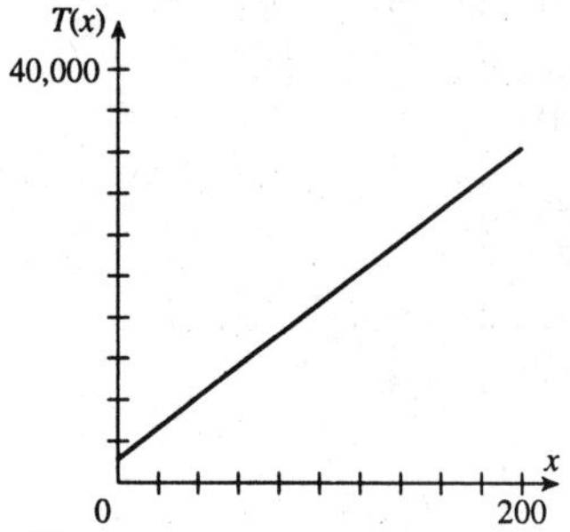

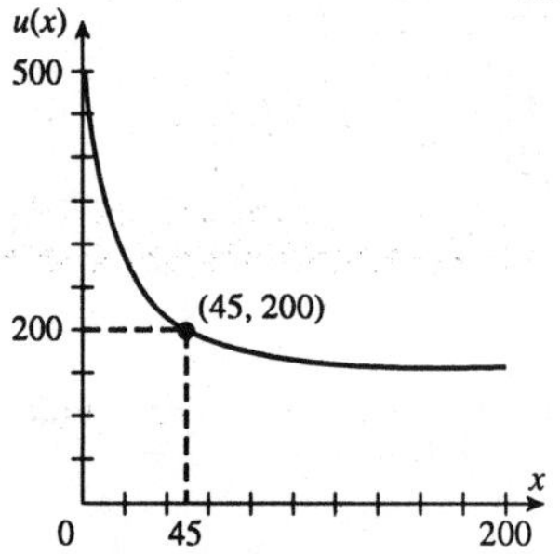

$151+\dfrac{2200}{x}=200$

$\dfrac{2200}{x}=49$

$x\approx 45$

19. $E(x)=x-x^3$

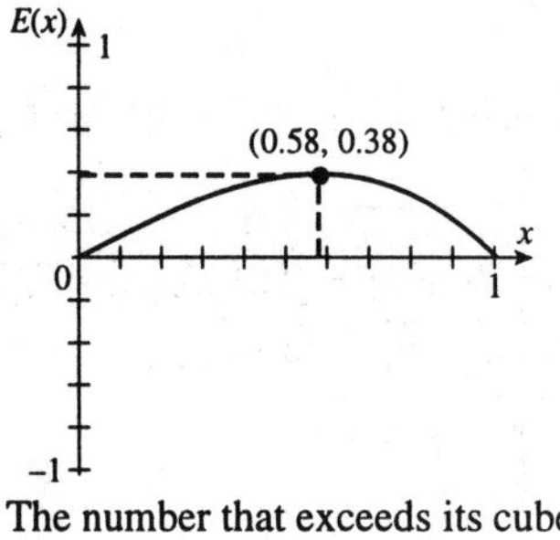

The number that exceeds its cube by the maximum amount is about $x = 0.58$.

21. a. $E(x)=24+0.40x$

b. $120=24+0.40x$

$0.40x=96$

$x=240$

23. The length of a side is $\dfrac{1}{2}(1-\pi d)$.

The area of the ends is $\pi\left(\dfrac{d}{2}\right)^2=\dfrac{\pi d^2}{4}$.

The area of the middle is

$d\left[\dfrac{1}{2}(1-\pi d)\right]=\dfrac{d}{2}-\dfrac{\pi d^2}{2}$.

Total area is

$$A(d) = \frac{\pi d^2}{4} + \frac{d - \pi d^2}{2} = \frac{d}{4}(2 - \pi d)..$$

Domain: $0 < d < \frac{2}{\pi}$

25. $\frac{3}{13} = 0.\overline{230769}$;

The range of $f(n)$ is $\{0, 2, 3, 6, 7, 9\}$.

27.

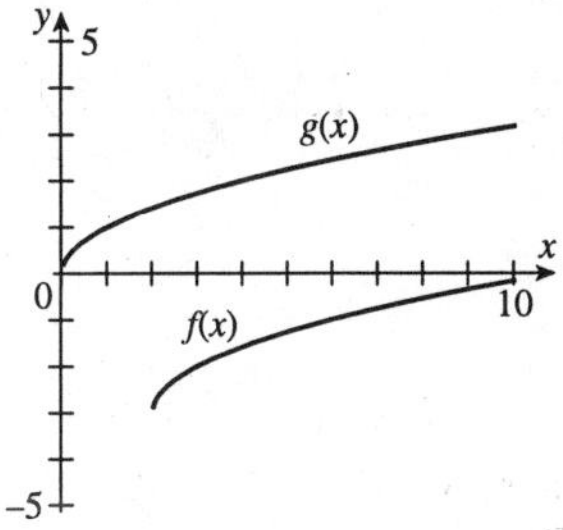

$f(x)$ is $g(x)$ shifted 3 units down and 2 units right.

29.

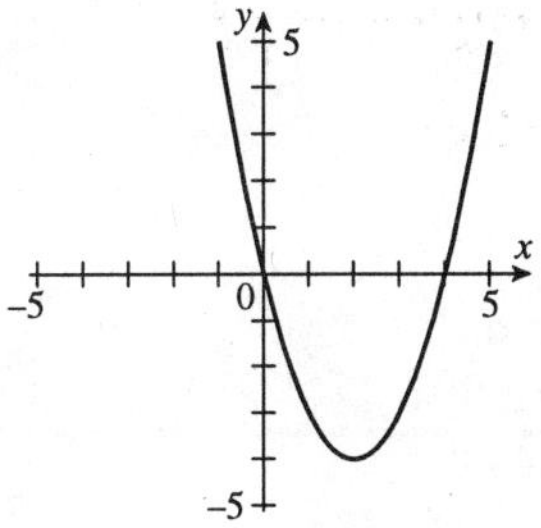

$f(x)$ is the graph of x^2 shifted 2 units right and 4 units down.

31. a. $P(t) = \sqrt{29 - 3(2+\sqrt{t}) + (2+\sqrt{t})^2}$

$= \sqrt{29 - 6 - 3\sqrt{t} + 4 + 4\sqrt{t} + t}$

$P(t) = \sqrt{27 + \sqrt{t} + t}$

b. $P(15) = \sqrt{27 + \sqrt{15} + 15} \approx 6.8$

c.

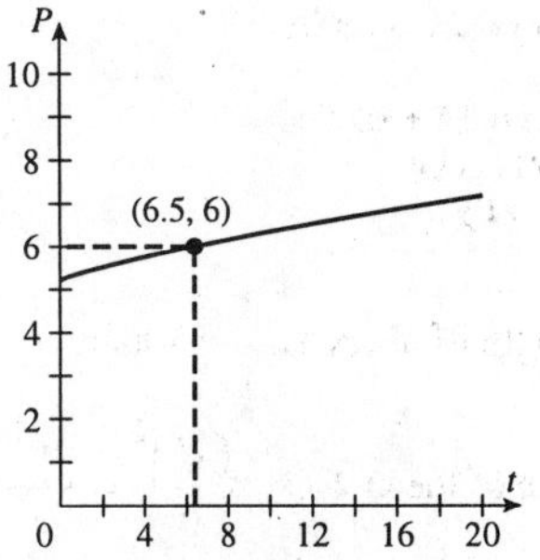

$P = 6$ when $t \approx 6.5$

33. When $0 \le t \le 1$, $D(t) = 400t$.

Use the Pythagorean Theorem to find $D(t)$ for $t > 1$.

$D(t) = \sqrt{(400t)^2 + [300(t-1)]^2}$

$= \sqrt{160,000t^2 + 90,000t^2 - 180,000t + 90,000}$

$= \sqrt{250,000t^2 - 180,000t + 90,000}$

$D(t) = 100\sqrt{25t^2 - 18t + 9}$

35.

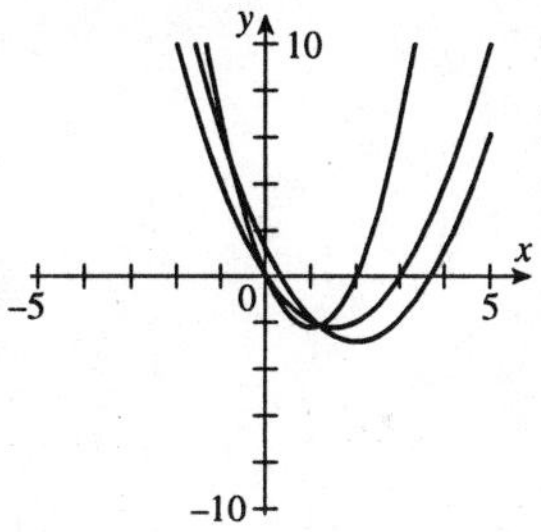

37.

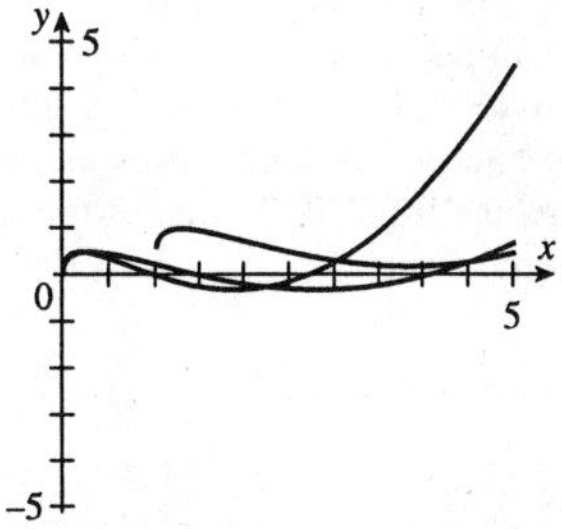

39. a.

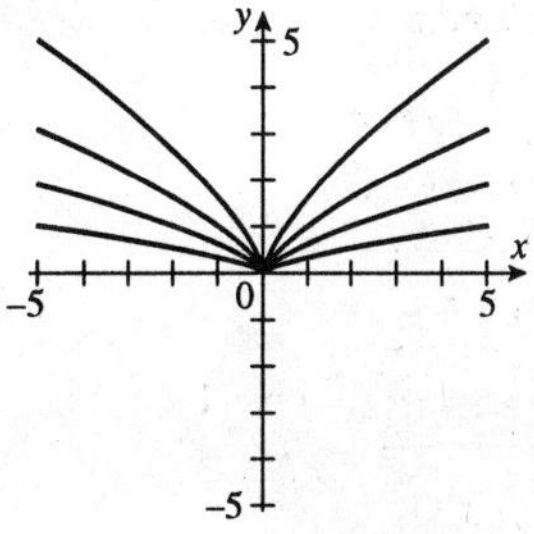

b.

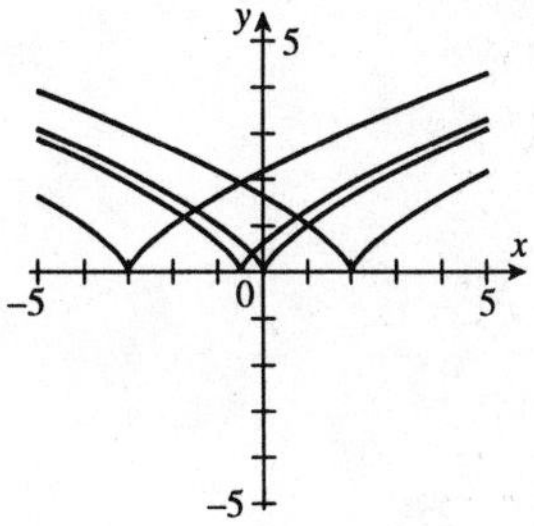

c.

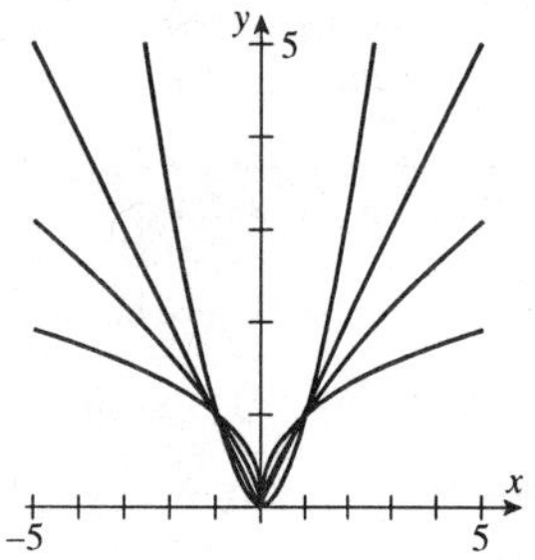

Section 0.3

Concepts Review

1. $\dfrac{d-b}{c-a}$

3. $y = mx + b$; $x = k$

5. The various forms make it easy to write linear equations when given various kinds of information (slope and intercept, slope and another point, etc.). Likewise, putting the equation of a line into one of the standard forms makes it easy to determine information about the line (slope, y-intercept, etc.).

Problem Set 0.3

1. $m = \dfrac{8-3}{4-2} = \dfrac{5}{2}$

3. $m = \dfrac{0-2}{3+4} = -\dfrac{2}{7}$

5. $m = \dfrac{5-0}{0-3} = -\dfrac{5}{3}$

7. $m = \dfrac{6.175-5.014}{4.315+1.732} \approx 0.192$

9. $y - 3 = 4(x - 2)$
$y - 3 = 4x - 8$
$4x - y - 5 = 0$

11. $y = -2x + 4$
$2x + y - 4 = 0$

13. $m = \dfrac{8-3}{4-2} = \dfrac{5}{2}$
$y - 3 = \dfrac{5}{2}(x-2)$
$2y - 6 = 5x - 10$
$5x - 2y - 4 = 0$

15. $m = \dfrac{5+3}{2-2}$ is undefined
$x = 2$; $x + 0y - 2 = 0$

17. $3y = 2x - 4$
$y = \dfrac{2}{3}x - \dfrac{4}{3}$
slope $= \dfrac{2}{3}$
y-intercept $= -\dfrac{4}{3}$

19. $2x + 3y = 6$
$3y = -2x + 6$
$y = -\dfrac{2}{3}x + 2$
slope $= -\dfrac{2}{3}$
y-intercept $= 2$

21. a. $m = 2$
$y + 3 = 2(x - 3)$
$y = 2x - 9$

b. $m = -\dfrac{1}{2}$
$y + 3 = -\dfrac{1}{2}(x-3)$
$y = -\dfrac{1}{2}x - \dfrac{3}{2}$

c. $2x + 3y = 6$
$3y = -2x + 6$
$y = -\dfrac{2}{3}x + 2$
$m = -\dfrac{2}{3}$
$y + 3 = -\dfrac{2}{3}(x-3)$
$y = -\dfrac{2}{3}x - 1$

d. $m = \dfrac{3}{2}$
$y + 3 = \dfrac{3}{2}(x-3)$
$y = \dfrac{3}{2}x - \dfrac{15}{2}$

e. $m = \dfrac{-1-2}{3+1} = -\dfrac{3}{4}$
$y + 3 = -\dfrac{3}{4}(x-3)$
$y = -\dfrac{3}{4}x - \dfrac{3}{4}$

f. $x = 3$

g. $y = -3$

23. At $x = 3$, $y = 3(3) - 1 = 8$.
The point (3, 9) lies above the line $y = 3x - 1$.

25. $m = -120{,}000(0.08) = -9600$
$V = 120{,}000 - 9600t$
$120{,}000 - 9600t = 0$
$t = \frac{120{,}000}{9600} = 12.5$ years

27. (0, 700,000), (10, 820,000)
$m = \frac{820{,}000 - 700{,}000}{10 - 0} = 12{,}000$
$N = 12{,}000n + 700{,}000$
At $n = 20$,
$N = 12{,}000(20) + 700{,}000$
$N = 940{,}000$
$\Delta N = m\Delta n$
$1{,}000{,}000 - 700{,}000 = 12{,}000\Delta n$
$\Delta n = \frac{300{,}000}{12{,}000} = 25$ years
Egg production will reach 1,000,000 cases in 2005.

29. a. When $x = 0$, $P = -2000$.
When the company sells no items, it loses \$2000.

b. slope = marginal profit = 450
The marginal profit is measured in dollars per item sold. It is the increase in profit that the company receives with each additional item sold.

31. When $B = 0$, $Ax + 0y + C = 0$
$Ax = -C$
$x = -\frac{C}{A}$ which is a vertical line.
When $B \neq 0$, $Ax + By + C = 0$
$By = -Ax - C$
$y = -\frac{A}{B}x - \frac{C}{B}$ which is a straight line with slope $-\frac{A}{B}$ and y-intercept $-\frac{C}{B}$.

33. The required distance is the same as that between $y = mx$ and $y = mx + B - b$. (0, 0) is on $y = mx$.
$y = -\frac{1}{m}x$ is perpendicular to $y = mx$ at (0, 0) and will intersect $y = mx + B - b$ at a point (x_0, y_0).
$y_0 = -\frac{1}{m}x_0$ and $y_0 = mx_0 + B - b$ since (x_0, y_0) is on both $y = -\frac{1}{m}x$ and $y = mx + B - b$. Thus, $-\frac{1}{m}x_0 = mx_0 + B - b$ or
$x_0 = \frac{m(b-B)}{m^2+1}$. $y_0 = -\frac{1}{m}x_0 = -\frac{b-B}{m^2+1}$.

The perpendicular distance between the lines is
$\sqrt{(x_0-0)^2 + (y_0-0)^2}$
$= \sqrt{\left[\frac{m(b-B)}{m^2+1}\right]^2 + \left[-\frac{b-B}{m^2+1}\right]^2}$
$= \frac{1}{m^2+1}\sqrt{m^2(b-B)^2 + (b-B)^2}$
$= \frac{|b-B|}{\sqrt{m^2+1}}$

35. Consider points (a, b), (c, d), (e, f), and (g, h).
The midpoints of adjacent sides are
$\left(\frac{a+c}{2}, \frac{b+d}{2}\right), \left(\frac{a+g}{2}, \frac{b+h}{2}\right)$,
$\left(\frac{c+e}{2}, \frac{d+f}{2}\right)$, and $\left(\frac{e+g}{2}, \frac{f+h}{2}\right)$.
Pairs of opposite sides must have equal slope for the quadrilateral to be a parallelogram.
$m_1 = \frac{\frac{b+h}{2} - \frac{b+d}{2}}{\frac{a+g}{2} - \frac{a+c}{2}} = \frac{h-d}{g-c}$
$m_2 = \frac{\frac{f+h}{2} - \frac{d+f}{2}}{\frac{e+g}{2} - \frac{c+e}{2}} = \frac{h-d}{g-c}$
$m_3 = \frac{\frac{d+f}{2} - \frac{b+d}{2}}{\frac{c+e}{2} - \frac{a+c}{2}} = \frac{f-b}{e-a}$
$m_4 = \frac{\frac{f+h}{2} - \frac{b+h}{2}}{\frac{e+g}{2} - \frac{a+g}{2}} = \frac{f-b}{e-a}$
Since $m_1 = m_2$ and $m_3 = m_4$, the line segments joining the midpoints form a parallelogram.

Section 0.4

Concepts Review

1. $(-\infty, \infty)$; $[-1, 1]$

3. -1; 1; 0; π

5. If the function is even, it will be symmetric with respect to the y-axis. If the function is odd, it will be symmetric with respect to the origin.

Problem Set 0.4

1. a. $240\left(\frac{\pi}{180}\right) = \frac{4\pi}{3}$

b. $-60\left(\frac{\pi}{180}\right) = -\frac{\pi}{3}$

c. $-135\left(\frac{\pi}{180}\right) = -\frac{3\pi}{4}$

d. $540\left(\frac{\pi}{180}\right) = 3\pi$

e. $600\left(\frac{\pi}{180}\right) = \frac{10\pi}{3}$

f. $720\left(\frac{\pi}{180}\right) = 4\pi$

g. $33.3\left(\frac{\pi}{180}\right) = 0.185\pi$

h. $471.5\left(\frac{\pi}{180}\right) \approx 2.62\pi$

i. $-391.4\left(\frac{\pi}{180}\right) \approx -2.17\pi$

j. $14.9\left(\frac{\pi}{180}\right) \approx 0.083\pi$

k. $4.02\left(\frac{\pi}{180}\right) \approx 0.022\pi$

l. $-1.52\left(\frac{\pi}{180}\right) \approx -0.0084\pi$

3. a. $\frac{234.1\sin(1.56)}{\cos(0.34)} \approx 248.3$

b. $\sin^2(2.51) + \sqrt{\cos(0.51)} \approx 1.283$

5. a. $\tan\left(\frac{\pi}{6}\right) = \frac{\sqrt{3}}{3} \approx 0.577$

b. $\sec(\pi) = -1$

c. $\cot\left(\frac{3\pi}{4}\right) = -1$

d. $\csc\left(\frac{\pi}{2}\right) = 1$

e. $\cot\left(\frac{\pi}{4}\right) = 1$

f. $\tan\left(-\frac{\pi}{4}\right) = -1$

7. a. $y = \sin\left(t - \frac{\pi}{4}\right)$

t	y
$-\pi$	$\frac{\sqrt{2}}{2}$
$-\frac{3\pi}{4}$	0
$-\frac{\pi}{2}$	$-\frac{\sqrt{2}}{2}$
$-\frac{\pi}{4}$	-1
0	$-\frac{\sqrt{2}}{2}$
$\frac{\pi}{4}$	0
$\frac{\pi}{2}$	$\frac{\sqrt{2}}{2}$
$\frac{3\pi}{4}$	1
π	$\frac{\sqrt{2}}{2}$
$\frac{5\pi}{4}$	0
$\frac{3\pi}{2}$	$-\frac{\sqrt{2}}{2}$
$\frac{7\pi}{4}$	-1
2π	$-\frac{\sqrt{2}}{2}$

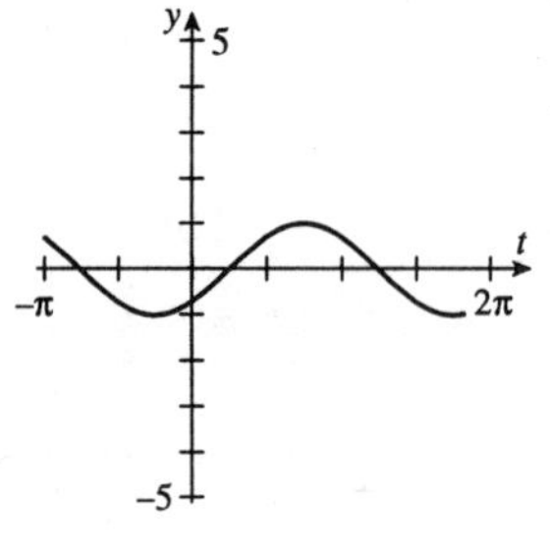

b. $y = 3\sin t$

t	y
$-\pi$	0
$-\frac{\pi}{2}$	-3
0	0
$\frac{\pi}{2}$	3
π	0
$\frac{3\pi}{2}$	-3
2π	0

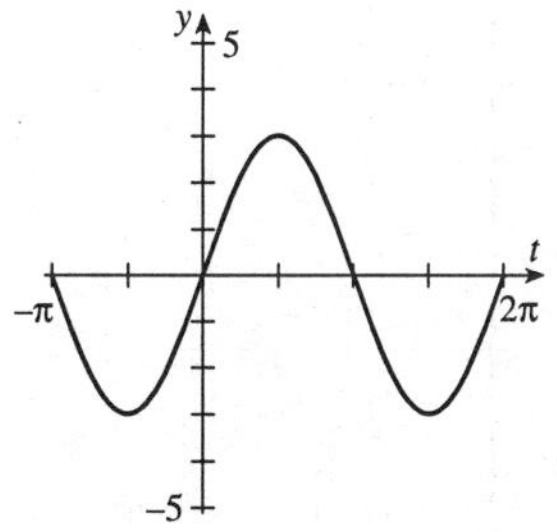

c. $y = \sin 2t$

t	y
$-\pi$	0
$-\frac{3\pi}{4}$	1
$-\frac{\pi}{2}$	0
$-\frac{\pi}{4}$	−1
0	0
$\frac{\pi}{4}$	1
$\frac{\pi}{2}$	0
$\frac{3\pi}{4}$	−1
π	0
$\frac{5\pi}{4}$	1
$\frac{3\pi}{2}$	0
$\frac{7\pi}{4}$	−1
2π	0

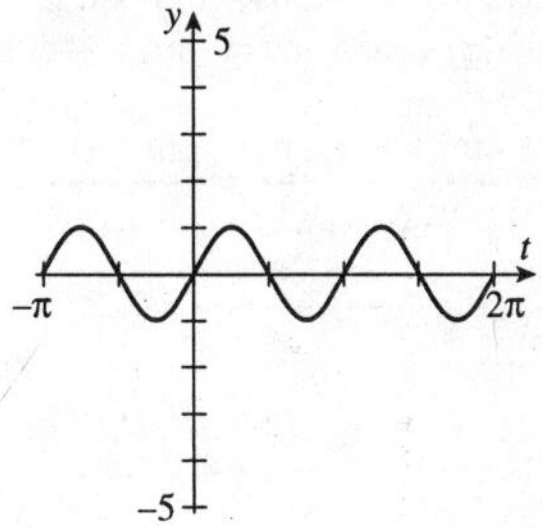

d. $y = \sec t$

t	y
$-\pi$	−1
$-\frac{\pi}{2}$	undefined
0	1
$\frac{\pi}{2}$	undefined
π	−1
$\frac{3\pi}{2}$	undefined
2π	1

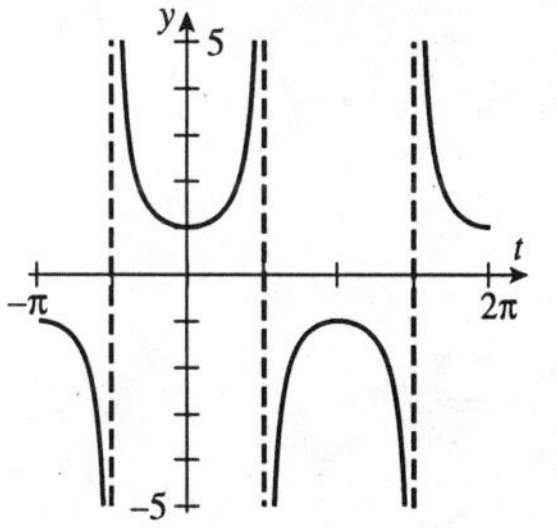

9. a. $\sin\left(t - \frac{\pi}{4}\right) = 0.5$

$t - \frac{\pi}{4} = \frac{\pi}{6}$

$t = \frac{5\pi}{12}$

$t - \frac{\pi}{4} = \pi - \frac{\pi}{6}$

$t = \frac{13\pi}{12}$

$t - \frac{\pi}{4} = -\pi - \frac{\pi}{6}$

$t = -\frac{11\pi}{12}$

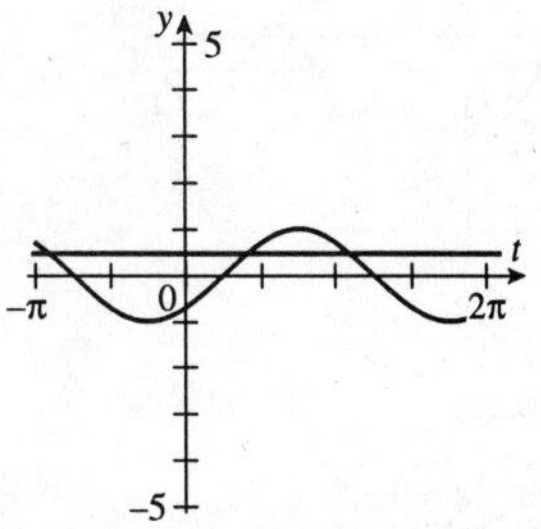

The solutions are $t = -\frac{11\pi}{12}, \frac{5\pi}{12}, \frac{13\pi}{12}$.

b. $3\sin t = 1$

$\sin t = \frac{1}{3}$

$t \approx 0.3398$

$t \approx \pi - 0.3398 \approx 2.802$

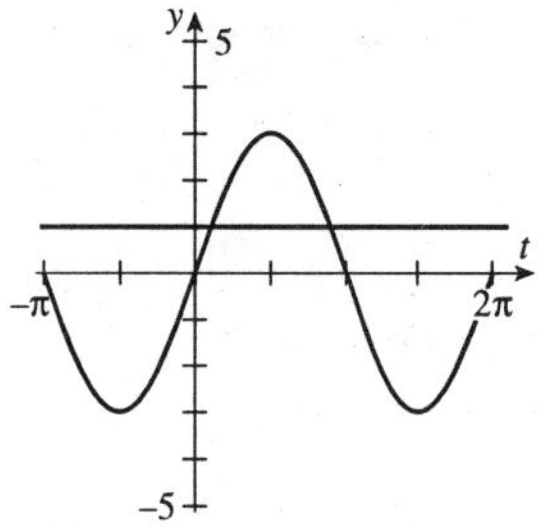

c. $\sin 2t = \frac{\sqrt{2}}{2}$

$2t = \frac{\pi}{4}$

$t = \frac{\pi}{8}$

$\pi + \frac{\pi}{8} = \frac{9\pi}{8}$

$2t = \frac{3\pi}{4}$

$t = \frac{3\pi}{8}$

$\frac{3\pi}{8} + \pi = \frac{11\pi}{8}$

$-\pi + \frac{\pi}{8} = -\frac{7\pi}{8}$

$-\frac{3\pi}{4} + \frac{\pi}{8} = -\frac{5\pi}{8}$

The solutions are

$t = -\frac{7\pi}{8}, -\frac{5\pi}{8}, \frac{\pi}{8}, \frac{3\pi}{8}, \frac{9\pi}{8}, \frac{11\pi}{8}$

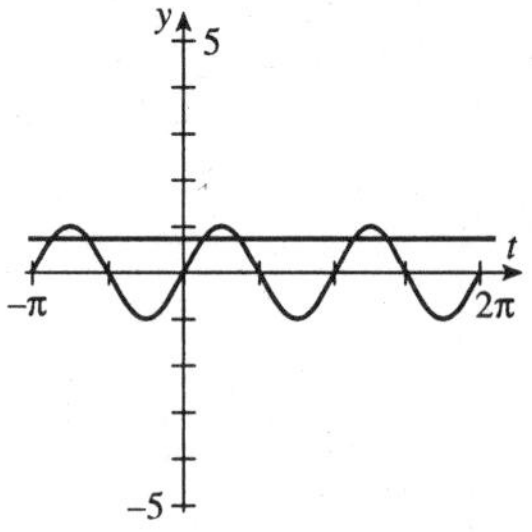

d. $\sec t = 2$

$\frac{1}{\cos t} = 2$

$\cos t = \frac{1}{2}$

$t = \frac{\pi}{3};\ 0 - \frac{\pi}{3} = -\frac{\pi}{3}$

$2\pi - \frac{\pi}{3} = \frac{5\pi}{3}$

The solutions are $t = -\frac{\pi}{3}, \frac{\pi}{3}, \frac{5\pi}{3}$

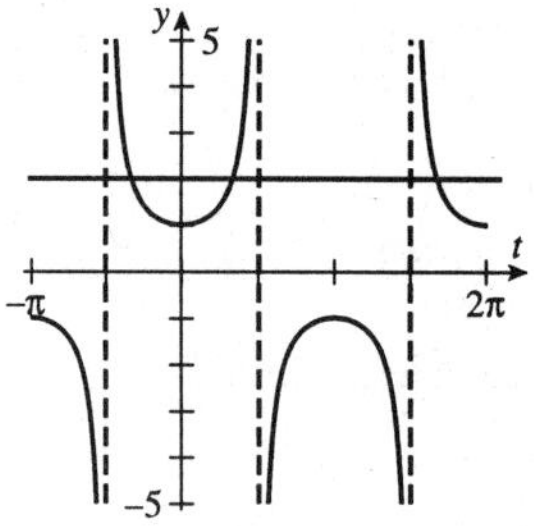

11. a. $(1 + \sin z)(1 - \sin z) = 1 - \sin^2 z$

$= \cos^2 z = \frac{1}{\sec^2 z}$

b. $(\sec t - 1)(\sec t + 1) = \sec^2 t - 1 = \tan^2 t$

13. a. Quadrant IV, cos t is positive

b. Quadrant II, cos t is negative

c. Quadrant II, cos t is negative

15. a. Even; $\sec(-t) = \sec t$

b. Odd; $\csc(-t) = -\csc t$

c. Even; $(-t)\sin(-t) = t \sin t$

d. Odd; $(-x)\cos(-x) = -x \cos x$

e. Even; $\sin^2(-x) = \sin^2 x$

f. Neither; $\sin(-x) + \cos(-x) \neq \sin x + \cos x$ and $\sin(-x) + \cos(-x) \neq -\sin x - \cos x$

17. $\tan(t + \pi) = \frac{\tan t + \tan \pi}{1 - \tan t \tan \pi} = \frac{\tan t + 0}{1 - 0} = \tan t$

19. a. $s = 2.5(6) = 15$

b. $t = 225\left(\frac{\pi}{180}\right) = \frac{5\pi}{4}$

$s = 2.5\left(\frac{5\pi}{4}\right) \approx 9.82$

21. 60 mph = 316,800 ft/h = 5,280 ft/min

$s = rt$

$5{,}280 = (2.5)(2\pi)x;\ x \approx 336$

23. a. $F\left(\frac{\pi}{4}\right) = \frac{50\mu}{\mu\sin\left(\frac{\pi}{4}\right) + \cos\left(\frac{\pi}{4}\right)}$

$= \frac{50\mu}{\frac{\sqrt{2}}{2}\mu + \frac{\sqrt{2}}{2}} = \frac{100\mu}{\sqrt{2}\mu + \sqrt{2}}$

b. $F(0) = \frac{50\mu}{\mu\sin 0 + \cos 0} = \frac{50\mu}{0\mu + 1} = 50\mu$

c. $F(1) = \frac{50\mu}{\mu\sin 1 + \cos 1} \approx \frac{50\mu}{0.89\mu + 0.54}$

d. $90\left(\frac{\pi}{180}\right) = \frac{\pi}{2}$

$F\left(\frac{\pi}{2}\right) = \frac{50\mu}{\mu\sin\left(\frac{\pi}{2}\right) + \cos\left(\frac{\pi}{2}\right)} = \frac{50\mu}{\mu} = 50$

25. a. $\tan\alpha = \sqrt{3};\ \alpha = \frac{\pi}{3}$

b. $\sqrt{3}x + 3y = 6$

$3y = -\sqrt{3}x + 6$

$y = -\frac{\sqrt{3}}{3}x + 2$

$\tan\alpha = -\frac{\sqrt{3}}{3}$

$\alpha = -\frac{\pi}{6}$

27. a. $\tan\theta = \frac{3-2}{1+2(3)}$

$\theta = \tan^{-1}\left(\frac{1}{7}\right) \approx 0.1419$

b. $\tan\theta = \frac{-1-\frac{1}{2}}{1+(-1)\left(\frac{1}{2}\right)}$

$\theta = \tan^{-1}(-3) \approx -1.2490$

c. $2x - 6y = 12$

$-6y = -2x + 12$

$y = \frac{1}{3}x - 2$

$m_1 = \frac{1}{3}$

$2x + y = 0$

$y = -2x$

$m_2 = -2$

$\tan\theta = \frac{-2-\frac{1}{3}}{1+(-2)\left(\frac{1}{3}\right)}$

$\theta = \tan^{-1}(-7) \approx -1.4289$

29. $A = \frac{1}{2}(5)^2(2) = 25$ cm

31. In 18 minutes, the record made $18 \times \left(33\frac{1}{3}\right) = 600$ revolutions. If the radius of the groove was 6 inches, the length of the groove would be $600(6)(2\pi) \approx 22{,}619.5$ inches. If the radius was 3 inches, the length would be $600(3)(2\pi) \approx 11{,}309.7$ inches. The approximate length of the groove is $\frac{1}{2}(22{,}619.5 + 11{,}309.7) \approx 16{,}964.6$ inches.

33. Let the base of the triangle be the side opposite the angle t. Then the base has length $2r\sin\frac{t}{2}$, the radius of the semicircle is $r\sin\frac{t}{2}$, and the height of the triangle is $r\cos\frac{t}{2}$.

$A = r^2\sin\frac{t}{2}\cos\frac{t}{2} + \frac{1}{2}\pi r^2\sin^2\frac{t}{2}$

35. For the following 6 graphs, the graph window is $-\pi \le x \le 2\pi$, $-3 \le y \le 3$

a.

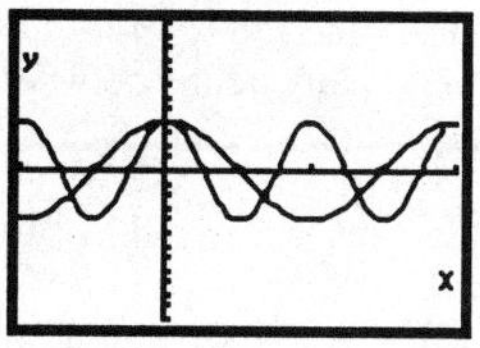

b.

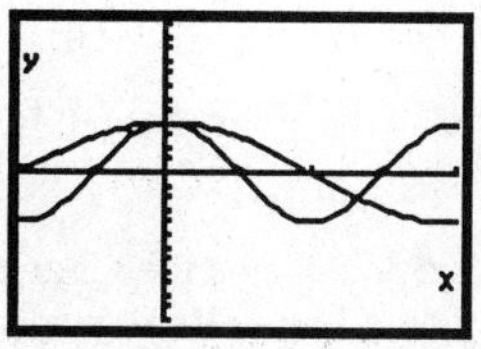

c.

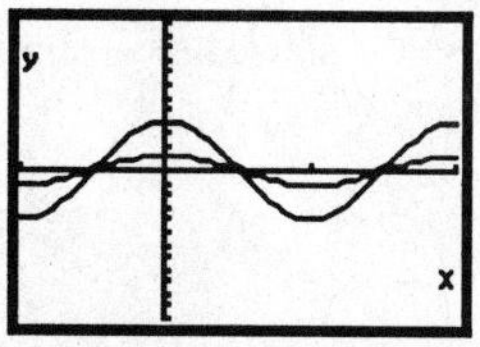

d.

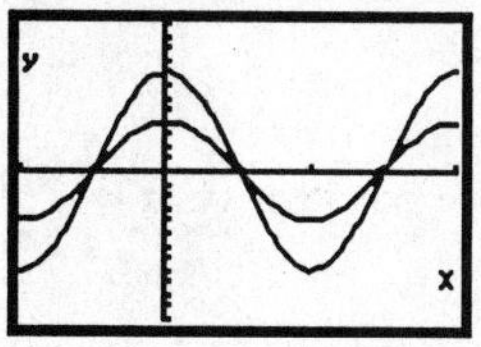

e. 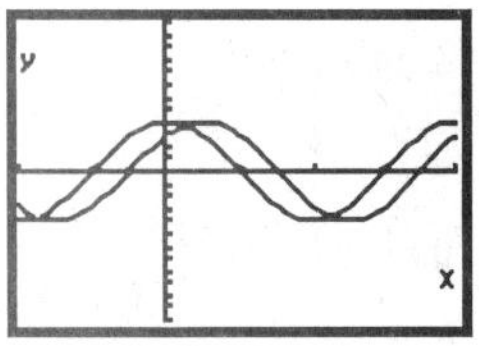

f. 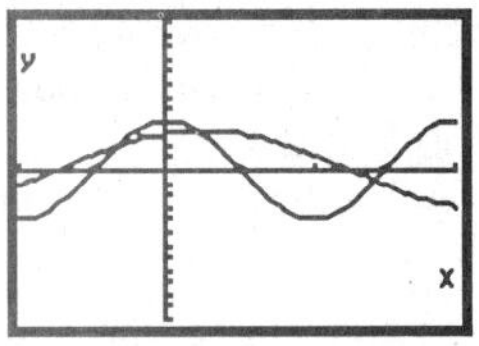

37. 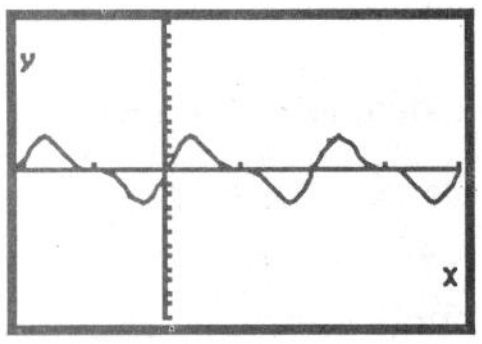

The graph window is $-\pi \le x \le 2\pi$, $-3 \le y \le 3$

Section 0.5 Chapter Review

Concepts Test

1. False. p and q must be integers and $q \neq 0$.

3. True. If two open intervals have a point in common, such as (a, b) and (c, d) if $b > c$, then there are infinitely many points between b and c that the intervals share.

5. False. The y-intercepts are found by setting $x = 0$.

7. False. The equation describes a circle with radius 2 centered at (–1, –3).

9. False. The intersection could contain four points, three points, one point, or no points at all.

11. True. If $ab > 0$, a and b have the same sign, so (a, b) is in either the first or third quadrant.

13. True. $y_1 = y_2$, so the points lie on a horizontal line.

15. True. $xy + x^2 = 3y$
$3y - xy = x^2$
$y = \dfrac{x^2}{3-x}$

17. True. $f(x)$ only exists when $0 \le x < 4$.

19. True. $\sqrt{x^2} = |x|$, so if $x < 0$, $\sqrt{x^2} = -x$.

21. False. Numerical or graphing methods may also be utilized.

23. False. Consider the function $y = 5$ which has domain $(-\infty, \infty)$ but range 5.

25. True. If $f(a) = 0$, $f(a - h + h) = f(a) = 0$.

27. False. The equation of a vertical line cannot be written in point-slope form.

29. False. Slopes of perpendicular lines are negative reciprocals.

31. True. $\Delta y = m\Delta x$
$12 = 3\Delta x$
$\Delta x = 4$

33. True. $\cot(-x) = -\cot x$

35. False. The tangent function is undefined at any point where the cosine is zero.

37. False. $\cos^{-1} 0.3$ only gives one solution to $\cos t = 0.3$.

Sample Test Problems

1. Answers will vary. Possible answer: $\dfrac{5\pi}{31}$

3. Center $= \left(\dfrac{2+10}{2}, \dfrac{0+4}{2}\right) = (6, 2)$
radius $= \dfrac{1}{2}\sqrt{(10-2)^2 + (4-0)^2} = \dfrac{1}{2}\sqrt{80} = 2\sqrt{5}$
circle: $(x-6)^2 + (y-2)^2 = 20$

5. $x^2 + y^2 = 3$

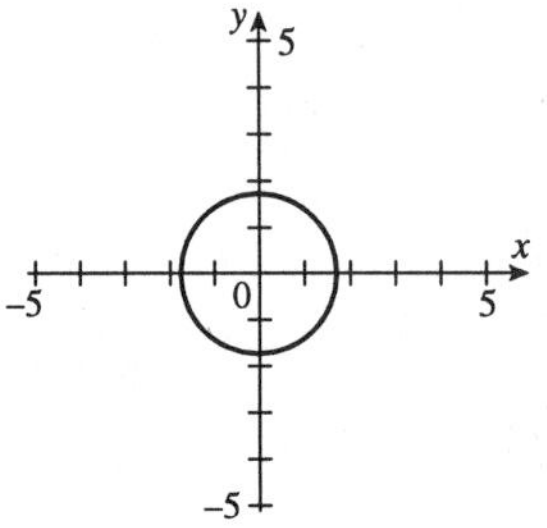

7. $x = y^2 - 3$

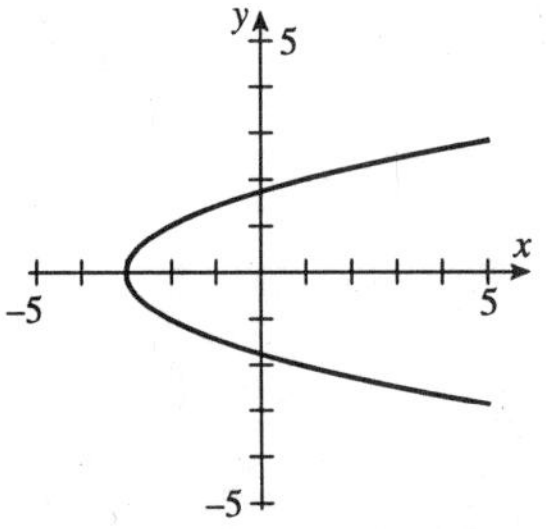

9. a. $f(1) = -\dfrac{1}{2}$

b. $f\left(-\dfrac{1}{2}\right) = 4$

c. $f(-1)$ is undefined

11. a. $(-\infty, \infty)$

b. $[-2, 2]$

c. $x \neq -\dfrac{3}{2}$

13. $V(x) = (32 - 2x)(24 - 2x)(x)$
$= 4x^3 - 112x^2 + 768x$ for $0 < x < 12$
$V(x) = 1000$ when $x \approx 1.7$ and $x \approx 8.1$

15. a. $m = \dfrac{3-1}{7+2} = \dfrac{2}{9}$

$y - 1 = \dfrac{2}{9}(x+2)$

$y = \dfrac{2}{9}x + \dfrac{13}{9}$

b. $3x - 2y = 5$
$-2y = -3x + 5$

$y = \dfrac{3}{2}x - \dfrac{5}{2}$

$m = \dfrac{3}{2}$

$y - 1 = \dfrac{3}{2}(x+2)$

$y = \dfrac{3}{2}x + 4$

c. $x = -2$

d. contains (0, 3)

$m = \dfrac{3-1}{0+2} = 1$

$y = x + 3$

17. Let d be the distance to her house and t the time in hours.
$d = 4 - 3t$
$2.5 = 4 - 3t$
$-3t = -1.5$
$t = 0.5$
$\Delta d = m\Delta t$
$2.5 - 4 = (-3)\Delta t$
$\Delta t = 0.5$

19. a. $\sin(-t) = -\sin t = -0.8$

b. $\cos^2 t + \sin^2 t = 1$
$\cos^2 t = 1 - (0.8)^2 = 0.36$
Since $\cos t < 0$, $\cos t = -0.6$

c. $\sin 2t = 2\sin t \cos t = 2(0.8)(-0.6) \approx -0.96$

d. $\tan t = \dfrac{\sin t}{\cos t} = \dfrac{0.8}{-0.6} \approx -1.33$

e. $\cos\left(\dfrac{\pi}{2} - t\right) = \sin t = 0.8$

f. $\sin(\pi + t) = -\sin t = -0.8$

21. The wheel spins at $\dfrac{20}{60} = \dfrac{1}{3}$ rev/sec

$s = 9\left(\dfrac{1}{3}\right)(2\pi) = 6\pi \approx 18.8$ in.

Chapter 1

Section 1.1

Concepts Review

1. L; c

3. L; right

Problem Set 1.1

1. $\lim_{x\to 3}(2x-8)=-2$

x	$2x-8$	x	$2x-8$
4	0	2	–4
3.1	–1.8	2.9	–2.2
3.01	–1.98	2.99	–2.02
3.001	–1.998	2.999	–2.002

3. $\lim_{x\to -2}(x^2-3x+1)=4+6+1=11$

x	x^2-3x+1	x	x^2-3x+1
–3	19	–1	5
–2.1	11.71	–1.9	10.31
–2.01	11.0701	–1.99	10.9301
–2.001	11.007001	–1.999	10.993001

5. $\lim_{x\to\sqrt{3}}\frac{\sqrt{12-x^2}}{x^4}=\frac{\sqrt{12-3}}{9}=\frac{1}{3}$

x	$\frac{\sqrt{12-x^2}}{x^4}$	x	$\frac{\sqrt{12-x^2}}{x^4}$
0.732	12	2.732	0
1.632	0.4	1.832	0.3
1.722	0.34	1.742	0.33
1.731	0.334	1.733	0.333

7. $\lim_{x\to 1}\frac{x^2+3x-4}{x-1}=\lim_{x\to 1}(x+4)=5$

x	$\frac{x^2+3x-4}{x-1}$	x	$\frac{x^2+3x-4}{x-1}$
0	4	2	6
0.9	4.9	1.1	5.1
0.99	4.99	1.01	5.01
0.999	4.999	1.001	5.001

9. $\lim_{x\to -3}\frac{2x^2+5x-3}{x+3}=\lim_{x\to -3}2x-1=-7$

x	$\frac{2x^2+5x-3}{x+3}$	x	$\frac{2x^2+5x-3}{x+3}$
–4	–9	–2	–5
–3.1	–7.2	–2.9	–6.8
–3.01	–7.02	–2.99	–6.98
–3.001	–7.002	–2.999	–6.998

11. $\lim_{x\to 9}\frac{x-9}{\sqrt{x}-3}=\lim_{x\to 9}\sqrt{x}+3=6$

x	$\frac{x-9}{\sqrt{x}-3}$	x	$\frac{x-9}{\sqrt{x}-3}$
8	6	10	6
8.9	6.0	9.1	6.0
8.99	6.00	9.01	6.00
8.999	6.000	9.001	6.000

13. $\lim_{x\to 2}\frac{x^2+x-6}{x+2}=\lim_{x\to 2}\frac{0}{4}=0$

x	$\frac{x^2+x-6}{x+2}$	x	$\frac{x^2+x-6}{x+2}$
1	–1	3	1
1.9	–0.1	2.1	0.1
1.99	–0.01	2.01	0.01
1.999	–0.001	2.001	0.001

15. $\lim_{t\to 2}\frac{t^2-5t+6}{t^2-t-2}=\lim_{t\to 2}\frac{t-3}{t+1}=-\frac{1}{3}$

t	$\frac{t^2-5t+6}{t^2-t-2}$	t	$\frac{t^2-5t+6}{t^2-t-2}$
1	–1	3	0
1.9	–0.4	2.1	–0.3
1.99	–0.34	2.01	–0.33
1.999	–0.334	2.001	–0.333

17.

x	$\frac{\tan x}{2x}$	x	$\frac{\tan x}{2x}$
–1	1	1	1
–0.1	0.5	0.1	0.5
–0.01	0.50	0.01	0.50
–0.001	0.500	0.001	0.500

$\lim_{x\to 0}\frac{\tan x}{2x}=0.5$

19.

x	$\frac{x-\sin x}{x^3}$	x	$\frac{x-\sin x}{x^3}$
–1	0	1	0
–0.1	0.2	0.1	0.2
–0.01	0.17	0.01	0.17
–0.001	0.167	0.001	0.167

$\lim_{x\to 0}\frac{x-\sin x}{x^3}\approx 0.167$

21.

t	$\frac{\sin t}{t^2}$	t	$\frac{\sin t}{t^2}$
–1	–1	1	1
–0.1	–10.0	0.1	10.0
–0.01	–100.00	0.01	100.00
–0.001	–1000.000	0.001	1000.000

$\lim_{x\to 0}\frac{\sin t}{t^2}$ does not exist.

23.

x	$\frac{1+\cos x}{\sin 2x}$	x	$\frac{1+\cos x}{\sin 2x}$
2.142	–0.5050	4.142	0.5061
3.042	–0.0250	3.242	0.0253
3.132	–0.0024	3.152	0.0026
3.141	–0.0001	3.143	0.0003

$\lim_{x\to\pi}\frac{1+\cos x}{\sin 2x}=0$

25.

x	$\frac{\sin 3x}{x-1}$	x	$\frac{\sin 3x}{x-1}$
0	0	2	0
0.9	–4.3	1.1	–1.6
0.99	–17.08	1.01	11.14
0.999	–144.089	1.001	138.149

$\lim_{x\to 1}\frac{\sin 3x}{x-1}$ does not exist.

27. $\lim_{x\to 0}\sqrt{x}$ does not exist.

$\sqrt{x}$ is not defined for $x<0$, so $\lim_{x\to 0^-}\sqrt{x}$ does not exist.

29. $\lim_{x\to 0}\sqrt{|x|}=0$

A table will verify this.

31. $\lim_{h\to 0}(1+2h)^{1/h}=\lim_{h\to 0}[(1+2h)^{1/(2h)}]^2=e^2$
≈ 7.389

h	$(1+2h)^{1/h}$	h	$(1+2h)^{1/h}$
–1	–1	1	3
–0.1	9.313	0.1	6.192
–0.01	7.540	0.01	7.245
–0.001	7.404	0.001	7.374

33. $\lim_{x\to 0}\cos\frac{1}{x}$ does not exist.

As $x\to 0$, $\frac{1}{x}$ grows rapidly, and $\cos\frac{1}{x}$ oscillates between –1 and 1 without approaching any particular number.

35. $\lim_{x\to 1}\frac{x^3-1}{\sqrt{2x+2}-2}=\lim_{x\to 1}\frac{(x^3-1)\left(\sqrt{2x+2}+2\right)}{2x-2}$

$=\lim_{x\to 1}\frac{(x^2+x+1)\left(\sqrt{2x+2}+2\right)}{2}=\frac{3(2+2)}{2}=6$

37. $\lim\limits_{x \to 2^-} \dfrac{x^2 - x - 2}{|x-2|} = \lim\limits_{x \to 2^-} \dfrac{x^2 - x - 2}{-(x-2)}$

$= \lim\limits_{x \to 2^-} \dfrac{x+1}{-1} = -3$

Since if $x < 2$, $x - 2 < 0$ so $|x-2| = -(x-2)$.

39. **a.** $\lim\limits_{x \to -3} f(x) = 2$

b. $f(-3) = 1$

c. $f(-1)$ does not exist.

d. $\lim\limits_{x \to -1} f(x) = 2.5$

e. $f(1) = 2$

f. $\lim\limits_{x \to 1} f(x)$ does not exist.

g. $\lim\limits_{x \to 1^-} f(x) = 2$

h. $\lim\limits_{x \to 1^+} f(x) = 1$

41.

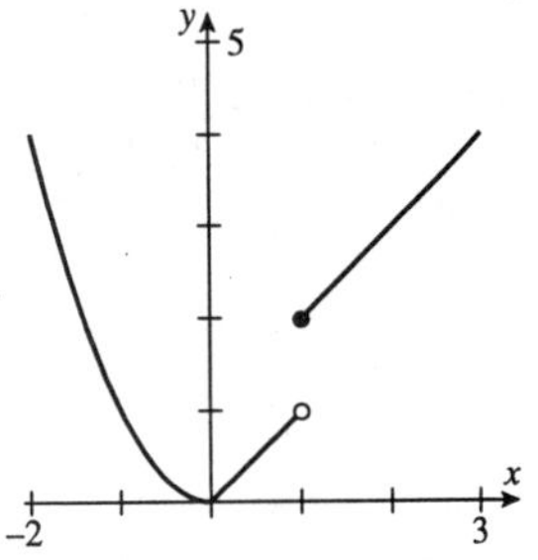

a. $\lim\limits_{x \to 0} f(x) = 0$

b. $f(1) = 2$

c. $\lim\limits_{x \to 1} f(x)$ does not exist.

d. $\lim\limits_{x \to 1^-} f(x) = 1$

43.

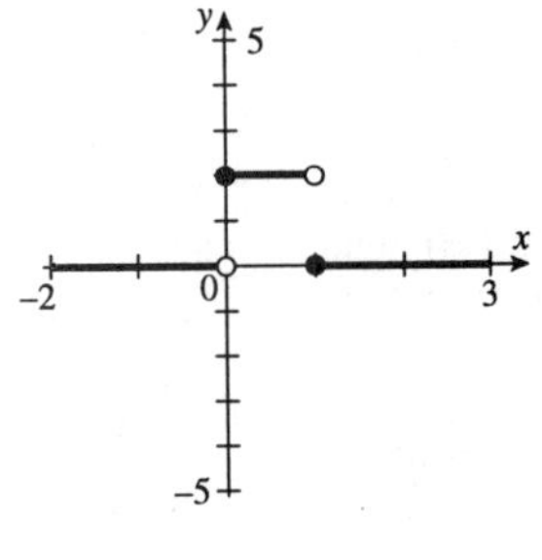

a. $f(1) = 0$

b. $\lim\limits_{x \to 1} f(x)$ does not exist.

c. $\lim\limits_{x \to 1^-} f(x) = 1$

d. $\lim\limits_{x \to 1^+} f(x) = 0$

45. $\lim\limits_{x \to 1} \dfrac{x^2 - 1}{|x-1|}$ does not exist.

x	$\dfrac{x^2-1}{\lvert x-1 \rvert}$	x	$\dfrac{x^2-1}{\lvert x-1 \rvert}$
0	–1	2	3
0.9	–1.9	1.1	2.1
0.99	–1.99	1.01	2.01
0.999	–1.999	1.001	2.001

47. **a.** $\lim\limits_{x \to 1} f(x)$ does not exist.

x	$f(x)$	x	$f(x)$
1.1	1.1	$1 + \dfrac{\sqrt{2}}{10}$	–1.14
1.01	1.01	$1 + \dfrac{\sqrt{2}}{100}$	–1.014
1.001	1.001	$1 + \dfrac{\sqrt{2}}{1000}$	–1.0014
0.9	0.9	$1 - \dfrac{\sqrt{2}}{10}$	–0.86
0.99	0.99	$1 - \dfrac{\sqrt{2}}{100}$	–0.986
0.999	0.999	$1 - \dfrac{\sqrt{2}}{1000}$	–0.9986

b. $\lim_{x \to 0} f(x) = 0$

x	$f(x)$	x	$f(x)$
0.1	0.1	$\frac{\sqrt{2}}{10}$	–0.14
0.01	0.001	$\frac{\sqrt{2}}{100}$	–0.014
0.001	0.001	$\frac{\sqrt{2}}{1000}$	–0.0014
–0.1	–0.1	$-\frac{\sqrt{2}}{10}$	0.14
–0.01	–0.01	$-\frac{\sqrt{2}}{100}$	0.014
–0.001	–0.001	$-\frac{\sqrt{2}}{1000}$	0.0014

49. $\lim_{x \to a} f(x)$ exists for $a = -1, 0, 1$ since for these values $a^2 = a^4$.

51. a. $\lim_{x \to 1} \frac{|x-1|}{x-1}$ does not exist since

$$\lim_{x \to 1^-} \frac{|x-1|}{x-1} = -1 \text{ and } \lim_{x \to 1^+} = -1$$

b. $\lim_{x \to 1^-} \frac{|x-1|}{x-1} = -1$

c. $\lim_{x \to 1^-} \frac{x^2 - |x-1| - 1}{|x-1|}$

$$= \lim_{x \to 1^-} \left(\frac{x-1}{|x-1|}(x+1) - 1 \right) = -3$$

d. $\lim_{x \to 1^-} \left[\frac{1}{x-1} - \frac{1}{|x-1|} \right] = \lim_{x \to 1^-} \frac{2}{x-1}$ does not exist since $\frac{2}{x-1}$ decreases without bound as x approaches 1 from the left.

Section 1.2

Concepts Review

1. x increases without bound; $f(x)$ gets close to L as x increases without bound; $f(x)$ increases without bound as x approaches a from the right.

3. $\lim_{x \to c} f(x)$

Problem Set 1.2

1. a. $f(x)$ increases without bound as x approaches a from the left

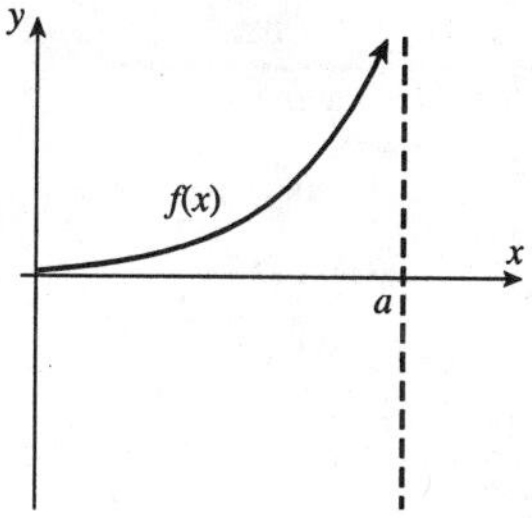

b. $f(x)$ decreases without bound as x approaches a from the right.

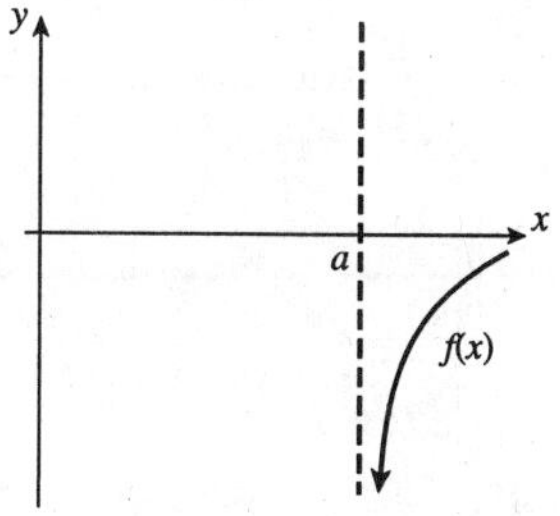

c. $f(x)$ decreases without bound as x approaches a from the left.

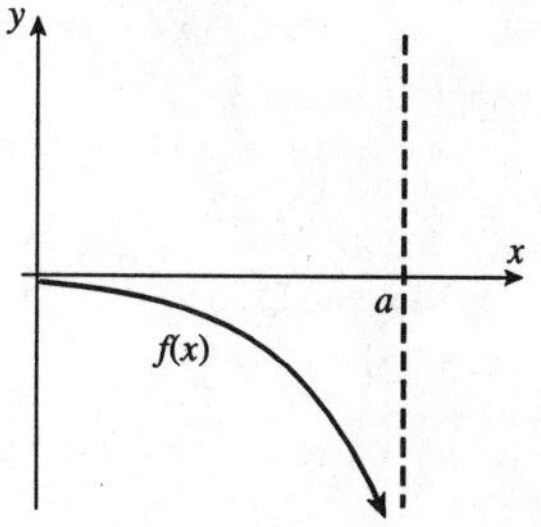

3.

x	$y = \frac{3-2x}{x+5}$
10	–1.13333
100	–1.87619
1000	–1.98706
10,000	–1.99870

$$\lim_{x \to \infty} \frac{3-2x}{x+5} = -2$$

5.

x	$y = \dfrac{2x+7}{x^2 - x}$
0.9	–97.77778
0.99	–907.07071
0.999	–9007.00701
0.9999	–90007.00070

$\lim_{x \to 1^-} \dfrac{2x+7}{x^2 - x} = -\infty$

7.

x	$y = \dfrac{1+\cos x}{\sin x}$
–1	–1.83049
–0.1	–19.98333
–0.01	–199.99833
–0.001	–1999.99983

$\lim_{x \to 0^-} \dfrac{1+\cos x}{\sin x} = -\infty$

9. Vertical asymptote at $x = -1$, $\lim_{x \to 1^-} f(x) = -\infty$, $\lim_{x \to 1^+} f(x) = \infty$. horizontal asymptote at $y = 0$ as $x \to \infty$ and $x \to -\infty$. Plot (–4, –1), (0, 3), and (2, 1).

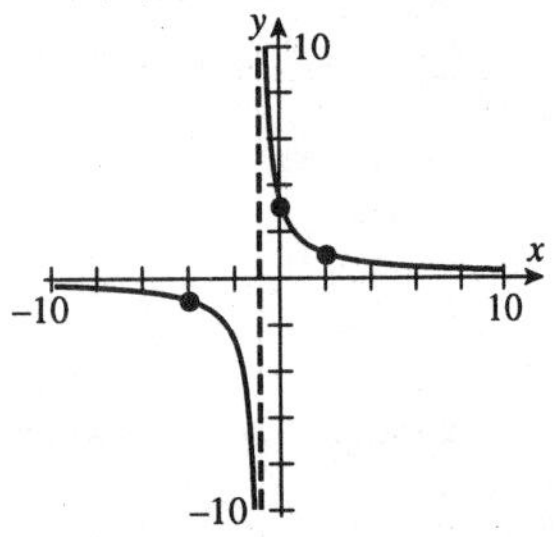

11. Vertical asymptote at $x = 3$, $\lim_{x \to 3^-} f(x) = -\infty$, $\lim_{x \to 3^+} f(x) = \infty$; horizontal asymptote at $y = 2$ as $x \to \infty$ and $x \to -\infty$. Plot (0, 0), (1, –1), and (5, 5).

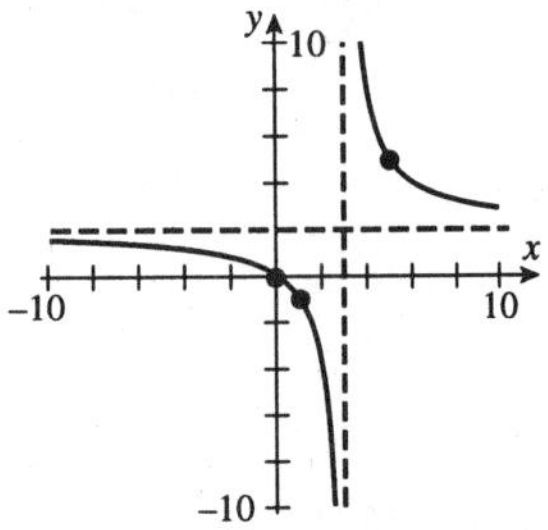

13. No vertical asymptotes; horizontal asymptote at $y = 0$ as $x \to -\infty$ and $x \to \infty$.

Plot (0, 2), $\left(2, \dfrac{14}{15}\right)$, and $\left(-2, \dfrac{14}{15}\right)$.

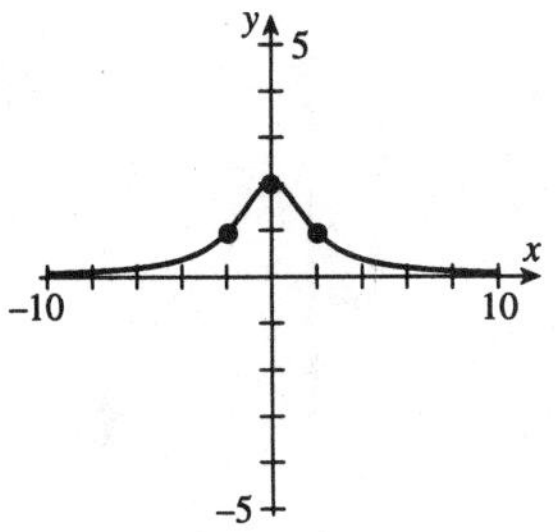

15. No vertical asymptotes; horizontal asymptotes at $y = \sqrt{\dfrac{2}{5}} \approx 0.632$ as $x \to -\infty$ and $x \to \infty$.

Plot (0, 0), $\left(\dfrac{3}{2}, 0\right)$ and $\left(-1, \sqrt{\dfrac{5}{6}}\right)$. Note that the function is not defined for $0 < x < \dfrac{3}{2}$.

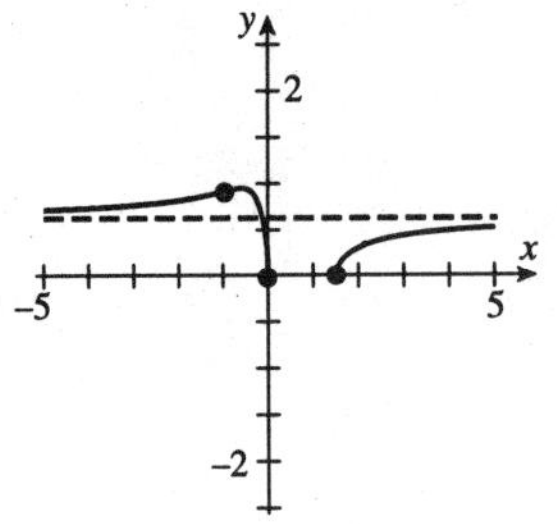

17. Vertical asymptotes at $x = \dfrac{3}{2}$ and $x = -1$,

$\lim_{x \to \frac{3}{2}^-} H(x) = -\infty$, $\lim_{x \to \frac{3}{2}^+} H(x) = \infty$,

$\lim_{x \to -1^-} H(x) = -\infty$, $\lim_{x \to -1^+} H(x) = \infty$; horizontal asymptote at $y = 0$ as $x \to -\infty$ and $x \to \infty$. Plot $\left(-2, -\dfrac{3}{7}\right)$, $\left(0, -\dfrac{1}{3}\right)$, $\left(1, -\dfrac{3}{2}\right)$, and $\left(2, \dfrac{5}{3}\right)$.

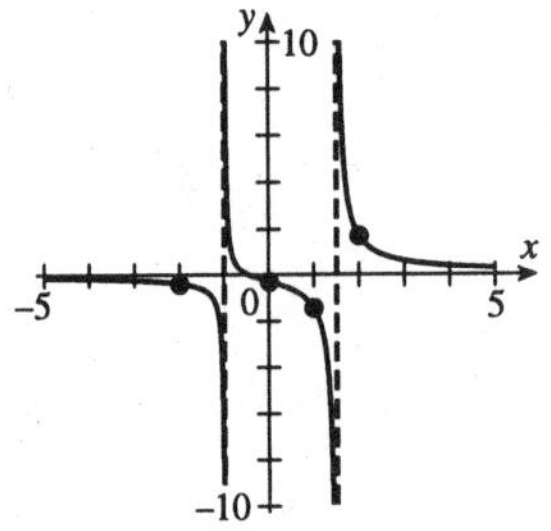

19. No vertical asymptotes; no horizontal asymptotes. Plot (–1, 4) and (1, 8). Note that $h(x)$ is not defined at $x = 3$.

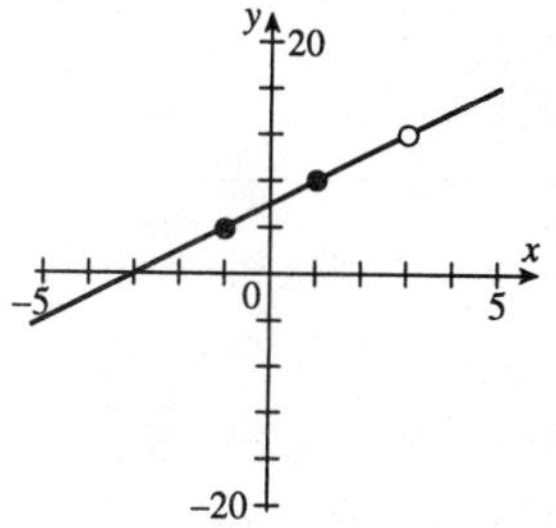

21. $\lim_{x \to \infty} \frac{2x}{x+4} = 2$

23. $\lim_{t \to 3^+} \frac{3+t}{3-t} = -\infty$

25. $\lim_{x \to -\infty} \frac{x^2 - 2x + 5}{8x^2 - 6x} = \frac{1}{8} = 0.125$

27. $\lim_{x \to 2^+} \frac{x^2 + 2x - 8}{x^2 - 4} = \lim_{x \to 2^+} \frac{(x+4)(x-2)}{(x+2)(x-2)} = \frac{3}{2}$
$= 1.5$

29. $\lim_{x \to \infty} \frac{(3x-2)(2x+4)}{(2x+1)(x+2)} = \lim_{x \to \infty} \frac{6x^2 + 8x - 8}{2x^2 + 5x + 2}$
$= 3$

31. $\lim_{x \to \frac{3}{2}^-} \frac{4x+1}{2x-3} = -\infty$

33. $\lim_{x \to \infty} \frac{3x\sqrt{x} + 3x + 11}{x^2 - x + 11} = 0$

35. $\lim_{x \to \infty} \frac{\cos(x-3)}{x-3} = 0$
As $x \to \infty$, $\cos(x-3)$ oscillates between –1 and 1 while $\frac{1}{x-3} \to 0$.

37. $\lim_{x \to 0^-} \left(\frac{1 + \cos x}{\sin x} \right) = -\infty$; See problem 7.

39. $\lim_{x \to \infty} \left(1 + \frac{1}{x} \right)^{10} = 1$

41. $\lim_{x \to 0^+} \left(1 + \sqrt{x}\right)^{1/\sqrt{x}} \approx 2.718$

x	$\left(1+\sqrt{x}\right)^{1/\sqrt{x}}$
1	2
0.1	2.384
0.01	2.594
0.001	2.677
0.0001	2.705
0.00001	2.714
0.000001	2.717
0.0000001	2.718
0.00000001	2.718

43. $\lim_{x \to \infty} \left(1 + \frac{1}{x} \right)^{x^2} = \infty$

45. $\lim_{x \to \infty} \frac{2x+1}{\sqrt{x^2+3}} = \lim_{x \to \infty} \frac{2 + \frac{1}{x}}{\sqrt{1 + \frac{3}{x^2}}} = 2$

47. Divide both numerator and denominator by x^n.

$$\lim_{x \to \infty} \frac{a_0 x^n + a_1 x^{n-1} + \cdots + a_n}{b_0 x^n + b_1 x^{n-1} + \cdots + b_n}$$

$$= \lim_{x \to \infty} \frac{a_0 + \frac{a_1}{x} + \cdots + \frac{a_n}{x^n}}{b_0 + \frac{b_1}{x} + \cdots + \frac{b_n}{x_n}} = \frac{a_0}{b_0}$$

49. Vertical asymptote at $x = 2$; horizontal asymptote at $y = 0$; discontinuity at $x = 2$.

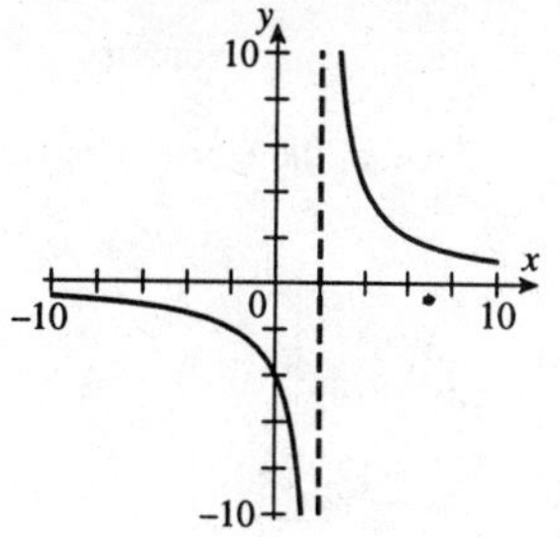

51. No asymptotes; no discontinuities.

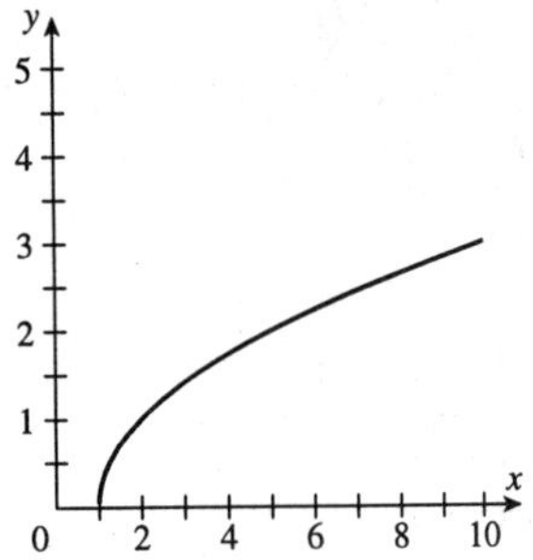

53. No vertical asymptotes; horizontal asymptote of $y = -2$ as $x \to -\infty$; no discontinuities.

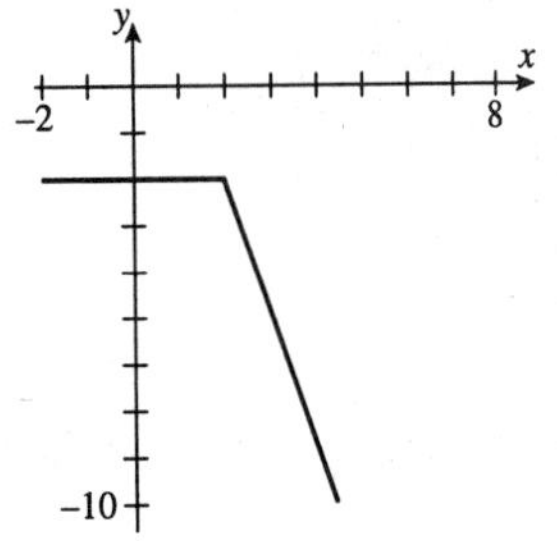

55. No asymptotes; discontinuity at $x = 1$.

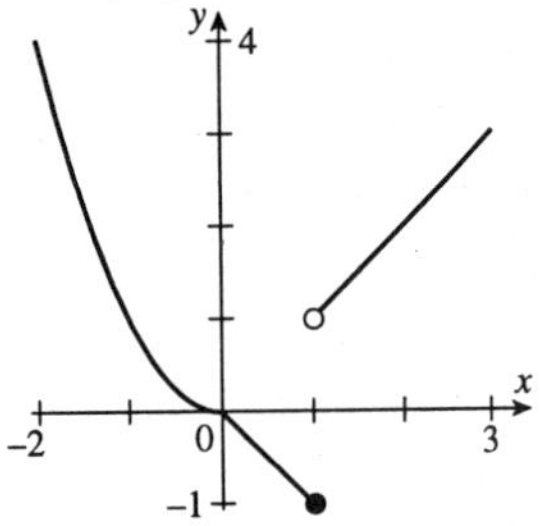

57. **a.** Suppose the block rotates to the left. Using geometry, $f(x) = -\frac{\sqrt{3}}{4}$. Suppose the block rotates to the right. Using geometry, $f(x) = \frac{\sqrt{3}}{4}$. If $x = 0$, the block does not rotate, so $f(x) = 0$.

Domain: $\left[-\frac{\sqrt{3}}{4}, \frac{\sqrt{3}}{4}\right]$;

Range: $\left\{-\frac{\sqrt{3}}{4}, 0, \frac{\sqrt{3}}{4}\right\}$

b. $x = 0$

c.

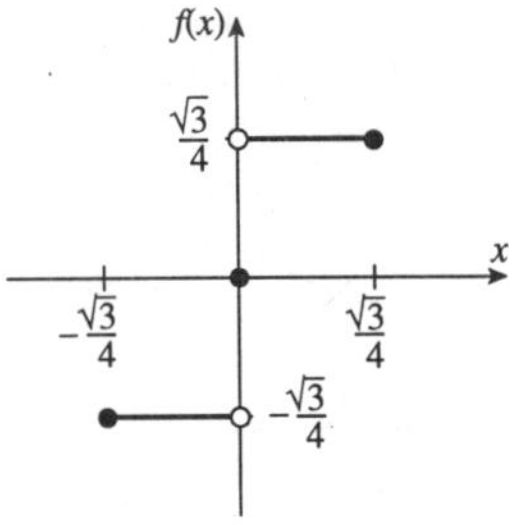

59. Possible answer:

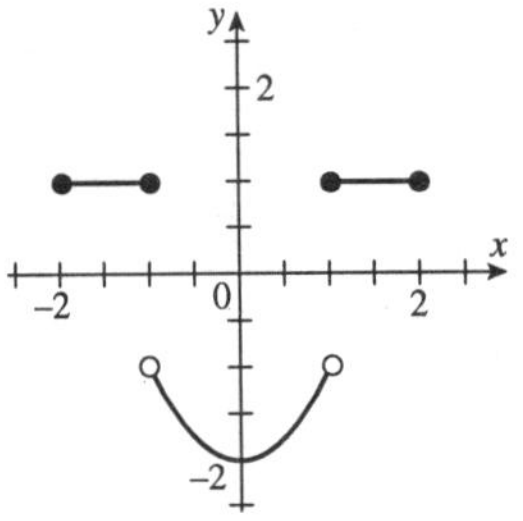

61. Use the hint from Problem 60 and divide the denominator into the numerator.

$$f(x) = \frac{3x^3 + 4x^2 - x + 1}{x^2 - 1} = 3x + 4 + \frac{2x + 5}{x^2 - 1}$$

$$\lim_{x \to \infty} \frac{2x + 5}{x^2 - 1} = \lim_{x \to \infty} \frac{\frac{2}{x} + \frac{5}{x^2}}{1 - \frac{1}{x^2}} = 0$$

and $\lim_{x \to -\infty} \frac{2x + 5}{x^2 - 1} = 0$;,

thus, $\lim_{x \to \infty} [f(x) - (3x + 4)] = 0$ and $\lim_{x \to -\infty} [f(x) - (3x + 4)] = 0$ so $y = 3x + 4$ is the oblique asymptote for $f(x)$.

Section 1.3

Concepts Review

1. Tangent line

3. $\frac{f(c+h) - f(c)}{h}$

Problem Set 1.3

1. slope $\approx \dfrac{1}{\left(\frac{1}{4}\right)} = 4$

3.

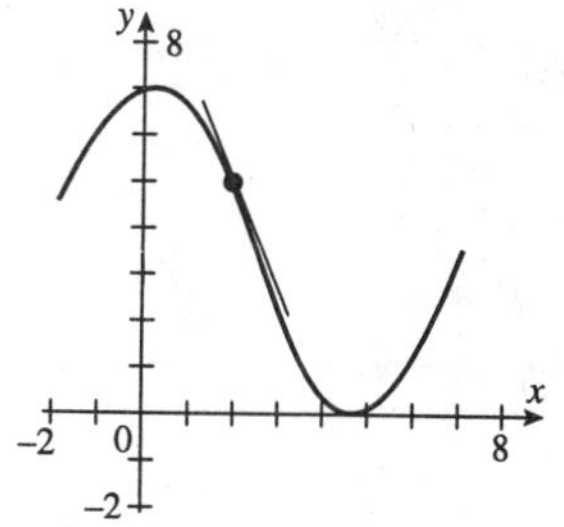

Possible answer: slope $\approx -\dfrac{7}{3}$

5.

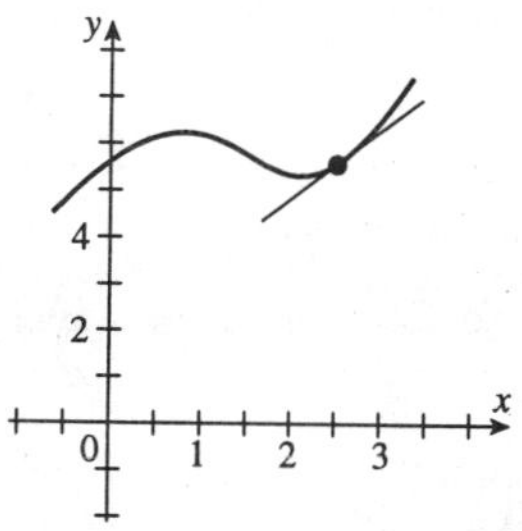

Possible answer: slope $\approx \dfrac{4}{2} = 2$

9. $m_{\tan} = \lim\limits_{h\to 0} \dfrac{(x+h)^2 - 3(x+h) + 2 - (x^2 - 3x + 2)}{h}$

$= \lim\limits_{h\to 0} \dfrac{2xh + h^2 - 3h}{h} = \lim\limits_{h\to 0} 2x + h - 3$

$= 2x - 3$

$x = -2,\ m_{\tan} = -7;\ x = 1.5,\ m_{\tan} = 0;$

$x = 2,\ m_{\tan} = 1;\ x = 5,\ m_{\tan} = 7$

11. $f'(c) = \lim\limits_{h\to 0} \dfrac{\cos(c+h) - \cos c}{h}$

For $c = 0$, we get $f'(0) = \lim\limits_{h\to 0} \dfrac{\cos h - 1}{h} \approx 0.$

For $c = \dfrac{\pi}{2}$, we get $f'\left(\dfrac{\pi}{2}\right) = \lim\limits_{h\to 0} \dfrac{\cos\left(\frac{\pi}{2} + h\right)}{h} \approx -1.$

For $c = \pi$, we get $f'(\pi) = \lim\limits_{h\to 0} \dfrac{\cos(\pi + h) + 1}{h} \approx 0.$

7. a.b.

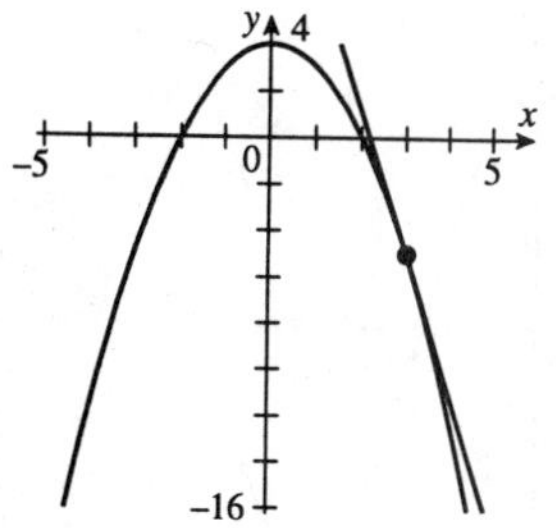

c. Written response

d. $m_{\sec} = \dfrac{-5 - (4 - 3.01^2)}{3 - 3.01} = -6.01$

e. $m_{\tan} = \lim\limits_{h\to 0} \dfrac{4 - (3+h)^2 - (4 - 3^2)}{h}$

$= \lim\limits_{h\to 0} -\dfrac{6h + h^2}{h} = \lim\limits_{h\to 0} -6 - h = -6$

f. $m_{\sec}$ is close to $m_{\tan}$ since 3.01 is close to 3.

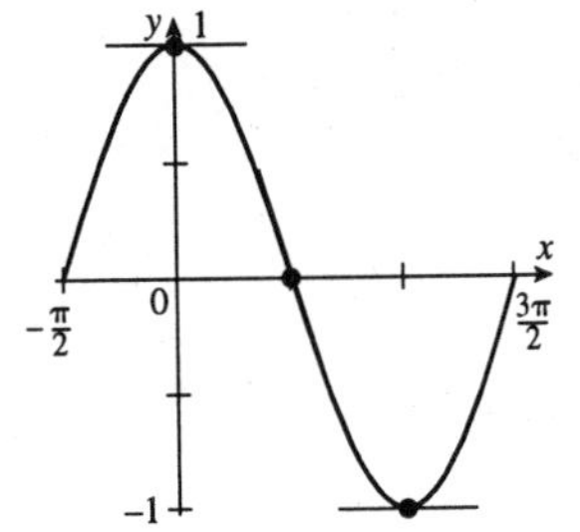

13. $f'(c) = \lim\limits_{h\to 0} \dfrac{\frac{c+h}{c+h+1} - \frac{c}{c+1}}{h}$

$= \lim\limits_{h\to 0} \dfrac{1}{(c+h+1)(c+1)} = \dfrac{1}{(c+1)^2}$

$f'(0) = 1;\ f'(1) = \dfrac{1}{4},\ f'(2) = \dfrac{1}{9}$

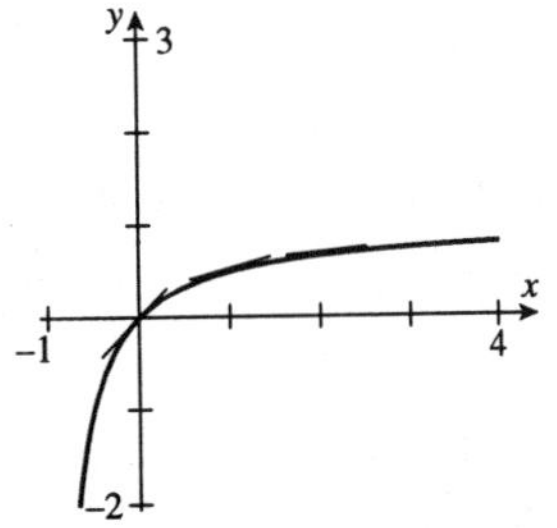

15. a. Let $f(t) = 16t^2$ be the distance the body falls in t seconds. $f(3) = 144, f(4) = 256$; so the body will fall 112 feet between $t = 3$ and $t = 4$.

b. $\dfrac{256 - 144}{4 - 3} = 112$ feet per second

c. $\dfrac{16(3.02)^2 - 144}{3.02 - 3} = 96.32$ feet per second

d. instantaneous velocity

$= \lim\limits_{h\to 0} \dfrac{16(3+h)^2 - 144}{h}$

$= \lim\limits_{h\to 0} 96 + 16h = 96$ feet per second

e. Written response

17. a. $\lim\limits_{h\to 0} \dfrac{16(c+h)^2 - 16c^2}{h} = \lim\limits_{h\to 0} 32c + 16h$

$= 32c$ feet per second

b. $32c = 100;\ c = \dfrac{100}{32} = 3.125$ seconds

19. a. $\lim\limits_{h\to 0} \dfrac{\sqrt{c+h} - \sqrt{c}}{h}$

$= \lim\limits_{h\to 0} \dfrac{\left(\sqrt{c+h} - \sqrt{c}\right)\left(\sqrt{c+h} + \sqrt{c}\right)}{h\left(\sqrt{c+h} + \sqrt{c}\right)}$

$= \lim\limits_{h\to 0} \dfrac{h}{h\left(\sqrt{c+h} + \sqrt{c}\right)}$

$= \lim\limits_{h\to 0} \dfrac{1}{\sqrt{c+h} + \sqrt{c}} = \dfrac{1}{2\sqrt{c}}$

b. $\dfrac{1}{2\sqrt{c}} = \dfrac{1}{6};\ 3 = \sqrt{c};\ c = 9$ seconds

21. a. Let $f(t) = 10 - \dfrac{9}{t+1}$ be the mass after t hours.

$f(2) = 7$ grams, $f(2.01) \approx 7.00997$ grams

It grew $f(2) - f(2.01) \approx 0.00997$ grams.

b. The average growth rate is approximately $\dfrac{0.00997}{0.01} = 0.997$ grams per hour.

c. The instantaneous growth rate is

$\lim\limits_{h\to 0} \dfrac{3 - \frac{9}{3+h}}{h} = \lim\limits_{h\to 0} \dfrac{3}{3+h} = 1$ gram per hour.

23. a. Mass $= 5^3 - 3^3 = 98$ g

Length = 2 cm

Average density = 49 g/cm

b. $\lim\limits_{h\to 0} \dfrac{(3+h)^3 - (3)^3}{h} = \lim\limits_{h\to 0} 27 + 9h + h^2$

$= 27$ g/cm

25. $\lim\limits_{h\to 0} \dfrac{0.2(10+h)^2 + 0.09(10+h)^3 - 0.2(10)^2 - 0.009(10)^3}{h}$

$= \lim\limits_{h\to 0} 6.7 + 0.47h + 0.009h^2 = 6.7$ g/wk

27. 700,000 gallons used in 24 hours, so average usage was $\frac{700,000}{24} \approx 29,167$ gallons per hour.
At 8, the average usage was approximately $\frac{200,000}{1.5} \approx 133,333$ gallons per hour.

29. $r = 2t;\ A = \pi r^2 = 4\pi t^2$

$$A'(3) = \lim_{h\to 0} \frac{4\pi(3+h)^2 - 4\pi(3)^2}{h}$$
$$= \lim_{h\to 0}(24\pi + 4\pi h) = 24\pi \approx 75.4 \text{ km}^2 \text{ per day}$$

31.

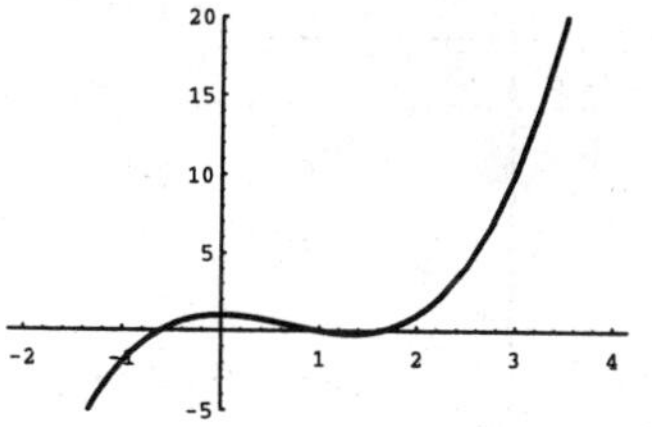

a. Using a table of values, $f'(-1) = 7$

b. Using a table of values, $f'(0) = 0$

c. Using a table of values, $f'(1) = -1$

d. Using a table of values, $f'(3.2) = 17.92$

33. Using a table of values, $s'(3) \approx 2.8$

35. Instantaneous velocity at $t = 1$ is

$$\lim_{h\to 0} \frac{16.09(1+h)^2 - 16.09}{h} = \lim_{h\to 0} 32.18 + 16.09h$$
$= 32.18$ feet per second

Section 1.4

Concepts Review

1. $\frac{1}{a^n}$; $\sqrt[m]{a^n}$; $\left(\sqrt[m]{a}\right)^n$ if m and n are positive integers.

3. 0

Problem Set 1.4

1. a. 1000

b. $\frac{1}{1000}$

c. 1

d. $8^{2/3} = \left(\sqrt[3]{8}\right)^2 = 4$

e. $8^{-2/3} = \left(\sqrt[3]{8}\right)^{-2} = \frac{1}{4}$

f. 1

3.

x	10^x
–2	0.01
–1	0.1
0	1
1	10
2	100

x	10^{-x}
–2	100
–1	10
0	1
1	0.1
2	0.01

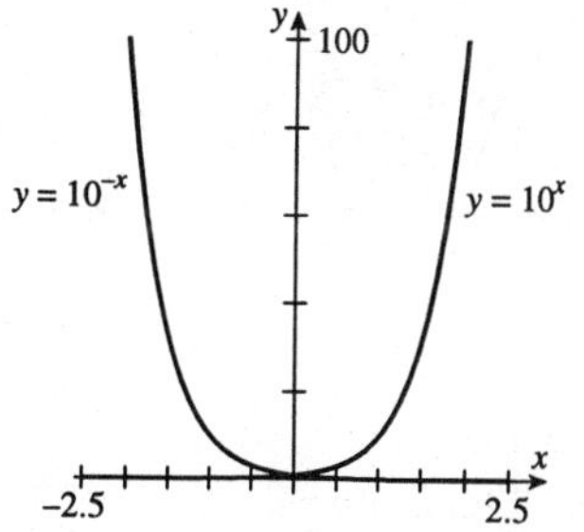

5.

x	x^2e^{-x}
–2	29.6
–1	2.718
0	0
1	0.368
2	0.541
3	0.448
4	0.293

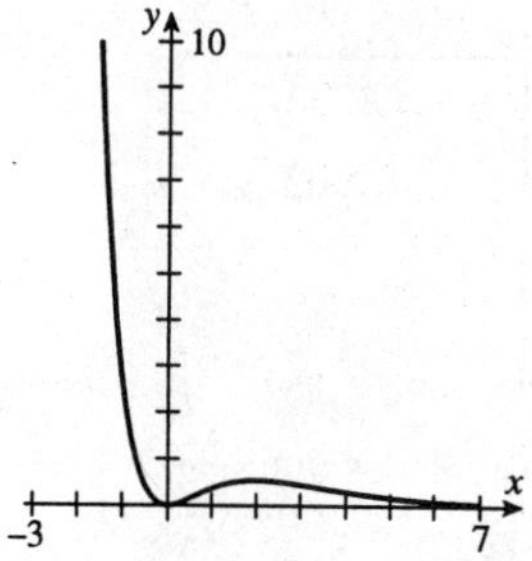

From left to right, the graph descends rapidly to (0, 0), then rises briefly near $x = 2$.
$$\lim_{x\to\infty} x^2e^{-x} = 0$$

7.

x	e^{-x^2}
–3	1.23E – 4
–2	0.0183
–1	0.368
0	1
1	0.368
2	0.0183
3	1.23E – 4

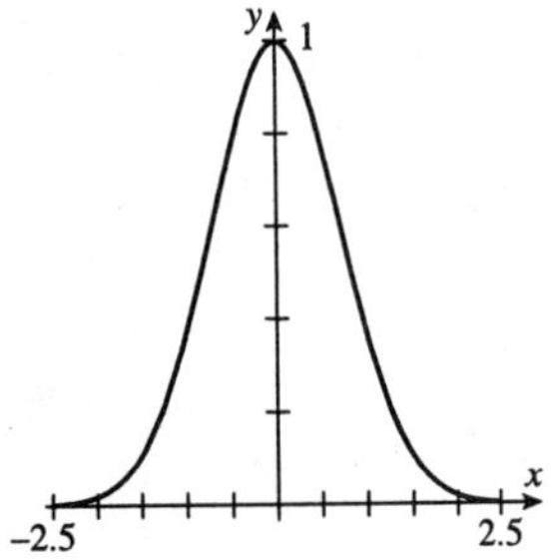

The graph is close to the x-axis except near $x = 0$ when it rises briefly. $\lim_{x\to\infty} e^{-x^2} = 0$

$\lim_{x\to-\infty} e^{-x^2} = 0$

9. a. $\lim_{x\to\infty} 1+\frac{x}{3^x} = 1$

x	$1+\frac{x}{3^x}$
1	1.3333
10	1.0002
100	1
1000	1

b. $\lim_{x\to\infty} 10^{-x}(x^2+1) = \lim_{x\to\infty} \frac{x^2}{10^x}+\frac{1}{10^x} = 0$

x	$10^{-x}(x^2+1)$
1	0.2
10	1.01E – 8
100	1.0001E – 96
1000	0

c. $\lim_{x\to\infty} x^{-1}(x^2+1) = \lim_{x\to\infty} x+\frac{1}{x} = \infty$

x	$x^{-1}(x^2+1)$
1	2
10	10.1
100	100.01
1000	1000.001

d. $\lim_{x\to\infty} xe^{-x}+5 = 5$

x	$xe^{-x}+5$
1	5.3679
10	5.0005
100	5
1000	5

11. $f'(c) = \lim_{h\to 0}\frac{e^{3(c+h)}-e^{3c}}{h} \approx 3e^{3c}$

$f'(-1) \approx 3e^{-3} \approx 0.149$

$f'(0) \approx 3e^{0} = 3$

$f'(1) \approx 3e^{3} \approx 60.3$

The slope of the curve $y = e^{3x}$ is 3 times the y-coordinate at any point.

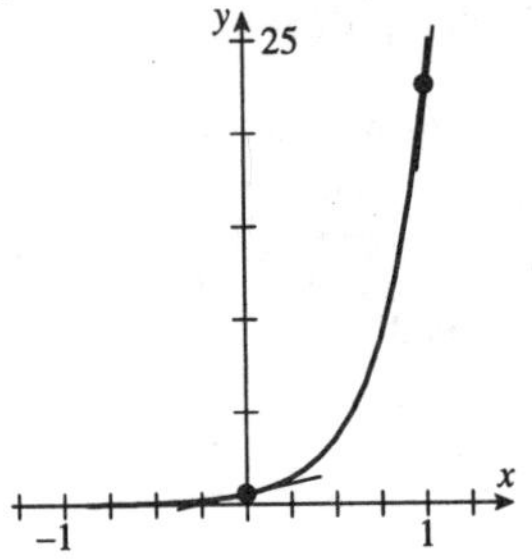

13. $f'(c) = \lim_{h\to 0}\frac{2^{c+h}-2^c}{h} = \lim_{h\to 0} 2^c\frac{2^h-1}{h}$

$\approx 0.693(2^c)$

$f'(-1) \approx 0.3465$

$f'(0) \approx 0.693$

$f'(1) \approx 1.386$

The slope of the curve $y = 2^x$ is approximately 0.693 times the y-coordinate at any point.

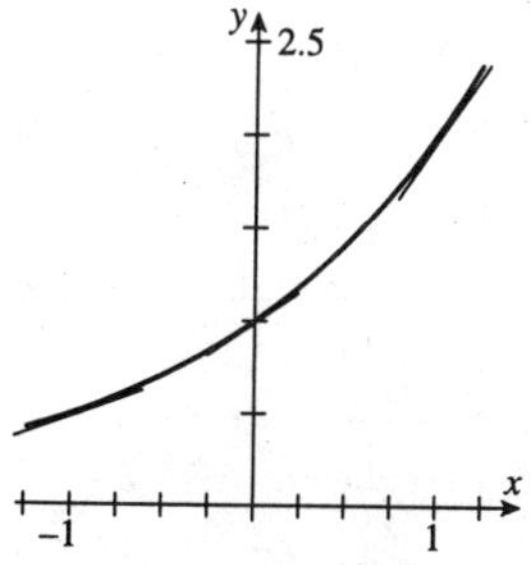

15. $A(n) = (1.03)^n(500)$

$A(10) = (1.03)^{10}(500) = 671.96$

To find n when $A(n) = 600$ check values n: $A(6) = 597.03$, $A(7) = 614.94$. Since interest is paid once each year, $A(n)$ reaches \$600 after 7 years.

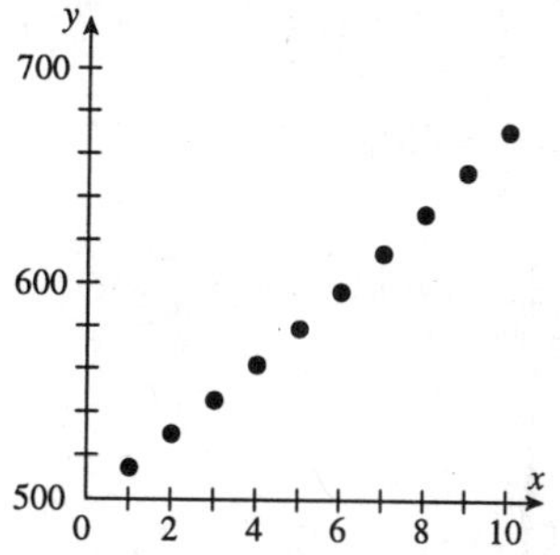

17. If $a < b$, then $e^a < e^b$ and dividing both sides by e^{ab} gives $\frac{1}{e^b} < \frac{1}{e^a}$ or $e^{-b} < e^{-a}$.

19. $e^{0.3} \approx 1.34985880758$

$$\left\{\left[\left(\frac{0.3}{4}+1\right)\frac{0.3}{3}+1\right]\frac{0.3}{2}+1\right\}0.3+1 = 1.3498375$$

21.

x	$\frac{e^{1/x}+e^{2/x}+\cdots+e^{x/x}}{x}$
1	$e \approx 2.718$
2	2.184
5	1.896
10	1.806
100	1.727
1000	1.719
10,000	1.718

$$\lim_{x\to\infty} \frac{e^{1/n}+e^{2/n}+\cdots+e^{n/n}}{n} \approx 1.72$$

Section 1.5

Concepts Review

1. $\ln y;\ e^y$

3. $\ln x;\ \ln y;\ y;\ \ln x$

Problem Set 1.5

1. a. Let $f(x) = 2e^x$.

$f(-4) = 2e^{-4} \approx 0.0366;$

$f(-2) = 2e^{-2} \approx 0.271;$

$f(0) = 2e^0 = 2;\ f(2) = 2e^2 \approx 14.8;$

$f(4) = 2e^4 \approx 109$

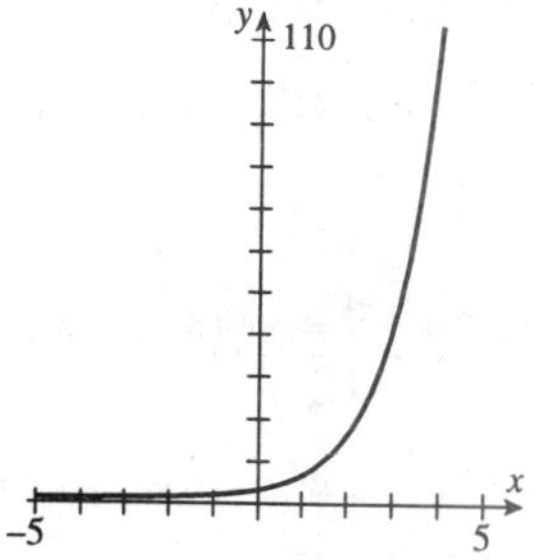

$2 = 2e^x;\ 1 = e^x;\ x = \ln 1 = 0$

b. Let $f(x) = e^{2x}$.

$f(-4) = e^{-8} \approx 0.0003;$

$f(-2) = e^{-4} \approx 0.0183;$

$f(0) = e^0 = 1;\ f(2) = e^4 \approx 54.6$

$f(4) = e^8 \approx 2981$

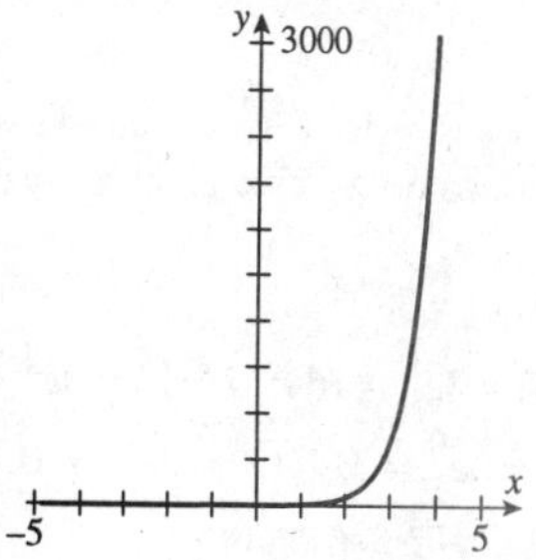

$2 = e^{2x};\ \ln 2 = 2x;\ x = \frac{1}{2}\ln 2 \approx 0.347$

c. Let $f(x) = 1 + e^x$.
$f(-4) = 1 + e^{-4} \approx 1.018$;
$f(-2) = 1 + e^{-2} \approx 1.135$;
$f(0) = 1 + e^0 = 2$; $f(2) = 1 + e^2 \approx 8.389$;
$f(4) = 1 + e^4 \approx 55.6$

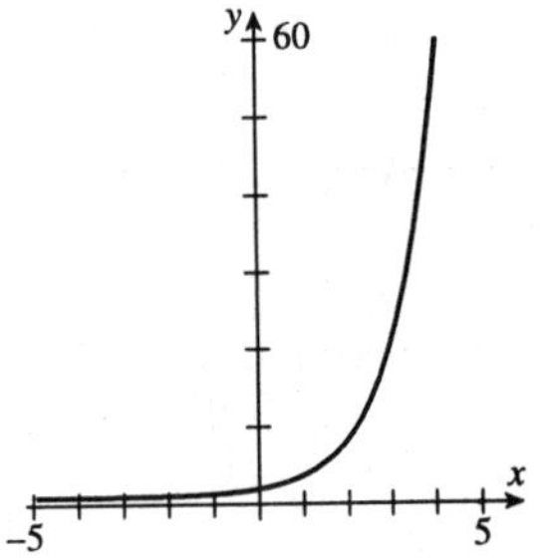

$2 = 1 + e^x$; $1 = e^x$; $x = \ln 1 = 0$

d. Let $f(x) = 3 + e^x$.
$f(-4) = 3 + e^{-4} \approx 3.018$;
$f(-2) = 3 + e^{-2} \approx 3.135$;
$f(0) = 3 + e^0 = 4$; $f(2) = 3 + e^2 \approx 10.389$;
$f(4) = 3 + e^4 \approx 57.6$

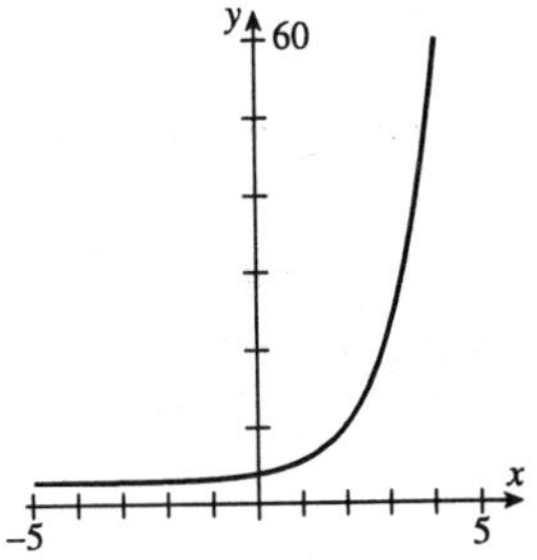

$2 = 3 + e^x$; $-1 = e^x$; $x = \ln(-1)$ which is not defined, so no x-value gives a y-value of 2.

e. Let $f(x) = 3e^{-x}$.
$f(-4) = 3e^4 \approx 164$; $f(-2) = 3e^2 \approx 22.2$;
$f(0) = 3e^0 = 3$; $f(2) = 3e^{-2} \approx 0.406$;
$f(4) = 3e^{-4} \approx 0.0549$

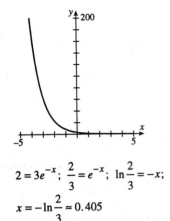

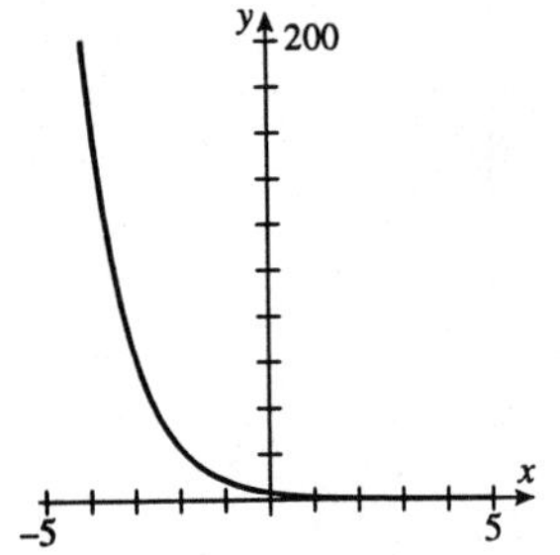

$2 = 3e^{-x}$; $\frac{2}{3} = e^{-x}$; $\ln\frac{2}{3} = -x$;
$x = -\ln\frac{2}{3} \approx 0.405$

f. Let $f(x) = e^{-3x}$.
$f(-4) = e^{12} \approx 162,755$;
$f(-2) = e^6 \approx 403$; $f(0) = e^0 = 1$;
$f(2) = e^{-6} \approx 0.00248$;
$f(4) = e^{-12} \approx 6.14\text{E}-6$

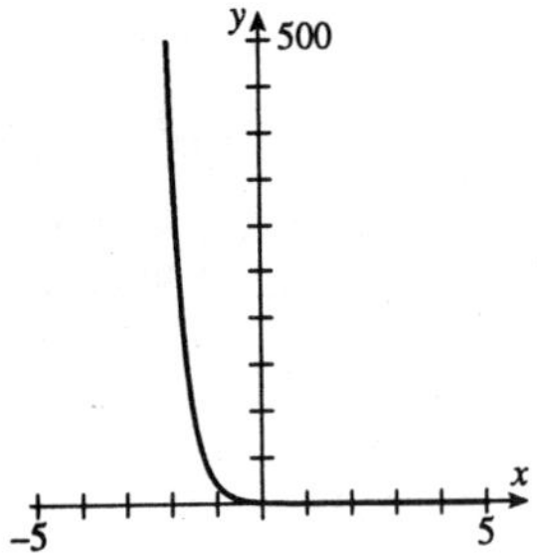

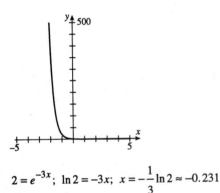

$2 = e^{-3x}$; $\ln 2 = -3x$; $x = -\frac{1}{3}\ln 2 \approx -0.231$

g. Let $f(x) = 3 - e^{2x}$.
$f(-4) = 3 - e^{-8} \approx 2.9997$;
$f(-2) = 3 - e^{-4} \approx 2.982$; $f(0) = 3 - e^0 = 2$;
$f(2) = 3 - e^4 \approx -51.6$;
$f(4) = 3 - e^8 \approx -2978$

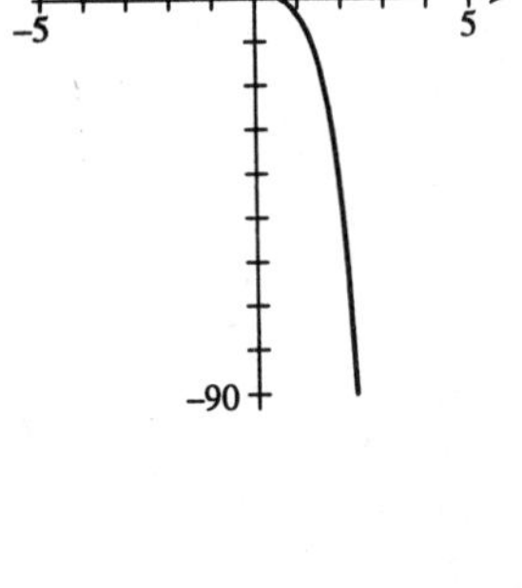

$2 = 3 - e^{2x};\ 1 = e^{2x};\ \ln 1 = 2x;$
$x = \frac{1}{2}\ln 1 = 0$

h. Let $f(x) = 1 - e^{-2x}$.
$f(-4) = 1 - e^{8} \approx -2980;$
$f(-2) = 1 - e^{4} \approx -53.6;$
$f(0) = 1 - e^{0} = 0;\ f(2) = 1 - e^{-4} \approx 0.982;$
$f(4) = 1 - e^{-8} \approx 0.9997$

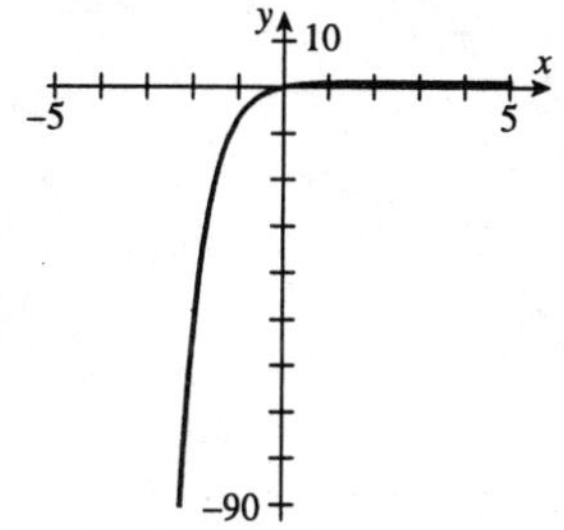

$2 = 1 - e^{-2x};\ -1 = e^{-2x};\ -2x = \ln(-1)$
which is not defined, so no x-value gives a y-value of 2.

3. a. $10^x = (e^{\ln 10})^x = e^{(\ln 10)x};$
$y = e^{(\ln 10)x} \approx e^{2.3026x}$

b. $3^x = (e^{\ln 3})^x = e^{(\ln 3)x};$
$y = e^{(\ln 3)x} \approx e^{1.0986x}$

c. $2^{-x} = (e^{\ln 2})^{-x} = e^{-(\ln 2)x};$
$y = e^{-(\ln 2)x} \approx e^{-0.6931x}$

d. $5^{-x} = (e^{\ln 5})^{-x} = e^{-(\ln 5)x};$
$y = e^{-(\ln 5)x} \approx e^{-1.6094x}$

e. $5^x = (e^{\ln 5})^x = e^{(\ln 5)x};$
$y = 4e^{(\ln 5)x} \approx 4e^{1.6094x}$

f. $2^{-x} = (e^{\ln 2})^{-x} = e^{-(\ln 2)x};$
$y = 5e^{-(\ln 2)x} \approx 5e^{-0.6931x}$

g. $2^{-3x} = (e^{\ln 2})^{-3x} = e^{-3(\ln 2)x};$
$y = e^{-3(\ln 2)x} \approx e^{-2.0794x}$

h. $4^{5x} = (e^{\ln 4})^{5x} = e^{5(\ln 4)x};$
$y = -e^{5(\ln 4)x} \approx -e^{6.9315x}$

5. a. $f(0) = 2e^{0} = 2;$ \$2000

b. $f(5) = 2e^{0.35} \approx 2.83814;$ \$2838.14

c. $5 = 2e^{0.07t};\ 2.5 = e^{0.07t};$
$\ln 2.5 = \ln e^{0.07t} = 0.07t \ln e;$
$t = \frac{1}{0.07}\ln 2.5 \approx 13.09$. It takes a little bit longer than 13 years and 1 month.

7. a. $\ln 6 = \ln(2\cdot 3) = \ln 2 + \ln 3$
$= 0.693 + 1.099 = 1.792$

b. $\ln 1.5 = \ln\frac{3}{2} = \ln 3 - \ln 2$
$= 1.099 - 0.693 = 0.406$

c. $\ln 81 = \ln 3^4 = 4\ln 3 = 4(1.099) = 4.396$

d. $\ln\sqrt{2} = \ln 2^{1/2} = \frac{1}{2}\ln 2 = \frac{1}{2}(0.693)$
$= 0.3465$

e. $\ln\left(\frac{1}{36}\right) = \ln 1 - \ln 36 = -\ln(9\cdot 4)$
$= -\ln 3^2 - \ln 2^2 = -2\ln 3 - 2\ln 2$
$= -2(1.099) - 2(0.693) = -3.584$

f. $\ln 48 = \ln(3\cdot 16) = \ln 3 + \ln 2^4$
$= \ln 3 + 4\ln 2 = 1.099 + 4(0.693) = 3.871$

9. $3^x = 9;\ x = 2$

11. $9^{3/2} = x;\ 3^3 = x;\ x = 27$

13. $10^{1/2} = \frac{x}{3};\ x = 3\sqrt{10}$

15. $\log_2 \frac{x+1}{x} = 2;\ \frac{x+1}{x} = 2^2;$
$x + 1 = 4x;\ 1 = 3x;\ x = \frac{1}{3}$

17. $x\ln 2 = \ln 19;\ x = \frac{\ln 19}{\ln 2} \approx 4.248$

19. $(3x-1)\ln 4 = \ln 5;\ 3x = \frac{\ln 5}{\ln 4} + 1;$
$x = \frac{1}{3}\left(\frac{\ln 5}{\ln 4} + 1\right) \approx 0.7203$

21. $e^{2\ln x} = e^{\ln x^2} = x^2$

23. $\ln e^{\sin x} = \sin x$

25. $\ln(x^2e^{-2x}) = \ln x^2 + \ln e^{-2x}$
$= 2\ln x - 2x\ln e = -2x + 2\ln x$

27. $e^{\ln 2 + \ln x} = e^{\ln 2x} = 2x$

29. $2\ln(x+1) - \ln x = \ln(x+1)^2 - \ln x$
$= \ln \dfrac{(x+1)^2}{x}$

31. $\ln(x-2) - \ln(x+2) + 2\ln x$
$= \ln \dfrac{x-2}{x+2} + \ln x^2 = \ln \dfrac{x^2(x-2)}{x+2}$

33.

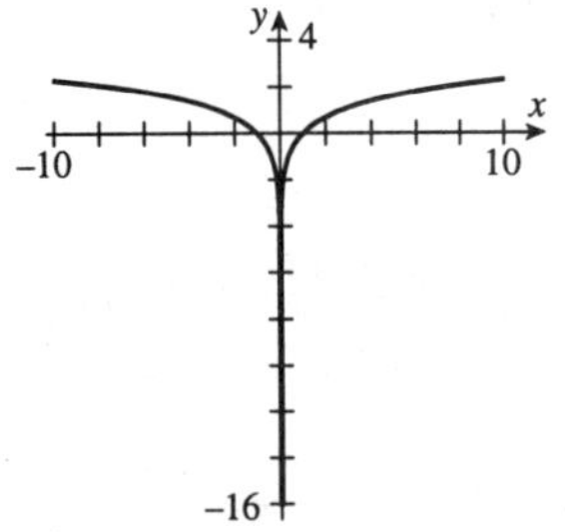

35.

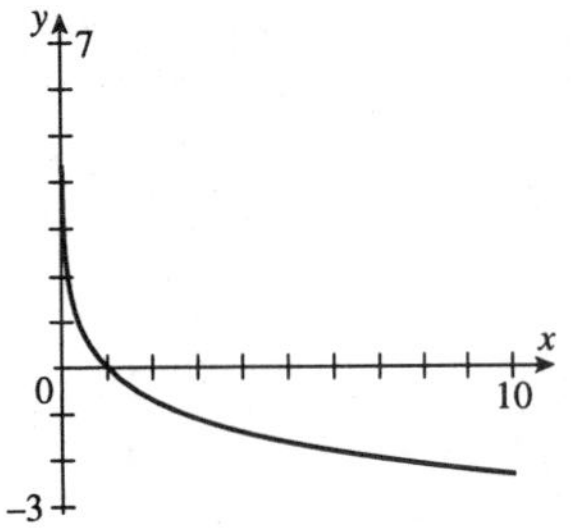

37.

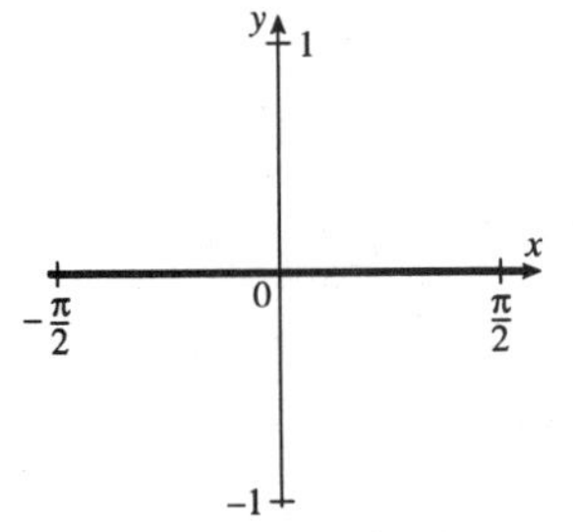

$\ln \cos x + \ln \sec x = \ln(\cos x \sec x) = \ln 1 = 0$

39. $\log_6(0.12) = \dfrac{\ln(0.12)}{\ln 6} \approx -1.183$

41. $\log_{10}(91.2)^3 = 3\log_{10}(91.2) = \dfrac{3\ln(91.2)}{\ln 10}$
≈ 5.88

43. a. $f'(c) = \lim\limits_{h\to 0} \dfrac{\ln 2(c+h) - \ln 2c}{h}$
$= \lim\limits_{h\to 0} \dfrac{\ln\left(1+\frac{h}{c}\right)}{h} = \lim\limits_{h\to 0} \ln\left(1+\dfrac{h}{c}\right)^{1/h}$
$= \lim\limits_{h\to 0} \ln\left[\left(1+\dfrac{h}{c}\right)^{c/h}\right]^{1/c} = \ln e^{1/c}$
$= \dfrac{1}{c}$
$f'(0.5) = 2;\ f'(3) = \dfrac{1}{3}$

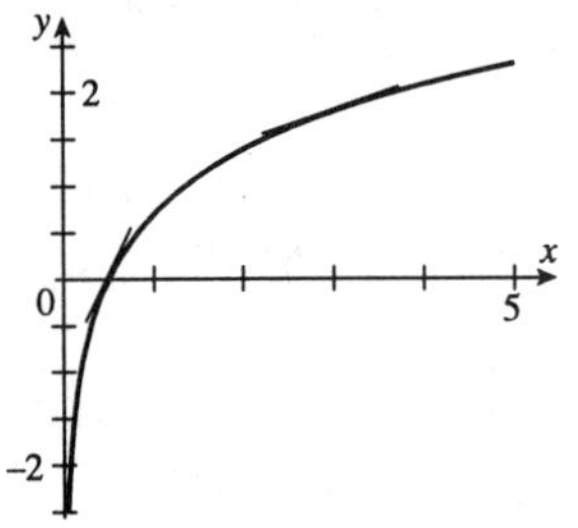

b. $f'(c) = \lim\limits_{h\to 0} \dfrac{\ln(2+c+h) - \ln(2+c)}{h}$
$= \lim\limits_{h\to 0} \ln\left(1+\dfrac{h}{2+c}\right)^{1/h}$
$= \lim\limits_{h\to 0} \ln\left[\left(1+\dfrac{h}{2+c}\right)^{\frac{2+c}{h}}\right]^{\frac{1}{2+c}}$
$= \ln e^{\frac{1}{2+c}} = \dfrac{1}{2+c}$
$f'(0.5) = 0.4;\ f'(3) = 0.2$

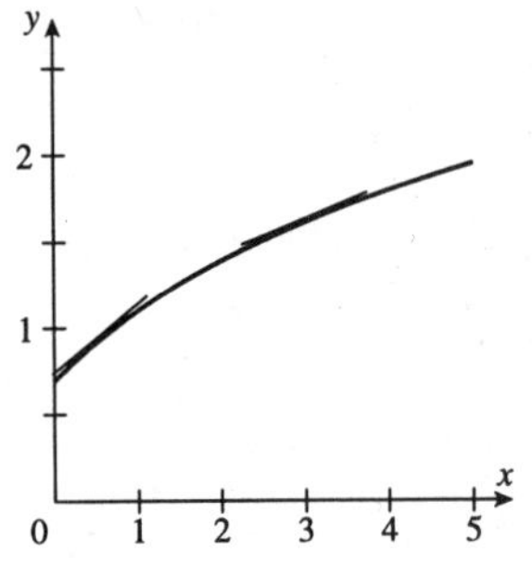

c. $f'(c) = \lim_{h \to 0} \frac{\ln(c+h-1) - \ln(c-1)}{h}$

$= \lim_{h \to 0} \ln\left(1 + \frac{h}{c-1}\right)^{1/h}$

$= \lim_{h \to 0} \ln\left[\left(1 + \frac{h}{c-1}\right)^{\frac{c}{1-h}}\right]^{\frac{1}{c-1}}$

$= \ln e^{\frac{1}{c-1}} = \frac{1}{c-1}$

$x - 1 > 0$ only if $x > 1$ so $f(0.5)$ is not defined; $f'(3) = 0.5$.

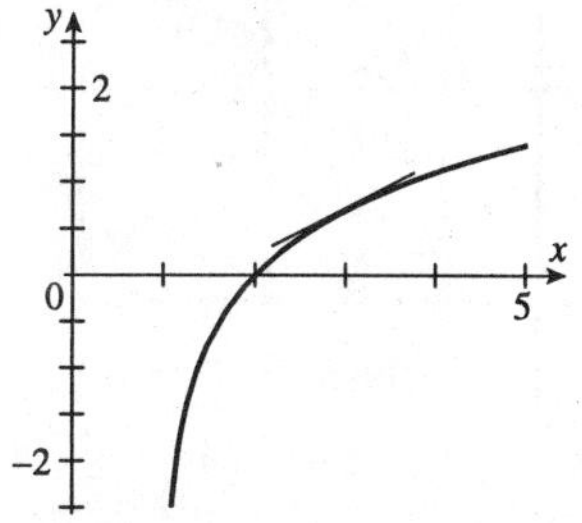

d. $f'(c) = \lim_{h \to 0} \frac{2\ln(c+h) - 2\ln c}{h}$

$= 2 \lim_{h \to 0} \ln\left(1 + \frac{h}{c}\right)^{1/h}$

$= 2 \lim_{h \to 0} \ln\left[\left(1 + \frac{h}{c}\right)^{c/h}\right]^{1/c} = 2\ln e^{1/c}$

$= \frac{2}{c}$

$f'(0.5) = 4;\ f'(3) = 0.667$

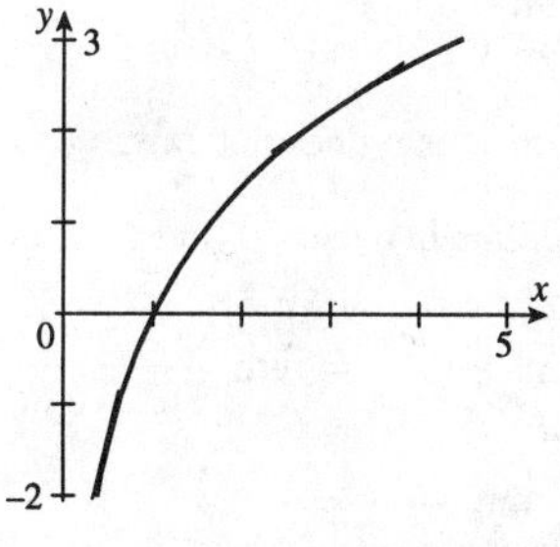

45. $a^{\log_a x} = x;\ \ln a^{\log_a x} = \log_a x \ln a$, so $\log_a x \ln a = \ln x$ and $\log_a x = \frac{\ln x}{\ln a}$.

47.

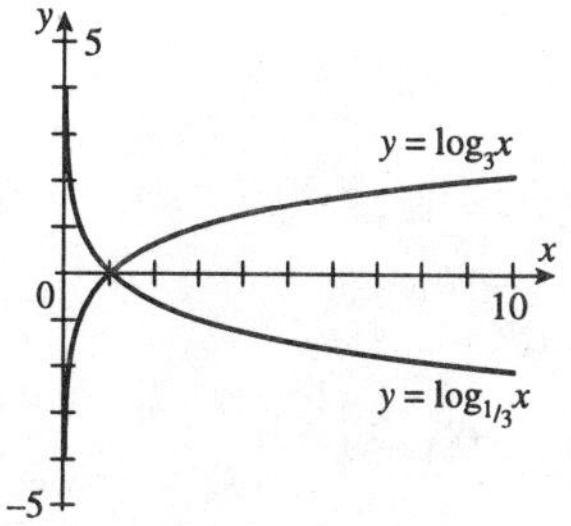

49. $\ln 4^m = m \ln 4 > m$ since $\ln 4 > 1$.
We can conclude $\lim_{x \to \infty} \ln x = \infty$.

51. $\frac{1,000,000}{\ln 1,000,000} \approx 72,382$. There are approximately 72,382 primes less than 1,000,000.

53. $115 = 20 \log_{10}(121.3P);\ 5.75 = \log_{10}(121.3P);$

$P = \frac{1}{121.3} 10^{5.75} \approx 4636$

The variation in pressure is approximately 4636 pounds per square inch.

55. Assume $\log_2 3 = \frac{m}{n}$. $2^{m/n} = 3;\ 2^m = 3^n$. But 2^m is divisible only by powers of 2, while 3^n is divisible only by powers of 3, so the only way $2^m = 3^n$ is possible if $m = n = 0$. But then $\frac{m}{n}$ is undefined.

57.

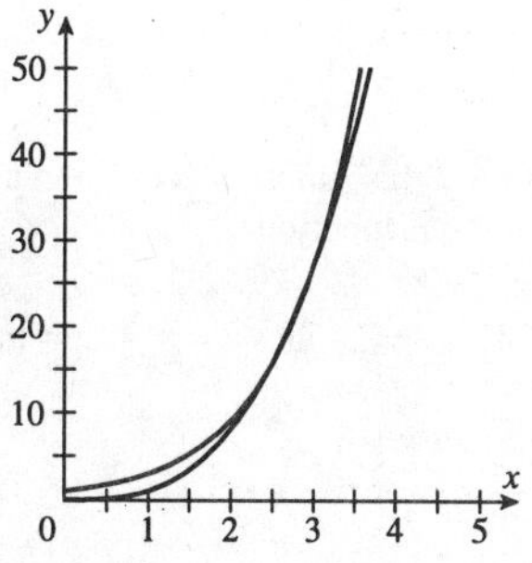

$x^3 = 3^x$ when $x = 3$.

Section 1.6 Chapter Review

Concepts Test

1. False. Consider $f(x) = \begin{cases} x^2 & x \neq 2 \\ 2 & x = 2 \end{cases}$.

3. True; $\lim_{x \to 5} f(x) = 10$

5. False. Consider $f(x)=\begin{cases} x^2 & x\neq 2 \\ 2 & x=2 \end{cases}$.

$\lim_{x\to 2^-} f(x)=4=\lim_{x\to 2^+} f(x)$, but $f(x)$ is not continuous at $x = 2$.

7. False. Consider $f(x)=\begin{cases} 10-|x-2| & x\neq 2 \\ 8 & x=2 \end{cases}$.

$f(x) = 10$ for all x and $\lim_{x\to 2} f(x)$ exists and is 10.

9. True. Some limits cannot be evaluated algebraically with the techniques available to us.

11. False. See Example 5 in Section 1.1.

13. False. The function is discontinuous at $x = 3$, however $\lim_{x\to 3} f(x)=5$.

15. False. look at the line tangent to $y=\sin x$ at $x = 0$.

17. True. $f'(c)$ is the slope of the tangent line to $f(x)$ at $x = c$.

19. False. If the directed distance is given by $f(t)=t^3$, then the average velocity between $t = 1$ and $t = 3$ is $\frac{27-1}{3-1}=13$ while the instantaneous velocity at $t = 2$ is $f'(2)=12$.

21. True. $\left(\frac{1}{2}\right)^x=2^{-x}=\frac{1}{2^x}$

23. True. This is why this function is defined as the natural exponential function.

25. True. $2e^x=4;\ e^x=2;\ \ln e^x=\ln 2;\ x=\ln 2$

27. True. $\log_2 x=\frac{\ln x}{\ln 2}$

Sample Test Problems

1. $\lim_{u\to 1}\frac{u^2-1}{u+1}=0,\ 1^2-1=0,\ 1+1=2$

3. $\lim_{u\to 1}\frac{u+1}{u^2-1}=\lim_{u\to 1}\frac{u+1}{(u+1)(u-1)}$
$=\lim_{u\to 1}\frac{1}{u-1}$ does not exist since
$\lim_{u\to 1^-}\frac{1}{u-1}=-\infty;\ \lim_{u\to 1^+}\frac{1}{u-1}=\infty$

5. $\lim_{z\to 2}\frac{z^2-4}{z^2+z-6}=\lim_{z\to 2}\frac{z+2}{z+3}=\frac{4}{5}$

7. $\lim_{y\to 1}\frac{y^3-1}{y^2-1}=\lim_{y\to 1}\frac{y^2+y+1}{y+1}=\frac{3}{2}$

9. $\lim_{x\to 0}\frac{\cos x}{x}$ does not exist. $\lim_{x\to 0^-}\frac{\cos x}{x}=-\infty$; $\lim_{x\to 0^+}\frac{\cos x}{x}=\infty$

x	$\frac{\cos x}{x}$	x	$\frac{\cos x}{x}$
1	0.540	–1	–0.540
0.1	9.950	–0.1	–9.950
0.01	99.995	–0.01	–99.995
0.001	999.9995	–0.001	–999.9995

11. a. $\lim_{x\to\infty}\frac{3x^2-2x+7}{2x^2+5x+9}=\lim_{x\to\infty}\frac{3-\frac{2}{x}+\frac{7}{x^2}}{2+\frac{5}{x}+\frac{9}{x^2}}=\frac{3}{2}$

b. $\lim_{x\to\infty}\frac{3x+9}{\sqrt{2x^2+1}}=\lim_{x\to\infty}\frac{3+\frac{9}{x}}{\sqrt{2+\frac{1}{x^2}}}=\frac{3}{\sqrt{2}}$

c. $\lim_{x\to-\infty}\frac{\sin x}{x}=0$ since $|\sin x|\le 1$
$-\frac{1}{x}\le\frac{\sin x}{x}\le\frac{1}{x}$. $\lim_{x\to-\infty}-\frac{1}{x}=0$ and
$\lim_{x\to\infty}\frac{1}{x}=0$.

d. $\lim_{x\to-\infty}\cos x$ does not exist. $\cos x$ continues to oscillate between –1 and 1 as $x\to-\infty$.

e. $\lim_{x\to 3^+}\frac{x+3}{x^2-9}=\lim_{x\to 3^+}\frac{x+3}{(x+3)(x-3)}$
$=\lim_{x\to 3^+}\frac{1}{x-3}=\infty$

f. $\lim_{x\to 2}\frac{x+2}{x^2-4}=\lim_{x\to 2}\frac{1}{x-2}$ does not exist since
$\lim_{x\to 2^+}\frac{x+2}{x^2-4}=\infty$ and $\lim_{x\to 2^-}\frac{x+2}{x^2-4}=-\infty$.

g. $\lim_{x\to 1^-}\frac{|x-1|}{x-1}=-1$ since for $x<1$, $x-1<0$ so $|x-1|=-(x-1)$.

h. $\lim_{x\to\infty}\cos\left(\frac{1}{x}\right)=1$. As $x\to\infty$, $\frac{1}{x}\to 0$, so $\cos\left(\frac{1}{x}\right)\to 1$.

13. **a.** f is discontinuous at $x=-1$, since $f(-1)$ is not defined. f is discontinuous at $x=1$, since $\lim_{x\to 1^-} f(x)=1$ and $f(1)=0$.

b. Define $f(-1)=-1$, or change either of the first two inequalities to allow $x=-1$.

15. Possible answer:

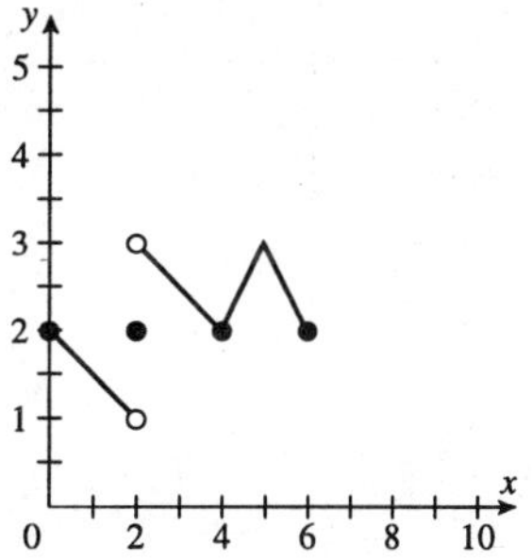

17. **a.** $f'(c)=\lim_{h\to 0}\frac{(c+h)^2-5(c+h)-[c^2-5c]}{h}$
$=\lim_{h\to 0} 2c+h-5=2c-5$
$f'(2)=-1$

b. $f'(c)=\lim_{h\to 0}\frac{\frac{1}{c+h-3}-\frac{1}{c-3}}{h}$
$=\lim_{h\to 0}\frac{-1}{(c+h-3)(c-3)}=-\frac{1}{(c-3)^2}$
$f'(2)=-\frac{1}{(-1)^2}=-1$

c. $f'(c)=\lim_{h\to 0}\frac{\sqrt{9-c-h}-\sqrt{9-c}}{h}$
$=\lim_{h\to 0}\frac{-1}{\sqrt{9-c-h}+\sqrt{9-c}}=-\frac{1}{2\sqrt{9-c}}$
$f'(2)=-\frac{1}{2\sqrt{7}}$

d. $f'(c)=\lim_{h\to 0}\frac{\ln(c+h)-\ln c}{h}$
$=\lim_{h\to 0}\ln\left(1+\frac{h}{c}\right)^{1/h}$
$=\lim_{h\to 0}\ln\left[\left(1+\frac{h}{c}\right)^{c/h}\right]^{1/c}=\ln e^{1/c}$
$=\frac{1}{c}$
$f'(2)=\frac{1}{2}$

19.

x	xe^{-x}
–1	–2.718
0	0
1	0.368
2	0.271
3	0.149
4	0.733

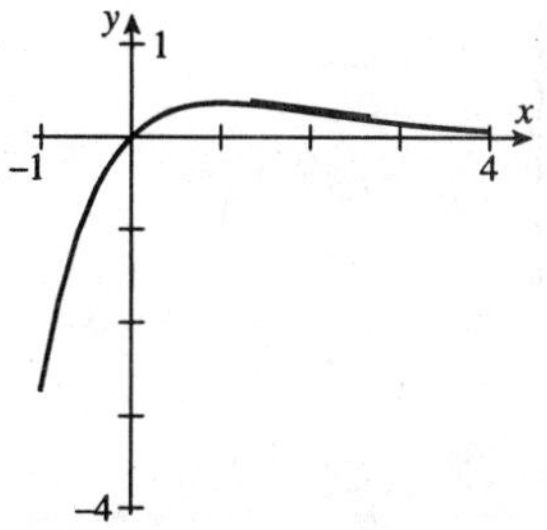

$f'(c)=\lim_{h\to 0}\frac{(c+h)e^{-c-h}-ce^{-c}}{h}$
$=\lim_{h\to 0}\left(e^{-c}e^{-h}+ce^{-c}\frac{e^{-h}-1}{h}\right)$
$=\lim_{h\to 0}\left(e^{-c}e^{-h}-ce^{-c}\frac{e^{-h}-1}{-h}\right)$
$=e^{-c}-ce^{-c}$
$f'(2)=e^{-2}-2e^{-2}\approx -0.135$

21. **a.** $f'(c)=\lim_{h\to 0}\frac{e^{-c-h}-e^{-c}}{h}$
$=\lim_{h\to 0}-e^{-c}\frac{e^{-h}-1}{-h}=-e^{-c}$
$f(-1)\approx 2.718;\ f'(-1)\approx -2.718$
$f(0)=1;\ f'(0)=-1$
$f(1)\approx 0.368;\ f'(1)\approx -0.368$
The slope is the negative of the y-coordinate at any point.

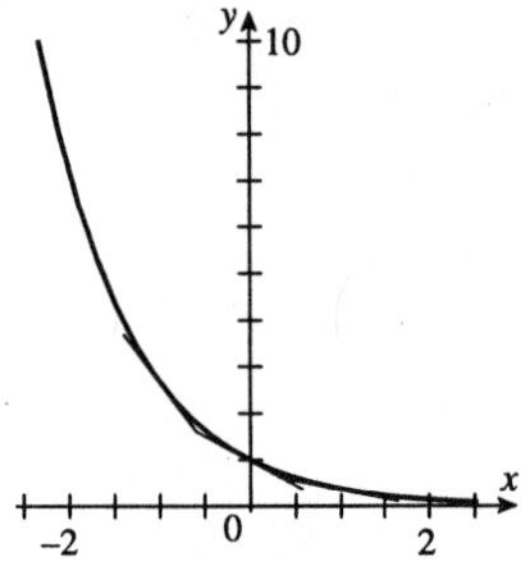

b. $f'(c) = \lim_{h \to 0} \frac{10^{-c-h} - 10^{-c}}{h}$

$= \lim_{h \to 0} 10^{-c} \frac{10^{-h} - 1}{h} = 10^{-c} \lim_{h \to 0} \frac{10^{-h} - 1}{h}$

$\approx (-2.3)10^{-c}$

h	$\frac{10^{-h} - 1}{h}$
1	–0.9
0.1	–2.057
0.01	–2.276
0.001	–2.3

$f(-1) = 10;\ f'(-1) \approx 23$
$f(0) = 1;\ f'(0) \approx -2.3$
$f(1) = 0.1;\ f'(1) \approx -0.23$
The slope is –2.3 times the y-coordinate at any point.

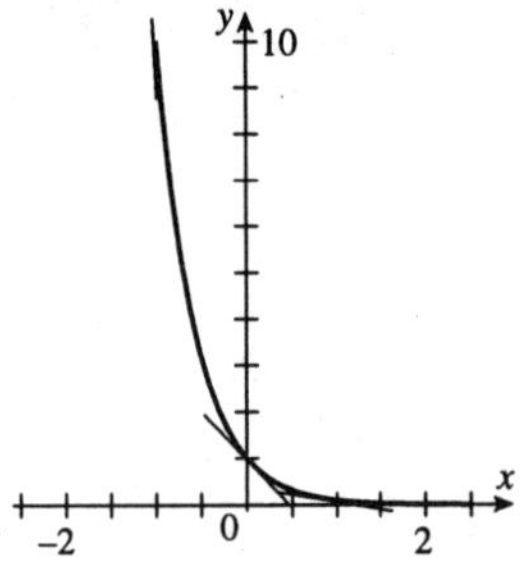

23. a. $\ln e^x = \ln 19;\ x = \ln 19 \approx 2.944$

b. $\ln e^x = \ln 4;\ x = \ln 4 \approx 1.386$

c. $\ln e^{3x-1} = \ln 5;\ 3x - 1 = \ln 5;$

$x = \frac{1}{3}(1 + \ln 5) \approx 0.870$

d. $\ln 12^x = \ln 3;\ x = \frac{\ln 3}{\ln 12} \approx 0.442$

Chapter 2

Section 2.1

Concepts Review

1. Pixels

3. Sometimes

Problem Set 2.1

1. a.

x	y	
–2.00	4.00	Plot at (–2, 4).
–1.00	1.00	Plot at (–1, 0).
0.00	0.00	Plot at (0, 0).
1.00	1.00	Plot at (1, 0).
2.00	4.00	Plot at (2, 4).

b.

x	y	
–2.00	–8.00	Plot at (–2, 8).
–1.00	–1.00	Plot at (–1, 0).
0.00	0.00	Plot at (0, 0).
1.00	1.00	Plot at (1, 0).
2.00	8.00	Plot at (2, 8).

Fill in graph with lit pixels at (2, 4) and (–2, –4) in order to give an appearance of a continuous curve.

c.

x	y	
–2.00	0.14	Plot at (–2, 0).
–1.00	0.37	Plot at (–1, 0).
0.00	1.00	Plot at (0, 0).
1.00	2.72	Plot at (1, 4).
2.00	7.39	Plot at (2, 8).

d.

x	y	
–2.00	–0.91	Plot at (–2, 0).
–1.00	–0.84	Plot at (–1, 0).
0.00	0.00	Plot at (0, 0).
1.00	0.84	Plot at (1, 0).
2.00	0.91	Plot at (2, 0).

3. The following graphs were done using Mathematica. The last graph in each part is generated with a default y-range.

a. $f(x) = 5 \sin x$

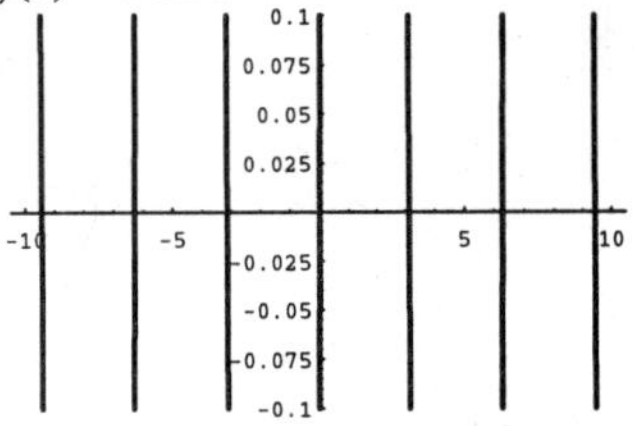

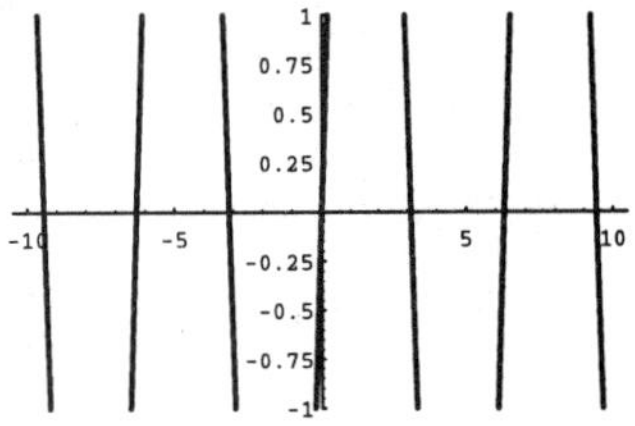

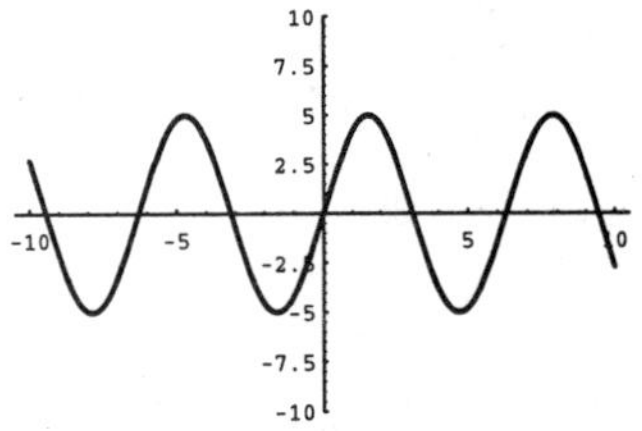

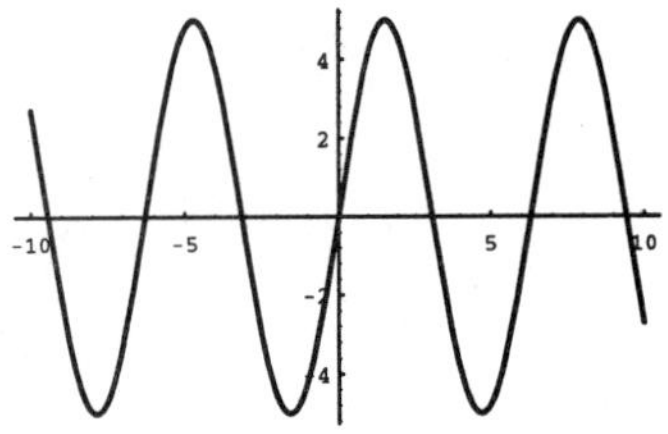

b. $f(x) = \dfrac{5}{x}$

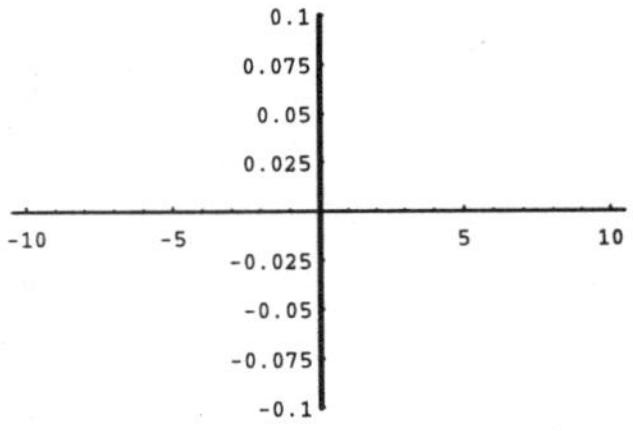

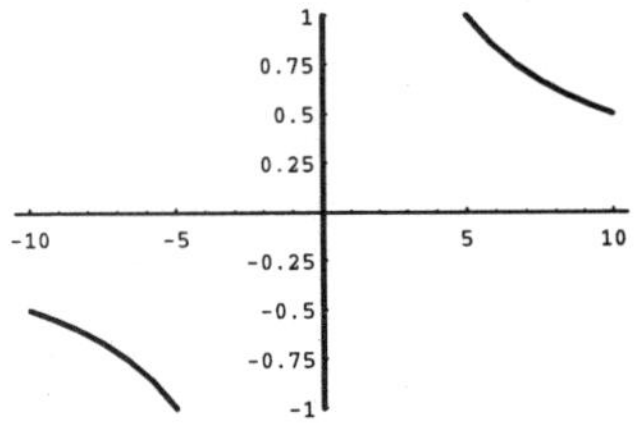

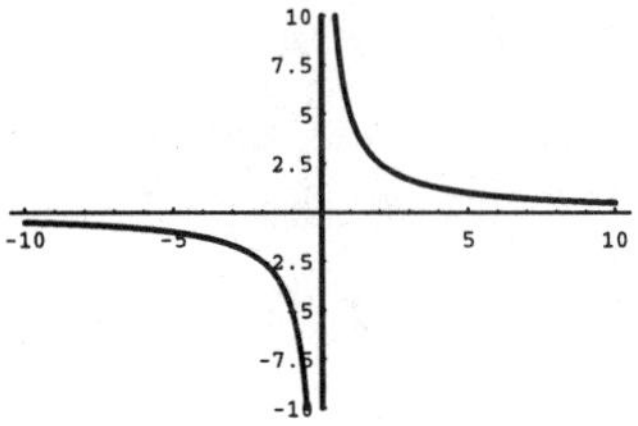

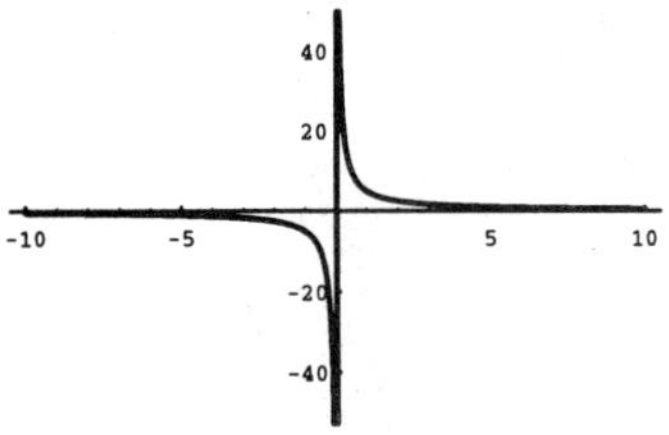

c. $f(x) = x^2$

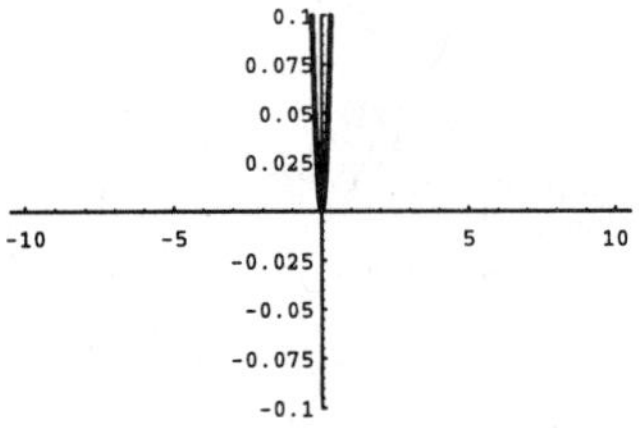

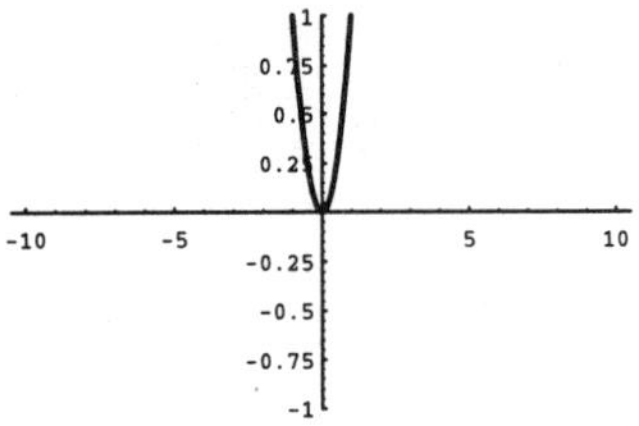

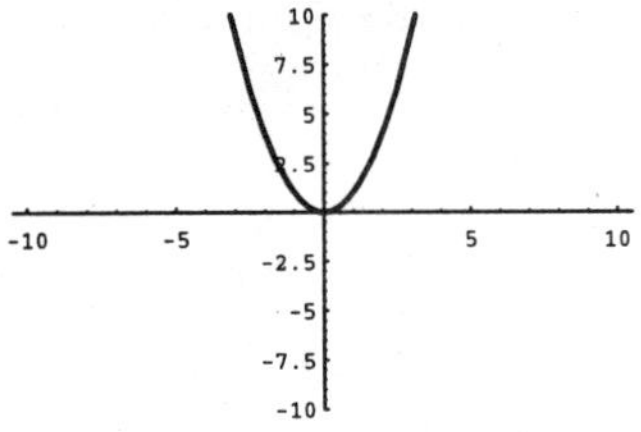

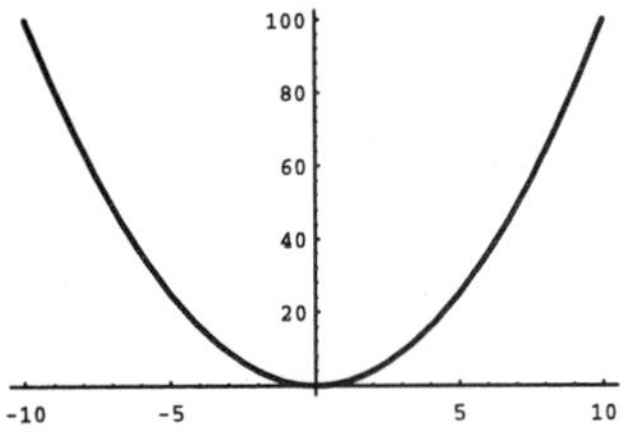

d. $f(x) = 0.01x + 0.001x^2$

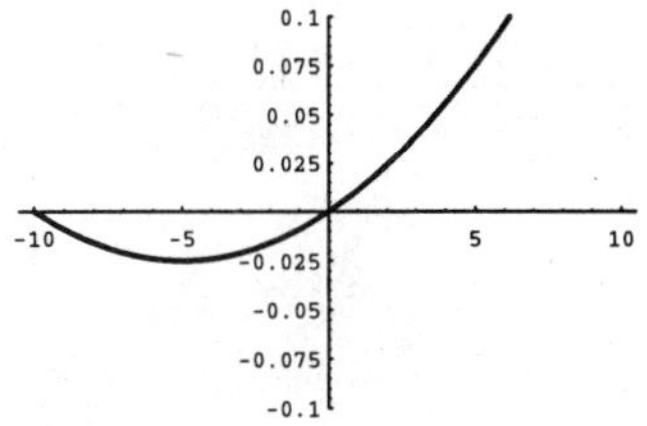

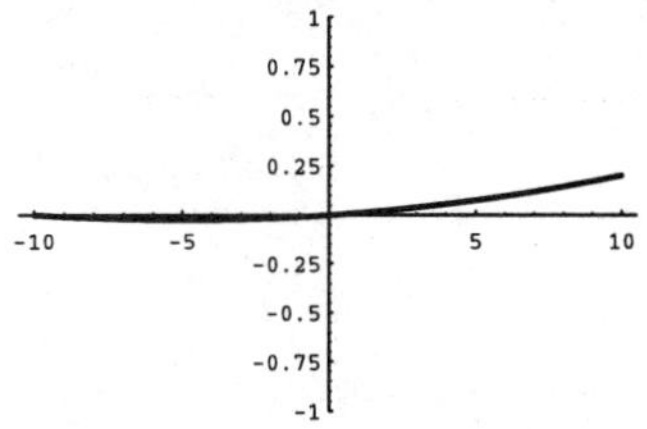

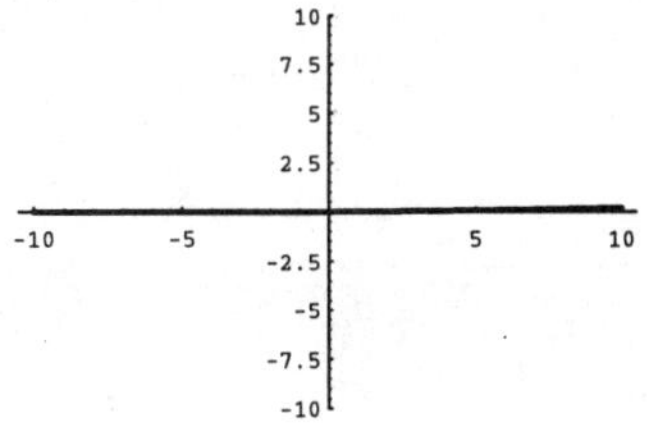

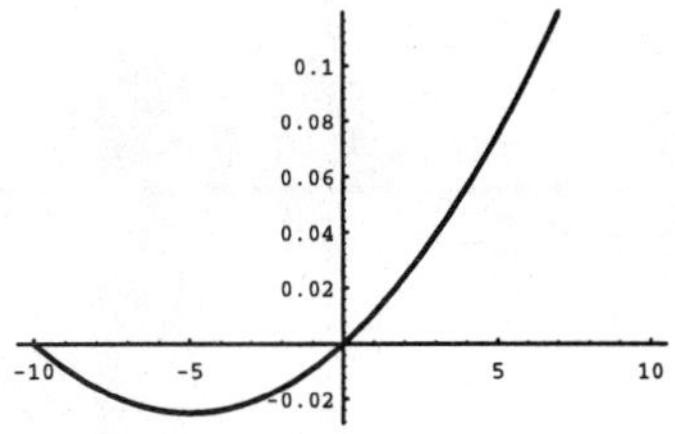

5. Written response. Possible intervals: $-5 \le x \le 5$, $-200 \le y \le 200$.

7. Answers will vary. For example, on the TI-82 set $-47 \le x \le 47$ and $-31 \le y \le 31$.

Section 2.2

Concepts Review

1. Table

3. 0; 1

Problem Set 2.2

1. a. $f(x) = 2e^{-x} \cos x + 1$

x	$f(x)$
0	3.0000
0.5	2.0646
1	1.3975
1.5	1.0316
2	0.8874
2.5	0.8685
3	0.9014
3.5	0.9434
4	0.9761
4.5	0.9953
5	1.0038

Interval estimate for maximum y-value: $0 \le x \le 0.5$, no interval estimates for which $f(x) = 0$.

b. $f(x) = 2x^2 e^{-x} + 1$

x	$f(x)$
0	1.0000
0.5	1.3033
1	1.7358
1.5	2.0041
2	2.0827
2.5	2.0261
3	1.8962
3.5	1.7398
4	1.5861
4.5	1.4499
5	1.3369

Interval estimate for maximum y-value: $1.5 \le x \le 2.5$, no interval estimates for which $f(x) = 0$.

c. $f(x) = 1 + 47x - 12x^2 + x^3$

x	$f(x)$
0	1.000
0.5	21.625
1	37.000
1.5	47.875
2	55.000
2.5	59.125
3	61.000
3.5	61.375
4	61.000
4.5	60.625
5	61.000

Interval estimate for maximum y-value: $3 \le x \le 4$, no interval estimates for which $f(x) = 0$.

d. $f(x) = \dfrac{3\ln(x+1)}{x+1} - 1$

x	$f(x)$
0	−1.0000
0.5	−0.1891
1	0.0397
1.5	0.0995
2	0.0986
2.5	0.0738
3	0.0397
3.5	0.0027
4	−0.0343
4.5	−0.0701
5	−0.1041

Interval estimate for maximum y-value: $1 \le x \le 2$

Interval estimate for which $f(x) = 0$: $0.5 \le x \le 1$, $3.5 \le x \le 4$

3. $y = -5e^{-10t}(\cos 10t + \sin 10t)$

t	y
1.21	−0.00001233
1.22	−0.00001447
1.23	−0.00001597
1.24	−0.00001690
1.25	−0.00001736
1.26	−0.00001742
1.27	−0.00001715
1.28	−0.00001662
1.29	−0.00001589
1.30	−0.00001500

Estimate the second rebound ratio to be $\dfrac{0.0000174}{5} = 0.000348\%$.

5.

V	P
80	54.102
90	41.802
100	41.355
110	44.158
120	47.436
130	50.311
140	52.558
150	54.186
160	55.276
170	55.922
180	56.213
190	56.224
200	56.018
210	55.644
220	55.142

Estimate a maximum y-value of 56.224 and a minimum y-value of 41.355. From Problem 6, Section 2.1, the maximum y-value is approximately 56.25 and the minimum y-value is approximately 40.85.

Section 2.3

Concepts Review

1. Parameter; family

3. Good fit

Problem Set 2.3

1. a. $y = x^3 - ax$

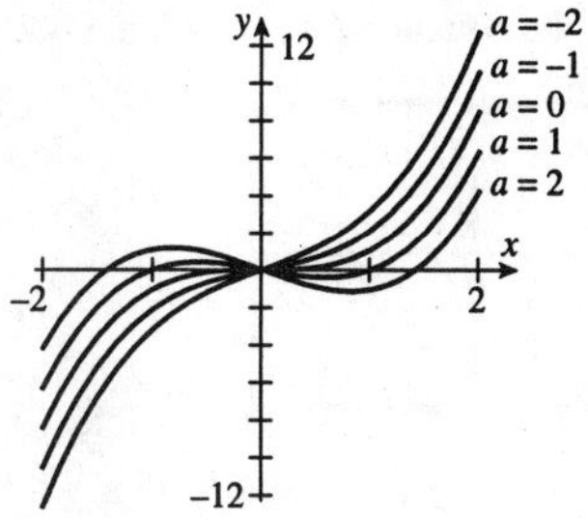

Written response

b. $y = xe^{-ax}$

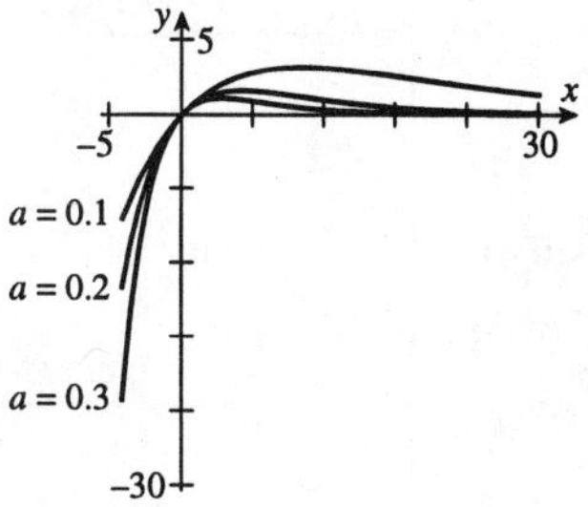

Written response

c. $y = x^n e^{-x}$

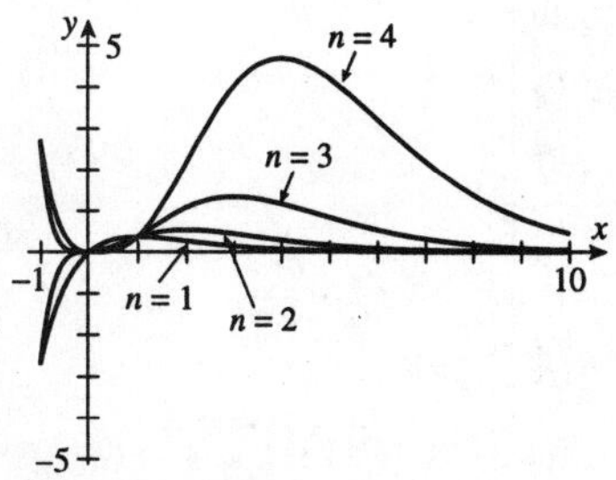

Written response

3. a. $y = Ae^{kt}$

When $t = 0$, $A = 100$. The graph of $y = 100e^{kt}$ fits the data for $k = 0.04$.

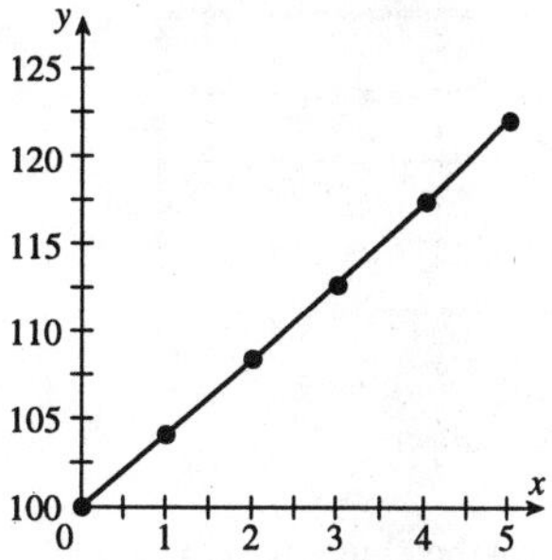

b. $y = Ae^{kt}$

When $t = 0$, $A = 500{,}000$. The graph of $y = 500{,}000e^{kt}$ fits the data for $k = 0.2$.

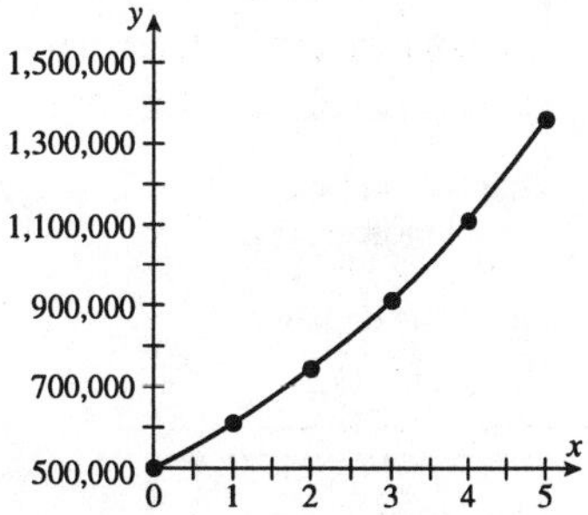

5. Written response

7. Estimate the middle temperature to be $\frac{40+66}{2} = 53$. A is about $66 - 53 = 13$, and d is about 4.5. $T = 12$, so $w = \frac{1}{T} = \frac{1}{12}$. The model is $y = 13\sin\left(\frac{\pi}{6}(x-4.5)\right) + 53$.

Section 2.4

Concepts Review

1. Zoom in; zoom out

3. Local linearity; tangent line

Problem Set 2.4

1. a. Graph the equation $f(x) = 2e^{-x}\cos x + 1$.

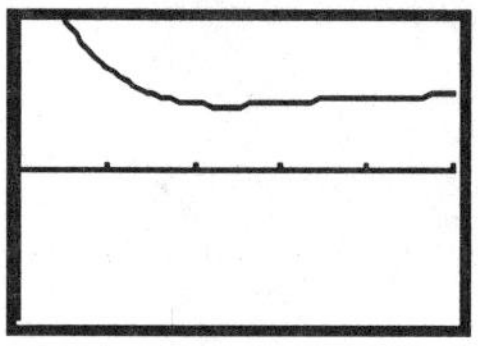

Graph window is $0 \le x \le 5, -2 \le y \le 2$.
No solution.

b. Graph the equation $f(x) = 2x^2 e^{-x} + 1$.

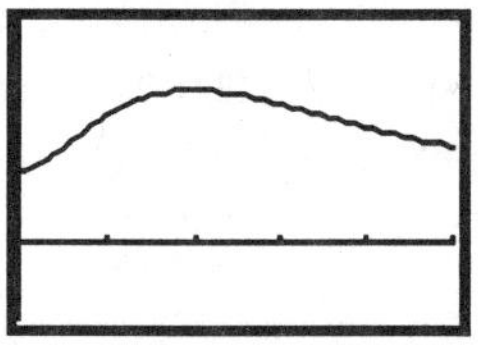

Graph window is $0 \le x \le 5, -1 \le y \le 3$.
No solution.

c. Graph the equation
$f(x) = 1 + 47x + 12x^2 + x^3$.

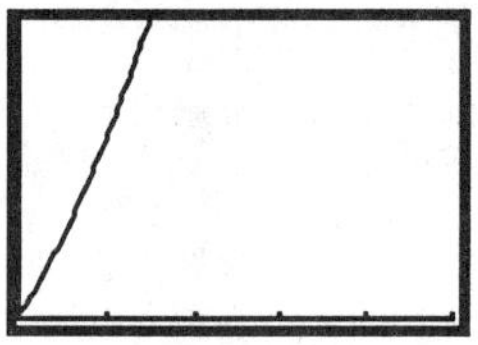

Graph window is $0 \le x \le 5, 0 \le y \le 100$.
No solution.

d. Graph the equation $f(x) = \dfrac{3\ln(x+1)}{x+1} - 1$.

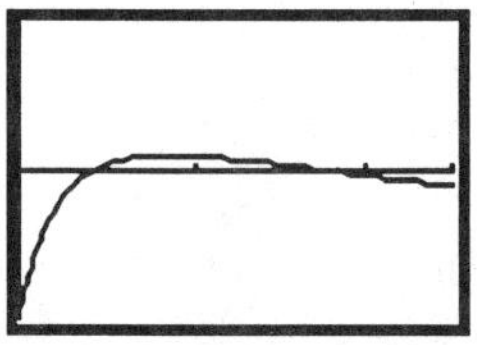

Graph window is $0 \le x \le 5, -1 \le y \le 1$.
There are two solutions.

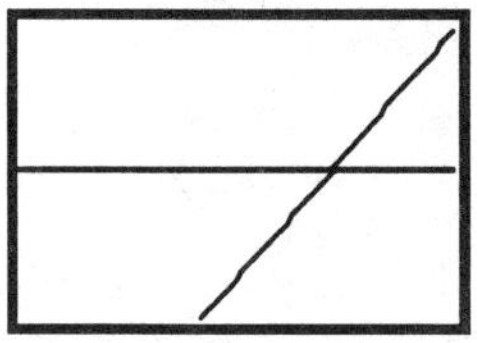

Graph window is $0.85 \le x \le 0.86$,
$-0.001 \le y \le 0.001$.

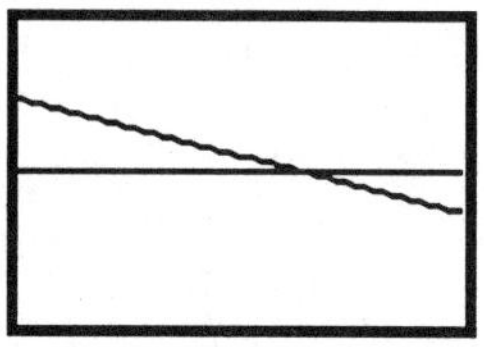

Graph window is $3.53 \le x \le 3.54$,
$-0.001 \le y \le 0.001$.
The two solutions are $x = 0.86$ and $x = 3.54$.

3. Graph the equation $f(x) = x + \sin x - 2$.

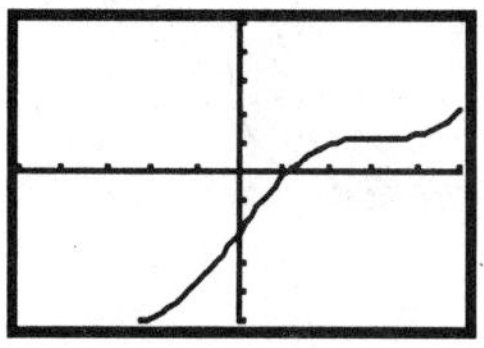

Graph window is $-5 \le x \le 5, -5 \le y \le 5$.
By zooming, $x = 1.106$ to three decimal places.
Then $\Delta x \approx 1.106 - 1 = 0.106$.

5. a.

h	$\dfrac{g(2+h) - g(2)}{h}$
0.1	1.7123
0.01	1.6489
0.001	1.6427
0.0001	1.6421
0.00001	1.6420

$g'(2) \approx 1.642$

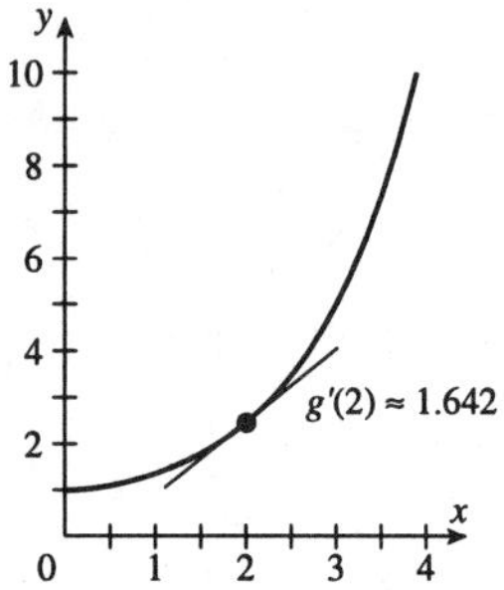

b.

h	$\dfrac{g(2+h) - g(2)}{h}$
0.1	0.1075
0.01	0.1107
0.001	0.1111
0.0001	0.1111

$g'(2) \approx 0.111$

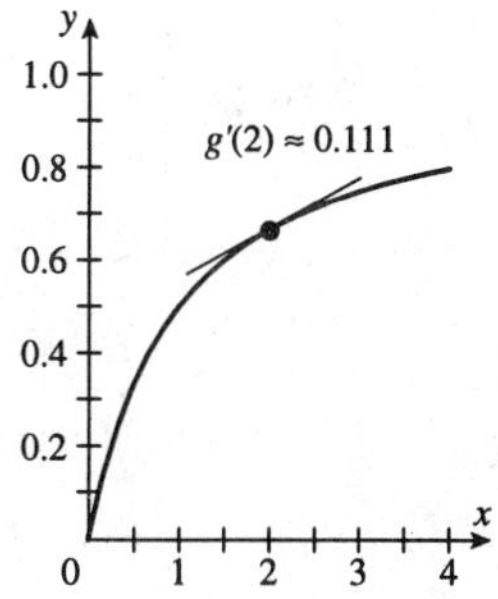

c.

h	$\frac{g(2+h)-g(2)}{h}$
0.1	1.7177
0.01	1.6956
0.001	1.6934
0.0001	1.6932

$g'(2) \approx 1.693$

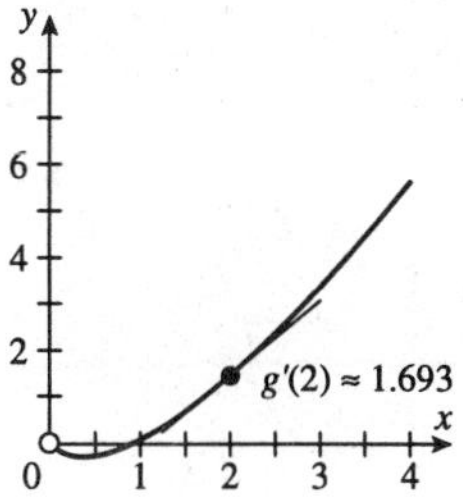

d.

h	$\frac{g(2+h)-g(2)}{h}$
0.1	–0.0586
0.01	0.0637
0.001	0.0757
0.0001	0.0769
0.00001	0.0770

$g'(2) \approx 0.077$

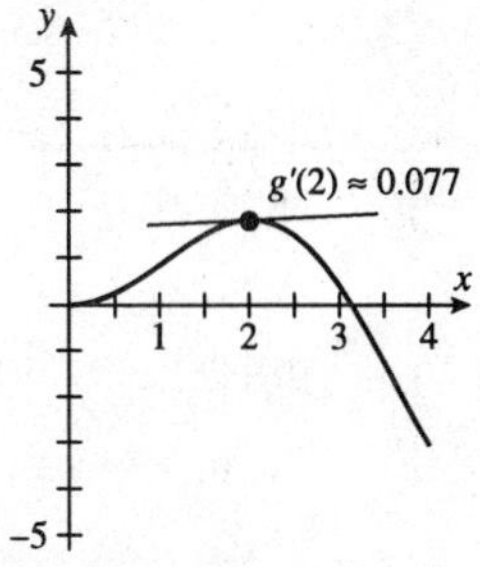

7. a. Let x change from 2 to 2.05. $\Delta x = 0.05$

$\Delta y \approx g'(2)\Delta x \approx (1.642)(0.05) = 0.0821$

$\Delta y = g(2.05) - g(2) \approx 0.0838$

Differentials give a good approximation.
Let x change from 2 to 3. $\Delta x = 1$

$\Delta y \approx g'(2)\Delta x \approx (1.642)(1) = 1.642$

$\Delta y = g(3) - g(2) \approx 2.558$

Differentials do not give a good approximation. The tangent line is close to the graph for small values of Δx.

b. Let x change from 2 to 2.05 $\Delta x = 0.05$

$\Delta y \approx g'(2)\Delta x \approx (0.111)(0.05) = 0.00555$

$\Delta y = g(2.05) - g(2) \approx 0.00546$

Differentials give a good approximation.
Let x change from 2 to 3. $\Delta x = 1$

$\Delta y \approx g'(2)\Delta x = (0.111)(1) = 0.111$

$\Delta y = g(3) - g(2) \approx 0.0833$

Differentials do not give a good approximation. The tangent line is close to the graph for small values of Δx.

c. Let x change from 2 to 2.05. $\Delta x = 0.05$

$\Delta y \approx g'(2)\Delta x \approx (1.693)(0.05) = 0.08465$

$\Delta y = g(2.05) - g(2) = 0.08528$

Differentials give a good approximation.
Let x change from 2 to 3. $\Delta x = 1$

$\Delta y \approx g'(2)\Delta x \approx (1.693)(1) = 1.693$

$\Delta y = g(3) - g(2) \approx 1.910$

Differentials do not give a good approximation. The tangent line is close to the graph for small values of Δx.

d. Let x change from 2 to 2.05. $\Delta x = 0.05$

$\Delta y \approx g'(2)\Delta x \approx (0.077)(0.05) = 0.00385$

$\Delta y = g(2.05) - g(2) = 0.000498$

Differentials do not give a good approximation.
Let x change from 2 to 3. $\Delta x = 1$

$\Delta y \approx g'(2)\Delta x \approx (0.077)(1) = 0.077$

$\Delta y = g(3) - g(2) \approx -1.395$

Differentials do not give a good approximation. The tangent line does not stay close to the graph.

9. a. $\Delta y = 1$

$$\Delta x = \frac{\Delta y}{g'(2)} \approx \frac{1}{1.642} \approx 0.609$$

By examining the graph, the approximation appears to be a good one.

b. $\Delta y = 1$

$$\Delta x = \frac{\Delta y}{g'(2)} \approx \frac{1}{0.111} \approx 9.01$$

Since there is no x such that $g(x) = g(2) + 1$, the approximation is not a good one.

c. $\Delta y = 1$

$\Delta x = \frac{\Delta y}{g'(2)} \approx \frac{1}{1.693} \approx 0.591$

By examining the graph, the approximation appears to be a good one.

d. $\Delta y = 1$

$\Delta x = \frac{\Delta y}{g'(2)} \approx \frac{1}{0.077} \approx 12.987$

Since there is no x such that $g(x) = g(2) + 1$, the approximation is not a good one.

Section 2.5

Concepts Review

1. Root (or zero)

3. a. Exact

b. Numerical

c. Numerical

Problem Set 2.5

1. a. Let $f(x) = x^4 - x - 1$.

$f(0) = -1, f(2) = 13$

Interval	f(midpoint)	Signs	Root Lies
$0 \le x < 2$	$f(1) = -1$	$- - +$	Right half
$1 \le x \le 2$	$f(1.5) \approx 2.56$	$- + +$	Left half
$1 \le x \le 1.5$	$f(1.25) \approx 0.19$	$- + +$	Left half
$1 \le x \le 1.25$	$f(1.13) \approx -0.50$	$- - +$	Right half
$1.13 \le x \le 1.25$	$f(1.19) \approx -0.18$	$- - +$	Right half

The root lies between 1.19 and 1.25. $x = 1.2$ to one decimal place.

b. Let $f(x) = x^4 - x - 1$.

From part (a), one solution on $0 \le x \le 2$ is $x = 1.2$ to one decimal place.

$f(-2) = 17, f(0) = -1$

Interval	f(midpoint)	Signs	Root Lies
$-2 \le x < 0$	$f(-1) = 1$	$+ + -$	Right half
$-1 \le x \le 0$	$f(-0.5) \approx -0.44$	$+ - -$	Left half
$-1 \le x \le -0.5$	$f(-0.75) \approx 0.07$	$+ + -$	Right half
$-0.75 \le x \le -0.5$	$f(-0.62) \approx -0.23$	$+ - -$	Left half
$-0.75 \le x \le -0.62$	$f(-0.68) \approx -0.11$	$+ - -$	Left half

The root lies between –0.75 and –0.68. $x = -0.7$ and $x = 1.2$ to one decimal place.

c. Let $f(x) = \tan x - x$

$f(1.571) \approx -4911.4, f(4.71) \approx 413.9$

Interval	f(midpoint)	Signs	Root Lies
$1.58 \le x \le 4.71$	$f(3.14) = -3.14$	$- - +$	Right side
$3.14 \le x \le 4.71$	$f(3.93) \approx -2.92$	$- - +$	Right side
$3.93 \le x \le 4.71$	$f(4.32) \approx -1.90$	$- - +$	Right side
$4.32 \le x \le 4.71$	$f(4.52) \approx 0.61$	$- + +$	Left side
$4.32 \le x \le 4.52$	$f(4.42) \approx -1.10$	$- - +$	Right side
$4.42 \le x \le 4.52$	$f(4.47) \approx -0.43$	$- - +$	Right side

The root lies between 4.47 and 4.52. $x = 4.5$ to one decimal place.

d. Let $f(x) = e^{-x} - x$.
$f(0) = 1,\ f(2) \approx -1.86$

Interval	f(midpoint)	Signs	Root Lies
$0 \le x \le 2$	$f(1) = -0.63$	+ – –	Left side
$0 \le x \le 1$	$f(0.5) \approx 0.11$	+ + –	Right side
$0.5 \le x \le 1$	$f(0.75) \approx -0.28$	+ – –	Left side
$0.5 \le x \le 0.75$	$f(0.63) \approx -0.10$	+ – –	Left side
$0.5 \le x \le 0.63$	$f(0.57) \approx -0.004$	+ – –	Left side
$0.5 \le x \le 0.57$	$f(0.54) \approx 0.04$	+ + –	Right side
$0.54 \le x \le 0.57$	$f(0.56) \approx 0.01$	+ + –	Right side

The root lies between 0.56 and 0.57. $x = 0.6$ to one decimal place.

3. a. Let $f(x) = x^3 - 3^x$.

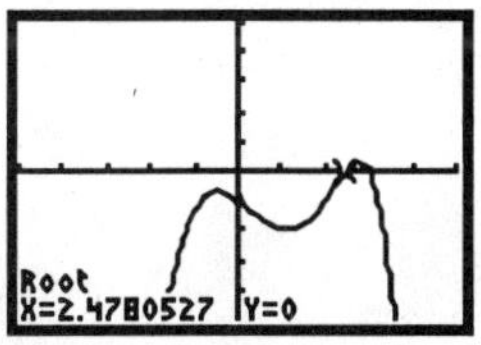

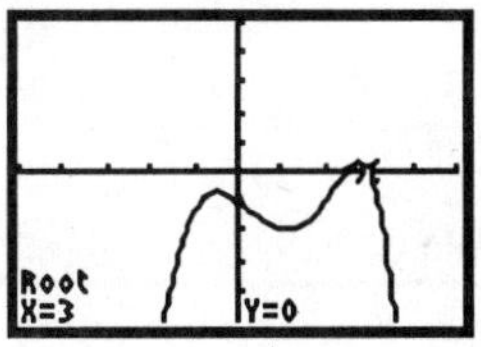

Graph windows are $-5 \le x \le 5, -5 \le y \le 5$.
$x = 2.4780527$ and $x = 3$ (exact).

b. Let $f(x) = x^4 - 4^x$.

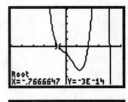

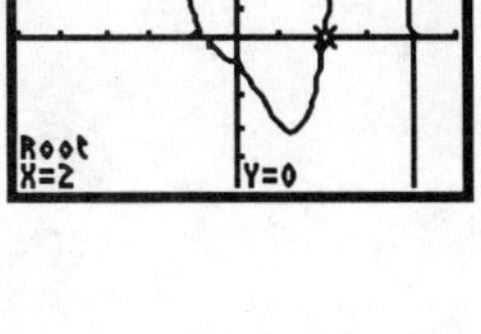

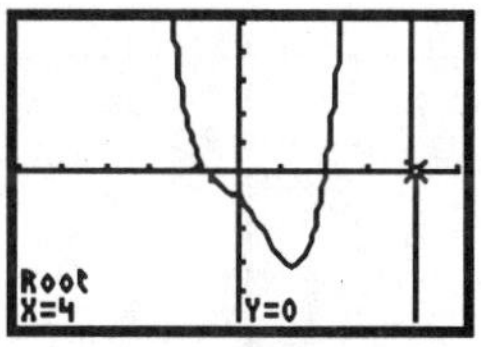

Graph windows are $-5 \le x \le 5, -5 \le y \le 5$.
$x = -0.7666647$, $x = 2$ (exact), and
$x = 4$ (exact).

c. Let $f(x) = x^4 - x - 1$.

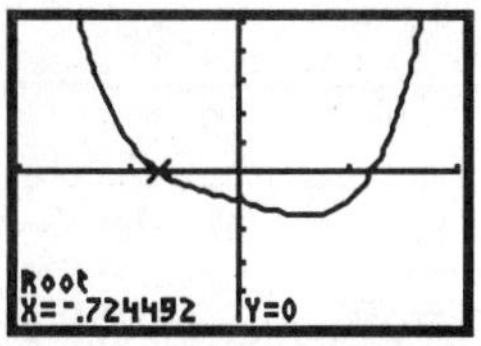

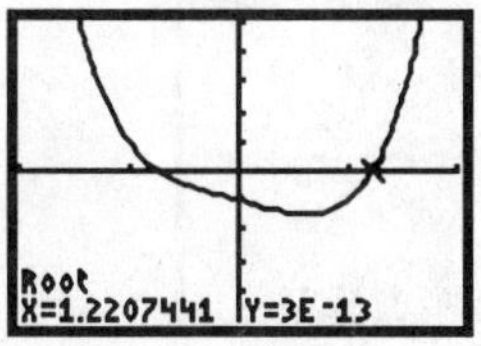

Graph windows are $-2 \le x \le 2, -5 \le y \le 5$.
$x = -0.724492$ and $x = 1.2207441$

d. Let $f(x) = e^{-x} - x$.

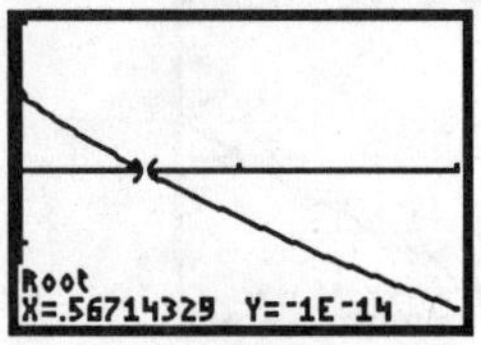

Graph window is $0 \le x \le 2, -2 \le y \le 2$.
$x = 0.56714329$

5. **a.** Exact results are $x = 0$ and $x = \ln(2)$. The numerical approximation of $x = \ln(2)$ is $x \approx 0.693$. To solve algebraically, rewrite the equation as $e^{2x} - 3e^{x} + 2 = 0$ and then factor the left-hand side to get $(e^{x} - 1)(e^{x} - 2) = 0$. The solutions are from solving $e^{x} - 1 = 0$ and $e^{x} - 2 = 0$, so $x = 0$ and $x = \ln(2)$ are solutions.

b. No solutions are found using a computer algebra system. From Problem 3(a), an exact solution is $x = 3$, and a numerical approximation is $x \approx 2.478$. There is no algebraic procedure to solve this equation.

7. **a.** $P(0) = 50e^{0.03(0)} + 2e^{-0}\sin(2\pi(0)) + 3 = 53$, so the current population is 53,000.

b. $$P'(0) = \lim_{h\to\infty} \frac{50e^{0.03h} + 2e^{-h}\sin(2\pi h) + 3 - 53}{h}$$

h	$\frac{50e^{0.03h} + 2e^{-h}\sin(2\pi h) + 3 - 53}{h}$
0.1	12.139254
0.01	13.933374
0.001	14.053750
0.0001	14.065115
0.00001	14.066245
0.000001	14.066358

The current growth rate is 14,066 people per year.

c. $P\left(\frac{1}{12}\right) = 50e^{0.03(1/12)} + 2e^{-1/12}\sin\left(2\pi\left(\frac{1}{12}\right)\right) + 3 \approx 54.045$ or 54,045 people

d. $P(3) = 50e^{0.03(3)} + 2e^{-3}\sin(2\pi(3)) + 3 \approx 57.709$ or 57,709 people

e. Let $f(t) = P(t) - 54 = 50e^{0.03t} + 2e^{-t}\sin(2\pi t) - 51$.

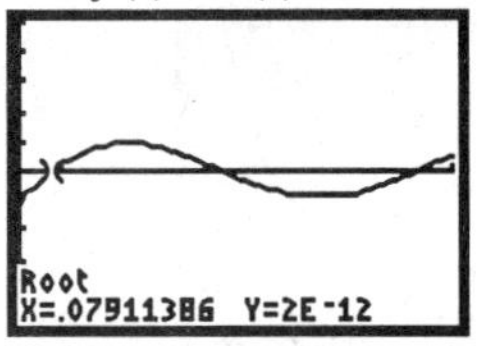

Graph windows is $0 \le x \le 1, -5 \le y \le 5$.
Solving for the first root, $t \approx 0.079$ years.

f. Let $f(t) = P(t) - 60 = 50e^{0.03t} + 2e^{-t}\sin(2\pi t) - 57$.

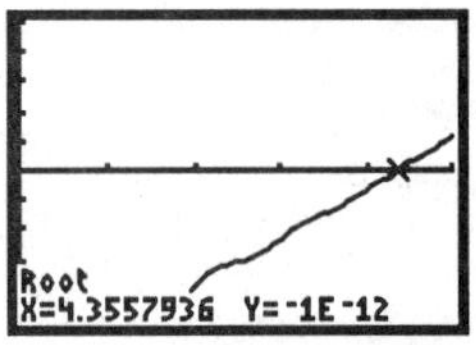

Graph window is $0 \le x \le 5, -5 \le y \le 5$.
Solving for the root, $t \approx 4.36$ years.

g. Written response. Short-term predictions are usually more reliable.

Section 2.6

Concepts Review

1. Sum of the squares

3. Outside the range of the data points.

Problem Set 2.6

1. a. Use the data points (1, 10) and (2, 8).

$$m = \frac{8-10}{2-1} = -2$$

This gives the linear equation $y = -2x + 12$.

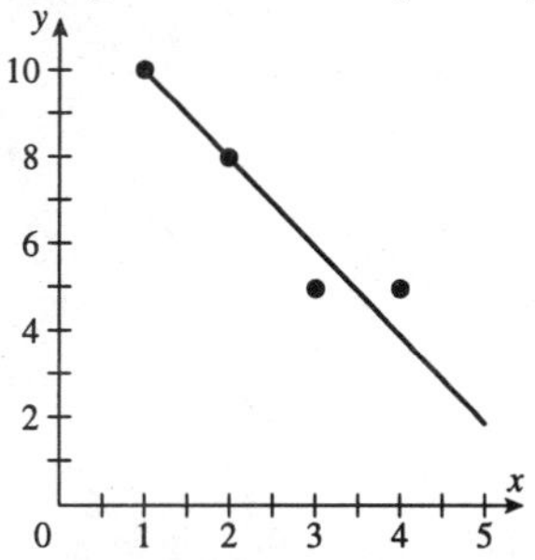

b. Use the data points (10, 0) and (40, 80).

$$m = \frac{80-0}{40-10} = \frac{8}{3}$$

This gives the linear equation $y = \frac{8}{3}x - \frac{80}{3}$.

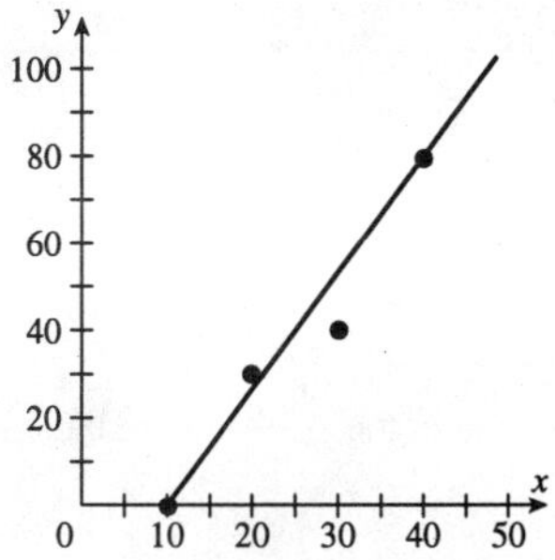

3. a.

	x_i	y_i	x_i^2	$x_i y_i$
	1	10	1	10
	2	8	4	16
	3	5	9	15
	4	5	16	20
Σ	10	28	30	61

$$m = \frac{(4)(61)-(10)(28)}{(4)(30)-(10)^2} = -1.8$$

$$b = \frac{(30)(28)-(10)(61)}{(4)(30)-(10)^2} = 11.5$$

The equation of the least squares line is $y = -1.8x + 11.5$.

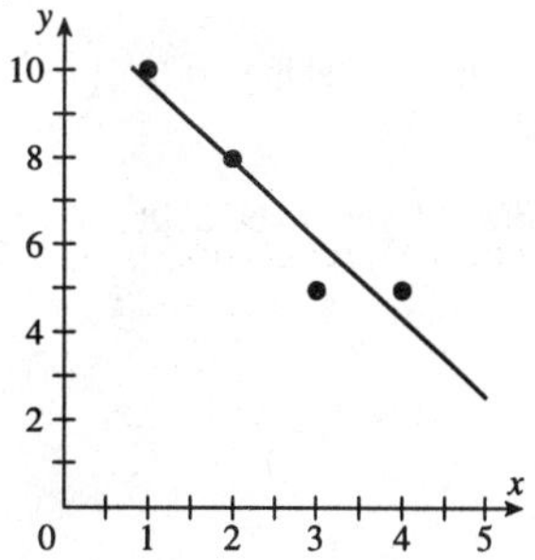

b.

	x_i	y_i	x_i^2	$x_i y_i$
	10	0	100	0
	20	30	400	600
	30	40	900	1200
	40	80	1600	3200
Σ	100	150	3000	5000

$$m = \frac{(4)(5000)-(100)(150)}{(4)(3000)-(100)^2} = 2.5$$

$$b = \frac{(3000)(150)-(100)(5000)}{(4)(3000)-(100)^2} = -25$$

The equation of the least squares line is $y = 2.5x - 25$.

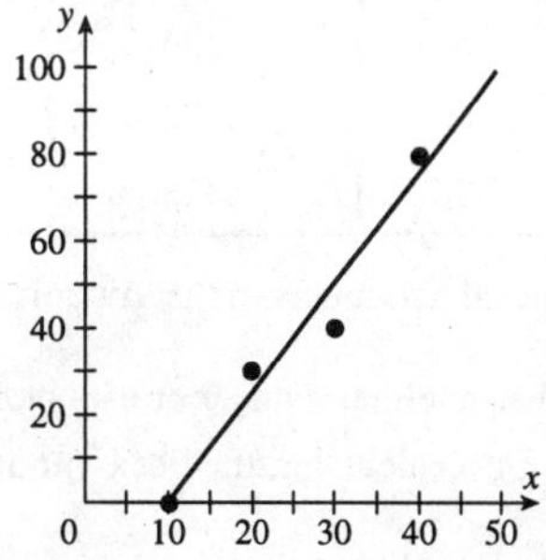

5. a. The first and last data points are (1984, 3777.2) and (1993, 6377.9).

$$m = \frac{6377.9-3777.2}{1993-1984} \approx 288.97$$

This gives the linear equation $y = 288.97x - 569{,}539.28$.
Since the slope is 288.97, the GDP increases at a rate of 289 billion dollars per year.

b. Using a calculator, the equation of the least squares line is $y = 291.18x - 573{,}973.77$. The slope of the line is the rate at which the GDP increases in billions of dollars per year.

c. Using the model in part (a), predict the GDP (in billions of dollars) to be 6666.9 in 1994, 6955.9 in 1995, and 7244.8 in 1996.

Using the model in part (b), predict the GDP (in billions of dollars) to be 6639.2 in 1994, 6930.3 in 1995, and 7221.5 in 1996.

d. Predict the year to be 2006. Written response.

7. Written response.

Chapter 2.7

Concepts Review

1. Linear

3. $y = ax^b$

Problem Set 2.7

1. Written response.

3. a. Let $x = 0$ represent 1985.

Year x_i	Population y_i	$\ln y_i$	$\Delta y = y_{i+1} - y_i$	$\frac{y_{i+1}}{y_i}$
0	4850	8.4867	86	1.0177
1	4936	8.5043	88	1.0178
2	5024	8.5220	88	1.0175
3	5112	8.5393	90	1.0176
4	5202	8.5568	92	1.0177
5	5294	8.5743	90	1.0170
6	5384	8.5912	94	1.0175
7	5478	8.6085		

Since the numbers in the Δy column are nearly constants and the numbers in the $\frac{y_{i+1}}{y_i}$ column are nearly constant either a linear or exponential model will fit the data.
Using a calculator, the linear model is $y = 89.714x + 4846$, and the exponential model is $y = 4851e^{0.0174x}$.

b. Graph for linear model:

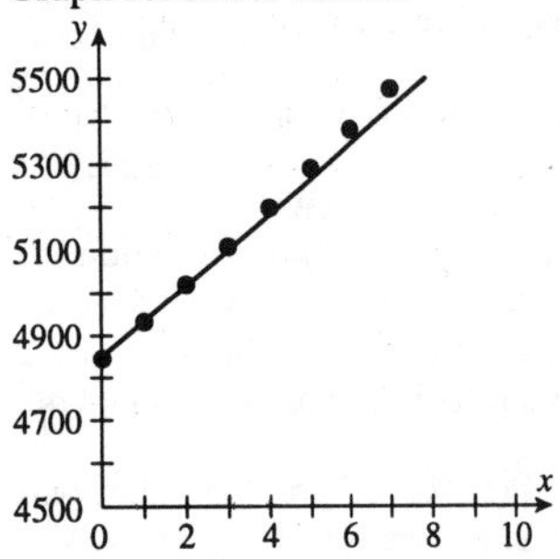

Graph for exponential model:

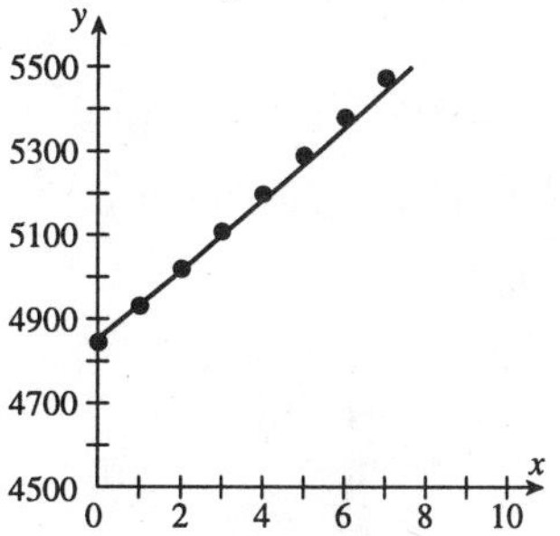

Year x_i	Population y_i	$y = 89.714x + 4846$	$y = 4851e^{0.0174x}$
0	4850	4846	4851
1	4936	4935.7	4936.1
2	5024	5025.4	5022.8
3	5112	5115.1	5110.9
4	5202	5204.9	5200.7
5	5294	5294.6	5291.9
6	5384	5384.3	5384.8
7	5478	5474	5479.3

c. Using the linear model, the predicted population in 1993 ($x = 8$) is $89.714(8) + 4846 \approx 5563.7$ million people and the predicted population in 2000 ($x = 15$) is $89.714(15) + 4846 \approx 6191.7$ million people.
Using the exponential model, the predicted population in 1993 ($x = 8$) is $4851e^{0.0174(8)} \approx 5575.5$ million people and the predicted population in 2000 ($x = 15$) is $4851e^{0.0174(15)} \approx 6297.7$ million people.
The world population in 1993 was approximately 5574.5 million people.

d. Using the linear model, solve $10{,}000 = 89.714x + 4846$. $x \approx 57.4$, or in 2043. Using the exponential model, solve $10{,}000 = 4851e^{0.0174x}$; $x \approx 41.6$ or in 2027. Written response.

Section 2.8 Chapter Review

Concepts Test

1. False. The machine's choice will sometimes be the best one.

3. True. Computers and calculators have a difficult time dealing with asymptotes.

5. False. The choice of graph window does have an effect on the appearance of the graph of a function.

7. False. A computer or calculator uses essentially the same process you would go through when graphing by hand.

9. True. Both methods can be used to estimate roots.

11. False. Parameters are often varied to provide a good fit for given data.

13. True.

15. False. The family of curves are lines since y is linear with respect to a.

17. False. For example, $x^4 - 4^x = 0$ cannot be solved algebraically.

19. True.

21. False. For example, a computer algebra system will not find the exact solution $x = 2$ of the equation $x^2 - 2^x = 0$.

23. False. A linear model that does not seem to follow a straight line will not fit.

25. True. The logarithm of a nonpositive number is not defined.

Sample Test Problems

1. a.

x	y	
–2.00	not defined	
–1.50	not defined	
–1.00	not defined	
–0.50	not defined	
0.00	not defined	
0.50	–0.69	Plot at (0.5, 0).
1.00	0	Plot at (1, 0).
1.50	0.41	Plot at (1.5, 0).
2.00	0.69	Plot at (2, 0).

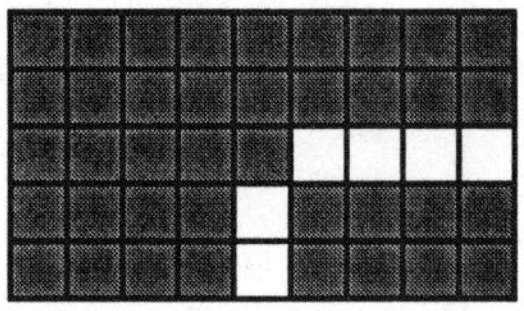

Fill in graph with lit pixels at (0, –4) and (0, –8) to give an appearance of an asymptote at $x = 0$.

b.

x	y	
–2.00	–0.50	Plot at (–2, 0).
–1.50	–0.67	Plot at (–1.5, 0)
–1.00	–1.00	Plot at (–1, 0).
–0.50	–2.00	Plot at (–0.5, –4) or at (–0.5, 0).
0.00	not defined	
0.50	2.00	Plot at (0.5, 4) or at (0.5, 0).
1.00	1.00	Plot at (1, 0).
1.50	0.67	Plot at (1.5, 0).
2.00	0.50	Plot at (2, 0).

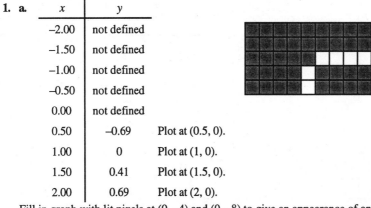

Fill in graph with lit pixels at (0, 8) and (0, –8) (or (0, 8), (0, 4), (0, –4) and (0, –8)) to give an appearance of an asymptote at $x = 0$.

3. a. $f(x) = 1 + x + x^3 - x^4$

x	y
–2.5	–56.1875
–2	–25.0000
–1.5	–8.9375
–1	–2.0000
–0.5	0.3125
0	1.0000
0.5	1.5625
1	2.0000
1.5	0.8125
2	–5.0000
2.5	–19.9375

Interval estimate for maximum y-value:
$0.5 \le x \le 1.5$
Interval estimate for which $f(x) = 0$:
$-1 \le x \le -0.5$, $1.5 \le x \le 2$

b. $f(x) = \sin(2x) + x + 1$

x	y
–2.5	–0.5411
–2	–0.2432
–1.5	–0.6411
–1	–0.9093
–0.5	–0.3415
0	1.0000
0.5	2.3415
1	2.9093
1.5	2.6411
2	2.2432
2.5	2.5411

Interval estimate for maximum y-value:
$0.5 \le x \le 1.5$
Interval estimate for which $f(x) = 0$:
$-0.5 \le x \le 0$.

5. a.

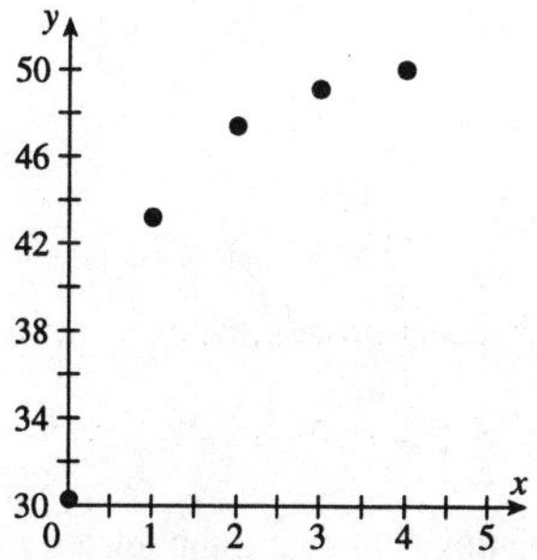

From the graph, it appears that as $x \to \infty$, y approaches 51, and the function is 30.29 at $x = 0$. A function of the form $y = 51 - 20.71e^{-kx}$ has these properties. $y = 51 - 20.71e^{-0.9x}$ provides a good fit. Other curves are possible.

b.

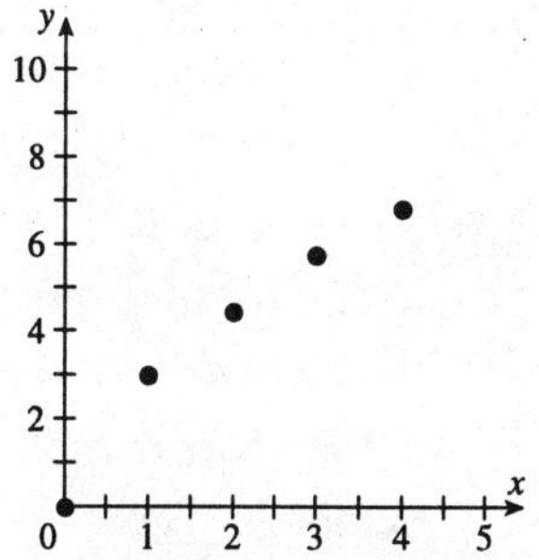

Use a model of the form $y = ax^b$. When $x = 1$, $y = 3.02$ so let $a = 3$. $y = 3x^{0.6}$ provides a good fit.

c.

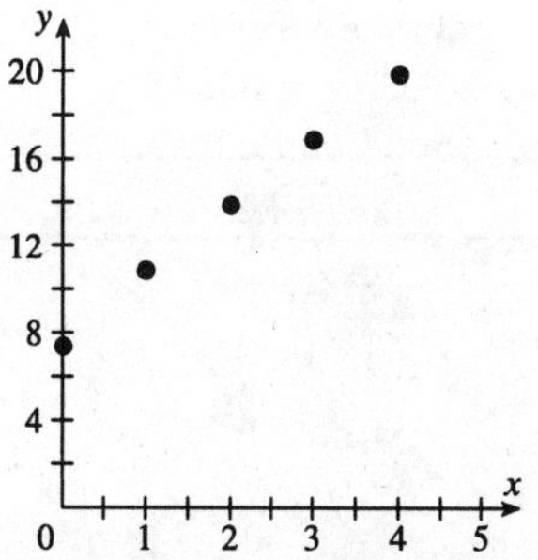

The points appear to be related by a linear equation. Using a calculator, the least squares regression line is $y = 3.086x + 7.658$.

d.

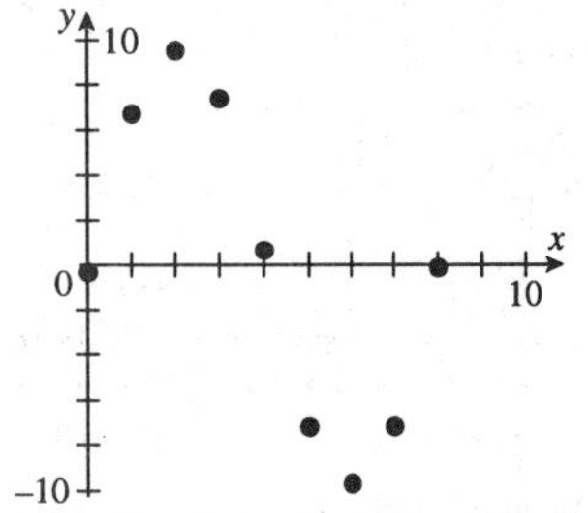

The points appear to be related by a trigonometric equation. Use the family $y = A\sin(2\pi w(x-d))$.

Let $A = 9.6$, $d = 0$, and $T = 8$ so $w = \frac{1}{8}$. Therefore $y = 9.6\sin\left(\frac{\pi}{8}x\right)$.

7. a. Let $f(x) = x^5 + x + 1$

$f(-1) = -1,\ f(4) = 1029$

Interval	f(midpoint)	Signs	Root lies
$-1 \le x \le 4$	$f(1.5) \approx 10.094$	$-++$	Left half
$-1 \le x \le 1.5$	$f(0.25) \approx 1.251$	$-++$	Left half
$-1 \le x \le 0.25$	$f(-0.375) \approx 0.618$	$-++$	Left half
$-1 \le x \le -0.375$	$f(-0.687) \approx 0.160$	$-++$	Left half
$-1 \le x \le -0.687$	$f(-0.843) \approx -0.269$	$--+$	Right half
$-0.843 \le x \le -0.687$	$f(-0.765) \approx -0.270$	$--+$	Right half
$-0.765 \le x \le -0.687$	$f(-0.726) \approx 0.072$	$-++$	Left half
$-0.765 \le x \le -0.726$	$f(-0.745) \approx 0.0255$	$-++$	Left half
$-0.765 \le x < -0.745$	$f(-0.755) \approx -0.0003$	$--+$	Right half

The root lies between −0.755 and −0.745. $x = -0.75$ to two decimal places.

b. Let $f(x) = x + 2\sin(2x) + 2$

$f(-1) \approx -0.819, f(4) \approx 7.979$

Interval	f(midpoint)	Signs	Root Lies
$-1 \le x \le 4$	$f(1.5) \approx 3.782$	$-++$	Left half
$-1 \le x \le 1.5$	$f(0.25) \approx 3.209$	$-++$	Left half
$-1 \le x \le 0.25$	$f(-0.375) \approx 0.262$	$-++$	Left half
$-1 \le x \le -0.375$	$f(-0.687) \approx -0.648$	$--+$	Right half
$-0.687 \le x \le -0.375$	$f(-0.531) \approx -0.278$	$--+$	Right half
$-0.531 \le x \le -0.375$	$f(-0.453) \approx -0.027$	$--+$	Right half
$-0.453 \le x \le -0.375$	$f(-0.414) \approx 0.113$	$-++$	Left half
$-0.453 \le x \le -0.414$	$f(-0.433) \approx 0.044$	$-++$	Left half
$-0.453 \le x \le -0.433$	$f(-0.443) \approx 0.008$	$-++$	Left half
$-0.453 \le x \le -0.443$	$f(-0.448) \approx -0.010$	$--+$	Right half
$-0.448 \le x \le -0.443$	$f(-0.445) \approx 0.0009$	$-++$	Left half

The root lies between −0.448 and −0.445. $x = -0.45$ to two decimal places.

9. a.

h	$\frac{f(1+h)-f(1)}{h}$
0.1	0.3489
0.01	0.3660
0.001	0.3677
0.0001	0.3679

$f'(1) \approx 0.368$

b. Let x change from 1 to 1.05. $\Delta x = 0.05$

$\Delta y \approx f'(1)\Delta x \approx (0.368)(0.05) = 0.0184$

$\Delta y = f(1.05) - f(1) \approx 0.0179$

The graph below shows that the approximation is a good one.

Let x change from 1 to 2. $\Delta x = 1$

$\Delta y \approx f'(1)\Delta x \approx (0.368)(1) = 0.368$

$\Delta y = f(2) - f(1) \approx 0.173$

The graph below shows the approximation is a poor one.

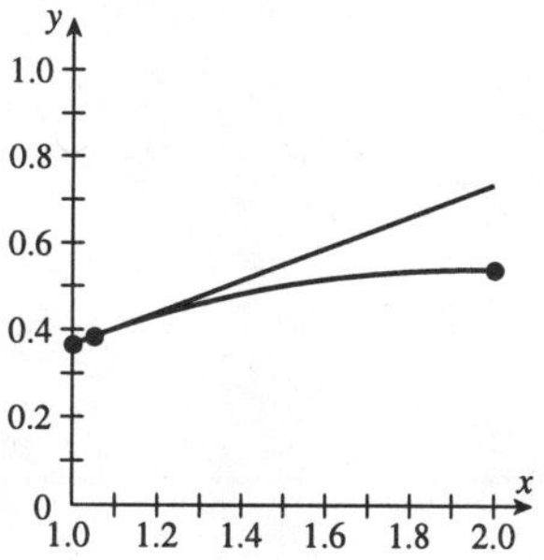

c. $\Delta y = 1$

$$\Delta x = \frac{\Delta y}{f'(1)} \approx \frac{1}{0.368} \approx 2.717$$

$f(1+\Delta x) = f(3.717) \approx 0.336$

Since $f(3.717) \neq f(1) + 1 \approx 1.368$, the estimate is not a good one.

11. Plot the data.

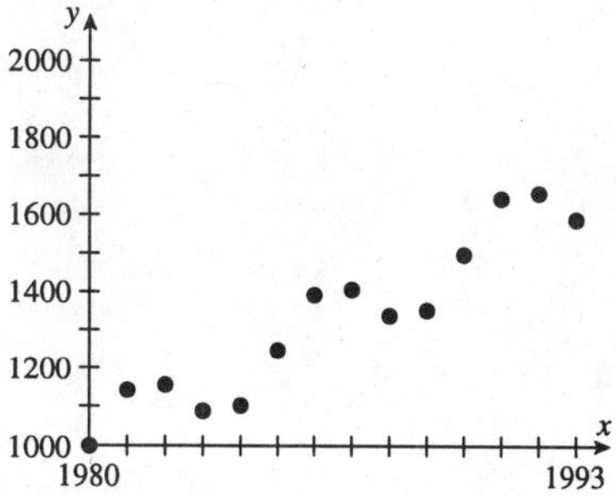

From the plot, observe that $P(x)$ should have trigonometric and linear terms. Note that as x increases by 5, y increases by 250. Then the linear term has slope $\frac{250}{5} = 50$ and passes through the point (1980, 1000), so the linear term is $50(x - 1980) + 1000$ or $50x - 98{,}000$.

The period of the trigonometric term appears to be 5. The trigonometric term is zero at 1980, so the term has the form $A\sin\left(\frac{2\pi}{5}(x-1980)\right)$.

$$P(x) = 50x - 98{,}000 + A\sin\left(\frac{2\pi}{5}(x-1980)\right)$$

To find A, substitute $x = 1981$ and $y = 1145$.

$$1145 = 50(1981) - 98{,}000 + A\sin\left(\frac{2\pi}{5}\right)$$

$$A = \frac{95}{\sin\left(\frac{2\pi}{5}\right)} \approx 100$$

Therefore,

$$P(x) = 50x - 98{,}000 + 100\sin 6\left(\frac{2\pi}{5}(x-1980)\right)$$

is a mathematical model.

$P(1994) = 1605$

$P(1995) = 1750$

$P(1996) = 1895$

Chapter 3

Section 3.1

Concepts Review

1. $\frac{f(x+h)-f(x)}{h}$

3. Continuous; $f(x)=|x|$

Problem Set 3.1

1. $f'(x)=\lim_{h\to 0}\frac{f(x+h)-f(x)}{h}=\lim_{h\to 0}\frac{[(x+h)^2-(x+h)]-(x^2-x)}{h}$
$=\lim_{h\to 0}\frac{(x^2+2xh+h^2-x-h)-(x^2-x)}{h}=\lim_{h\to 0}\frac{2xh+h^2-h}{h}$
$=\lim_{h\to 0} 2x+h-1=2x-1$
$f'(3)=2(3)-1=5$
$f'(3)=\lim_{h\to 0}\frac{f(3+h)-f(3)}{h}=\lim_{h\to 0}\frac{[(3+h)^2-(3+h)]-[3^2-3]}{h}$
$=\lim_{h\to 0}\frac{[(3+h)^2-(3+h)]-6}{h}$

h	$\frac{[(3+h)^2-(3+h)]-6}{h}$
1	6
0.1	5.1
0.01	5.01
0.001	5.001
0.0001	5.0001
0.00001	5.00001

$f'(3)=5.000$, accurate to three decimal places.

3. $f'(x)=\lim_{h\to 0}\frac{f(x+h)-f(x)}{h}=\lim_{h\to 0}\frac{[(x+h)^3+2(x+h)^2]-(x^3+2x^2)}{h}$
$=\lim_{h\to 0}\frac{[(x^3+3x^2h+3xh^2+h^3)+(2x^2+4xh+2h^2)]-x^3-2x^2}{h}$
$=\lim_{h\to 0}\frac{3x^2h+3xh^2+h^3+4xh+2h^2}{h}=\lim_{h\to 0} 3x^2+3xh+h^2+4x+2h$
$=3x^2+4x$
$f'(-1)=3(-1)^2+4(-1)=-1$
$f'(-1)=\lim_{h\to 0}\frac{f(-1+h)-f(-1)}{h}=\lim_{h\to 0}\frac{[(-1+h)^3+2(-1+h)^2]-[(-1)^3+2(-1)^2]}{h}$
$=\lim_{h\to 0}\frac{[(-1+h)^3+2(-1+h)^2]-1}{h}$

h	$\dfrac{[(-1+h)^3+2(-1+h)^2]-1}{h}$
1	−1
0.1	−1.09
0.01	−1.0099
0.001	−1.000999
0.0001	−1.00009999

$f'(-1) = -1.000$, accurate to three decimal places.

5. $f(x) = 5x - 4$

$$f'(x) = \lim_{h\to 0} \frac{f(x+h)-f(x)}{h} = \lim_{h\to 0} \frac{[5(x+h)-4]-(5x-4)}{h}$$

$$= \lim_{h\to 0} \frac{5h}{h} = \lim_{h\to 0} 5 = 5$$

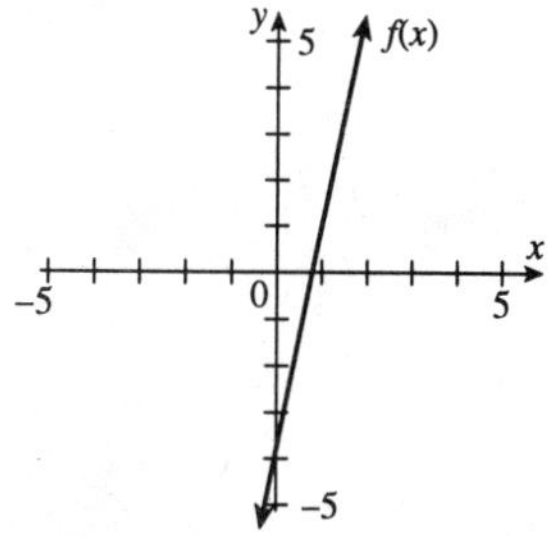

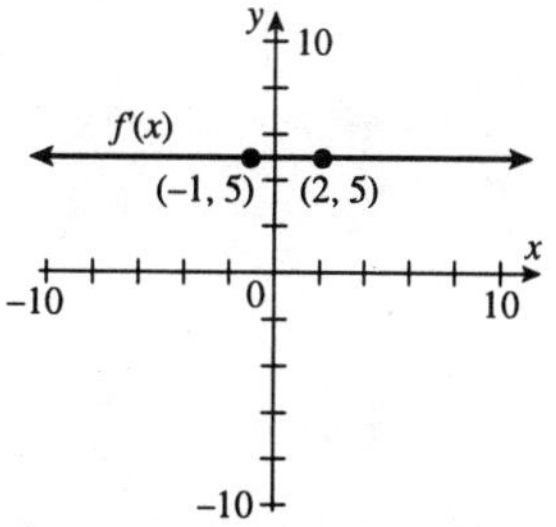

7. $f(x) = 8x^2 - 1$

We use the derivative formula from Example 3 with $a = 8$, $b = 0$, and $c = -1$. $f'(x) = 2(8)x + 0 = 16x$

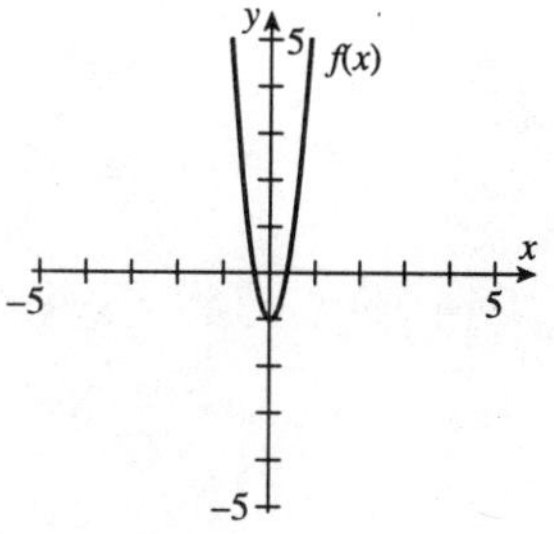

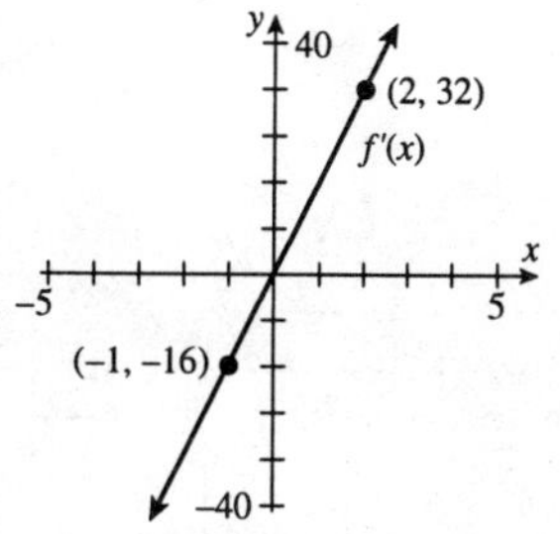

9. $f(x) = ax^3 + bx^2 + cx + d$

$$f'(x) = \lim_{h\to 0} \frac{f(x+h)-f(x)}{h} = \lim_{h\to 0} \frac{[a(x+h)^3 + b(x+h)^2 + c(x+h) + d] - (ax^3 + bx^2 + cx + d)}{h}$$

$$= \lim_{h\to 0} \frac{ax^3 + 3ax^2h + 3axh^2 + ah^3 + bx^2 + 2bxh + bh^2 + cx + ch + d - ax^3 - bx^2 - cx - d}{h}$$

$$= \lim_{h\to 0} \frac{3ax^2h + 3axh^2 + ah^3 + 2bxh + bh^2 + ch}{h} = \lim_{h\to 0} 3ax^2 + 3axh + ah^2 + 2bx + bh + c$$

$$= 3ax^2 + 2bx + c$$

11. $f(x) = x^3 - 2x$

We use the derivative formula from Problem 9 with $a = 1$, $b = 0$, $c = -2$, and $d = 0$.

$f'(x) = 3(1)x^2 + 2(0)x + (-2) = 3x^2 - 2$

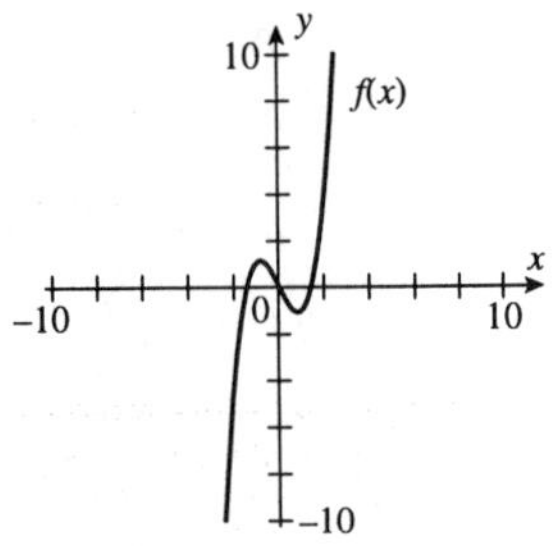

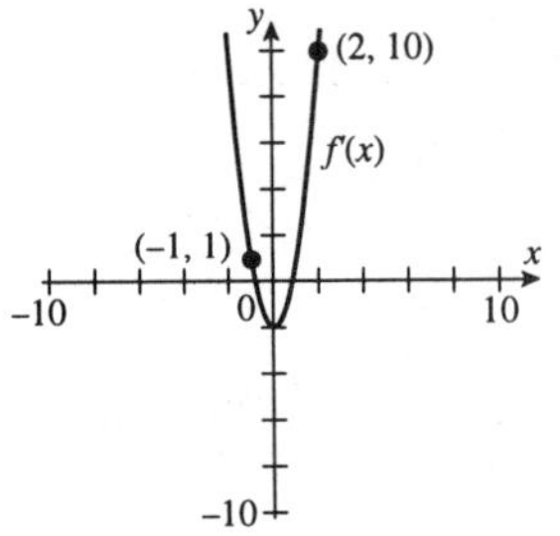

13. $g(x) = \dfrac{3}{5x}$

$$g'(x) = \lim_{h\to 0} \frac{g(x+h) - g(x)}{h} = \lim_{h\to 0} \frac{\frac{3}{5(x+h)} - \frac{3}{5x}}{h}$$

$$= \lim_{h\to 0} \left[\frac{3x - 3(x+h)}{5(x+h)x} \cdot \frac{1}{h} \right] = \lim_{h\to 0} \left[\frac{-3h}{5x(x+h)} \cdot \frac{1}{h} \right] = \lim_{h\to 0} \frac{-3}{5x(x+h)}$$

$$= -\frac{3}{5x^2}$$

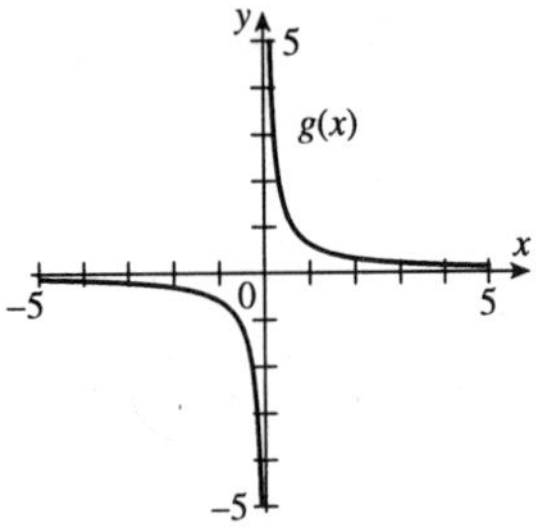

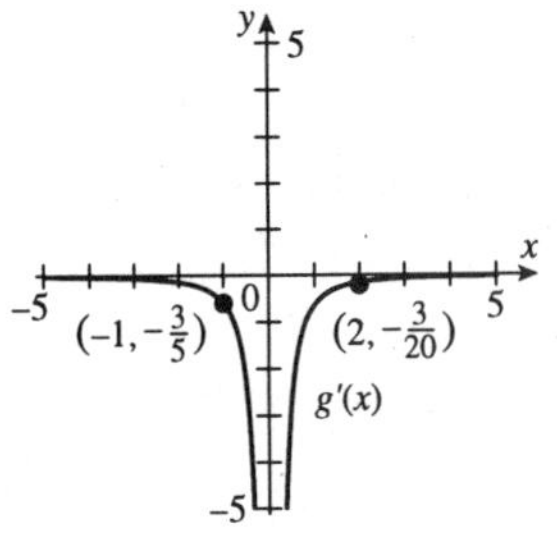

15. $F(x) = \dfrac{6}{x^2 + 1}$

$$F'(x) = \lim_{h\to 0} \frac{F(x+h) - F(x)}{h} = \lim_{h\to 0} \frac{\frac{6}{(x+h)^2+1} - \frac{6}{x^2+1}}{h} = \lim_{h\to 0} \left[\frac{6(x^2+1) - 6(x^2 + 2xh + h^2 + 1)}{(x^2 + 2xh + h^2 + 1)(x^2+1)} \cdot \frac{1}{h} \right]$$

$$= \lim_{h\to 0} \left[\frac{6x^2 + 6 - 6x^2 - 12xh - 6h^2 - 6}{(x^2 + 2xh + h^2 + 1)(x^2+1)} \cdot \frac{1}{h} \right] = \lim_{h\to 0} \left[\frac{-12xh - 6h^2}{(x^2 + 2xh + h^2 + 1)(x^2+1)} \cdot \frac{1}{h} \right]$$

$$= \lim_{h\to 0} \frac{-12x - 6h}{(x^2 + 2xh + h^2 + 1)(x^2+1)} = \frac{-12x}{(x^2+1)^2} = -\frac{12x}{x^4 + 2x^2 + 1}$$

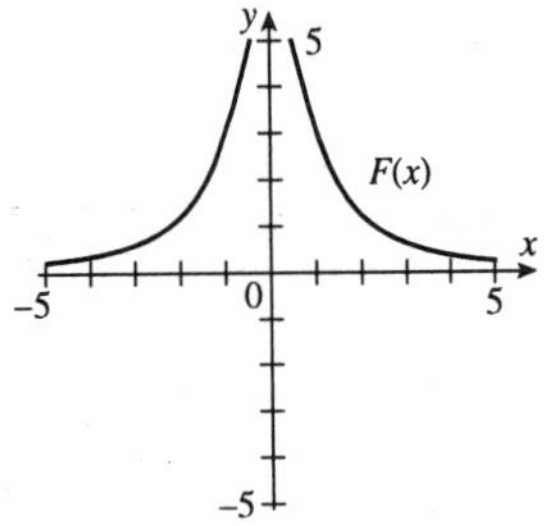

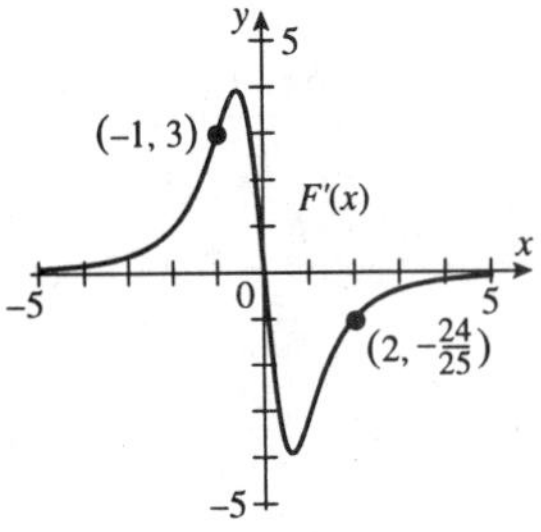

17. $G(x) = \dfrac{2x-1}{x-4}$

$$G'(x) = \lim_{h \to 0} \frac{G(x+h) - G(x)}{h} = \lim_{h \to 0} \frac{\frac{2(x+h)-1}{(x+h)-4} - \frac{2x-1}{x-4}}{h}$$

$$= \lim_{h \to 0} \left[\frac{(2x+2h-1)(x-4) - (2x-1)(x+h-4)}{(x+h-4)(x-4)} \cdot \frac{1}{h} \right]$$

$$= \lim_{h \to 0} \left[\frac{2x^2 - 8x + 2hx - 8h - x + 4 - 2x^2 - 2xh + 8x + x + h - 4}{(x+h-4)(x-4)} \cdot \frac{1}{h} \right]$$

$$= \lim_{h \to 0} \left[\frac{-7h}{(x+h-4)(x-4)} \cdot \frac{1}{h} \right] = \lim_{h \to 0} \frac{-7}{(x+h-4)(x-4)} = \frac{-7}{(x-4)^2} = -\frac{7}{x^2 - 8x + 16}$$

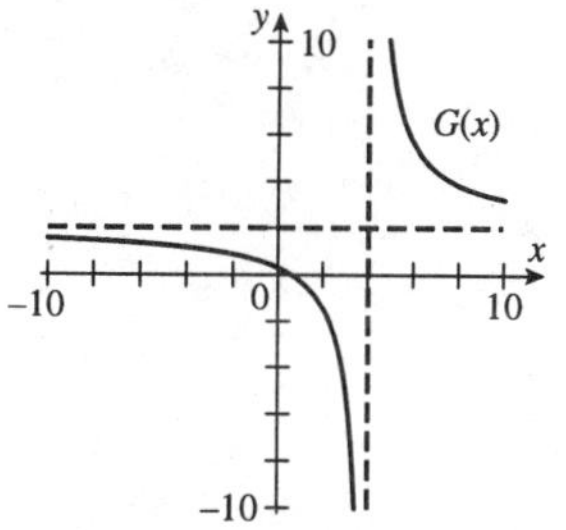

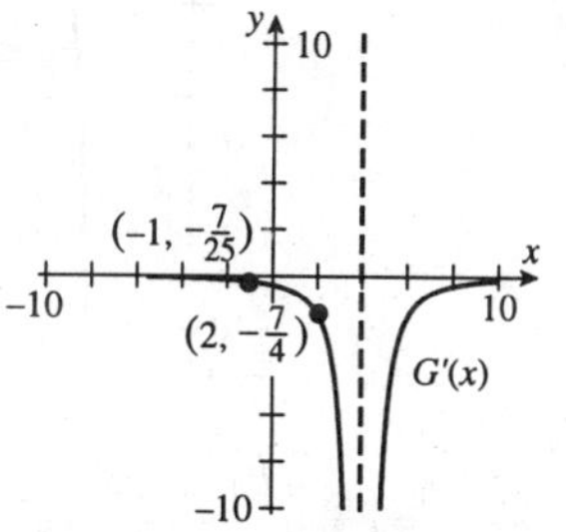

19. $g(x) = \sqrt{3x}$

$$g'(x) = \lim_{h \to 0} \frac{g(x+h) - g(x)}{h} = \lim_{h \to 0} \frac{\sqrt{3(x+h)} - \sqrt{3x}}{h}$$

$$= \lim_{h \to 0} \left[\frac{\sqrt{3x+3h} - \sqrt{3x}}{h} \cdot \frac{\sqrt{3x+3h} + \sqrt{3x}}{\sqrt{3x+3h} + \sqrt{3x}} \right]$$

$$= \lim_{h \to 0} \left[\frac{3h}{h(\sqrt{3x+3h} + \sqrt{3x})} \right] = \lim_{h \to 0} \frac{3}{\sqrt{3x+3h} + \sqrt{3x}}$$

$$= \frac{3}{2\sqrt{3x}}$$

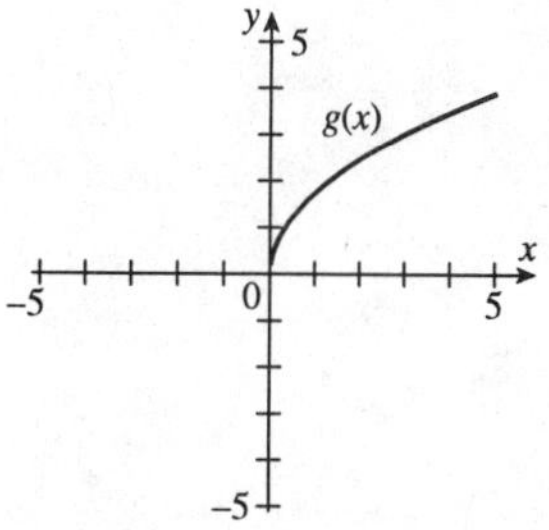

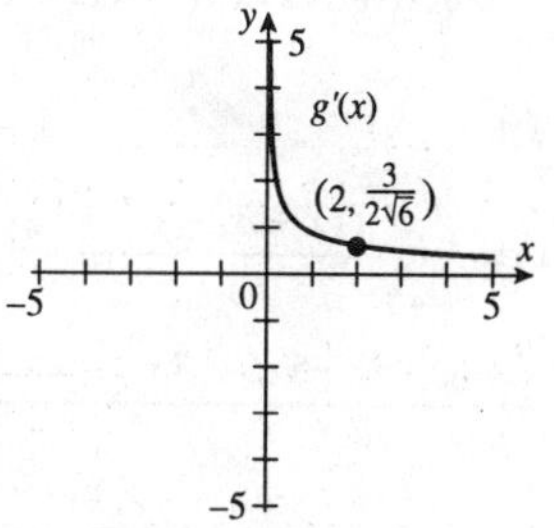

21. $H(x) = \dfrac{3}{\sqrt{x-2}}$

$$H'(x) = \lim_{h \to 0} \frac{H(x+h) - H(x)}{h} = \lim_{h \to 0} \frac{\frac{3}{\sqrt{(x+h)-2}} - \frac{3}{\sqrt{x-2}}}{h} = \lim_{h \to 0} \left[\frac{3\sqrt{x-2} - 3\sqrt{x+h-2}}{\sqrt{(x+h-2)(x-2)}} \cdot \frac{1}{h} \right]$$

$$= \lim_{h \to 0} \left[\frac{3\left(\sqrt{x-2} - \sqrt{x+h-2}\right)}{\sqrt{(x+2h-2)(x-2)}} \cdot \frac{1}{h} \cdot \frac{\sqrt{x-2} + \sqrt{x+h-2}}{\sqrt{x-2} + \sqrt{x+h-2}} \right]$$

$$= \lim_{h\to 0}\left[\frac{3(-h)}{\sqrt{(x+h-2)(x-2)}\left(\sqrt{x-2}+\sqrt{x+h-2}\right)}\cdot\frac{1}{h}\right]$$

$$= \lim_{h\to 0}\left[\frac{-3}{\sqrt{(x+h-2)(x-2)}\left(\sqrt{x-2}+\sqrt{x+h-2}\right)}\right]$$

$$= -\frac{3}{2\sqrt{(x-2)^3}}$$

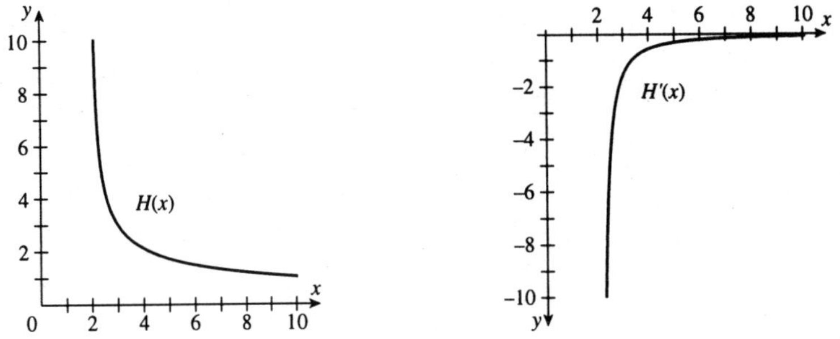

23. $f'(1) = \lim_{h\to 0}\frac{e^{1+h}-e}{h} = \lim_{h\to 0} e\left(\frac{e^h-1}{h}\right) = e\lim_{h\to 0}\left(\frac{e^h-1}{h}\right) = e$

$\lim_{h\to 0}\left(\frac{e^h-1}{h}\right) = 1$ from Theorem A of Section 1.4.

25. $f'(1) = \lim_{h\to 0}\frac{\ln(1+h)-\ln(1)}{h} = \lim_{h\to 0}\frac{\ln(1+h)}{h} = \lim_{h\to 0}\ln(1+h)^{1/h} = \ln e = 1$

27. $\lim_{h\to 0}\frac{2(5+h)^3-2(5)^3}{h}$. This is the derivative of $f(x) = 2x^3$ at $x = 5$.

29. $\lim_{h\to 0}\frac{\cos(x+h)-\cos(x)}{h}$. This is the derivative of $f(x) = \cos x$ at x.

31. $f(x) = 5x^7 - 3x^2 + x + 1$

For the limit command, use $f'(x) = \lim_{h\to 0}\frac{f(x+h)-f(x)}{h}$

$= \lim_{h\to 0}\frac{[5(x+h)^7 - 3(x+h)^2 + (x+h)+1] - 5x^7 + 3x^2 - x - 1}{h}$

$f'(x) = 35x^6 - 6x + 1$

For the Derivative command, use $f(x) = 5x^7 - 3x^2 + x + 1$.

$f'(x) = 35x^6 - 6x + 1$

33. $f(x) = e^{x^2}$

For the Limit command, use $f'(x) = \lim_{h\to 0}\frac{f(x+h)-f(x)}{h} = \lim_{h\to 0}\frac{e^{(x+h)^2}-e^{x^2}}{h} = 2xe^{x^2}$.

For the Derivative command, use $f(x) = e^{x^2}$. $f'(x) = 2xe^{x^2}$

35. $f'(0) \approx -\frac{1}{2}$; $f'(2) \approx \frac{4}{3}$; $f'(5) \approx \frac{2}{3}$; $f'(7) \approx -3$

37.

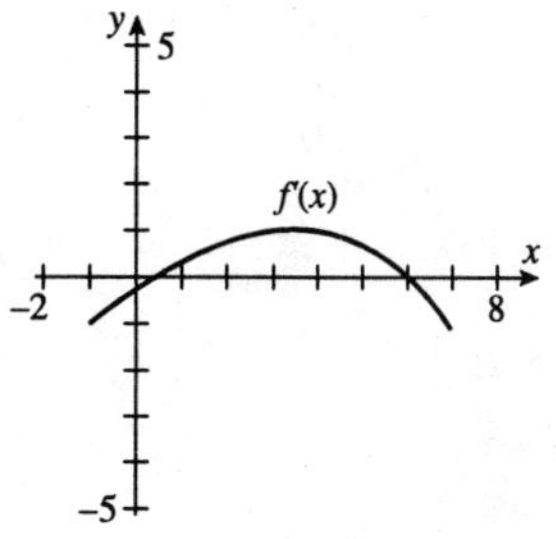

39. a. $f(2) \approx 2.5$, $f'(2) \approx 1$, $f(0.5) \approx 1.8$, $f'(0.5) \approx -1$

b. Average rate of change ≈ 1

c. At $x = 5$.

d. At $x = 3$, $x = 5$.

e. At $x = 1$, $x = 3$, and $x = 5$.

f. At $x = 0$.

g. At $x = -0.6$ and $x = 2$.

41.

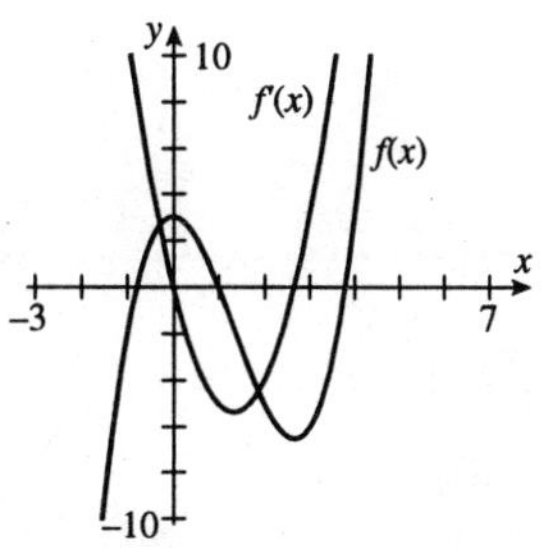

a. $0 < x < \frac{8}{3}$

b. $0 < x < \frac{8}{3}$

c. When $f(x)$ is decreasing as x increases, the derivative is negative. So, on intervals where $f(x)$ decreases, $f'(x) < 0$.

Section 3.2

Concepts Review

1. The derivative of the second; second; $f(x)g'(x) + g(x)f'(x)$

3. $nx^{n-1}h$; nx^{n-1}

Problem Set 3.2

1. $\frac{d}{dx}(2x^3) = 2\frac{d}{dx}(x^3) = 2\cdot 3x^2 = 6x^2$

3. $\frac{d}{dx}(\pi x^2) = \pi\frac{d}{dx}(x^2) = \pi\cdot 2x^1 = 2\pi x$

5. $\frac{d}{dx}(-3x^{-3}) = -3\frac{d}{dx}(x^{-3}) = -3\cdot(-3x^{-4}) = 9x^{-4} = \frac{9}{x^4}$

7. $\frac{d}{dx}\left(-\frac{2}{x^4}\right) = \frac{d}{dx}(-2x^{-4}) = -2\frac{d}{dx}(x^{-4}) = 8x^{-5} = \frac{8}{x^5}$

9. $\frac{d}{dx}\left(-8\sqrt{x}\right) = \frac{d}{dx}(-8x^{1/2}) = -8\frac{d}{dx}(x^{1/2}) = -8\cdot\frac{1}{2}x^{-1/2} = -4x^{-1/2} = -\frac{4}{\sqrt{x}}$

11. $\frac{d}{dx}(-x^3 + 2x) = \frac{d}{dx}(-x^3) + \frac{d}{dx}(2x) = -3x^2 + 2$

13. $\frac{d}{dx}(-x^4 + 3x^2 - 6x + 1) = \frac{d}{dx}(-x^4) + \frac{d}{dx}(3x^2) - \frac{d}{dx}(6x) + \frac{d}{dx}(1)$
$= -4x^3 + 3\cdot 2x - 6\cdot 1 + 0 = -4x^3 + 6x - 6$

15. $\frac{d}{dx}\left(5\sqrt{x} - 3\sqrt[3]{x} + 11\sqrt[3]{x^2} - 9\right) = \frac{d}{dx}(5x^{1/2}) - 3\frac{d}{dx}(x^{1/3}) + 11\frac{d}{dx}(x^{2/3}) - \frac{d}{dx}(9)$
$= \frac{5}{2}x^{-1/2} - 3\left(\frac{1}{3}\right)x^{-2/3} + 11\left(\frac{2}{3}\right)x^{-1/3} - 0$
$= \frac{5}{2}x^{-1/2} - x^{-2/3} + \frac{22}{3}x^{-1/3} = \frac{5}{2\sqrt{x}} - \frac{1}{\sqrt[3]{x^2}} + \frac{22}{3\sqrt[3]{x}}$

17. $\frac{d}{dx}(3x^{-5} + 2x^{-3}) = 3\frac{d}{dx}(x^{-5}) + 2\frac{d}{dx}(x^{-3}) = 3(-5)x^{-6} + 2(-3)x^{-4}$
$= -\frac{15}{x^6} - \frac{6}{x^4}$

19. $\frac{d}{dx}\left(\frac{2}{\sqrt{x}} - \frac{1}{x^2}\right) = \frac{d}{dx}(2x^{-1/2}) - \frac{d}{dx}(x^{-2}) = 2\left(-\frac{1}{2}\right)x^{-3/2} - (-2)x^{-3}$
$= -\frac{1}{\sqrt{x^3}} + \frac{2}{x^3}$

21. $\frac{d}{dx}\left(\frac{1}{2x} + 2x\right) = \frac{1}{2}\frac{d}{dx}(x^{-1}) + 2\frac{d}{dx}(x) = \frac{1}{2}(-1)x^{-2} + 2(1) = -\frac{1}{2x^2} + 2$

23. $\frac{d}{dx}[x(x^2+1)] = x\frac{d}{dx}(x^2+1) + (x^2+1)\frac{d}{dx}(x) = x(2x+0) + (x^2+1)(1)$
$= 2x^2 + x^2 + 1 = 3x^2 + 1$

25. $\frac{d}{dx}[(2x+1)^2] = \frac{d}{dx}[(2x+1)(2x+1)] = (2x+1)\frac{d}{dx}(2x+1) + (2x+1)\frac{d}{dx}(2x+1)$
$= (2x+1)(2) + (2x+1)(2) = 8x + 4$

27. $\frac{d}{dx}[(x^2+2)(x^3+1)] = (x^2+2)\frac{d}{dx}(x^3+1) + (x^3+1)\frac{d}{dx}(x^2+2)$
$= (x^2+2)(3x^2+0) + (x^3+1)(2x+0)$
$= 3x^4 + 6x^2 + 2x^4 + 2x = 5x^4 + 6x^2 + 2x$

29. $\frac{d}{dx}\left[\left(\sqrt{x}+17\right)\left(\sqrt[3]{x}+1\right)\right] = (x^{1/2}+17)\frac{d}{dx}(x^{1/3}+1) + (x^{1/3}+1)\frac{d}{dx}(x^{1/2}+17)$
$= (x^{1/2}+17)\left(\frac{1}{3}x^{-2/3}\right) + (x^{1/3}+1)\left(\frac{1}{2}x^{-1/2}\right)$
$= \frac{1}{3}x^{-1/3} + \frac{17}{3}x^{-2/3} + \frac{1}{2}x^{-1/6} + \frac{1}{2}x^{-1/2} = \frac{1}{3\sqrt[3]{x}} + \frac{17}{3\sqrt[3]{x^2}} + \frac{1}{2\sqrt[6]{x}} + \frac{1}{2\sqrt{x}}$

31. $\frac{d}{dx}[(5x^2-7)(3x^2-2x+1)] = (5x^2-7)\frac{d}{dx}(3x^2-2x+1) + (3x^2-2x+1)\frac{d}{dx}(5x^2-7)$
$= (5x^2-7)(6x-2) + (3x^2-2x+1)(10x)$
$= 30x^3 - 10x^2 - 42x + 14 + 30x^3 - 20x^2 + 10x = 60x^3 - 30x^2 - 32x + 14$

33. $\frac{d}{dx}\left(\frac{1}{3x^2+1}\right) = \frac{(3x^2+1)\frac{d}{dx}(1) - 1\frac{d}{dx}(3x^2+1)}{(3x^2+1)^2} = \frac{(3x^2+1)(0)-(6x)}{(3x^2+1)^2} = \frac{-6x}{(3x^2+1)^2}$

35. $\frac{d}{dx}\left(\frac{1}{4x^2-3x+9}\right) = \frac{(4x^2-3x+9)\frac{d}{dx}(1) - 1\frac{d}{dx}(4x^2-3x+9)}{(4x^2-3x+9)^2}$
$= \frac{(4x^2-3x+9)(0)-(8x-3)}{(4x^2-3x+9)^2} = \frac{-8x+3}{(4x^2-3x+9)^2}$

37. $\frac{d}{dx}\left(\frac{x-1}{x+1}\right) = \frac{(x+1)\frac{d}{dx}(x-1) - (x-1)\frac{d}{dx}(x+1)}{(x+1)^2} = \frac{(x+1)(1)-(x-1)(1)}{(x+1)^2}$
$= \frac{x+1-x+1}{(x+1)^2} = \frac{2}{(x+1)^2}$

39. $\frac{d}{dx}\left(\frac{2x^2-1}{3\sqrt{x}+5}\right) = \frac{(3x^{1/2}+5)\frac{d}{dx}(2x^2-1) - (2x^2-1)\frac{d}{dx}(3x^{1/2}+5)}{(3x^{1/2}+5)^2}$
$= \frac{(3x^{1/2}+5)(4x) - (2x^2-1)\left(\frac{3}{2}x^{-1/2}\right)}{(3x^{1/2}+5)^2} = \frac{12x^{3/2} + 20x - 3x^{3/2} + \frac{3}{2}x^{-1/2}}{(3x^{1/2}+5)^2}$
$= \frac{9x^{3/2} + 20x + \frac{3}{2}x^{-1/2}}{(3x^{1/2}+5)^2} = \frac{9\sqrt{x^3} + 20x + \frac{3}{2\sqrt{x}}}{\left(3\sqrt{x}+5\right)^2}$

41. $\frac{d}{dx}\left(\frac{2x^2-3x+1}{2x+1}\right) = \frac{(2x+1)\frac{d}{dx}(2x^2-3x+1) - (2x^2-3x+1)\frac{d}{dx}(2x+1)}{(2x+1)^2}$
$= \frac{(2x+1)(4x-3)-(2x^2-3x+1)(2)}{(2x+1)^2} = \frac{8x^2-2x-3-4x^2+6x-2}{(2x+1)^2}$
$= \frac{4x^2+4x-5}{(2x+1)^2}$

43. $\dfrac{d}{dx}\left(\dfrac{x^2-x+1}{x^2+1}\right)=\dfrac{(x^2+1)\frac{d}{dx}(x^2-x+1)-(x^2-x+1)\frac{d}{dx}(x^2+1)}{(x^2+1)^2}$

$=\dfrac{(x^2+1)(2x-1)-(x^2-x+1)(2x)}{(x^2+1)^2}=\dfrac{2x^3-x^2+2x-1-2x^3+2x^2-2x}{(x^2+1)^2}$

$=\dfrac{x^2-1}{(x^2+1)^2}$

45. a. $(f\cdot g)'(0)=f(0)g'(0)+g(0)f'(0)=4(5)+(-3)(-1)=23$

b. $(f+g)'(0)=f'(0)+g'(0)=-1+5=4$

c. $\left(\dfrac{f}{g}\right)'(0)=\dfrac{g(0)f'(0)-f(0)g'(0)}{g^2(0)}=\dfrac{-3(-1)-4(5)}{(-3)^2}=-\dfrac{17}{9}$

47. $\dfrac{d}{dx}[f(x)]^2=\dfrac{d}{dx}[f(x)\cdot f(x)]=f(x)\cdot\dfrac{d}{dx}f(x)+f(x)\cdot\dfrac{d}{dx}f(x)$

$=2\cdot f(x)\cdot\dfrac{d}{dx}f(x)$

49. $y=3x^2-6x+1$; slope $=\dfrac{d}{dx}(3x^2-6x+1)=6x-6$

Slope at (1, –2): 6(1) – 6 = 0; equation: $y-(-2)=0(x-1)$; $y=-2$

51. $y=x^3-x^2$; slope of the tangent line $=\dfrac{d}{dx}(x^3-x^2)=3x^2-2x$

The tangent line is horizontal when its slope is equal to zero:

$3x^2-2x=0$

$x(3x-2)=0$

$x=0$ or $3x-2=0$

$x=\dfrac{2}{3}$

$y=(0)^3-(0)^2=0$; $y=\left(\dfrac{2}{3}\right)^3-\left(\dfrac{2}{3}\right)^2=\dfrac{8}{27}-\dfrac{4}{9}=-\dfrac{4}{27}$

The tangent line is horizontal at (0, 0) and $\left(\dfrac{2}{3},-\dfrac{4}{27}\right)$.

53. a. $\dfrac{d}{dt}(-16t^2+40t+100)=-32t+40$; at $t=2$: $-32(2)+40=-24$

b. At $t=0$: $-32(0)+40=40$.

55. The slope of the tangent line at $(x_0,\ y_0)$ is $2x_0$, so the tangent line has equation $y-y_0=2x_0(x-x_0)$ or $y=2xx_0-2x_0^2+y_0=2xx_0-x_0^2$ since $y_0=x_0^2$. This line goes through (4, 15), so we can solve for x_0: $15=8x_0-x_0^2$; $x_0^2-8x_0+15=0$; $x_0=3$ or $x_0=5$. Since she is moving left to right along the curve, she should shut off the engines at (3, 9).

57. Let x represent the radius of the watermelon. The volume of the whole watermelon is $\frac{4}{3}\pi x^3$. The edible part of the watermelon is a sphere of radius $\frac{4}{3}\pi\left(x-\frac{x}{10}\right)^3=\frac{4}{3}\pi\left(\frac{9x}{10}\right)^3$, so the volume of the rind is

$$V=\frac{4}{3}\pi\left(x^3-\left(\frac{9x}{10}\right)^3\right)=\frac{4}{3}\pi\cdot\frac{271x^3}{1000}=\frac{271}{750}\pi x^3$$

$$\frac{d}{dx}\left(\frac{271}{750}\pi x^3\right)=\frac{271\pi}{750}\frac{d}{dx}(x^3)=\frac{271\pi}{750}(3x^2)=\frac{271\pi}{250}x^2.$$

At the end of the fifth week, the radius is $x = 2(5) = 10$ cm. The volume of the rind is growing at the rate of $\frac{271\pi}{250}(10)^2=\frac{542\pi}{5}\approx 340\text{ cm}^3$ per cm of radius growth. Since the radius is growing 2 cm per week, the volume of the rind is growing at the rate of $\frac{542\pi}{5}(2)\approx 681\text{ cm}^3$ per week.

59. $$\frac{d}{dx}[f(x)-g(x)]=\lim_{h\to 0}\frac{[f(x+h)-g(x+h)]-[f(x)-g(x)]}{h}$$

$$=\lim_{h\to 0}\left[\frac{f(x+h)-f(x)}{h}-\frac{g(x+h)-g(x)}{h}\right]$$

$$=\lim_{h\to 0}\frac{f(x+h)-f(x)}{h}-\lim_{h\to 0}\frac{g(x+h)-g(x)}{h}=\frac{d}{dx}f(x)-\frac{d}{dx}g(x)$$

Section 3.3

Concepts Review

1. $\cos x;\ -\sin x$

3. $e^x;\ \frac{1}{x}$

Problem Set 3.3

1. $\frac{d}{dx}(3\sin x-5\cos x)=3\frac{d}{dx}(\sin x)-5\frac{d}{dx}(\cos x)=3\cos x+5\sin x$

3. $\frac{d}{dx}(e^x-\ln x)=\frac{d}{dx}(e^x)-\frac{d}{dx}(\ln x)=e^x-\frac{1}{x}$

5. $$\frac{d}{dx}(\cot x)=\frac{d}{dx}\left(\frac{\cos x}{\sin x}\right)=\frac{\sin x\frac{d}{dx}(\cos x)-\cos x\frac{d}{dx}(\sin x)}{\sin^2 x}$$

$$=\frac{-\sin^2 x-\cos^2 x}{\sin^2 x}=\frac{-(\sin^2 x+\cos^2 x)}{\sin^2 x}=-\frac{1}{\sin^2 x}=-\csc^2 x$$

7. $\frac{d}{dx}(x^2e^x)=x^2\frac{d}{dx}(e^x)+e^x\frac{d}{dx}(x^2)=x^2e^x+2xe^x$

9. $\frac{d}{dx}[3\sin(2x+1)]=3\frac{d}{dx}[\sin(2x+1)]=3\cdot 2\cos(2x+1)=6\cos(2x+1)$

11. $\frac{d}{dx}(e^{2x}\cos\pi x)=e^{2x}\frac{d}{dx}\cos(\pi x)+\cos(\pi x)\frac{d}{dx}(e^{2x})=-\pi e^{2x}\sin\pi x+2e^{2x}\cos\pi x$

13. $\dfrac{d}{dx}\dfrac{\ln x}{e^{0.1x}+\cos x}=\dfrac{(e^{0.1x}+\cos x)\frac{d}{dx}(\ln x)-\ln x\frac{d}{dx}(e^{0.1x}+\cos x)}{(e^{0.1x}+\cos x)^2}$

$=\dfrac{\frac{e^{0.1x}+\cos x}{x}-0.1(\ln x)e^{0.1x}+(\ln x)\sin x}{(e^{0.1x}+\cos x)^2}$

15. $\dfrac{d}{dx}[3\cos(\pi x)]=-3\pi\sin(\pi x)$

At $x = 1$: $y = 3\cos\pi = -3$.

Slope $= -3\pi\sin\pi = 0$

$y-(-3)=0(x-1);\ y=-3$

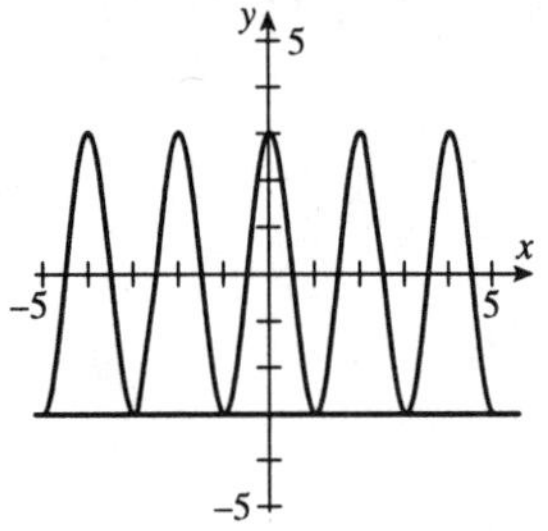

17. $\dfrac{d}{dx}\left(\dfrac{\cos x}{x}\right)=\dfrac{x\frac{d}{dx}(\cos x)-\cos x\frac{d}{dx}x}{x^2}=\dfrac{-x\sin x-\cos x}{x^2}$

At $x=1$: $y=\dfrac{\cos 1}{1}\approx 0.540$.

Slope $=\dfrac{-1\sin 1-\cos 1}{1^2}\approx -1.382$

$y-0.540=-1.382(x-1);\ y=-1.382x+1.922$

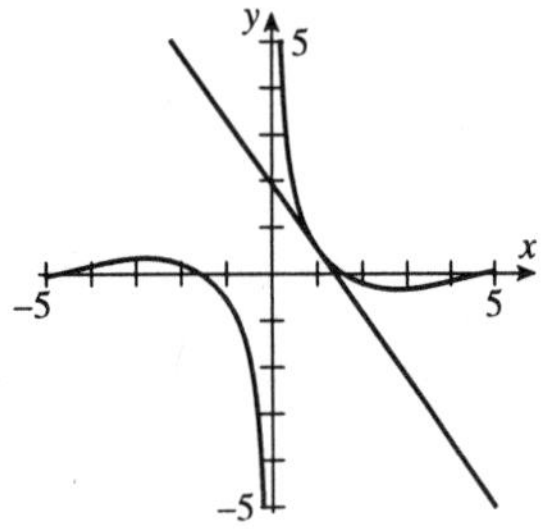

19. $f'(t)=\dfrac{d}{dt}[10e^{-0.1t}\cos(3t)]=10e^{-0.1t}\dfrac{d}{dt}\cos(3t)+\cos(3t)\dfrac{d}{dt}10e^{-0.1t}$

$=-30e^{-0.1t}\sin(3t)-e^{-0.1t}\cos(3t)$

$f'(2)=-30e^{-0.2}\sin 6-e^{-0.2}\cos 6\approx 6.077$

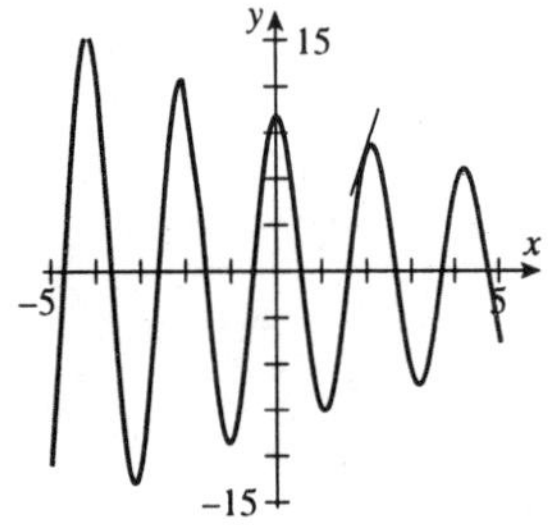

21. $h'(t) = \frac{d}{dt}\frac{100}{9e^{-0.1t}+1} = \frac{(9e^{-0.1t}+1)\frac{d}{dt}100 - 100\frac{d}{dt}(9e^{-0.1t}+1)}{(9e^{-0.1t}+1)^2}$

$= \frac{90e^{-0.1t}}{(9e^{-0.1t}+1)^2}$

$h'(50) = \frac{90e^{-5}}{(9e^{-5}+1)^2} \approx 0.539$

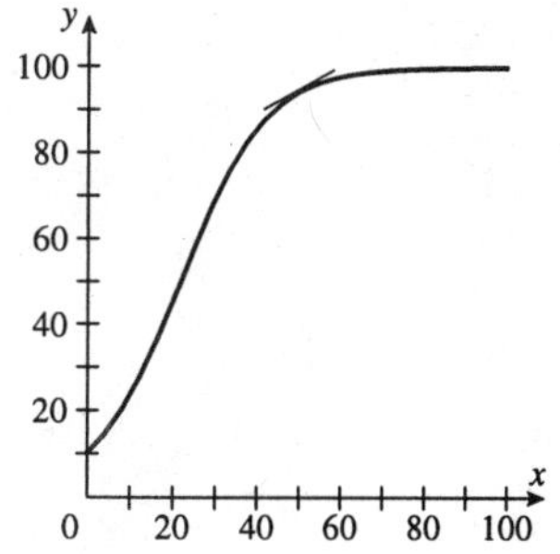

23. $\frac{d}{dt}(3.956e^{0.02951t}) = 3.956(0.02951)e^{0.02951t} = 0.11674156e^{0.02951t}$

At $t = 10$: $0.11674156e^{0.02951(10)} \approx 0.157$.

At $t = 60$: $0.11674156e^{0.02951(60)} \approx 0.686$.

These numbers represent the rate of population growth in 1800 and 1850 respectively.

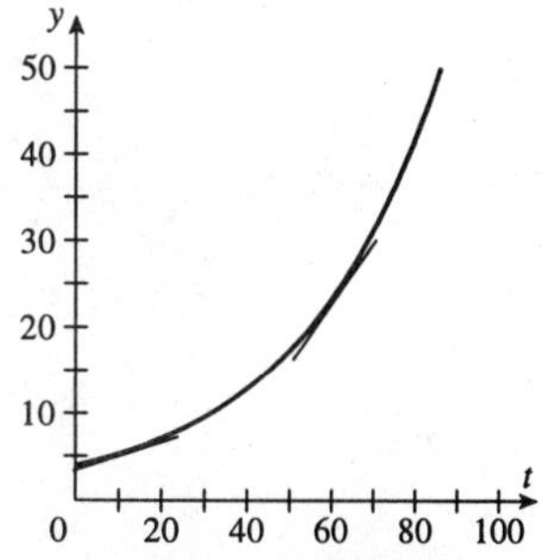

25. $y = \sqrt{2}\sin x;\ \frac{d}{dx}\left(\sqrt{2}\sin x\right) = \sqrt{2}\cos x$

$y = \sqrt{2}\cos x;\ \frac{d}{dx}\left(\sqrt{2}\cos x\right) = -\sqrt{2}\sin x$

Two lines intersect at right angles if their slopes are negative reciprocals of each other:

$\sqrt{2}\cos x = \frac{1}{\sqrt{2}\sin x};\ 2\sin x\cos x = 1;\ \sin 2x = 1;\ x = \frac{1}{2}\sin^{-1} 1.$

Two lines intersect at $x = \frac{\sin^{-1} 1}{2} = \frac{\pi}{4}$.

27. $\frac{d}{dx}(\sin x^2) = \lim_{h\to 0} \frac{\sin(x+h)^2 - \sin x^2}{h} = \lim_{h\to 0} \frac{\sin(x^2+2hx+h^2) - \sin x^2}{h}$

$= \lim_{h\to 0} \frac{\sin x^2 \cos(2hx+h^2) + \cos(x^2)\sin(2hx+h^2) - \sin x^2}{h}$

$= \lim_{h\to 0} \frac{\sin x^2 (\cos(2hx+h^2)-1) + \cos x^2 \sin(2hx+h^2)}{h}$

$= \lim_{h\to 0} (2x+h)\sin x^2 \frac{(\cos(2hx+h^2)-1)}{2hx+h^2} + \lim_{h\to 0} (2x+h)\cos x^2 \frac{\sin(2hx+h^2)}{2hx+h^2}$

$= 2x\sin x^2 \cdot 0 + 2x\cos x^2 \cdot 1 = 2x\cos x^2$

29. **a.** Written response.

b. $D = \frac{1}{2}(1-\cos t)(\sin t) = \frac{\sin t(1-\cos t)}{2}$

$E = \frac{t}{2} - \frac{\sin t \cos t}{2}$

$\frac{D}{E} = \frac{\sin t(1-\cos t)}{t - \sin t \cos t}$

c. $\lim_{t\to 0^+}\left(\frac{D}{E}\right) \approx 0.75$

31. $f(x) = x\sin x$

$f'(x) = \frac{d}{dx}(x\sin x) = x\frac{d}{dx}(\sin x) + \sin x \frac{d}{dx}(x) = x\cos x + \sin x$

a.

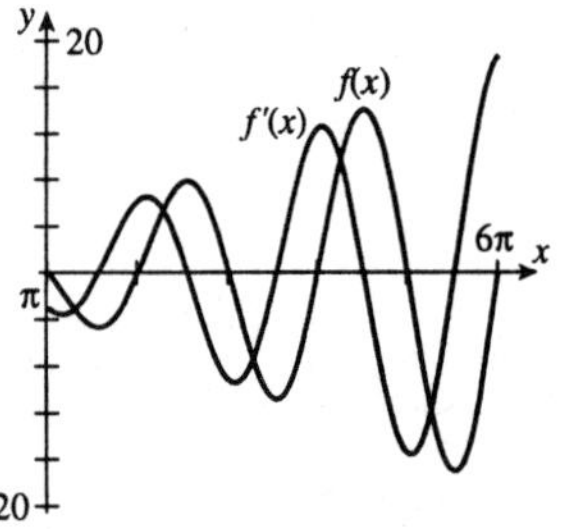

b. $f(x) = 0$ has 6 solutions on $[\pi, 6\pi]$.
$f'(x) = 0$ has 5 solutions on $[\pi, 6\pi]$.

c. $f'(x) = 0$ can have more than $n - 1$ solutions on $[a, b]$.
For example, look at $f(x) = 0$ and $f'(x) = 0$ for $f(x) = x^4 - 3x^2 - 4$ on $[-2, 2]$.

d. The maximum value of $|f(x) - f'(x)|$ occurs at $x \approx 18.149$. The value is 24.930.

Section 3.4

Concepts Review

1. $\frac{du}{dt}$

3. $2xe^{x^2}$

Problem Set 3.4

1. $y = u^{15}$ and $u = 2-9x$

$$\frac{dy}{dx} = \frac{dy}{du}\cdot\frac{du}{dx} = 15u^{14}(-9) = -135(2-9x)^{14}$$

3. $y = u^5$ and $u = 5x^2+2x-8$

$$\frac{dy}{dx} = \frac{dy}{du}\cdot\frac{du}{dx} = 5u^4(10x+2) = 5(5x^2+2x-8)^4(10x+2)$$

5. $y = u^9$ and $u = x^3 - 3e^x + 4\ln x$

$$\frac{dy}{dx} = \frac{dy}{du}\cdot\frac{du}{dx} = 9u^8\left(3x^2 - 3e^x + \frac{4}{x}\right) = 9(x^3 - 3e^x + 4\ln x)^8\left(3x^2 - 3e^x + \frac{4}{x}\right)$$

7. $y = \sqrt{u}$ and $u = 3x^4 + x - 8$

$$\frac{dy}{dx} = \frac{dy}{du}\cdot\frac{du}{dx} = \frac{1}{2\sqrt{u}}(12x^3+1) = \frac{12x^3+1}{2\sqrt{3x^4+x-8}}$$

9. $y = \ln u$ and $u = 4x^3 - 3x^2 + 11x - 1$

$$\frac{dy}{dx} = \frac{dy}{du}\cdot\frac{du}{dx} = \frac{1}{u}(12x^2 - 6x + 11) = \frac{12x^2-6x+11}{4x^3-3x^2+11x-1}$$

11. $y = \sin u$ and $u = 3x^2 + 11x$

$$\frac{dy}{dx} = \frac{dy}{du}\cdot\frac{du}{dx} = \cos u(6x+11) = (6x+11)\cos(3x^2+11x)$$

13. $y = u^3$ and $u = \sin x$

$$\frac{dy}{dx} = \frac{dy}{du}\cdot\frac{du}{dx} = 3u^2\cos x = 3\sin^2 x\cos x$$

15. $y = u^4$ and $u = \dfrac{e^x}{x+4}$

$$\frac{dy}{dx} = \frac{dy}{du}\cdot\frac{du}{dx} = 4u^3\frac{(x+4)\frac{d}{dx}(e^x) - e^x\frac{d}{dx}(x+4)}{(x+4)^2} = 4\left(\frac{e^x}{x+4}\right)^3\frac{(x+3)e^x}{(x+4)^2} = \frac{4e^{4x}(x+3)}{(x+4)^5}$$

17. $y = \sin u$ and $u = \dfrac{3x-1}{2x+5}$

$$\frac{dy}{dx} = \frac{dy}{du}\cdot\frac{du}{dx} = \cos u\frac{(2x+5)\frac{d}{dx}(3x-1) - (3x-1)\frac{d}{dx}(2x+5)}{(2x+5)^2} = \cos\left(\frac{3x-1}{2x+5}\right)\frac{6x+15-(6x-2)}{(2x+5)^2}$$

$$= \frac{17}{(2x+5)^2}\cos\left(\frac{3x-1}{2x+5}\right)$$

19. $$\frac{dy}{dx} = (4x-7)^2\frac{d}{dx}(2x+3) + (2x+3)\frac{d}{dx}(4x-7)^2 = 2(4x-7)^2 + (2x+3)\cdot 2(4x-7)\cdot 4$$

$$= 96x^2 - 128x - 70$$

21. $$\frac{dy}{dx} = (2x-1)^3\frac{d}{dx}(e^{5x}) + e^{5x}\frac{d}{dx}(2x-1)^3 = 5(2x-1)^3e^{5x} + e^{5x}\cdot 3(2x-1)^2\cdot 2$$

$$= 5e^{5x}(2x-1)^3 + 6e^{5x}(2x-1)^2$$

23. $\dfrac{dy}{dx} = \dfrac{e^{3x-4}\frac{d}{dx}(x+1)^2 - (x+1)^2\frac{d}{dx}(e^{3x-4})}{(e^{3x-4})^2} = \dfrac{e^{3x-4}\cdot 2(x+1)\cdot 1 - (x+1)^2\cdot 3e^{3x-4}}{e^{6x-8}}$

$= \dfrac{(2x+2)e^{3x-4} - 3(x+1)^2 e^{3x-4}}{e^{6x-8}}$

25. $\dfrac{dy}{dx} = \dfrac{(2x^2-5)\frac{d}{dx}(\ln x)^2 - (\ln x)^2\frac{d}{dx}(2x^2-5)}{(2x^2-5)^2} = \dfrac{(2x^2-5)\cdot 2(\ln x)\left(\frac{1}{x}\right) - (\ln x)^2\cdot 4x}{(2x^2-5)^2}$

$= \dfrac{\frac{2\ln x(2x^2-5)}{x} - 4x(\ln x)^2}{(2x^2-5)^2}$

27. $\dfrac{d}{dt}\left(\dfrac{3t-2}{t+5}\right)^3 = 3\left(\dfrac{3t-2}{t+5}\right)^2\dfrac{d}{dt}\left(\dfrac{3t-2}{t+5}\right) = 3\left(\dfrac{3t-2}{t+5}\right)^2\dfrac{(t+5)\frac{d}{dx}(3t-2) - (3t-2)\frac{d}{dx}(t+5)}{(t+5)^2}$

$= 3\left(\dfrac{3t-2}{t+5}\right)^2\dfrac{17}{(t+5)^2} = \dfrac{51(3t-2)^2}{(t+5)^4}$

29. $\dfrac{d}{dt}\left(\dfrac{(3t-2)^3}{e^{5t}}\right) = \dfrac{e^{5t}\frac{d}{dt}(3t-2)^3 - (3t-2)^3\frac{d}{dt}(e^{5t})}{(e^{5t})^2} = \dfrac{e^{5t}\cdot 3(3t-2)^2\cdot 3 - (3t-2)^3\cdot 5e^{5t}}{e^{10t}}$

$= \dfrac{9e^{5t}(3t-2)^2 - 5e^{5t}(3t-2)^3}{e^{10t}} = \dfrac{9(3t-2)^2 - 5(3t-2)^3}{e^{5t}}$

31. $\dfrac{d}{d\theta}(\sin^3\theta) = 3\sin^2\theta\dfrac{d}{d\theta}(\sin\theta) = 3\sin^2\theta\cos\theta$

33. $\dfrac{d}{dx}\left(\dfrac{\sin x}{\ln 2x}\right)^3 = 3\left(\dfrac{\sin x}{\ln 2x}\right)^2\dfrac{(\ln 2x)\frac{d}{dx}(\sin x) - (\sin x)\frac{d}{dx}(\ln 2x)}{(\ln 2x)^2}$

$= 3\left(\dfrac{\sin x}{\ln 2x}\right)^2\dfrac{\ln 2x\cos x - \sin x\left(\frac{1}{2x}\right)\cdot 2}{(\ln 2x)^2} = \dfrac{3\sin^2 x\left(\ln 2x\cos x - \frac{\sin x}{x}\right)}{(\ln 2x)^4}$

35. $\dfrac{d}{dx}[\sin^4(x^2+3x)] = 4\sin^3(x^2+3x)\dfrac{d}{dx}[\sin(x^2+3x)] = 4\sin^3(x^2+3x)\cos(x^2+3x)\dfrac{d}{dx}(x^2+3x)$

$= 4\sin^3(x^2+3x)\cos(x^2+3x)(2x+3)$

37. $\dfrac{d}{dt}[\sin^3(e^t)] = 3\sin^2(e^t)\dfrac{d}{dt}[\sin(e^t)] = 3\sin^2(e^t)\cos(e^t)\dfrac{d}{dt}(e^t)$

$= 3e^t\sin^2(e^t)\cos(e^t)$

39. $\dfrac{d}{d\theta}[\cos(\sin\theta^2)] = -\sin(\sin\theta^2)\dfrac{d}{d\theta}(\sin\theta^2) = -\sin(\sin\theta^2)\cos(\theta^2)\dfrac{d}{d\theta}(\theta^2)$

$= -2\theta\sin(\sin\theta^2)\cos(\theta^2)$

41. $\dfrac{d}{dx}e^{\cos(\sin 2x)} = e^{\cos(\sin 2x)}\dfrac{d}{dx}[\cos(\sin 2x)] = e^{\cos(\sin 2x)}(-\sin(\sin 2x))(\cos 2x)(2)$

$= -2e^{\cos(\sin 2x)}\sin(\sin 2x)\cos 2x$

43. $f'(x) = \dfrac{d}{dx}(e^{-x^2}) = e^{-x^2}\cdot\dfrac{d}{dx}(-x^2) = -2xe^{-x^2}$

$f'(1) = -2(1)e^{-(1)^2} \approx -0.736$

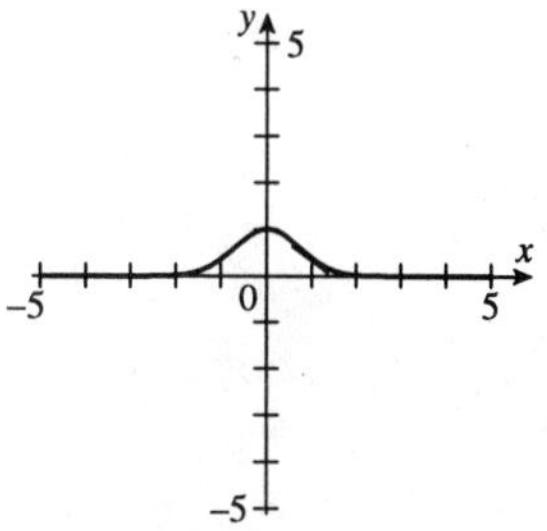

45. $F'(t) = \frac{d}{dt}\sin(t^2 + 3t + 1) = \cos(t^2 + 3t + 1)\cdot(2t + 3) = (2t + 3)\cos(t^2 + 3t + 1)$

$F'(1) = (2\cdot 1 + 3)\cos(1^2 + 3\cdot 1 + 1) \approx 1.418$

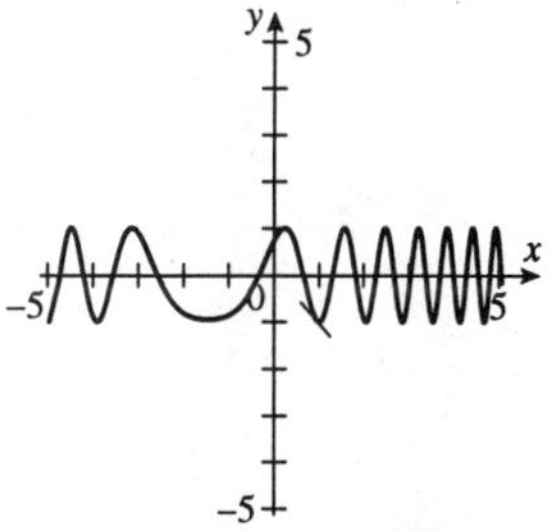

47. $\frac{dy}{dt} = -32 + 32e^{-t}$

a. At $t = 2$: $-32 + 32e^{-2} \approx -27.67$ m/s

b. $E = \frac{1}{2}m(-32 + 32e^{-t})^2$; at $t = 2$: $E \approx \frac{1}{2}(1)(-27.67)^2 \approx 382.81$

$\frac{dE}{dt} = m(-32 + 32e^{-t})\cdot -32e^{-t}$; at $t = 2$: $\frac{dE}{dt} = -32e^{-2}(1)(-32 + 32e^{-2}) \approx 119.83$

The numbers represent the kinetic energy and the rate of change of energy after 2 seconds.

49. a. $\left(\frac{4\cos 2t}{4}\right)^2 + \left(\frac{7\sin 2t}{7}\right) = \cos^2 2t + \sin^2 2t = 1$

b. $L = \sqrt{(4\cos 2t - 0)^2 + (7\sin 2t - 0)^2} = \sqrt{16\cos^2 2t + 49\sin^2 2t}$

c. $\frac{dL}{dt} = \frac{d}{dt}\sqrt{16\cos^2 2t + 49\sin^2 2t} = \frac{-64\cos 2t\sin 2t + 196\sin 2t\cos 2t}{2\sqrt{16\cos^2 2t + 49\sin^2 2t}} = \frac{66\sin 2t\cos 2t}{\sqrt{16\cos^2 2t + 49\sin^2 2t}}$

At $t = \frac{\pi}{8}$: $\frac{66\sin\frac{\pi}{4}\cos\frac{\pi}{4}}{\sqrt{16\cos^2\frac{\pi}{4} + 49\sin^2\frac{\pi}{4}}} \approx 5.789$ ft/sec.

51. a. $(\cos 2t, \sin 2t)$

b. $(0 - \cos 2t)^2 + (y - \sin 2t)^2 = 5^2$; $y = \sin 2t + \sqrt{25 - \cos^2 2t}$

c. $\frac{d}{dt}\left(\sin 2t + \sqrt{25 - \cos^2 2t}\right) = 2\cos 2t + \frac{4\cos 2t\sin 2t}{2\sqrt{25 - \cos^2 2t}} = 2\cos 2t\left(1 + \frac{\sin 2t}{\sqrt{25 - \cos^2 2t}}\right)$

53. a. $\frac{d}{dx}|x^2-1| = \frac{|x^2-1|}{x^2-1} \cdot 2x = \frac{2x|x^2-1|}{x^2-1}$

b. $\frac{d}{dx}|\sin x| = \frac{|\sin x|}{\sin x} \cdot \cos x = \cot x |\sin x|$

55. a. Possible answer: 0.68

b. $f'(x) = \cos(\sin(\sin(\sin(x))) \cdot \cos(\sin(\sin(x))) \cdot \cos(\sin(x)) \cdot \cos x$
Possible answer: 1

Section 3.5

Concepts Review

1. $f'''(x);\ y''';\ \frac{d^3y}{dx^3}$

3. 0; <0

Problem Set 3.5

1. $\frac{dy}{dx} = 3x^2 + 6x - 2;\ \frac{d^2y}{dx^2} = 6x + 6;\ \frac{d^3y}{dx^3} = 6$

3. $\frac{dy}{dx} = 4(2x+5)^3 \cdot 2 = 8(2x+5)^3$
$\frac{d^2y}{dx^2} = 24(2x+5)^2 \cdot 2 = 48(2x+5)^2$
$\frac{d^3y}{dx^3} = 96(2x+5) \cdot 2 = 384x + 960$

5. $\frac{dy}{dx} = 3\cos(3x);\ \frac{d^2y}{dx^2} = -9\sin(3x);\ \frac{d^3y}{dx^3} = -27\cos(3x)$

7. $\frac{dy}{dx} = \frac{1}{x-3};\ \frac{d^2y}{dx^2} = -\frac{1}{(x-3)^2};\ \frac{d^3y}{dx^3} = \frac{2}{(x-3)^3}$

9. $f'(x) = 6x^2;\ f''(x) = 12x$
$f''(2) = 12(2) = 24$

11. $f'(t) = -\frac{1}{t^2};\ f''(t) = \frac{2}{t^3}$
$f''(2) = \frac{2}{2^3} = \frac{1}{4}$

13. $f'(x) = x\frac{d}{dx}e^x + e^x\frac{d}{dx}x = xe^x + e^x$
$f''(x) = xe^x + e^x + e^x = xe^x + 2e^x$
$f''(2) = 2e^2 + 2e^2 \approx 29.556$

15. $f'(x) = 2\sin(\pi x)\cos(\pi x)\pi = 2\pi\sin(\pi x)\cos(\pi x)$

$f''(x) = 2\pi\left[\sin(\pi x)\frac{d}{dx}\cos(\pi x) + \cos(\pi x)\frac{d}{dx}\sin(\pi x)\right]$

$= 2\pi[-\pi\sin^2(\pi x) + \pi\cos^2(\pi x)]$

$f''(2) = 19.739$

17. $\frac{d}{dx}(x^n) = nx^{n-1}$; $\frac{d^2}{dx^2} = n(n-1)x^{n-2}$; $\frac{d^3}{dx^3} = n(n-1)(n-2)x^{n-3}$

$\frac{d^n}{dx^n} = n(n-1)(n-2)\ldots 2\cdot 1x^{n-n} = n(n-1)(n-2)\ldots 2\cdot 1x^0 = n!$

19. a. $\frac{d^4}{dx^4}(3x^3 + 2x - 19) = 0$

b. $\frac{d^{12}}{dx^{12}}(100x^{11} - 79x^{10}) = 0$

c. $\frac{d^{11}}{dx^{11}}(x^2 - 3)^5 = 0$

21. $f(x) = x^3 + 3x^2 - 45x - 6$

$f'(x) = 3x^2 + 6x - 45 = 0$; $3(x^2 + 2x - 15) = 0$; $3(x+5)(x-3) = 0$

$f'(c) = 0$ when $c = -5$ and $c = 3$.

$f''(x) = 6x + 6$

$f''(-5) = 6(-5) + 6 = -24$; $f''(3) = 6(3) + 6 = 24$

23. a. $v(t) = \frac{ds}{dt} = 12 - 4t$

$a(t) = \frac{d^2s}{dt^2} = -4$

b. $12 - 4t > 0$; $t < 3$

c. $12 - 4t < 0$; $t > 3$

d. $-\infty < t < \infty$

e.

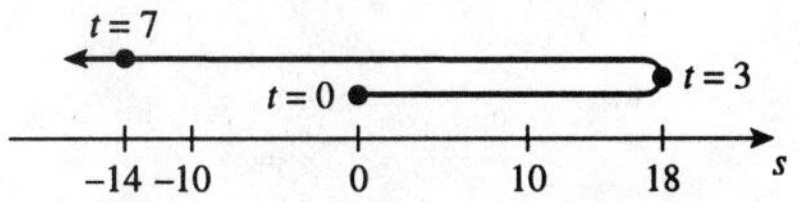

25. a. $v(t) = \frac{ds}{dt} = e^t - 5$

$a(t) = \frac{d^2s}{dt^2} = e^t$

b. $e^t - 5 > 0$; $t > \ln 5$

c. $e^t - 5 < 0$; $t < \ln 5$

d. $\varnothing$

e.

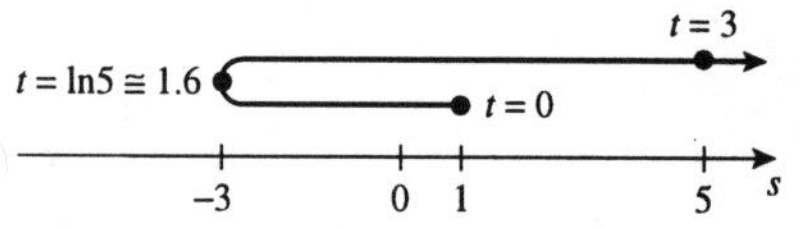

27. a. $v(t) = \frac{ds}{dt} = 2t - \frac{16}{t^2}$

$a(t) = \frac{d^2s}{dt^2} = 2 + \frac{32}{t^3}$

b. $2t - \frac{16}{t^2} > 0;\ \frac{2t^3 - 16}{t^2} > 0;\ \frac{2(t-2)(t^2+2t+4)}{t^2} > 0$

$t > 2$

c. $2t - \frac{16}{t^2} < 0;\ t < 2$

d. $2 + \frac{32}{t^3} < 0;\ \frac{2t^3 + 32}{t^3} < 0;\ \frac{2(t^3+16)}{t^3} < 0;\ a(t) > 0$ for all $t > 0$.

e.

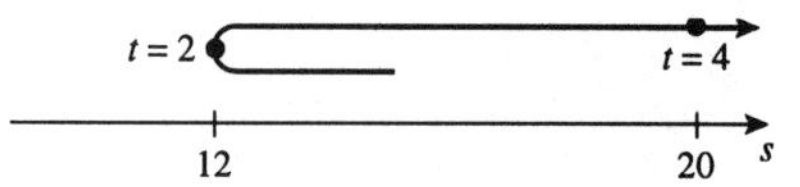

29. $\frac{ds}{dt} = 2t^3 - 15t^2 + 24t;\ \frac{d^2s}{dt^2} = 6t^2 - 30t + 24$

$6t^2 - 30t + 24 = 0;\ 6(t^2 - 5t + 4) = 0;\ 6(t-4)(t-1) = 0;\ t = 1,\ 4$

Velocity at $t = 1$: $2(1)^3 - 15(1)^2 + 24(1) = 11$

Velocity at $t = 4$: $2(4)^3 - 15(4)^2 + 24(4) = -16$

31. a. $\frac{ds_1}{dt} = 4 - 6t;\ \frac{ds_2}{dt} = 2t - 2;\ 4 - 6t = 2t - 2;\ 8t = 6;\ t = 0.75$

b. $t = 0.75$ and when $4 - 6t = -(2t - 2)$; $4t = 2$; $t = 0.5$

c. $4t - 3t^2 = t^2 - 2t;\ 4t^2 - 6t = 0;\ 2t(2t-3) = 0;\ t = 0,\ 1.5$

33. $s = -16t^2 + 48t + 256$

$v = \frac{ds}{dt} = -32t + 48$

a. $v = -32(0) + 48 = 48$ ft/sec

b. $-32t + 48 = 0$; $t = 1.5$ sec

c. $s = -16(1.5)^2 + 48(1.5) + 256 = 292$ ft

d. $-16t^2 + 48t + 256 = 0;\ -16(t^2 - 3t - 16) = 0;\ t = \frac{3 \pm \sqrt{(-3)^2 - 4(-16)}}{2}$

$t \approx 5.772$ sec

e. $v = -32(5.772) + 48 \approx -136.7$ ft/sec

35. $\frac{ds}{dt} = v_0 - 32t = 0;\ v_0 = 32t$

$s = v_0 t - 16t^2;\ 5280 = 32t^2 - 16t^2;\ t^2 = 330;\ t \approx 18.166$

$5280 = v_0(18.166) - 16(18.166)^2;\ v_0 \approx 581.3$ ft/sec.

37. $s = t^3 - 3t^2 - 24t - 6$

$v = 3t^2 - 6t - 24$

$a = 6t - 6;\ 6t - 6 < 0;\ t < 1$ sec. The point is slowing down when t is less than one second.

39. **a.** $\frac{ds}{dt} = ks$ for some k.

b. $\frac{d^2s}{dt^2} > 0$

c. $\frac{d^3s}{dt^3} < 0$

d. $\frac{d^2s}{dt^2} = 10$

e. $\frac{d^2s}{dt^2}$ and $\frac{ds}{dt}$ are approaching zero.

f. $\frac{ds}{dt}$ is a constant.

41. Possible answers

a. $c = f(t)$ is the cost at time t. $\frac{d^2c}{dt^2} > 0,\ \frac{dc}{dt} > 0$

b. $c = f(t)$ is the consumption at time t. $\frac{d^2c}{dt^2} < 0,\ \frac{dc}{dt} < 0$

c. $p = f(t)$ is the population at time t. $\frac{dp}{dt} > 0,\ \frac{d^2p}{dt^2} < 0$

d. $\frac{d^2s}{dt^2}$ is constant and positive.

e. $\theta = f(t)$ is the angle at time t. $\frac{d\theta}{dt} > 0,\ \frac{d^2\theta}{dt^2} > 0$

f. $p = f(t)$ is the profit at time t. $\frac{dp}{dt} > 0,\ \frac{d^2p}{dt^2} < 0$

g. $p = f(t)$ is the profit at time t. $p < 0,\ \frac{dp}{dt} > 0$

43. $\frac{d}{dx}(uv) = uv' + vu'$; $\frac{d^2}{dx^2}(uv) = uv'' + v'u' + vu'' + u'v' = uv'' + 2u'v' + u''v$

$\frac{d^3}{dx^3}(uv) = u'''v + 3u''v' + 3u'v'' + uv'''$

The coefficients are the same as those in a binomial expansion.

45. $f(x) = x\left[\sin x - \cos\left(\frac{x}{2}\right)\right]$

$f'(x) = x\left[\cos x + \frac{1}{2}\sin\left(\frac{x}{2}\right)\right] + \sin x - \cos\left(\frac{x}{2}\right)$

$f''(x) = x\left[-\sin x + \frac{1}{4}\cos\left(\frac{x}{2}\right)\right] + \cos x + \frac{1}{2}\sin\left(\frac{x}{2}\right) + \cos x + \frac{1}{2}\sin\left(\frac{x}{2}\right)$

$f'''(x) = x\left[-\cos x - \frac{1}{8}\sin\left(\frac{x}{2}\right)\right] + \left[-\sin x + \frac{1}{4}\cos\left(\frac{x}{2}\right)\right] + \left[-\sin x + \frac{1}{4}\cos\left(\frac{x}{2}\right)\right] + \left[-\sin x + \frac{1}{4}\cos\left(\frac{x}{2}\right)\right]$

$= x\left[-\cos x - \frac{1}{8}\sin\left(\frac{x}{2}\right)\right] - 3\sin x + \frac{3}{4}\cos\left(\frac{x}{2}\right)$

a.

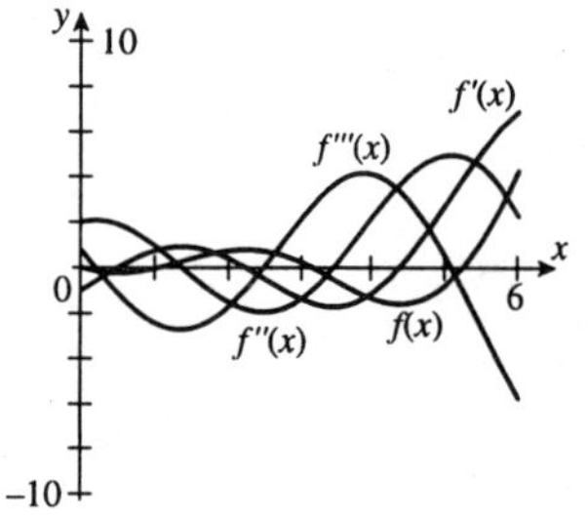

b. $f'''(2.13) \approx -1.283$

Section 3.6

Concepts Review

1. $\frac{9}{x^3 - 3}$

3. $\frac{du}{dt}$; $t = 2$

Problem Set 3.6

1. $\frac{d}{dx}(x^2 - y^2) = \frac{d}{dx}(9)$; $2x - 2y\frac{dy}{dx} = 0$; $\frac{dy}{dx} = \frac{-2x}{-2y} = \frac{x}{y}$

3. $\frac{d}{dx}(xy) = \frac{d}{dx}(4)$; $x\frac{dy}{dx} + y = 0$; $\frac{dy}{dx} = -\frac{y}{x}$

5. $\frac{d}{dx}\left(6x-\sqrt{2xy}+xy^3\right)=\frac{d}{dx}(y^2);\ 6-\frac{1}{\sqrt{2xy}}\left(x\frac{dy}{dx}+y\right)+3xy^2\frac{dy}{dx}+y^3=2y\frac{dy}{dx};$

$-\frac{x}{\sqrt{2xy}}\frac{dy}{dx}+3xy^2\frac{dy}{dx}-2y\frac{dy}{dx}=-6+\frac{y}{\sqrt{2xy}}-y^3;\ \frac{dy}{dx}=\frac{-6+\frac{y}{\sqrt{2xy}}-y^3}{-\frac{x}{\sqrt{2xy}}+3xy^2-2y}$

7. $\frac{d}{dx}(x^2-y^4)=\frac{d}{dx}(9);\ 2x-4y^3\frac{dy}{dx}=0;\ \frac{dy}{dx}=\frac{x}{2y^3}$

$\text{Slope}=\frac{5}{2(2)^3}=\frac{5}{16}$

$y-2=\frac{5}{16}(x-5)$

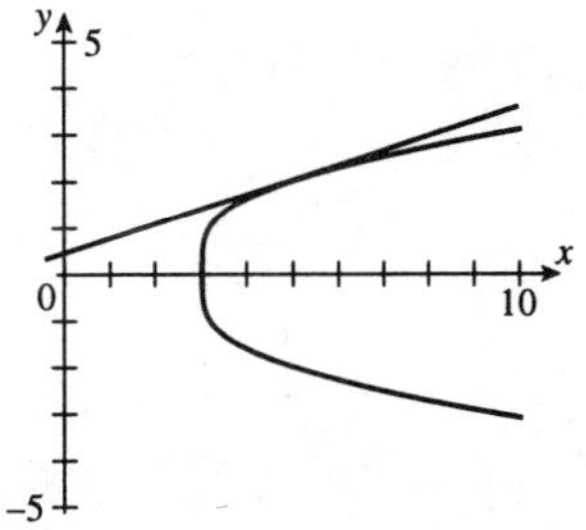

9. $\frac{d}{dx}\sin(xy)=\frac{d}{dx}(y);\ \cos(xy)\left(x\frac{dy}{dx}+y\right)=\frac{dy}{dx};\ y\cos(xy)=\frac{dy}{dx}-x\cos(xy)\frac{dy}{dx}$

$\frac{dy}{dx}=\frac{y\cos(xy)}{1-x\cos(xy)};\ \text{slope}=\frac{1\cos\left(\frac{\pi}{2}\right)}{1-\frac{\pi}{2}\cos\left(\frac{\pi}{2}\right)}=0$

$y=1$

The following graph was plotted using the Implicit Plot command on Mathematica. It is not possible to sketch the curve by hand because y cannot be solved in terms of x.

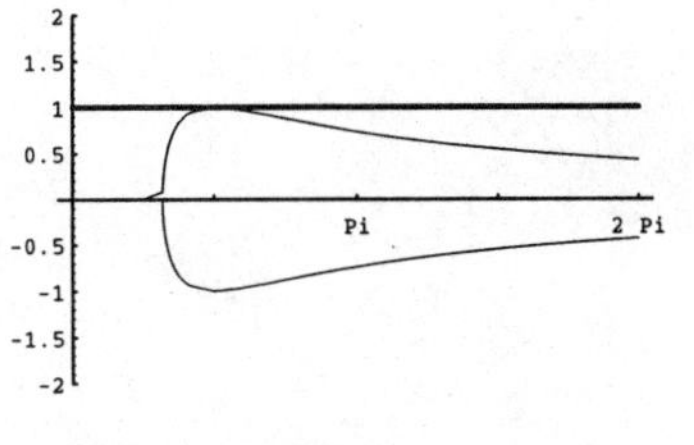

11. $\frac{d}{dt}(s^2t+t^3)=\frac{d}{dt}(1);\ s^2+t(2s)\frac{ds}{dt}+3t^2=0;\ \frac{ds}{dt}=-\frac{3t^2+s^2}{2ts}$

$\frac{d}{ds}(s^2t+t^3)=\frac{d}{ds}(1);\ s^2\frac{dt}{ds}+t(2s)+3t^2\frac{dt}{ds}=0;\ \frac{dt}{ds}=-\frac{2st}{s^2-3t^2}$

13. $\frac{d}{dt}m(v^2-v_0^2)=\frac{d}{dt}k(x_o^2-x^2);\ 2mv\frac{dv}{dt}=-2kx\frac{dx}{dt};\ m\frac{dv}{dt}=-\frac{2kvx}{2v}$

$m\frac{dv}{dt}=-kx$

15. $V = x^3$ and $\frac{dx}{dt} = 3$

$\frac{dV}{dt} = 3x^2 \frac{dx}{dt}$; $\frac{dV}{dt} = 3(10)^2(3) = 900$ in.^{3}/sec

17. $y^2 = x^2 + 1$ and $\frac{dx}{dt} = 240$. At $t = \frac{1}{120}$, $x = 2$, and $y = \sqrt{5}$.

$2y\frac{dy}{dt} = 2x\frac{dx}{dt}$; $\frac{dy}{dt} = \frac{x}{y}\frac{dx}{dt}$

At $t = \frac{1}{120}$: $\frac{dy}{dt} = \frac{2}{\sqrt{5}}(240) \approx 215$ mph.

19. t is the number of hours after noon. The westbound plane will have flown $\frac{1}{2}(400) = 200$ miles from the town at $t = 0$.

$s^2 = (x + 200)^2 + y^2$ and $\frac{dx}{dt} = 400$, $\frac{dy}{dt} = 500$.

$2s\frac{ds}{dt} = 2(x + 200)\frac{dx}{dt} + 2y\frac{dy}{dt}$; $s\frac{ds}{dt} = (x + 200)\frac{dx}{dt} + y\frac{dy}{dt}$

At $t = 1$; $x = 400$, $y = 500$, and $s = \sqrt{(400 + 200)^2 + (500)^2} \approx 781$.

$781\frac{ds}{dt} = (400 + 200)(400) + (500)(500)$; $\frac{ds}{dt} \approx 627$ mph

21. $y^2 = 20^2 - x^2$, $\frac{dx}{dt} = 2$. When $x = 4$, $y = \sqrt{20^2 - 4^2} \approx 19.6$.

$2y\frac{dy}{dt} = -2x\frac{dx}{dt}$; $\frac{dy}{dt} = -\frac{x}{y}\frac{dx}{dt} = -\frac{4}{19.6}(2) \approx -0.408$ ft/sec

23. $V = \frac{1}{3}\pi r^2 h$, but $h = \frac{1}{4}d = \frac{1}{2}r$, so $r = 2h$.

$V = \frac{1}{3}\pi(2h)^2 h = \frac{4\pi}{3}h^3$ and $\frac{dV}{dt} = 16$.

$\frac{dV}{dt} = 4\pi h^2 \frac{dh}{dt}$; $16 = 4\pi(4)^2 \frac{dh}{dt}$; $\frac{dh}{dt} \approx 0.080$ ft/sec

25. $V = \frac{1}{2}bhw$; $\frac{40}{b} = \frac{5}{h}$, so $b = 8h$. $\frac{dV}{dt} = 40$ ft^3/min and $w = 20$.

$V = \frac{1}{2}(8h)(h)(20) = 80h^2$

$\frac{dV}{dt} = 160h\frac{dh}{dt}$; $40 = 160(3)\frac{dh}{dt}$; $\frac{dh}{dt} \approx 0.083$ ft/min

27. $A = \pi r^2$ and $\frac{dr}{dt} = 0.02$.

$\frac{dA}{dt} = 2\pi r\frac{dr}{dt}$; $\frac{dA}{dt} = 2\pi(8.1)(0.02) \approx 1.018$ in.^{2}/sec

29. Let θ be the angle of the beam with the line from the lighthouse to the point opposite the shore as the light rotates. Let y be the distance along the shoreline from the point opposite the lighthouse to where the light passes the shoreline. See figure.

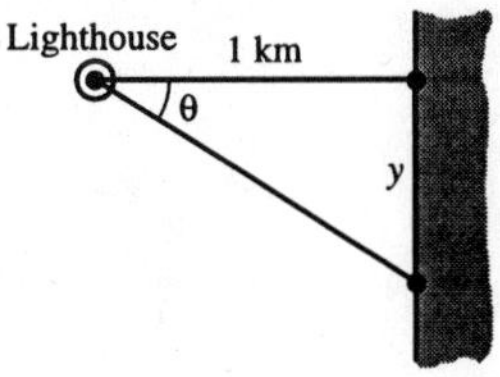

$$y = \tan\theta$$

$$\frac{dy}{dt} = \sec^2\theta \frac{d\theta}{dt}$$

At $y = \frac{1}{2}$ km, $\sec^2\theta = \left(\frac{\sqrt{1+\left(\frac{1}{2}\right)^2}}{1}\right)^2 = \frac{5}{4}$ and $\frac{d\theta}{dt} = \left(2\frac{\text{rev}}{\text{min}}\right)\left(2\pi\frac{\text{rad}}{\text{rev}}\right)\left(\sqrt{1+\left(\frac{1}{2}\right)^2}\frac{\text{km}}{\text{rad}}\right) = 2\pi\sqrt{5}\frac{\text{km}}{\text{min}}$.

$$\frac{dy}{dt} = \left(\frac{5}{4}\right)\left(2\pi\sqrt{5}\right) = \frac{5\pi\sqrt{5}}{2} \approx 17.56 \text{ km/min}$$

31. $V = \pi h^2\left[r - \left(\frac{h}{3}\right)\right]$ and $\frac{dV}{dt} = -2$. The radius is 8 feet, so $V = 8\pi h^2 - \frac{\pi}{3}h^3$.

$$\frac{dV}{dt} = 16\pi h\frac{dh}{dt} - \pi h^2\frac{dh}{dt} = (16\pi h - \pi h^2)\frac{dh}{dt}$$

$$-2 = (48\pi - 9\pi)\frac{dh}{dt};\ \frac{dh}{dt} \approx -0.016 \text{ ft/hr}$$

33. Let $y = x^{p/q}$, then $y^q = x^p$.

$$qy^{q-1}\frac{dy}{dx} = px^{p-1};\ \frac{dy}{dx} = \frac{px^{p-1}}{qy^{q-1}}$$

$$\frac{dy}{dx} = \frac{px^{p-1}}{q(x^{(p/q)})^{q-1}};\ \frac{dy}{dx} = \frac{px^{p-1}}{qx^{p-(p/q)}} = \frac{p}{q}x^{p-1-p+(p/q)} = \frac{p}{q}x^{(p/q)-1}$$

If $r = \frac{p}{q}$, then $\frac{d}{dx}x^{p/q} = \frac{p}{q}x^{(p/q)-1}$ can be written $\frac{d}{dx}x^r = rx^{r-1}$, proving the Power Rule for the case where r is rational.

Section 3.7

Concepts Review

1. $f'(x)dx$

3. Δx is small.

Problem Set 3.7

1. $\frac{dy}{dx} = 4x - 3;\ dy = (4x-3)dx$

3. $\frac{dy}{dx} = -4(3+2x^3)^{-5}\cdot 6x^2;\ dy = -24x^2(3+2x^3)^{-5}dx$

5. $\dfrac{dy}{dx}=\dfrac{1}{2}(4x^5+2x^4-5)^{-1/2}\cdot(20x^4+8x^3);\ \dfrac{dy}{dx}=\dfrac{10x^4+4x^3}{\sqrt{4x^5+2x^4-5}};\ dy=\dfrac{10x^4+4x^3}{\sqrt{4x^5+2x^4-5}}dx$

7. $\dfrac{ds}{dt}=\dfrac{3}{5}(t^2-3)^{-2/5}\cdot 2t;\ \dfrac{ds}{dt}=\dfrac{6t}{5\sqrt[5]{(t^2-3)^2}};\ ds=\dfrac{6t\,dt}{5\sqrt[5]{(t^2-3)^2}}$

9. $\dfrac{dy}{dx}=3x^2;\ dy=3x^2dx$

   a. $dy=3(0.5)^2(1)=0.75$

   b. $dy=3(-1)^2(0.75)=2.25$

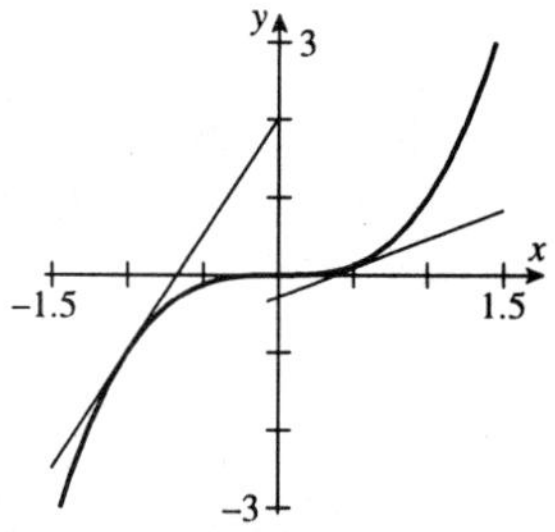

Written response.

11. $f(0.5+1)-f(0.5)=3.375-0.125=3.25$
$f(-1+0.75)-f(-1)=-0.15625-(-1)=0.984375$

13. $\dfrac{dy}{dx}=2x;\ dy=2x\,dx$

   a. $\Delta y=[(2.5)^2-3]-[(2)^2-3]=2.25$
   $dy=2(2)(0.5)=2$

   b. $\Delta y=[(2.88)^2-3]-[(3)^2-3]=-0.7056$
   $dy=2(3)(-0.12)=-0.72$

15. $\dfrac{d}{dx}(x^4+y^4)=\dfrac{d}{dx}(2);\ 4x^3+4y^3\dfrac{dy}{dx}=0;\ \dfrac{dy}{dx}=\dfrac{-4x^3}{4y^3};\ dy=-\dfrac{x^3}{y^3}dx$

$dy=-\dfrac{(1)^3}{(1)^3}dx=-dx$ or $dy=-\dfrac{(1)^3}{(-1)^3}dx=dx$.
$dx=\Delta x=-0.1$
$dy=0.1$ or $dy=-0.1$, so $y=1+0.1=1.1$ or $y=-1-0.1=-1.1$.
$x^4+y^4=2;\ (0.9)^4+y^4=2;\ y^4=1.3439;\ y\approx-1.077$ or $y\approx 1.077$.

17. a. $t=-\dfrac{\pi}{40}+\dfrac{\pi}{5}\approx 0.5498$

   b. $f'(t)=100e^{-10t}\sin(10t);\ f'(0.5498)\approx-0.2895$

   c. $0.1=-0.2895\,dt;\ dt\approx-0.3454$

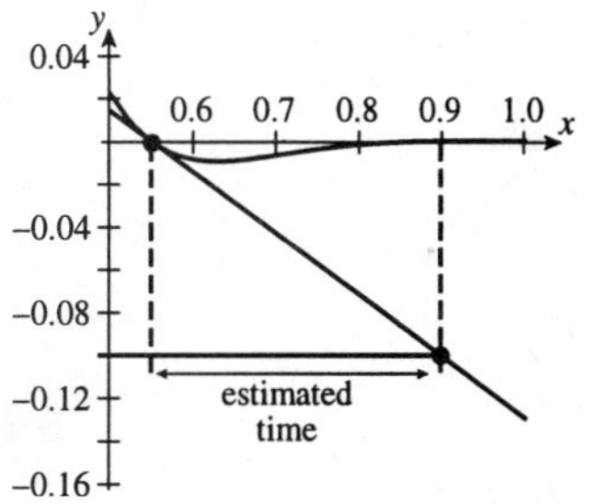

From the graph, we see that the bumper will not move 0.1 after reaching rest position for the second time.

19. $V = \frac{4}{3}\pi r^3;\ \frac{dV}{dr} = 4\pi r^2;\ dV = 4\pi r^2 dr$

$r = 5$ and $dr = 0.125$

$dV = 4\pi(5)^2(0.125) \approx 39.27\text{ cm}^3$

21. $V = \frac{4}{3}\pi r^3;\ dV = 4\pi r^2 dr$

$r = 6$ and $dr = -0.3$

$V = \frac{4}{3}\pi(6)^3 + 4\pi(6)^2(-0.3) \approx 769.06\text{ ft}^3$

23. $C = 2\pi r;\ dC = 2\pi dr$

$dr = 2$

$dC = 2\pi(2) = 4\pi \approx 12.57\text{ ft}$

25. $V = \frac{4}{3}\pi\left(\frac{d}{2}\right)^3;\ \frac{dV}{dd} = 4\pi\left(\frac{d}{2}\right)^2 \cdot \frac{1}{2};\ dV = \frac{\pi d^2}{2}dd$

$d = 20$ and $dd = 0.1$, so $V = \frac{4\pi}{3}\left(\frac{20}{2}\right)^3 \approx 4189\text{ cm}^3$

$dV = \frac{\pi}{2}(20)^2(0.1) \approx 62.8$

$V = 4189 \pm 62.8\text{ cm}^3$

27. $V = 100\pi r^2 + \frac{4}{3}\pi r^3;$

$dV = (200\pi r + 4\pi r^2)dr$

$r = 10$ and $dr = 0.1$

$dV = (200\pi(10) + 4\pi(10)^2)(0.1) = 240\pi \approx 753.98\text{ cm}^3$

Section 3.8 Chapter Review

Concepts Test

1. False. The y-coordinate on the curve $y = f'(x)$ is the same as the slope of the curve $y = f(x)$.

3. True. For example, if $s = t^2$, then $v = 2t$ and $\frac{dv}{dt} = 2$ so velocity is increasing, but speed $|v| = 2|t|$ is decreasing for $t < 0$.

5. False. For example, if $g(x) = f(x) + C$ for some nonzero constant C, $f'(x) = g'(x)$ but $f(x) \neq g(x)$.

7. True.

9. False. For example, $\frac{d}{dx}(x \cdot x) = \frac{d}{dx}(x^2) = 2x \neq \frac{dx}{dx} \cdot \frac{dx}{dx} = 1 \cdot 1 = 1$.

11. True. Let $f(x) = x^3 g(x)$. Then $f'(x) = 3x^2 g(x) + x^3 g'(x) = x^2(3g(x) + xg'(x))$.

13. False. For example, if $f(x) = x^2$ and $g(x) = x^2$, then $\frac{d^2y}{dx^2} = 12x^2 \neq x^2(2) + x^2(2) = 4x^2$.

15. True.

17. True. $h'(x) = f(x)g'(x) + g(x)f'(x)$, so $h'(c) = f(c)g'(c) + g(c)f'(c) = 0$.

19. False. $\frac{d}{dx}\ln(\sin x) = \frac{1}{\sin x}\frac{d}{dx}(\sin x) = \frac{\cos x}{\sin x} = \cot x$

21. True. $(f \circ g)'(x) = f'(g(x))g'(x)$, so $(f \circ g)'(2) = f'(g(2))g'(2) = f'(2)2 = 2 \cdot 2 = 4$.

23. False. $V = \frac{4}{3}\pi r^3$ so $\frac{dV}{dt} = 4\pi r^2 \frac{dr}{dt} = 12\pi r^2 \neq 27$ for $r \neq \frac{3}{2\sqrt{\pi}}$.

25. True. $\lim_{x\to 0} \frac{\tan x}{3x} = \lim_{x\to 0} \frac{\sin x}{x} \cdot \lim_{x\to 0} \frac{1}{3\cos x} = 1 \cdot \frac{1}{3} = \frac{1}{3}$

27. True. $V = \frac{4}{3}\pi r^3$ so $\frac{dV}{dt} = 4\pi r^2 \frac{dr}{dt}$ and $\frac{dr}{dt} = \frac{1}{4\pi r^2}\frac{dV}{dt} = \frac{3}{4\pi r^2}$. As the radius gets bigger, $\frac{dr}{dt}$ gets smaller.

29. True. $V = \frac{4}{3}\pi r^3$, so $dV = 4\pi r^2 dr$. Since $S = 4\pi r^2$, $dV = S\,dr$, so $\Delta V \approx S\Delta r$.

Sample Test Problems

1. $$f'(x) = \lim_{h\to 0} \frac{f(x+h) - f(x)}{h} = \lim_{h\to 0} \frac{(x+h)^n - x^n}{h}$$
$$= \lim_{h\to 0} \frac{x^n + nx^{n-1}h + \frac{n(n-1)}{2}x^{n-2}h^2 + \cdots + nxh^{n-1} + h^n - x^n}{h}$$
$$= \lim_{h\to 0} nx^{n-1} + \frac{n(n-1)}{2}x^{n-2}h^1 + \cdots + nxh^{n-2} + h^{n-1}$$
$$= nx^{n-1}$$

3. $f(x) = \frac{1}{x}$ at $x = 2$.

5. $\frac{d}{dx}(x^3 - 3x^2 + 2e^x) = 3x^2 - 6x + 2e^x$

7. $\frac{d^2}{dx^2}(e^{3x^2+2}) = \frac{d}{dx}(6xe^{3x^2+2}) = 6e^{3x^2+2} + 36x^2e^{3x^2+2}$

9. $\frac{d}{dx}5\ln(x^2+4) = \frac{5}{x^2+4} \cdot 2x = \frac{10x}{x^2+4}$

11. $\frac{d}{dx}(\cos^3 5x) = 3(\cos^2 5x) \cdot (-\sin 5x) \cdot 5 = -15\cos^2 5x \sin 5x$

13. $f'(x) = (x^2-1)^2(9x^2-4) + (3x^3-4x)2(x^2-1)(2x) = (x^2-1)^2(9x^2-4) + 4x(3x^3-4x)(x^2-1)$
$f'(2) = 672$

15. $V = \frac{4}{3}\pi r^3;\ \frac{dV}{dr} = \frac{4}{3}\pi(3r^2) = 4\pi r^2$
At $r = 5$: $4\pi(5)^2 = 100\pi \approx 314.16$.

17. $V = \frac{4}{3}\pi r^3;\ \frac{dV}{dt} = \frac{4}{3}\pi(3r^2)\frac{dr}{dt} = 4\pi r^2\frac{dr}{dt}$
$10 = 4\pi(5^2)\frac{dr}{dt};\ \frac{dr}{dt} \approx 0.032$ m/h

19. a. $\frac{ds}{dt} = 128 - 32t = 0;\ t = 4$ sec
$s = 128(4) - 16(4)^2 = 256$ ft

b. $128t - 16t^2 = 0;\ -16t(t-8) = 0;\ t = 0$ and $t = 8$.
The object hits the ground after 8 seconds with velocity $v = 128 - 32(8) = -128$ ft/sec

21. a. $\frac{d^{20}}{dx^{20}}(13x^{19} - 2x^{12} - 6x^5 + 18) = 0$

b. $\frac{d^{20}}{dx^{20}}(\ln x) = \frac{(-1)^{n-1}(n-1)!}{x^n} = \frac{(-1)^{19}(20-1)!}{x^{20}} = -\frac{19!}{x^{20}}$

c. $\frac{d^{20}}{dx^{10}}(3e^{2x}) = 3\cdot 2^{20}e^{2x}$

23. $\frac{d}{dx}(y^2) = \frac{d}{dx}(4x^3);\ 2y\frac{dy}{dx} = 12x^2;\ \frac{dy}{dx} = \frac{6x^2}{y}$
At (1, 2): $\frac{dy}{dx} = \frac{6(1)^2}{2} = 3$.
$\frac{d}{dx}(2x^2 + 3y^2) = \frac{d}{dx}(14);\ 4x + 6y\frac{dy}{dx} = 0;\ \frac{dy}{dx} = -\frac{2x}{3y}$
At (1, 2): $\frac{dy}{dx} = -\frac{2(1)}{3(2)} = -\frac{1}{3}$.
The tangents are perpendicular since the slopes are negative reciprocals.

25. a. $\frac{d}{dx}[f^2(x) + g^3(x)] = 2f(x)f'(x) + 3g^2(x)g'(x)$
At $x = 2$: $2(3)(4) + 3(2)^2(5) = 84$.

b. $\frac{d}{dx}[f(x) + g(x)] = f(x)g'(x) + f'(x)g(x)$
At $x = 2$: $3(5) + 4(2) = 23$.

c. $\frac{d}{dx}[f(g(x))] = f'(g(x))\cdot g'(x)$
At $x = 2$: $f'(2)\cdot 5 = 4\cdot 5 = 20$.

27. $h = x \sin 15°$

$\dfrac{dh}{dt} = \dfrac{dx}{dt} \sin 15°$

$\dfrac{dx}{dt} = 400$

$\dfrac{dh}{dt} = 400 \sin 15° \approx 103.53$ miles per hour

Chapter 4

Section 4.1

Concepts Review

1. Continuous; closed

3. Endpoints; stationary points; singular points

Problem Set 4.1

1. Maximum y-value $\approx$ 1.841 at $x = 1$; minimum y-value ≈ -0.232 between $x = -0.454$ and $x = -0.446$.

3. $f'(x) = -2x + 4$: $f'(x) = 0 = -2x + 4$; $x = 2$
Critical numbers: 0, 2, 3
$f(0) = -(0)^2 + 4(0) - 1 = -1$: $f(2) = -(2)^2 + 4(2) - 1 = 3$;
$f(3) = -(3)^2 + 4(3) - 1 = 2$
Min = –1; max = 3

5. $f'(x) = 4x^3 - 4x + 4 = 0$; $x \approx -1.325$
Critical numbers: –2, –1.325, 2
$f(-2) = (-2)^4 - 2(-2)^2 + 4(-2) = 0$; $f(-1.325) = (-1.325)^4 - 2(-1.325)^2 + 4(-1.325) \approx -5.729$
$f(2) = (2)^4 - 2(2)^2 + 4(2) = 16$
Min = –5.729, accurate to three decimals; max = 16

7. $g(x) = (1 + x^2)^{-1}$: $g'(x) = -\frac{1}{(1 + x^2)^2} \cdot 2x = -\frac{2x}{(1 + x^2)^2}$
$-\frac{2x}{x^4 + 2x^2 + 1} = 0$; $x = 0$
Critical numbers: –2, 0, 1
$g(-2) = \frac{1}{1 + (-2)^2} = \frac{1}{5}$; $g(0) = \frac{1}{1 + (0)^2} = 1$; $g(1) = \frac{1}{1 + (1)^2} = \frac{1}{2}$
Min $= \frac{1}{5}$; max = 1

9. $f'(x) = \frac{(x^4 + 1)(2x) - x^2(4x^3)}{(x^4 + 1)^2} = \frac{-2x^5 + 2x}{(x^4 + 1)^2}$; $-\frac{2x^5 - 2x}{(x^4 + 1)^2} = 0$; $x = -1,\ 0,\ 1$
Critical numbers: –1, 0, 1, 4
$f(-1) = \frac{1}{2}$, $f(0) = 0$, $f(1) = \frac{1}{2}$, $f(4) \approx 0.0623$
Min = 0; max $= \frac{1}{2}$

11. $f'(x) = \frac{|x - 2|}{x - 2}$; $f'(x)$ does not exist at $x = 2$.
Critical numbers: 1, 2, 5
$f(1) = 1, f(2) = 0, f(5) = 3$; min = 0; max = 3

13. $F'(t) = \cos t + \sin t$; $\cos t + \sin t = 0$; $t = \tan^{-1}(-1) = \frac{3\pi}{4}$
Critical numbers: 0, $\frac{3\pi}{4}$, π

$F(0) = -1,\ F\left(\frac{3\pi}{4}\right) = \sqrt{2},\ F(\pi) = 1$

min = –1; max = $\sqrt{2}$

15. $f'(x) = x^2(-e^{-x}) + e^{-x}(2x);\ 2xe^{-x} - x^2e^{-x} = 0;\ x = 0,\ 2$
Critical numbers: –1, 0, 2, 3
$f(-1) = e, f(0) = 0, f(2) \approx 0.5413, f(3) \approx 0.4481$
Min = 0; max = e

17. $f(x) = x(10 - x) = 10x - x^2;\ f'(x) = 10 - 2x;\ 10 - 2x = 0;\ x = 5$
$f(0) = 0, f(5) = 25, f(10) = 0$; the numbers are $x = 5$ and $10 - x = 5$.

19. Let x represent the width, then $100 - x$ represents the length.
$f(x) = x(100 - x) = 100x - x^2;\ f'(x) = 100 - 2x;\ 100 - 2x = 0;\ x = 50$
$f(50) = 50(50) = 2500$. The dimensions should be 50 ft by 50 ft.

21. Let x be the width of the square to be cut out and V the volume of the resulting open box.
$V = x(24 - 2x)^2 = 4x^3 - 96x^2 + 576x$
$\frac{dV}{dx} = 12x^2 - 192x + 576 = 12(x - 12)(x - 4);$
$12(x - 12)(x - 4) = 0; x = 12$ or $x = 4$.
If $x = 12$ the volume of the box is 0, so $x = 4$.
$V = 4(24 - 2 \cdot 4)^2 = 1024 \text{ in.}^3$

23. Let A be the area of the pen.
$A = x(80 - 2x) = 80x - 2x^2;\ \frac{dA}{dx} = 80 - 4x;\ 80 - 4x = 0;\ x = 20$
The dimensions are 20 ft by 80 – 2(20) = 40 ft, with the length along the barn being 40 ft.

25. Let A be the area of the pen. The perimeter is 100 + 180 = 280 ft.
$y + y - 100 + 2x = 180;\ y = 140 - x$
$A = x(140 - x) = 140x - x^2;\ \frac{dA}{dx} = 140 - 2x;\ 140 - 2x = 0;\ x = 70$
Since $0 \le x \le 40$, test endpoints. When $x = 0$, $A = 0$. When $x = 40$, $A = 4000$. The dimensions are 40 ft by 100 ft.

27. $A = 2x(12 - x^2) = 24x - 2x^3;\ \frac{dA}{dx} = 24 - 6x^2;\ 24 - 6x^2 = 0;\ x = -2,\ 2$
When $x = 2$: $y = 12 - (2)^2 = 8$; when $x = -2$: $y = 12 - (-2)^2 = 8$.
The dimensions are $2x = 2(2) = 4$ by 8.

29. The carrying capacity of the gutter is maximized when the area of the vertical end of the gutter is maximized. The area is $A = 3(3\sin\theta) + 2\left(\frac{1}{2}\right)(3\cos\theta)(3\sin\theta) = 9\sin\theta + 9\cos\theta\sin\theta$.

$\frac{dA}{d\theta} = 9\cos\theta + 9(-\sin\theta)\sin\theta + 9\cos\theta\cos\theta$
$= 9(\cos\theta - \sin^2\theta + \cos^2\theta)$
$= 9(2\cos^2\theta + \cos\theta - 1)$
$2\cos^2\theta + \cos\theta - 1 = 0;\ \cos\theta = -1,\ \frac{1}{2};\ \theta = \pi,\ \frac{\pi}{3}.$

Since $0 \le \theta \le \frac{\pi}{2}$, the critical points are 0, $\frac{\pi}{3}$, and $\frac{\pi}{2}$. $A(0) = 0,\ A\left(\frac{\pi}{3}\right) \approx 11.7,\ A = \left(\frac{\pi}{2}\right) = 9.$

The carrying capacity is maximized at $\theta = \frac{\pi}{3}$.

31. Let c be the cost of driving the truck.

$$c = \frac{400}{x}(1200) + 400\left(25 + \frac{x}{4}\right) = 480,000x^{-1} + 10,000 + 100x$$

$$\frac{dc}{dx} = -480,000x^{-2} + 100; \ -480,000x^{-2} + 100 = 0; \ x \approx 69$$

This is higher than the allowable speed, so we check endpoints.
At $x = 40$, $c = 26,000$. At $x = 55$, $c \approx 24,227$.
55 mph is the most economical allowed speed.

33. Let D be the square of the distance.

$$D = (x-0)^2 + (y-4)^2 = x^2 + \left(\frac{x^2}{4} - 4\right)^2 = \frac{x^4}{16} - x^2 + 16$$

$$\frac{dD}{dx} = \frac{x^3}{4} - 2x; \ \frac{x^3}{4} - 2x = 0; \ x(x^2 - 8) = 0$$

$$x = 0, \ x = \pm 2\sqrt{2}$$

At $x = 0$, $y = 0$, and $D = 16$. At $x = 2\sqrt{2}$, $y = 2$, and $D = 12$.

At $x = 2\sqrt{3}$, $y = 3$, and $D = 13$.

$P\left(2\sqrt{2}, \ 2\right)$, $Q(0, \ 0)$

35. Let V be the volume. $y = 4 - x$ and $z = 5 - 2x$.

$$V = x(4-x)(5-2x) = 20x - 13x^2 + 2x^3$$

$$\frac{dV}{dx} = 20 - 26x + 6x^2; \ 2(3x^2 - 13x + 10) = 0; \ 2(3x - 10)(x - 1) = 0;$$

$$x = 1, \ \frac{10}{3}$$

At $x = 1$, $V = 9$; at $x = \frac{10}{3}$, $V \approx -3.704$.

Maximum volume when $x = 1$, $y = 4 - 1 = 3$, and $z = 5 - 2(1) = 3$.

37. **a.** $f'(x) = -\sin x + x\cos x + \sin x = x\cos x$

$$x\cos x = 0; \ x = 0, \ \frac{\pi}{2}, \ \frac{3\pi}{2}$$

Critical numbers: $-1, 0, \frac{\pi}{2}, \frac{3\pi}{2}, 5$

$$f(-1) \approx 3.382, \ f(0) = 3, \ f\left(\frac{\pi}{2}\right) \approx 3.571$$

$$f\left(\frac{3\pi}{2}\right) \approx -2.712, \ f(5) \approx -2.511$$

Minimum ≈ -2.712, maximum ≈ 3.571

b. $g'(x) = \frac{|f(x)|}{f(x)} \cdot f'(x)$; $g'(x) = 0$ at the same points where $f'(x) = 0$; $x = 0, \ \frac{\pi}{2}, \ \frac{3\pi}{2}$.

$g'(x)$ fails to exist at the points where $f(x) = 0$ on $[-1, 5]$; using a graphing calculator, $x \approx 3.45$.

Critical numbers: $-1, 0 \ \frac{\pi}{2}, \ 3.45, \ \frac{3\pi}{2}, \ 5$

$$g(-1) \approx 3.382, \ g(0) = 3, \ g\left(\frac{\pi}{2}\right) \approx 3.571, \ g(3.45) \approx 0, \ g\left(\frac{3\pi}{2}\right) \approx 2.712, \ g(5) \approx 2.511$$

Minimum $= 0$, maximum ≈ 3.571

Section 4.2

Concepts Review

1. Increasing; concave up

3. An inflection point

Problem Set 4.2

1. $f'(x) = 2x - 4;\ 2x - 4 > 0;\ x > 2$
   $2x - 4 < 0;\ x < 2.$ $f(x)$ is increasing on $[2, \infty)$, decreasing on $(-\infty, 2]$.

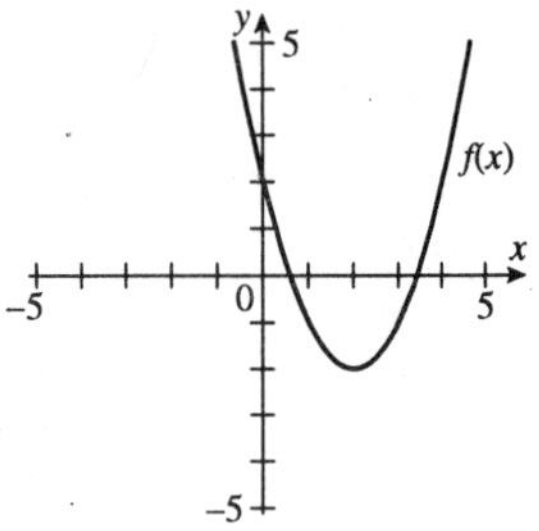

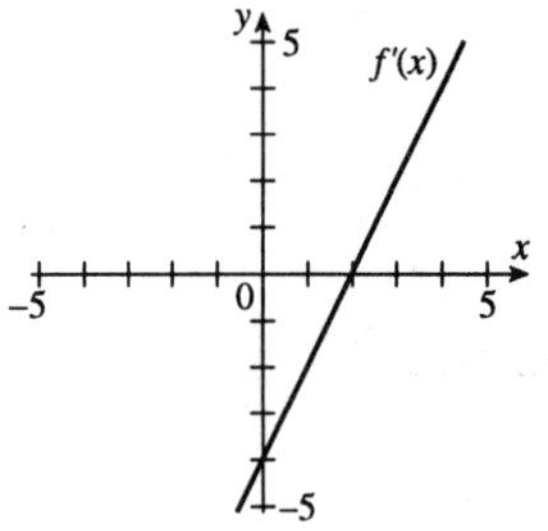

$f(x)$ is increasing when $f'(x) > 0$ and decreasing when $f'(x) < 0$.

3. $F'(x) = 3x^2;\ 3x^2 > 0;\ x < 0$ and $x > 0$.
   $F(x)$ is increasing on $(-\infty, \infty)$.

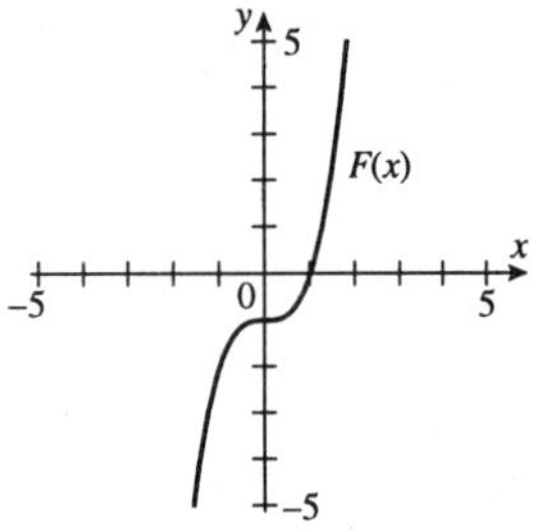

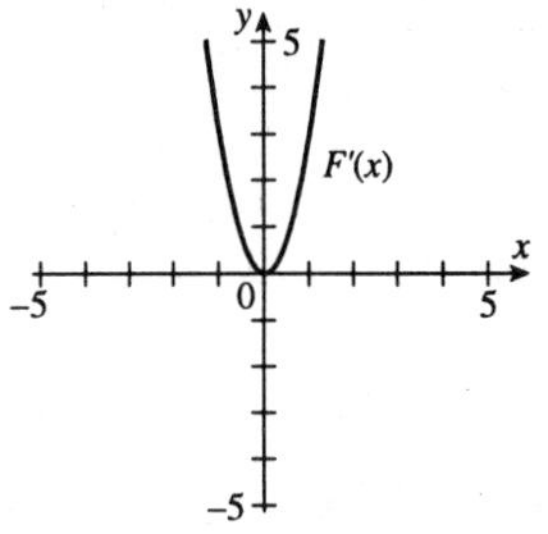

$F'(x)$ is above zero at all points except $x = 0$, and $F(x)$ is increasing for all x.

5. $g'(t) = 4t^3 + 12t^2 + 12;\ 4t^3 + 12t^2 + 12 > 0;\ t > -3.279$
   $4t^3 + 12t^2 + 12 < 0;\ t < -3.279$
   $g(t)$ is increasing on $[-3.279, \infty)$, decreasing on $(-\infty, -3.279]$.

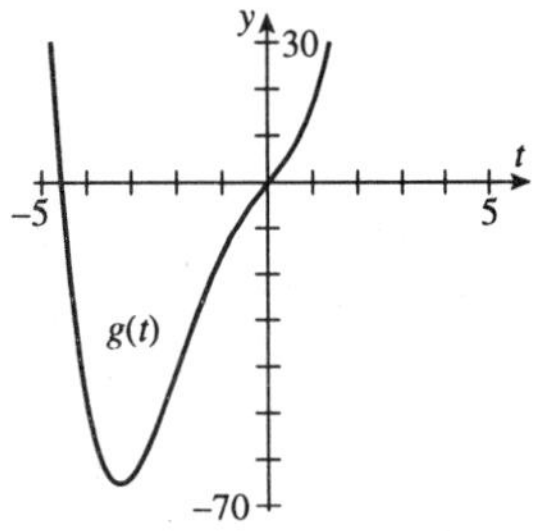

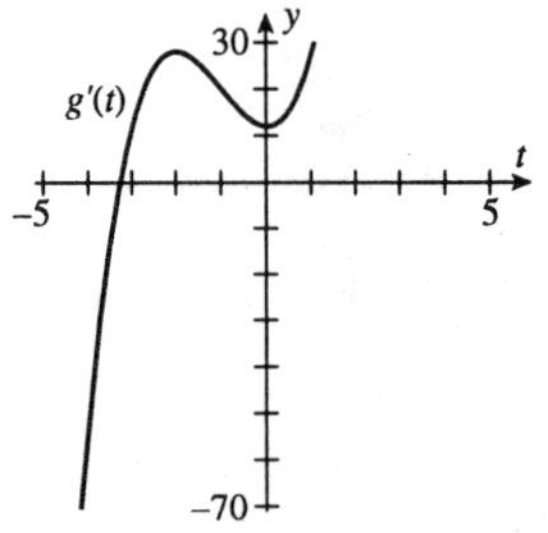

$g(t)$ is increasing when $g'(t)$ is positive and decreasing when $g'(t)$ is negative.

7. $f'(x)=10x^4-60x^3+90x^2-6$; $f'(x)>0$: $(-\infty, -0.24)\cup(0.29, 2.71)\cup(3.24, \infty)$;
$f'(x)<0$: $(-0.24, 0.29)\cup(2.71, 3.24)$
$f(x)$ is increasing on $(-\infty, -0.24]\cup[0.29, 2.71]\cup[3.24, \infty)$, decreasing on $[-0.24, 0.29]\cup[2.71, 3.24]$.

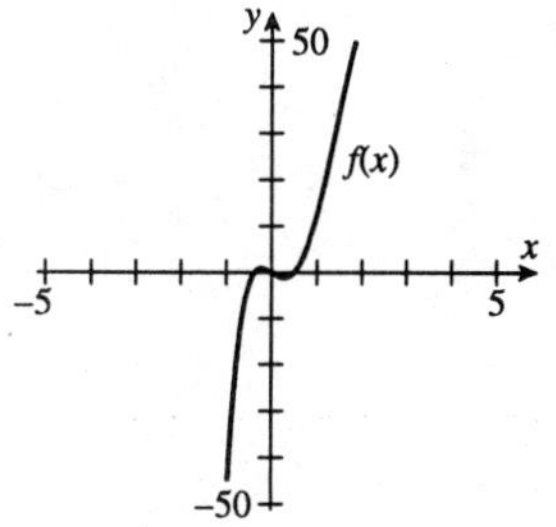

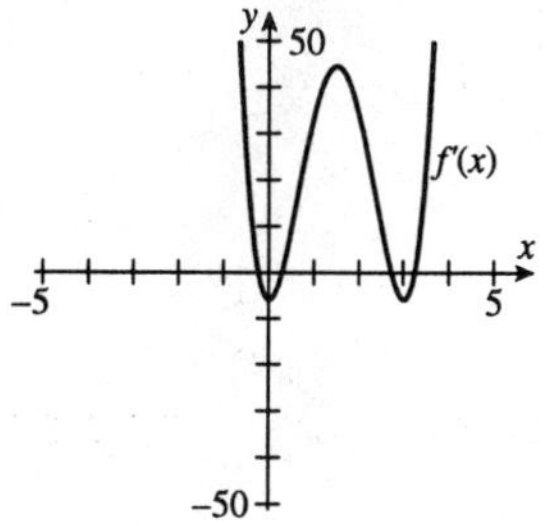

$f(x)$ is increasing when $f'(x)>0$ and decreasing when $f'(x)<0$.

9. $H'(t)=2\sin 2t(\cos 2t)(2)=2\sin 4t$
$H'(t)>0$: $\left(0, \frac{\pi}{4}\right)\cup\left(\frac{\pi}{2}, \frac{3\pi}{4}\right)$
$H'(t)<0$: $\left(\frac{\pi}{4}, \frac{\pi}{2}\right)\cup\left(\frac{3\pi}{4}, \pi\right)$
$H(t)$ is increasing on $\left[0, \frac{\pi}{4}\right]\cup\left[\frac{\pi}{2}, \frac{3\pi}{4}\right]$, decreasing on $\left[\frac{\pi}{4}, \frac{\pi}{2}\right]\cup\left[\frac{3\pi}{4}, \pi\right]$.

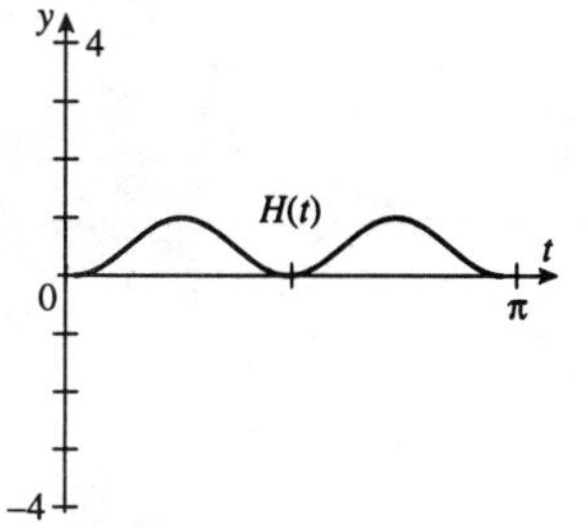

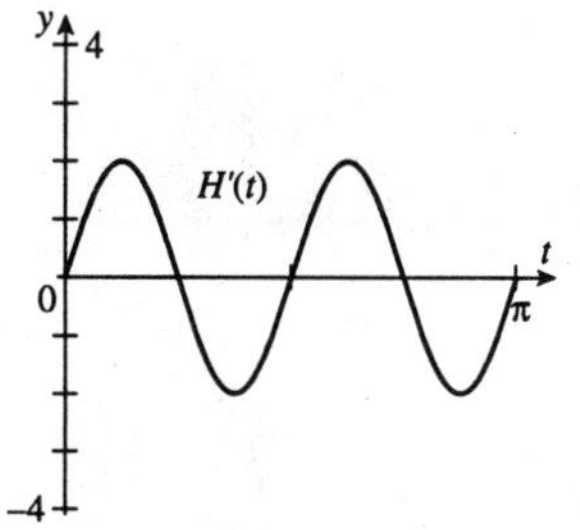

$H(t)$ increases when $H'(t)>0$ and decreases when $H'(t)<0$.

11. $f'(x)=2(x-3)=2x-6$; $f''(x)=2$
Since $f''(x)$ is always positive, $f(x)$ is always concave up, and has no inflection points.

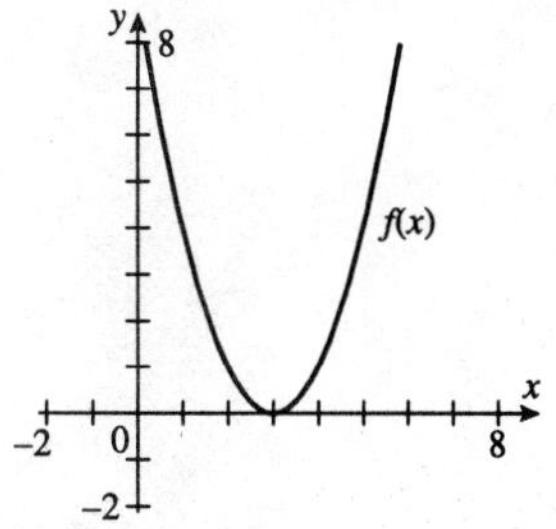

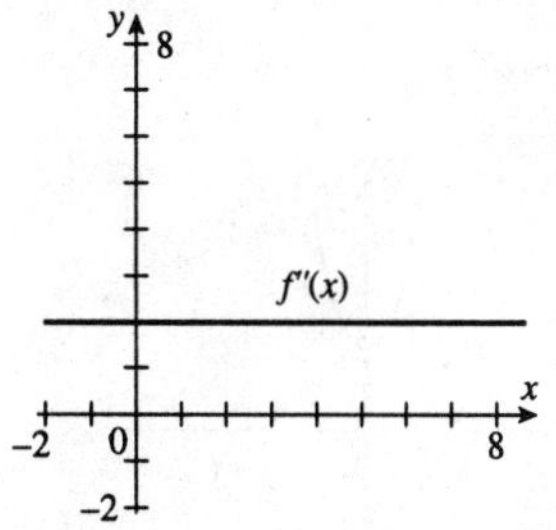

The function is concave up when $f''(x)$ is positive.

13. $F'(x) = 3x^2 - 12 + x^{-1}$; $F''(x) = 6x - x^{-2}$

Concave up: $6x - x^{-2} > 0$; $x > \frac{1}{\sqrt[3]{6}}$

Concave down: $6x - x^{-2} < 0$; $0 < x < \frac{1}{\sqrt[3]{6}}$

Inflection point: $x = \frac{1}{\sqrt[3]{6}} \approx 0.55$

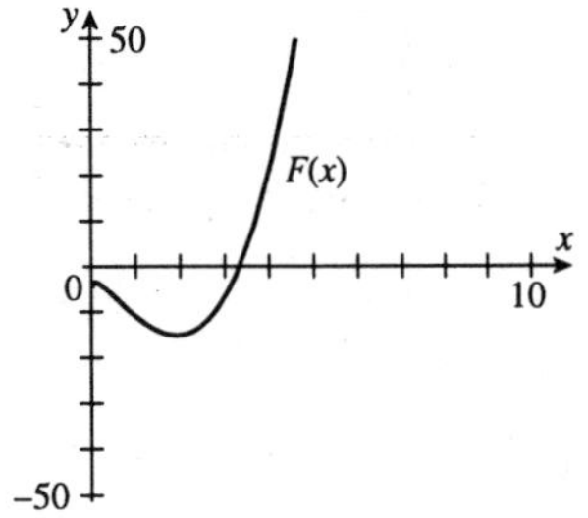

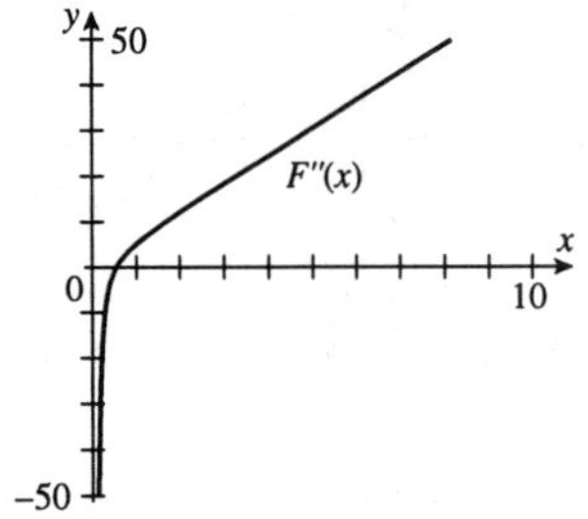

The domain of the function is $x > 0$, so even though $F''(x)$ has values for $x < 0$, these are not considered. The function is concave up when $F''(x) > 0$.

15. $g'(x) = 6x + \frac{2}{x^3}$; $g''(x) = 6 - \frac{6}{x^4}$

Concave up: $g''(x) > 0$; $(-\infty, -1) \cup (1, \infty)$

Concave down: $g''(x) < 0$; $(-1, 0) \cup (0, 1)$

Inflection points: $x = -1, x = 1$

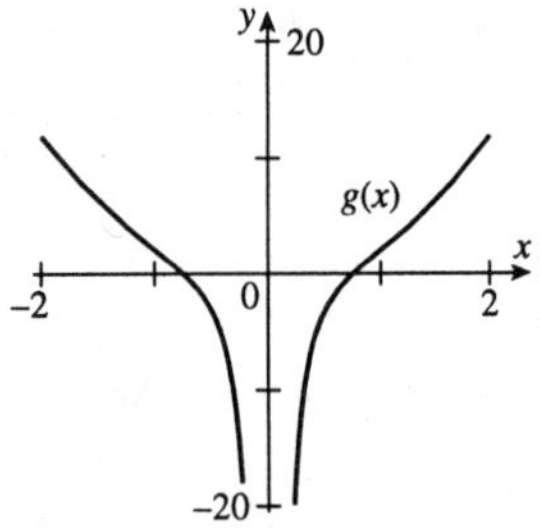

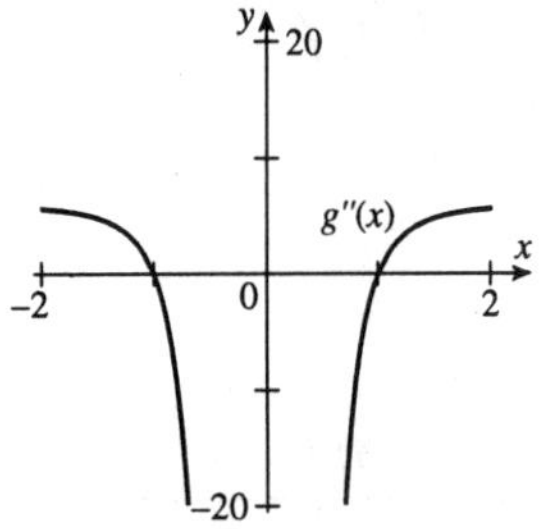

The function is undefined at $x = 0$, and this is not an inflection point.

17. $g'(x) = 12x^5 + 60x^3 + 180x + 120$; $g''(x) = 60x^4 + 180x^2 + 180$

Always concave up: $g''(x) > 0$ at all values of x

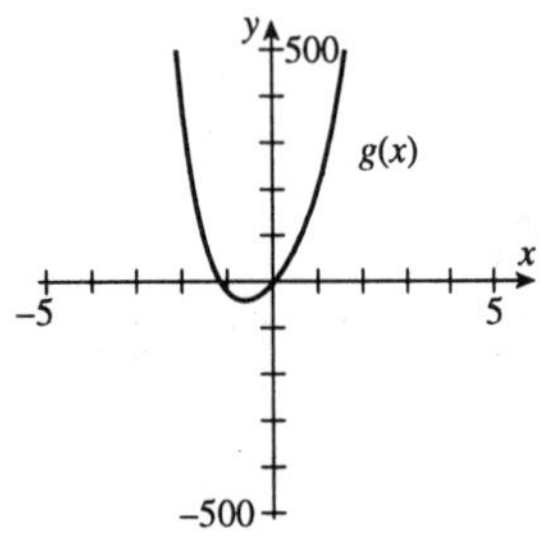

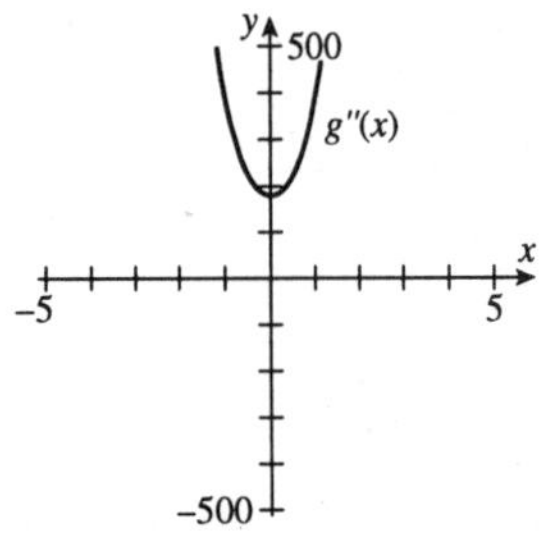

The function is always concave up as $g''(x)$ is always positive.

19. $f'(x) = -3x^2 - 3 = -3(x^2 + 1)$

Increasing: $f'(x) > 0$; no values of x

Decreasing: $f'(x) < 0$; all values of x

$f''(x) = -6x$

Concave up: $f''(x) > 0$; $(-\infty,\ 0)$

Concave down: $f''(x) < 0$; $(0,\ \infty)$

Inflection point: $x = 0$

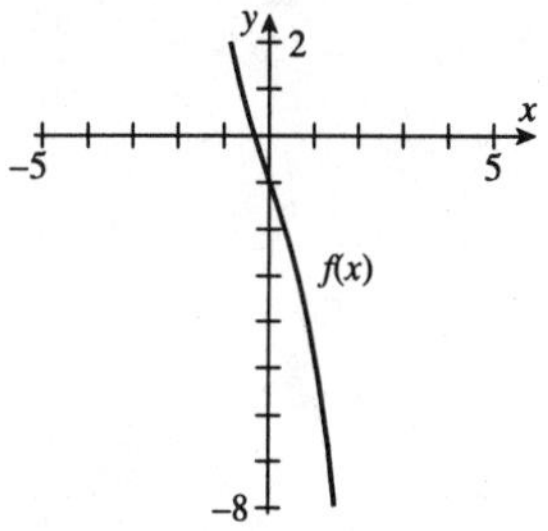

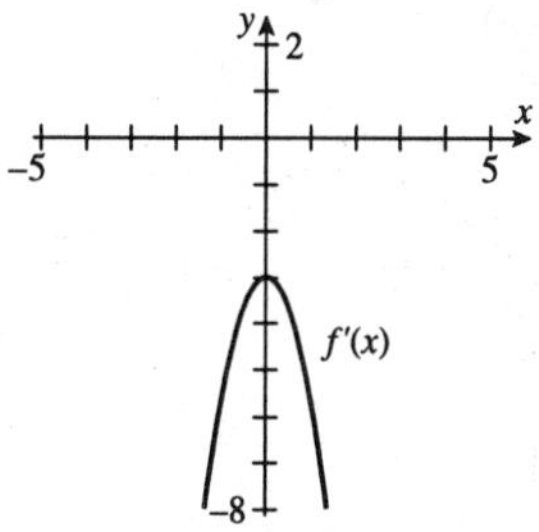

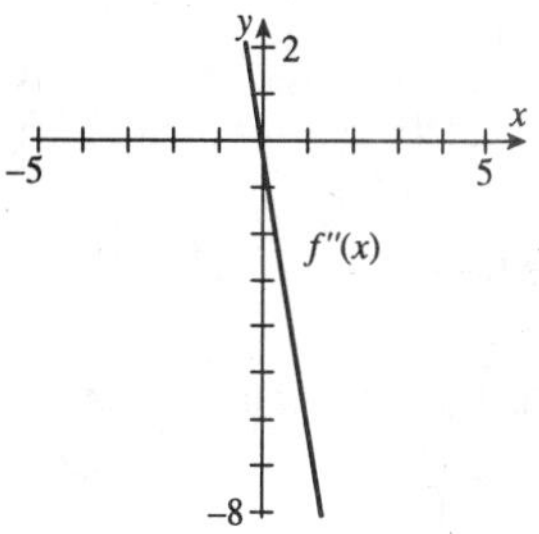

21. $g'(x) = 12x^3 - 12x^2 = 12x^2(x - 1)$

Increasing: $g'(x) > 0$; $[1,\ \infty)$

Decreasing: $g'(x) < 0$; $(-\infty,\ 1]$

$g''(x) = 36x^2 - 24x = 12x(3x - 2)$

Concave up: $g''(x) > 0$; $(-\infty,\ 0) \cup \left(\frac{2}{3},\ \infty\right)$

Concave down: $g''(x) < 0$; $\left(0,\ \frac{2}{3}\right)$

Inflection points: $g''(x) = 0$; $x = 0,\ x = \frac{2}{3}$

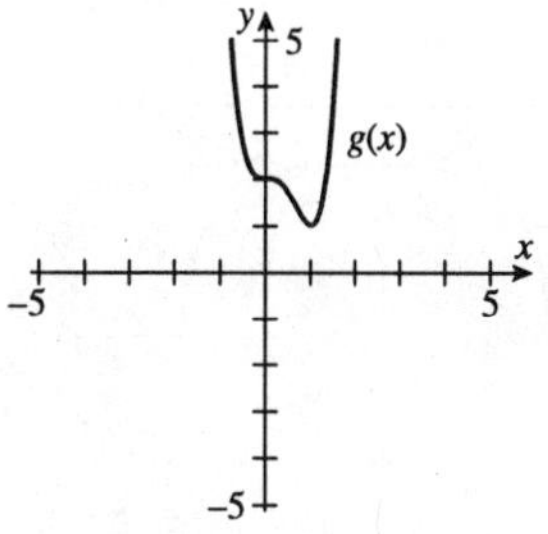

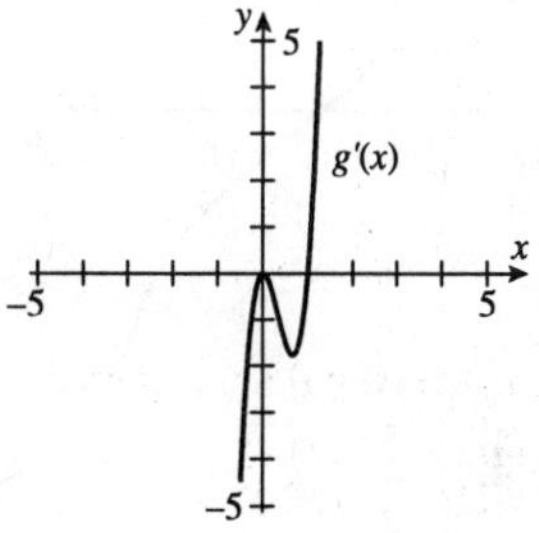

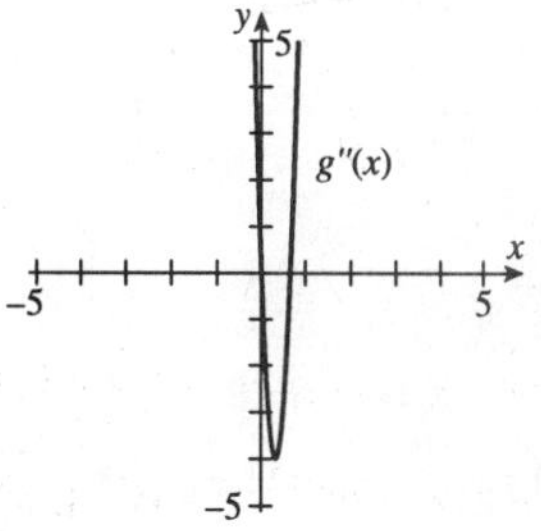

23. $G'(x) = 15x^4 - 15x^2 = 15x^2(x + 1)(x - 1)$

Increasing: $G'(x) > 0$; $(-\infty,\ -1] \cup [1,\ \infty)$

Decreasing: $G'(x) < 0$; $[-1,\ 1]$

$G''(x) = 60x^3 - 30x = 30x(2x^2 - 1)$

Concave up: $\left(-\frac{\sqrt{2}}{2},\ 0\right) \cup \left(\frac{\sqrt{2}}{2},\ \infty\right)$

Concave down: $G''(x) < 0$; $\left(-\infty,\ -\frac{\sqrt{2}}{2}\right) \cup \left(0,\ \frac{\sqrt{2}}{2}\right)$

Inflection points: $G''(x) = 0$; $x = -\frac{\sqrt{2}}{2},\ 0,\ \frac{\sqrt{2}}{2}$

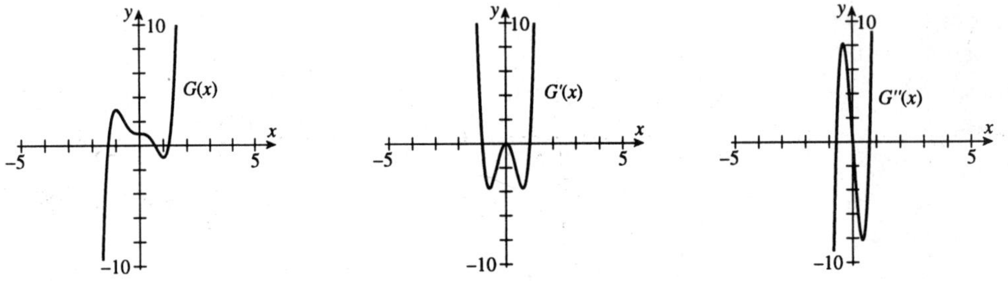

25. $f'(x) = \frac{1}{2}(\sin x)^{-1/2} \cos x = \frac{\cos x}{2\sqrt{\sin x}}$

Increasing: $f'(x) > 0$; $\left[0, \frac{\pi}{2}\right]$

Decreasing: $f'(x) < 0$; $\left[\frac{\pi}{2}, \pi\right]$

$$f''(x) = \frac{1}{2}\left[\frac{\sqrt{\sin x}(-\sin x) - \cos x\left(\frac{1}{2}(\sin x)^{-1/2}\cos x\right)}{\left(\sqrt{\sin x}\right)^2}\right]$$

$$= -\frac{\sqrt{\sin x}}{2} - \frac{\cos^2 x}{4\sqrt{\sin^3 x}}$$

Concave up: $f''(x) > 0$; no values of x in $(0, \pi)$

Concave down: $f''(x) < 0$; $(0, \pi)$

Inflection points: $f''(x) = 0$; no values of x in $(0, \pi)$

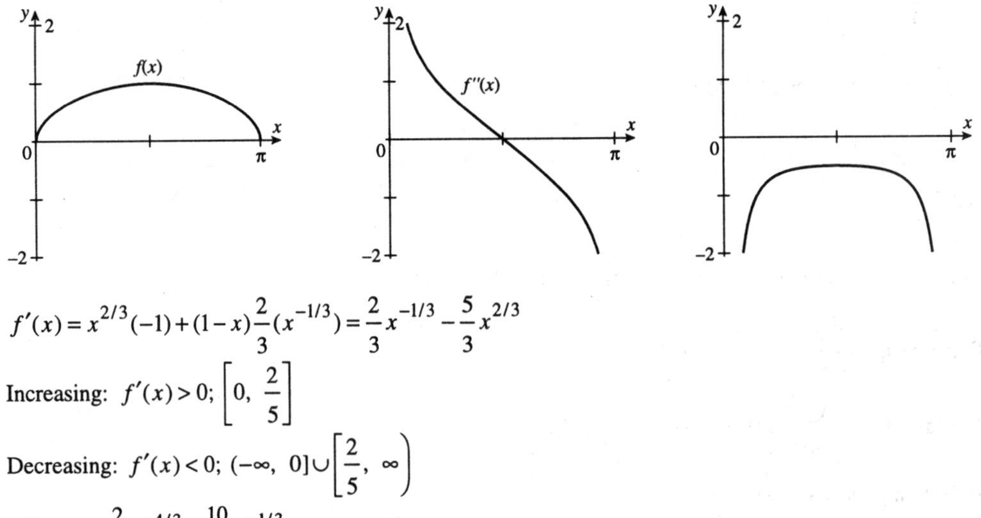

27. $f'(x) = x^{2/3}(-1) + (1 - x)\frac{2}{3}(x^{-1/3}) = \frac{2}{3}x^{-1/3} - \frac{5}{3}x^{2/3}$

Increasing: $f'(x) > 0$; $\left[0, \frac{2}{5}\right]$

Decreasing: $f'(x) < 0$; $(-\infty, 0] \cup \left[\frac{2}{5}, \infty\right)$

$f''(x) = -\frac{2}{9}x^{-4/3} - \frac{10}{9}x^{-1/3}$

Concave up: $f''(x) > 0$; $\left(-\infty, -\frac{1}{5}\right)$

Concave down: $f''(x) < 0$; $\left(-\frac{1}{5}, 0\right) \cup (0, \infty)$

Inflection point: $x = -\frac{1}{5}$

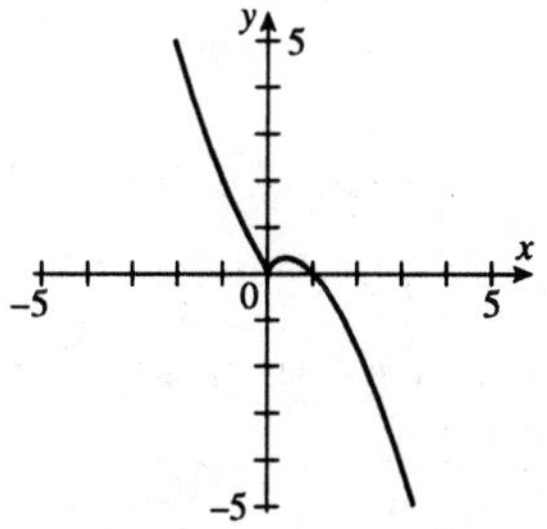

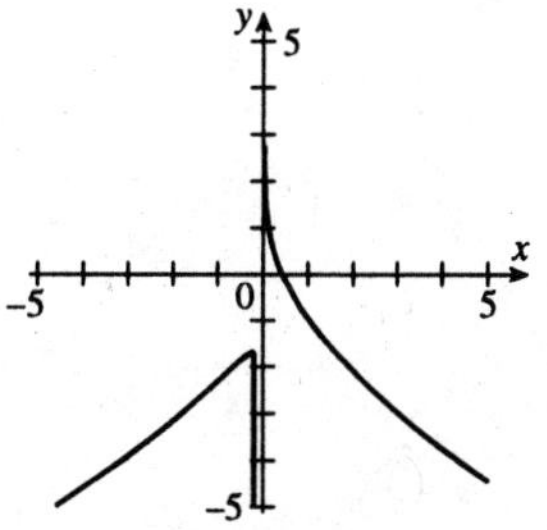

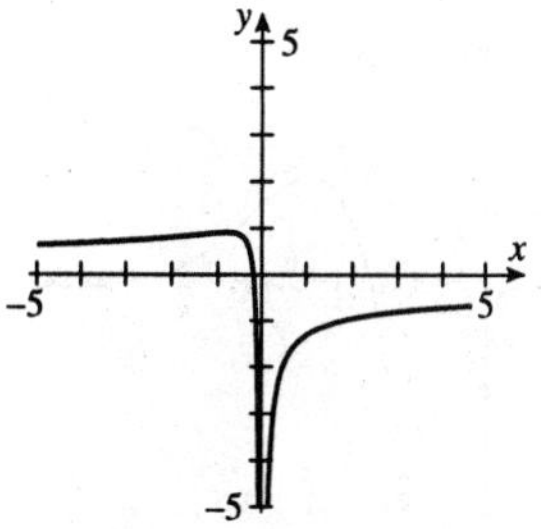

29. a.

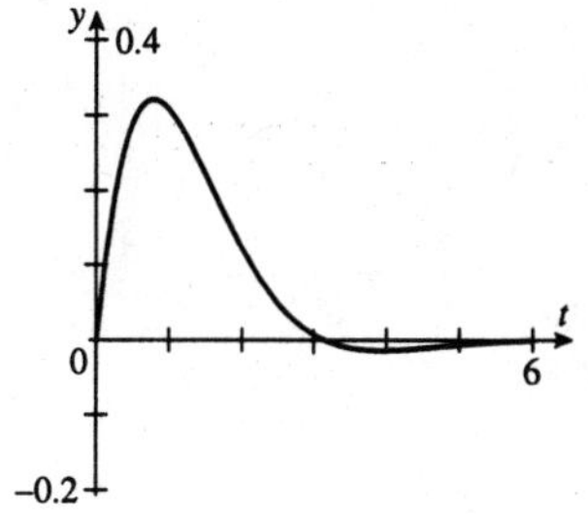

b. $\dfrac{dy}{dt} = \cos t\, e^{-t} - e^{-t}\sin t = e^{-t}(\cos t - \sin t)$

Increasing: $\dfrac{dy}{dt} > 0;\ \left[0, \dfrac{\pi}{4}\right] \cup \left[\dfrac{5\pi}{4}, 6\right]$

Decreasing: $\dfrac{dy}{dt} < 0;\ \left[\dfrac{\pi}{4}, \dfrac{5\pi}{4}\right]$

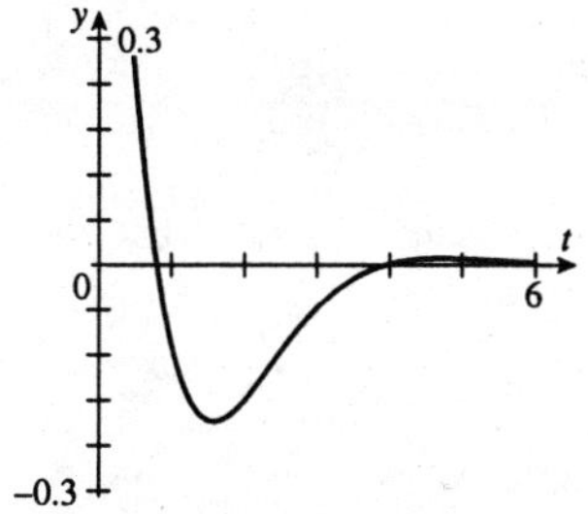

c. $\dfrac{d^2y}{dt^2} = e^{-t}(-\sin t - \cos t) + (\cos t - \sin t)(-e^{-t}) = -2e^{-t}\cos t$

Concave up: $\dfrac{d^2y}{dt^2} > 0;\ \left(\dfrac{\pi}{2}, \dfrac{3\pi}{2}\right)$

Concave down: $\dfrac{d^2y}{dt^2} < 0;\ \left[0, \dfrac{\pi}{2}\right) \cup \left(\dfrac{3\pi}{2}, 6\right]$

Inflection points: $x = \dfrac{\pi}{2}, \dfrac{3\pi}{2}$

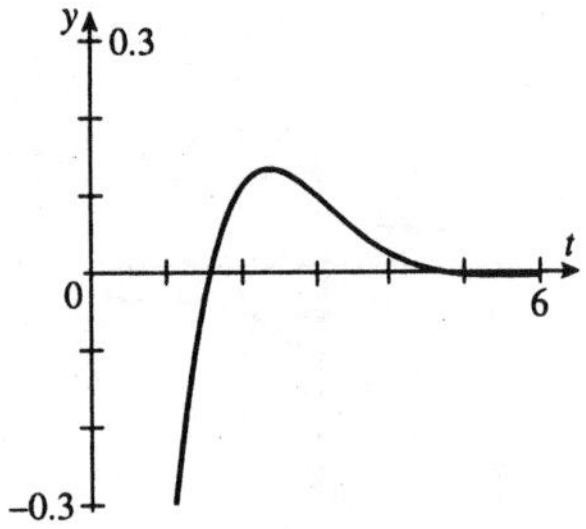

31.

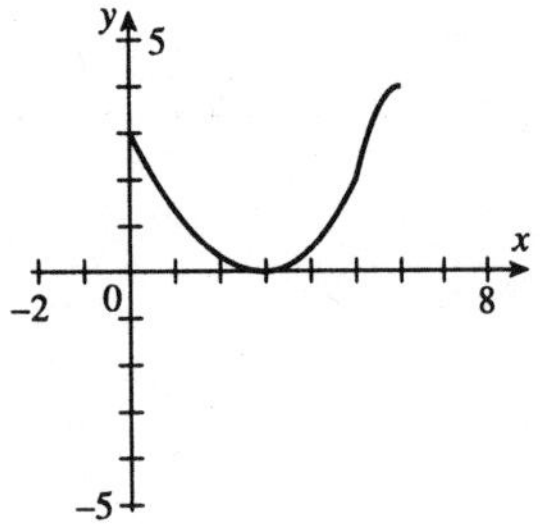

33.

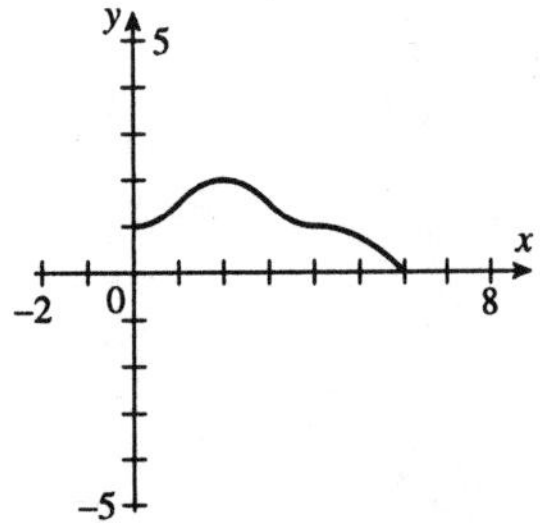

35. $f(x) = ax^2 + bx + c;\ f'(x) = 2ax + b;\ f''(x) = 2a$

An inflection point could occur where $f''(x) = 0$, or $2a = 0$. This could only occur when $a = 0$, but if $a = 0$, the equation is not quadratic. Thus, quadratic functions have no points of inflection.

37. a. Let $f(x) = x^2$ and let $I = [0, a]$, $a > y$.

$f'(x) = 2x > 0$ on I. Therefore, $f(x)$ is increasing on I, so $f(x) < f(y)$ for $x < y$.

b. Let $f(x) = \sqrt{x}$ and let $I = [0, a]$, $a > y$.

$f'(x) = \dfrac{1}{2\sqrt{x}} > 0$ on I. Therefore, $f(x)$ is increasing on I, so $f(x) < f(y)$ for $x < y$.

c. Let $f(x) = \dfrac{1}{x}$ and let $I = [0, a]$, $a > y$.

$f'(x) = -\dfrac{1}{x^2} < 0$ on I. Therefore $f(x)$ is decreasing on I, so $f(x) > f(y)$ for $x < y$.

39. $f(x) = a\sqrt{x} + \dfrac{\sqrt{b}}{\sqrt{x}};\ 13 = a\sqrt{4} + \dfrac{\sqrt{b}}{\sqrt{4}};\ 13 = 2a + \dfrac{\sqrt{b}}{2};\ 26 = 4a + \sqrt{b}$

$f'(x) = \dfrac{a}{2}x^{-1/2} - \dfrac{\sqrt{b}}{2}x^{-3/2};\ f''(x) = -\dfrac{a}{4}x^{-3/2} + \dfrac{3\sqrt{b}}{4}x^{-5/2}$

$f''(x) = 0;\ \dfrac{a}{4}x^{-3/2} = \dfrac{3\sqrt{b}}{4}x^{-5/2};\ a(4)^{-3/2} = 3\sqrt{b}(4)^{-5/2};\ \dfrac{a}{8} = \dfrac{3\sqrt{b}}{32};$

$$4a = 3\sqrt{b};\ \sqrt{b} = \frac{4a}{3}$$

$$26 = 4a + \left(\frac{4a}{3}\right);\ a = \frac{78}{16} = 4.875$$

$$b = \left(\frac{4(4.875)}{3}\right)^2 = 42.25$$

41. **a.** No conditions needed

b. $f(x)$ and $g(x) > 0$ for all x.

c. No conditions needed

43. **a.**

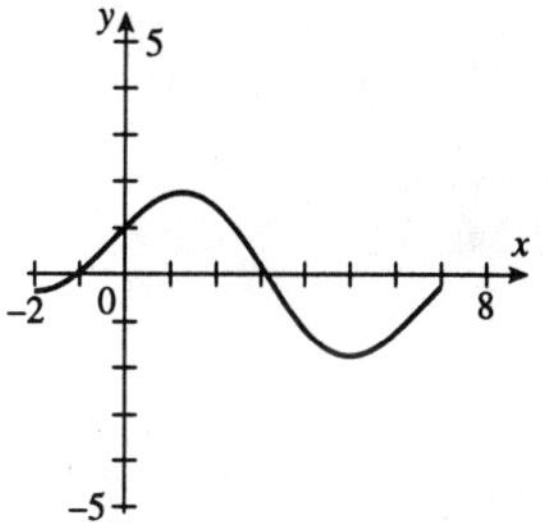

b. $f'(x) < 0;\ (1.3,\ 5)$

c. $f''(x) < 0;\ (-0.25,\ \pi) \cup (6.5,\ 7]$

d.

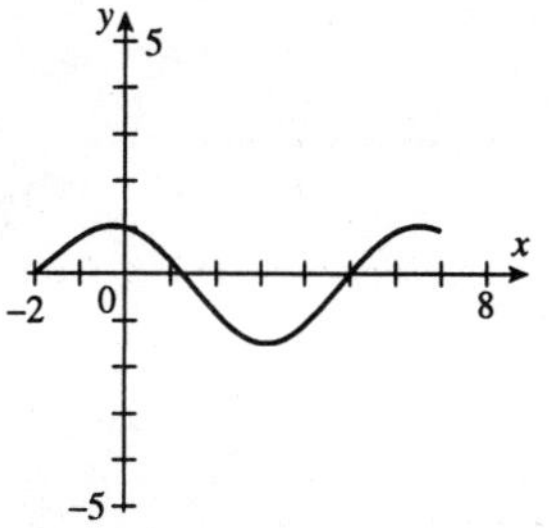

e.

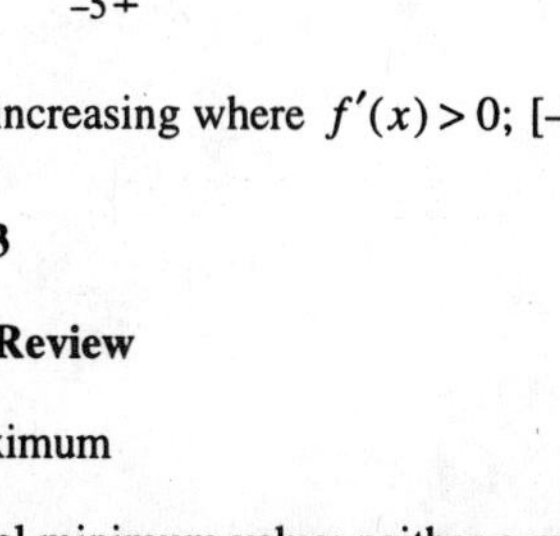

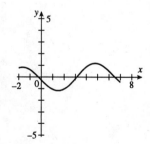

45. f is increasing where $f'(x) > 0$; $[-0.598, 0.680]$

Section 4.3

Concepts Review

1. Maximum

3. Local minimum value; neither a maximum nor minimum value

Problem Set 4.3

1. $f'(x) = 3x^2 - 6x$; $3x^2 - 6x = 0$; $3x(x-2) = 0$; $x = 0,\ 2$
Critical numbers: 0, 2
$f''(x) = 6x - 6$
$f''(0) = -6$; $f''(2) = 6$
Local minimum at $x = 2$; local maximum at $x = 0$

3. $f'(x) = \frac{1}{2} - \cos x$
Critical numbers: $\frac{\pi}{3}, \frac{5\pi}{3}$
$f''(x) = \sin x$
$f''\left(\frac{\pi}{3}\right) = \frac{\sqrt{3}}{2}$; $f''\left(\frac{5\pi}{3}\right) = -\frac{\sqrt{3}}{2}$;
Local minimum at $x = \frac{\pi}{3}$; local maximum at $x = \frac{5\pi}{3}$

5. $g'(x) = -x^2 e^{-x} + 2xe^{-x} = (2x - x^2)e^{-x} = x(2-x)e^{-x}$
Critical numbers: 0, 2
$g''(x) = x^2 e^{-x} - 2xe^{-x} - 2xe^{-x} + 2e^{-x} = e^{-x}(x^2 - 4x + 2)$
$g''(0) = 2$; $g''(2) \approx -0.27$
Local minimum at $x = 0$; local maximum at $x = 2$

7. $f'(x) = \frac{3}{4}x^2 - 3$
Critical numbers: –2, 2
$f''(x) = \frac{3}{2}x$
$f''(-2) = -3$, $f''(2) = 3$
Local minimum: $f(2) = -5$; local maximum: $f(-2) = 3$

9. $h'(x) = 4(x-a)^3 - 4$; $f'(x) = 0$ when $(x-a)^3 = 1$; $x - a = 1$; $x = a + 1$
Critical number: $a + 1$
$h''(x) = 12(x-a)^2$
$h''(a+1) = 12[(a+1) - a]^2 = 12$
Local minimum: $h(a+1) = (a+1-a)^4 - 4(a+1) = -4a - 3$

11. $g'(t) = 2t - 2e^{-2t}$; $g'(t) = 0$ when $t \approx 0.43$
Critical number: ≈ 0.43
$g''(t) = 2 + 4e^{-2t}$
$g''(0.43) \approx 3.69$
Local minimum: $g(0.43) \approx 0.61$

13. $f'(x) = 1 - \frac{1}{x^2}$; $f'(x) = 0$ when $x = -1,\ 1$
Critical numbers: –1, 1
$f''(x) = \frac{2}{x^3}$
$f''(-1) = -2$; $f''(1) = 2$
Local minimum: $f(1) = 2$; local maximum: $f(-1) = -2$

15. $f'(t) = \dfrac{(2+\cos t)(\cos t) - \sin t(-\sin t)}{(2+\cos t)^2} = \dfrac{2\cos t + 1}{(2+\cos t)^2}$

Critical numbers: $\dfrac{2\pi}{3}, \dfrac{4\pi}{3}$

$f'(t) > 0$ on $\left(0, \dfrac{2\pi}{3}\right) \cup \left(\dfrac{4\pi}{3}, 2\pi\right)$.

$f'(t) < 0$ on $\left(\dfrac{2\pi}{3}, \dfrac{4\pi}{3}\right)$.

Local minimum: $f\left(\dfrac{4\pi}{3}\right) = -\dfrac{\sqrt{3}}{3} \approx -0.58$; local maximum: $f\left(\dfrac{2\pi}{3}\right) = \dfrac{\sqrt{3}}{3} \approx 0.58,$

17. $F'(x) = \dfrac{3}{\sqrt{x}} - 3$; $F'(x) = 0$ when $x = 1$

Critical numbers: 0, 1, 9

$F''(x) = -\dfrac{3}{2}x^{-3/2}$

$F''(0)$ does not exist; $F''(1) = -\dfrac{3}{2}$; $F''(9) \approx -0.056$

Maximum: $F(1) = 3$, minimum: $F(9) = -9$

19. $f'(x) = 64(-1)(\sin x)^{-2}\cos x + 27(-1)(\cos x)^{-2}(-\sin x)$

$= -\dfrac{64\cos x}{\sin^2 x} + \dfrac{27\sin x}{\cos^2 x} = \dfrac{(3\sin x - 4\cos x)(9\sin^2 x + 12\cos x \sin x + 16\cos^2 x)}{\sin^2 x \cos^2 x}$

On $\left(0, \dfrac{\pi}{2}\right)$, $f'(x) = 0$ only where $3\sin x = 4\cos x$; $\tan x = \dfrac{4}{3}$;

$x = \tan^{-1}\dfrac{4}{3} \approx 0.9273$

Critical number: ≈ 0.9273

For $0 < x < 0.9273$, $f'(x) < 0$, while for $0.9273 < x < \dfrac{\pi}{2}$, $f'(x) > 0$.

Minimum: $f\left(\tan^{-1}\dfrac{4}{3}\right) = 125$.

21. $g'(x) = 2x + \dfrac{(8-x)^2(32x) - (16x^2)2(8-x)(-1)}{(8-x)^4} = 2x + \dfrac{256x}{(8-x)^3}$

$g'(x) = 0$; $x = 8 + \sqrt[3]{128} \approx 13.04$ for $x > 8$

Critical number: ≈ 13.04

$g(13.04) \approx 277$

$g'(13) = -0.624$;

$g'(14) \approx 11.407$;

$g(13.04) = 277$ is the minimum

23. Critical numbers: $x = -2, -1, 2, 3$

Use test values: $-3, -\dfrac{3}{2}, 0, \dfrac{5}{2}$, and 4 to find that $f'(x) < 0$ on $(-2, -1) \cup (-1, 2) \cup (2, 3)$ and $f'(x) > 0$ on $(-\infty, -2) \cup (3, \infty)$.

Local maximum at $x = -2$; local minimum at $x = 3$.

25. Written response (graph)

a. Increasing: $(-\infty, -3] \cup [-1, 0)$
Decreasing: $[-3, -1] \cup [0, \infty)$

b. Concave up: $(-2, 0) \cup (0, 2)$
Concave down: $(-\infty, -2) \cup (2, \infty)$

c. Local maximum at $x = -3$; local minimum at $x = -1$

d. Inflection points: $x = -2, x = 2$

27. $f'(x) = 5x^4 - 15x^2 = 5x^2(x^2 - 3)$
Critical numbers: $-2, -\sqrt{3}, 0, \sqrt{3}, 2.5$
$f''(x) = 20x^3 - 30x$
$f''(-\sqrt{3}) \approx -52;\ f''(0) = 0;\ f''(\sqrt{3}) \approx 52$
$f(-2) = 12$ and $f(\sqrt{3}) \approx -6.4$ are local minima while $f(-\sqrt{3}) \approx 14.4$ and $f(2.5) = 23.53125$ are local maxima.

29. $g'(x) = 0$ when $x = 1, \approx 2.1$
Critical numbers: $1, \approx 2.1$
$g''(x) = 5x^4 - 15x^2$
$g''(1) = -10;\ g''(2.1) \approx 31.1$
g has a local minimum at $x \approx 2.1$; local maximum at $x = 1$.

Section 4.4

Concepts Review

1. $0 < x < 100$

3. $P = 3x + \frac{300}{x}$

Problem Set 4.4

1. Let x be one number and $-\frac{12}{x}$ be the other.

$f(x) = x^2 + \left(-\frac{12}{x}\right)^2 = x^2 + \frac{144}{x^2}, x \neq 0$

$f'(x) = 2x - \frac{288}{x^3};\ f'(x) = 0$ when $x = -2\sqrt{3},\ 2\sqrt{3}$

Critical numbers: $-2\sqrt{3},\ 2\sqrt{3}$

$f''(x) = 2 + \frac{864}{x^4}$

$f''(-2\sqrt{3}) = 8, ;\ f''(2\sqrt{3}) = 8$

$f(x)$ has a minimum value at $-2\sqrt{3}$ and $2\sqrt{3}$, and $-\frac{12}{-2\sqrt{3}} = 2\sqrt{3}$ so one number is $-2\sqrt{3}$ and the other is $2\sqrt{3}$.

3. $Q = (x+10)^2 + (y-0)^2 = (2y^2+10)^2 + y^2 = 4y^4 + 41y^2 + 100$

$\frac{dQ}{dy} = 16y^3 + 82y = 2y(8y^2+41)$; $Q'(y) = 0$ when $y = 0$

$\frac{d^2Q}{dy^2} = 48y^2 + 82$; at $y = 0$: $48(0)^2 + 82 = 82$

Q has a minimum at $y = 0$, $x = 2(0)^2 = 0$.

5. $A = (2x)y = 1800$; $y = \frac{900}{x}$

$Q = 4x + 3y = 4x + 3\left(\frac{900}{x}\right) = 4x + \frac{2700}{x}$

$\frac{dQ}{dx} = 4 - \frac{2700}{x^2}$; $Q'(x) = 0$ when $x \approx -25.98,\ 25.98$

Critical number: 25.98

$\frac{d^2Q}{dx^2} = \frac{5400}{x^3}$; at $x = 25.98$: $\frac{5400}{25.98^3} \approx 0.31$

Q has a minimum at $x \approx 25.98$, $y \approx 34.64$.

7. $A = 3xy = 2700$; $y = \frac{900}{x}$

$Q = 6x + 4y = 6x + 4\left(\frac{900}{x}\right) = 6x + \frac{3600}{x}$

$\frac{dQ}{dx} = 6 - \frac{3600}{x^2}$; $Q'(x) = 0$ when $x \approx -24.5,\ 24.5$

Critical number: 24.5

$\frac{d^2Q}{dx^2} = \frac{7200}{x^3}$; at $x = 24.5$: $\frac{7200}{24.5^3} \approx 0.49$

Q has a minimum at $x \approx 24.5$, $y \approx 36.7$.

9. Let x be the length of a side of the base and y the height.

$V = x^2y = 12{,}000$; $y = \frac{12{,}000}{x^2}$

Let c be the cost per square foot of the concrete sides and base.

$Q = (2c)x^2 + 4cxy + cx^2 = 3cx^2 + 4cx\left(\frac{12{,}000}{x^2}\right) = 3cx^2 + \frac{48{,}000c}{x}$

$\frac{dQ}{dx} = 6cx - \frac{48{,}000c}{x^2}$; $Q'(x) = 0$ when $x = 20$.

$\frac{d^2Q}{dx^2} = 6c + \frac{96{,}000c}{x^3}$; at $x = 20$: $6c + \frac{96{,}000c}{20^3} = 18c$

Q has a minimum at $x = 20$, $y = \frac{12{,}000}{20^2} = 30$.

11. There are two cases:

In the first case, let $2x$ be the width and $2y$ be the depth. Then

$S = k(2x)(2y)^2 = 8kxy^2 = 8kx\left(9 - \frac{9x^2}{8}\right) = 72kx - 9kx^3$

$\frac{dS}{dx} = 72k - 27kx^2 = 9k(8 - 3x^2)$

$9k(8-3x^2)=0;\ x=\frac{2\sqrt{2}}{\sqrt{3}}$ in $\left(0,\ 2\sqrt{2}\right)$.

$\frac{d^2S}{dx^2}=-54kx$; at $x=\frac{2\sqrt{2}}{\sqrt{3}}$: $-36k\sqrt{6}<0$

Therefore S is maximum at $x=\frac{2\sqrt{2}}{\sqrt{3}}$. $y^2=9-\frac{9}{8}\left(\frac{2\sqrt{2}}{\sqrt{3}}\right)^2=6$.

The width is $\frac{4\sqrt{2}}{\sqrt{3}}\approx 3.27$ and depth is $2\sqrt{6}\approx 4.90$.

In the second case, let $2y$ be the width and $2x$ be the depth. Then

$S=k(2y)(2x)^2=8kx^2y=8k\left(8-\frac{8y^2}{9}\right)y=64ky-\frac{64ky^3}{9}$.

$\frac{dS}{dy}=64k-\frac{64ky^2}{3}=64k\left(1-\frac{y^2}{3}\right)$

$64k\left(1-\frac{y^2}{3}\right)=0;\ y=\sqrt{3}$ on (0, 3).

$\frac{d^2S}{dy^2}=-\frac{128ky}{3}$; at $y=\sqrt{3}$: $-\frac{128k}{\sqrt{3}}<0$

Therefore S is maximum at $y=\sqrt{3}$. $x^2=8-\frac{8}{9}(\sqrt{3})^2=\frac{16}{3}$.

The width is $2\sqrt{3}\approx 3.46$ and the depth is $\frac{8}{\sqrt{3}}\approx 4.62$.

In the first case, $S=k\left(\frac{4\sqrt{2}}{\sqrt{3}}\right)\left(2\sqrt{6}\right)^2=32\sqrt{6}k\approx 78.38k$.

In the second case, $S=k\left(2\sqrt{3}\right)\left(\frac{8}{\sqrt{3}}\right)^2=\frac{128\sqrt{3}}{3}k\approx 73.90k$.

The first case gives the dimensions of the strongest beam.

13. $200=\frac{100m}{(m-100)e^{-k}+100};\ 200=\frac{100m}{me^{-k}-100e^{-k}+100};\ 200me^{-k}-20{,}000e^{-k}+20{,}000=100m$

$100m-200me^{-k}=-20{,}000e^{-k}+20{,}000;\ m(1-2e^{-k})=-200e^{-k}+200;\ m=\frac{200-200e^{-k}}{1-2e^{-k}}$

$350=\frac{100m}{(m-100)e^{-2k}+100};\ 350=\frac{100\left(\frac{200-200e^{-k}}{1-2e^{-k}}\right)}{\left(\frac{200-200e^{-k}}{1-2e^{-k}}-100\right)e^{-2k}+100};$

Using a computer algebra system,
$k=\ln 7-\ln 3\approx 0.847;\ m=800$

$y=\frac{100\cdot 800}{(800-100)e^{-0.847t}+100}=\frac{80{,}000}{700e^{-0.847t}+100}$

$\frac{dy}{dt}=\frac{(700e^{-0.847t}+100)(0)-80{,}000(-592.9e^{-0.847t})}{(700e^{-0.847t}+100)^2}=\frac{47{,}432{,}000e^{-0.847t}}{(700e^{-0.847t}+100)^2}$

The colony grows fastest at t = 1. In the long run, there will be 800 organisms.

15. Let x be the distance from I_1.

$$Q = \frac{kI_1}{x^2} + \frac{kI_2}{(s-x)^2}$$

$$\frac{dQ}{dx} = \frac{-2kI_1}{x^3} + \frac{2kI_2}{(s-x)^3}$$

$$-\frac{2kI_1}{x^3} + \frac{2kI_2}{(s-x)^3} = 0;\ \frac{x^3}{(s-x)^3} = \frac{I_1}{I_2};\ x = \frac{s\sqrt[3]{I_1}}{\sqrt[3]{I_1} + \sqrt[3]{I_2}}$$

17. Let the coordinates of the first ship at 7:00 a.m. be (0, 0). Thus, the coordinates of the second ship at 7:00 a.m. are (–60, 0). Let t be the time in hours since 7:00 a.m. The coordinates of the first and second ships at t are $(-20t, 0)$ and $\left(-60 + 15\sqrt{2}t,\ -15\sqrt{2}t\right)$ respectively. Let D be the square of the distances at t.

$$D = \left(-20t + 60 - 15\sqrt{2}t\right)^2 + \left(0 + 15\sqrt{2}t\right)^2 = \left(1300 + 600\sqrt{2}\right)t^2 - \left(2400 + 1800\sqrt{2}\right)t + 3600$$

$$\frac{dD}{dt} = 2\left(1300 + 600\sqrt{2}\right)t - \left(2400 + 1800\sqrt{2}\right)$$

$$2\left(1300 + 600\sqrt{2}\right)t - \left(2400 + 1800\sqrt{2}\right) = 0;\ t = \frac{12 + 9\sqrt{2}}{13 + 6\sqrt{2}} \approx 1.15 \text{ hours or approximately 1 hour, 9 minutes}$$

D is the minimum at $t = \dfrac{12 + 9\sqrt{2}}{13 + 6\sqrt{2}}$ since $\dfrac{d^2D}{dt^2} > 0$ for all t.

The ships are closest at 8:09 a.m.

19. Let x be the radius of the base of the cylinder and h the height.

$$V = \pi x^2 h;\ r^2 = x^2 + \left(\frac{h}{2}\right)^2;\ x^2 = r^2 - \frac{h^2}{4}$$

$$V = \pi\left(r^2 - \frac{h^2}{4}\right)h = \pi h r^2 - \frac{\pi h^3}{4}$$

$$\frac{dV}{dh} = \pi r^2 - \frac{3\pi h^2}{4};\ V' = 0 \text{ when } h = \pm\frac{2\sqrt{3}r}{3}$$

The volume is maximized when $h = \dfrac{2\sqrt{3}r}{3}$.

$$V = \pi\left(\frac{2\sqrt{3}}{3}r\right)r^2 - \frac{\pi\left(\frac{2\sqrt{3}}{3}r\right)^3}{4} = \frac{2\pi\sqrt{3}}{3}r^3 - \frac{2\pi\sqrt{3}}{9}r^3 = \frac{4\pi\sqrt{3}}{9}r^3$$

21. Let x be the radius of the cylinder, r the radius of the sphere, and h the height of the cylinder.

$$A = 2\pi xh;\ r^2 = x^2 + \frac{h^2}{4};\ x = \sqrt{r^2 - \frac{h^2}{4}}$$

$$A = 2\pi\sqrt{r^2 - \frac{h^2}{4}}h = 2\pi\sqrt{h^2r^2 - \frac{h^4}{4}}$$

$$\frac{dA}{dh} = \frac{\pi\left(2r^2h - h^3\right)}{\sqrt{h^2r^2 - \frac{h^4}{4}}};\ A' = 0 \text{ when } h = 0,\ \pm r\sqrt{2}$$

The dimensions are $h = \sqrt{2}r,\ x = \dfrac{r}{\sqrt{2}}$.

23. Let x be the length of a side of the square, so $\frac{100-4x}{3}$ is the side of the triangle, $0 \le x \le 25$.

$$A = x^2 + \frac{1}{2}\left(\frac{100-4x}{3}\right)\frac{\sqrt{3}}{2}\left(\frac{100-4x}{3}\right) = x^2 + \frac{\sqrt{3}}{4}\left(\frac{10{,}000-800x+16x^2}{9}\right)$$

$$\frac{dA}{dx} = 2x - \frac{200\sqrt{3}}{9} + \frac{8\sqrt{3}}{9}x;\ A' = 0 \text{ when } x \approx 10.87$$

Critical numbers: $x = 0, 10.87, 25$

At $x = 0$, $A \approx 481$; at $x = 10.87$, $A \approx 272$; at $x = 25$, $A = 625$.

a. For minimum area, the cut should be approximately $4(10.87) = 43.48$ cm from one end and the shorter length should be bent to form the square.

b. For maximum area, the wire should not be cut, it should be bent to form a square.

25. Let r be the radius of the cylinder and h the height of the cylinder.

$$V = \pi r^2 h + \frac{2}{3}\pi r^3;\ h = \frac{V - \frac{2}{3}\pi r^3}{\pi r^2} = \frac{V}{\pi r^2} - \frac{2}{3}r$$

Let k be the cost per square foot of the cylindrical wall.

$$C = k(2\pi rh) + 2k(2\pi r^2) = k\left(2\pi r\left(\frac{V}{\pi r^2} - \frac{2}{3}r\right) + 4\pi r^2\right) = k\left(\frac{2V}{r} + \frac{8\pi r^2}{3}\right)$$

$$\frac{dC}{dr} = k\left(-\frac{2V}{r^2} + \frac{16\pi r}{3}\right);\ k\left(-\frac{2V}{r^2} + \frac{16\pi r}{3}\right) = 0,\ r^3 = \frac{3V}{8\pi},\ r = \frac{1}{2}\left(\frac{3V}{\pi}\right)^{1/3}$$

$$h = \frac{4V}{\pi\left(\frac{3V}{\pi}\right)^{2/3}} - \frac{1}{3}\left(\frac{3V}{\pi}\right)^{1/3} = \frac{4}{3}\left(\frac{3V}{\pi}\right)^{1/3} - \frac{1}{3}\left(\frac{3V}{\pi}\right)^{1/3} = \left(\frac{3V}{\pi}\right)^{1/3}$$

For a given volume V, the height of the cylinder is $\left(\frac{3V}{\pi}\right)^{1/3}$ and the radius is $\frac{1}{2}\left(\frac{3V}{\pi}\right)^{1/3}$.

27. $A = \frac{r^2\theta}{2};\ s = r\theta;\ \theta = \frac{2A}{r^2}$

$$Q = 2r + r\theta = 2r + \frac{2Ar}{r^2} = 2r + \frac{2A}{r}$$

$$\frac{dQ}{dr} = 2 - \frac{2A}{r^2};\ Q' = 0 \text{ when } r = \sqrt{A}$$

$$\theta = \frac{2A}{(\sqrt{A})^2} = 2$$

29. Let d_1 be the distance that the light travels in medium 1 and let d_2 be the distance the light travels in medium 2.

$$d_1 = \sqrt{a^2 + x^2},\ d_2 = \sqrt{b^2 + (d-x)^2}$$

The time that it takes the light to travel from A to B is

$$t = \frac{d_1}{c_1} + \frac{d_2}{c_2} = \frac{\sqrt{a^2+x^2}}{c_1} + \frac{\sqrt{b^2+(d-x)^2}}{c_2}$$

$$\frac{dt}{dx} = \frac{x}{c_1\sqrt{a^2+x^2}} - \frac{d-x}{c_2\sqrt{b^2-(d-x)^2}} = \frac{\sin\theta_1}{c_1} - \frac{\sin\theta_2}{c_2}$$

$$\frac{dt}{dx} = 0 \text{ implies } \frac{\sin\theta_1}{c_1} = \frac{\sin\theta_2}{c_2}.$$

31. Consider the following sketch.

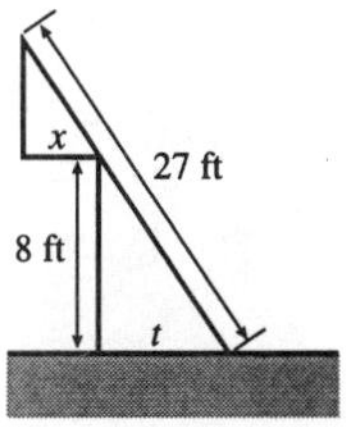

By similar triangles, $\dfrac{x}{27-\sqrt{t^2+64}} = \dfrac{t}{\sqrt{t^2+64}}$.

$$x = \frac{27t}{\sqrt{t^2+64}} - t$$

$$\frac{dx}{dt} = \frac{27\sqrt{t^2+64} - \frac{27t^2}{\sqrt{t^2+64}}}{t^2+64} - 1 = \frac{1728}{(t^2+64)^{3/2}} - 1$$

$$\frac{1728}{(t^2+64)^{3/2}} - 1 = 0;\ t = 4\sqrt{5}$$

$$\frac{d^2x}{dt^2} = \frac{-5184t}{(t^2+64)^{5/2}};\ \left.\frac{d^2x}{dt^2}\right|_{t=4\sqrt{5}} < 0$$

Therefore $x = \dfrac{27\left(4\sqrt{5}\right)}{\sqrt{\left(4\sqrt{5}\right)^2+64}} - 4\sqrt{5} = 5\sqrt{5} \approx 11.18$ ft is the maximum horizontal overhang.

33. Consider the figure below.

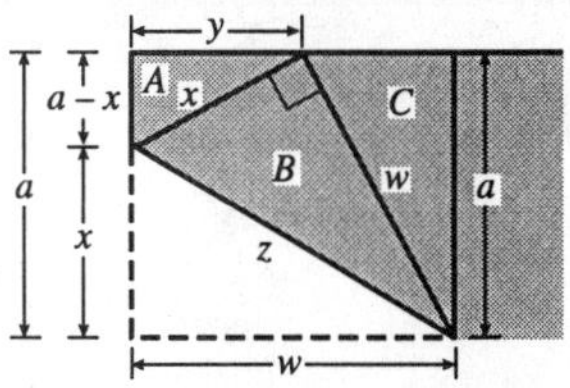

a. $y = \sqrt{x^2 - (a-x)^2} = \sqrt{2ax - a^2}$

Area of $A = A = \dfrac{1}{2}(a-x)y = \dfrac{1}{2}(a-x)\sqrt{2ax-a^2}$

$$\frac{dA}{dx} = -\frac{1}{2}\sqrt{2ax-a^2} + \frac{\frac{1}{2}(a-x)\left(\frac{1}{2}\right)(2a)}{\sqrt{2ax-a^2}} = \frac{a^2 - \frac{3}{2}ax}{\sqrt{2ax-a^2}}$$

$$\frac{a^2 - \frac{3}{2}ax}{\sqrt{2ax-a^2}} = 0,\ x = \frac{2a}{3}$$

b. Triangle A is similar to triangle C.

$$w = \frac{ax}{y} = \frac{ax}{\sqrt{2ax-a^2}}$$

area of $B = B = \dfrac{1}{2}xw = \dfrac{ax^2}{2\sqrt{2ax-a^2}}$

$$\frac{dB}{dx}=\frac{a}{2}\left(\frac{2x\sqrt{2ax-a^2}-x^2\frac{a}{\sqrt{2ax-a^2}}}{2ax-a^2}\right)=\frac{a}{2}\left(\frac{2x(2ax-a^2)-ax^2}{(2ax-a^2)^{3/2}}\right)=\frac{a}{2}\left(\frac{3ax^2-2xa^2}{(2ax-a^2)^{3/2}}\right)$$

$$\frac{a^2}{2}\left(\frac{3x^2-2xa}{(2ax-a^2)^{3/2}}\right)=0;\ x=0,\ \frac{2a}{3}$$

Since $x=0$ is not possible, $x=\frac{2a}{3}$.

c. $$z=\sqrt{x^2+w^2}=\sqrt{x^2+\frac{a^2x^2}{2ax-a^2}}=\sqrt{\frac{2ax^3}{2ax-a^2}}$$

$$\frac{dz}{dx}=\frac{1}{2}\sqrt{\frac{2ax-a^2}{2ax^3}}\left(\frac{6ax^2(2ax-a^2)-2ax^3(2a)}{(2ax-a^2)^2}\right)=\frac{4a^2x^3-3a^3x^2}{\sqrt{2ax^3(2ax-a^2)^3}}$$

$$\frac{dz}{dx}=0 \text{ when } x=0,\ \frac{3a}{4}$$

$$x=\frac{3a}{4}$$

35. a. $$L'(\theta)=15(9+25-30\cos\theta)^{-1/2}\sin\theta=15(34-30\cos\theta)^{-1/2}\sin\theta$$

$$L''(\theta)=-\frac{15}{2}(34-30\cos\theta)^{-3/2}(30\sin\theta)\sin\theta+15(34-30\cos\theta)^{-1/2}\cos\theta$$

$$=-225(34-30\cos\theta)^{-3/2}\sin^2\theta+15(34-30\cos\theta)^{-1/2}\cos\theta$$

$$=15(34-30\cos\theta)^{-3/2}[-15\sin^2\theta+(34-30\cos\theta)\cos\theta]$$

$$=15(34-30\cos\theta)^{-3/2}[-15\sin^2\theta+34\cos\theta-30\cos^2\theta]$$

$$=15(34-30\cos\theta)^{-3/2}[-15+34\cos\theta-15\cos^2\theta]$$

$$=-15(34-30\cos\theta)^{-3/2}[15\cos^2\theta-34\cos\theta+15]$$

$$L''=0 \text{ when } \cos\theta=\frac{34\pm\sqrt{(34)^2-4(15)(15)}}{2(15)}=\frac{5}{3},\ \frac{3}{5}$$

$$\theta=\cos^{-1}\left(\frac{3}{5}\right)$$

$$L'\left(\cos^{-1}\left(\frac{3}{5}\right)\right)=15\left(9+25-30\left(\frac{3}{5}\right)\right)^{-1/2}\left(\frac{4}{5}\right)=3$$

$$L\left(\cos^{-1}\left(\frac{3}{5}\right)\right)=\left(9+25-30\left(\frac{3}{5}\right)\right)^{1/2}=4$$

$\phi=90^\circ$ since the resulting triangle is a 3-4-5 right triangle.

b. $$L'(\theta)=65(25+169-130\cos\theta)^{-1/2}\sin\theta=65(194-130\cos\theta)^{-1/2}\sin\theta$$

$$L''(\theta)=-\frac{65}{2}(194-130\cos\theta)^{-3/2}(130\sin\theta)\sin\theta+65(194-130\cos\theta)^{-1/2}\cos\theta$$

$$=-4225(194-130\cos\theta)^{-3/2}\sin^2\theta+65(194-130\cos\theta)^{-1/2}\cos\theta$$

$$=65(194-130\cos\theta)^{-3/2}[-65\sin^2\theta+(194-130\cos\theta)\cos\theta]$$

$$=65(194-130\cos\theta)^{-3/2}[-65\sin^2\theta+194\cos\theta-130\cos^2\theta]$$

$$=65(194-130\cos\theta)^{-3/2}[-65\cos^2\theta+194\cos\theta-65]$$

$$=-65(194-130\cos\theta)^{-3/2}[65\cos^2\theta-194\cos\theta+65]$$

$L'' = 0$ when $\cos\theta = \dfrac{194 \pm \sqrt{(194)^2 - 4(65)(65)}}{2(65)} = \dfrac{13}{5}, \dfrac{5}{13}$

$\theta = \cos^{-1}\left(\dfrac{5}{13}\right)$

$L'\left(\cos^{-1}\left(\dfrac{5}{13}\right)\right) = 65\left(25 + 169 - 130\left(\dfrac{5}{13}\right)\right)^{1/2}\left(\dfrac{12}{13}\right) = 5$

$L\left(\cos^{-1}\left(\dfrac{5}{13}\right)\right) = \left(25 + 169 - 130\left(\dfrac{5}{13}\right)\right)^{1/2} = 12$

$\phi = 90°$ since the resulting triangle is a 5-12-13 right triangle.

c. When the tips are separating most rapidly, $\phi = 90°$, $L = \sqrt{m^2 - h^2}$, $L' = h$

d. $L'(\theta) = hm(h^2 + m^2 - 2hm\cos\theta)^{-1/2}\sin\theta$

$L''(\theta) = -h^2m^2(h^2 + m^2 - 2hm\cos\theta)^{-3/2}\sin^2\theta + hm(h^2 + m^2 - 2hm\cos\theta)^{-1/2}\cos\theta$

$= hm(h^2 + m^2 - 2hm\cos\theta)^{-3/2}[-hm\sin^2\theta + (h^2 + m^2)\cos\theta - 2hm\cos^2\theta]$

$= hm(h^2 + m^2 - 2hm\cos\theta)^{-3/2}[-hm\cos^2\theta + (h^2 + m^2)\cos\theta - hm]$

$= -hm(h^2 + m^2 - 2hm\cos\theta)^{-3/2}[hm\cos^2\theta - (h^2 + m^2)\cos\theta + hm]$

$L'' = 0$ when $hm\cos^2\theta - (h^2 + m^2)\cos\theta + hm = 0$

$(h\cos\theta - m)(m\cos\theta - h) = 0$

$\cos\theta = \dfrac{m}{h}, \dfrac{h}{m}$

Since $h < m$, $\cos\theta = \dfrac{h}{m}$ so $\theta = \cos^{-1}\left(\dfrac{h}{m}\right)$.

$L'\left(\cos^{-1}\left(\dfrac{h}{m}\right)\right) = hm\left(h^2 + m^2 - 2hm\left(\dfrac{h}{m}\right)\right)^{-1/2}\dfrac{\sqrt{m^2 - h^2}}{m}$

$= hm(m^2 - h^2)^{-1/2}\dfrac{\sqrt{m^2 - h^2}}{m} = h$

$L\left(\cos^{-1}\left(\dfrac{h}{m}\right)\right) = \left(h^2 + m^2 - 2hm\left(\dfrac{h}{m}\right)\right)^{1/2} = \sqrt{m^2 - h^2}$

Since $h^2 + L^2 = m^2$, $\phi = 90°$.

Section 4.5

Concepts Review

1. Continuous

3. Fixed; variable

Problem Set 4.5

1. $C(x) = 8000 + 110x$

3. $P(x) = R(x) - C(x) = (300x - 0.5x^2) - (8000 + 110x)$

   $P(x) = -8000 + 190x - 0.5x^2$

5. $C(x) = 1200 + (3.25)x - (0.002)x^2; \;\; x = 1800$

At $x = 1800$: $\dfrac{C(x)}{x} = \dfrac{1200 + (3.25)(1800) - (0.002)(1800)^2}{1800} \approx 0.32$

$\dfrac{dC}{dx} = 3.25 - 0.004x$; at $x = 1800$: $3.25 - 0.004(1800) = -3.95$

7. $A(x) = \dfrac{C(x)}{x} = \dfrac{80,000 - 400x^2 + x^3}{40,000x} = \dfrac{2}{x} - \dfrac{x}{100} + \dfrac{x^2}{40,000}$

$\dfrac{dA}{dx} = -\dfrac{2}{x^2} - \dfrac{1}{100} + \dfrac{x}{20,000}$; $\dfrac{dA}{dx} = 0$ when $x \approx 201$

Average cost is minimum at $x = 201$.

9. a. $R(x) = xp(x) = 20x + 4x^2 - \dfrac{x^3}{3}$

$\dfrac{dR}{dx} = 20 + 8x - x^2$

b. Increasing when $\dfrac{dR}{dx} > 0$

$20 + 8x - x^2 > 0$; $[0, \;10)$

c. $\dfrac{d^2R}{dx^2} = 8 - 2x$; $\dfrac{d^2R}{dx^2} = 0$ when $x = 4$

$\dfrac{d^3R}{dx^3} = -2$; $\dfrac{dR}{dx}$ is maximum at $x = 4$.

11. $R(x) = \dfrac{800x}{x+3} - 3x$

$\dfrac{dR}{dx} = \dfrac{(x+3)(800) - 800x}{(x+3)^2} - 3 = \dfrac{2400}{(x+3)^2} - 3$; $\dfrac{dR}{dx} = 0$ when $x \approx 25$

$x_1 = 25$; $R(25) \approx 639.29$

At x_1, $\dfrac{dR}{dx} = 0$.

13. $x = 4000 + \dfrac{6 - p(x)}{0.15}(250)$; $p(x) = 6 - (0.15)\dfrac{(x - 4000)}{250} = 8.4 - 0.0006x$

$R(x) = 8.4x - 0.0006x^2$

$\dfrac{dR}{dx} = 8.4 - 0.0012x$; $\dfrac{dR}{dx} = 0$ when $x = 7000$

Revenue is maximum when $p(x) = 8.4 - 0.0006(7000) = \$4.20$ per yard.

15. a. $C(x) = \begin{cases} 6000 + 1.40x & \text{if } 0 \le x \le 4500 \\ 6000 + 1.60x & \text{if } 4500 < x \end{cases}$

b. $x = 4000 + \dfrac{7 - p(x)}{0.10}(100)$; $p(x) = 7 - (0.10)\dfrac{x - 4000}{100}$

$p(x) = 11 - 0.001x$

c. $R(x) = 11x - 0.001x^2$

$P(x) = (11x - 0.001x^2) - (6000 + 1.40x) \qquad \text{if } 0 \le x \le 4500$

$P(x) = -6000 + 9.6x - 0.001x^2 \qquad \text{if } 0 \le x \le 4500$

$\frac{dP}{dx} = 9.6 - 0.002x; \; \frac{dP}{dx} = 0$ when $x = 4800$; this is not in the interval [0, 4500].

The critical numbers are 0 and 4500. $P(0) = -6000$ and $P(4500) = 16{,}950$

$P(x) = (11x - 0.001x^2) - (6000 + 1.60x) \qquad \text{if } 4500 < x$

$P(x) = -6000 + 9.4x - 0.001x^2 \qquad \text{if } 4500 < x$

$\frac{dP}{dx} = 9.4 - 0.002x; \; \frac{dP}{dx} = 0$ when $x = 4700$

$P(4700) = 16{,}090$

Therefore, the number of units for maximum profit is $x = 4500$.

17. $C(x) = \begin{cases} 200 + 4x - 0.01x^2 & \text{if } 0 \le x \le 300 \\ 800 + 3x - 0.01x^2 & \text{if } 300 < x \le 450 \end{cases}$

$P(x) = \begin{cases} -200 + 6x + 0.009x^2 & \text{if } 0 \le x \le 300 \\ -800 + 7x + 0.009x^2 & \text{if } 300 < x \le 450 \end{cases}$

There are no stationary points on the interval [0, 300].

On [300, 450]:

$\frac{dP}{dx} = 7 + 0.018x; \; \frac{dP}{dx} = 0$ when $x = -389$

The critical numbers are 0, 300, 450.

$P(0) = -200$, $P(300) = 2410$, $P(450) = 4172.5$

Monthly profit is maximized at $x = 450$, $P(450) = 4172.50$

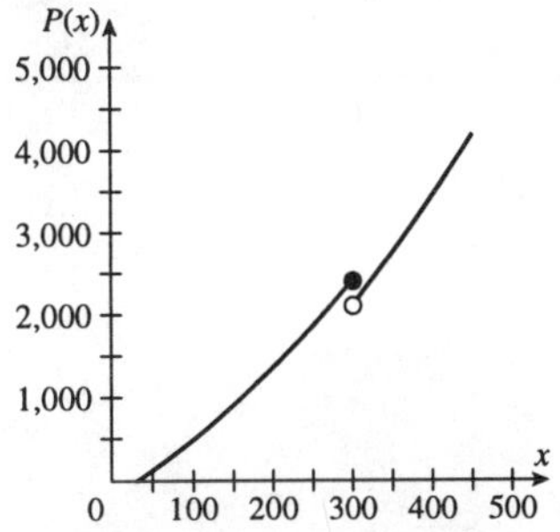

19. $x = \frac{5m^2}{\sqrt{m^2+13}}; \; p = 10x - 0.1x^2; \; R(x) = 10x^2 - 0.1x^3$

$$\frac{dR}{dm} = \frac{\sqrt{m^2+13}(10m) - (5m^2)\frac{2m}{2\sqrt{m^2+13}}}{\left(\sqrt{m^2+13}\right)^2}\left(10x - 0.1x^2 + \frac{5m^2}{\sqrt{m^2+13}}(10 - 0.2x)\right)$$

$$= \frac{10m(m^2+13) - 5m^3}{(m^2+13)^{3/2}}\left(10x - 0.1x^2 + \frac{5m^2(10-0.2x)}{\sqrt{m^2+13}}\right)$$

$$= \frac{5m^3 + 130m}{(m^2+13)^{3/2}}\left(10x - 0.1x^2 + \frac{5m^2(10-0.2x)}{\sqrt{m^2+13}}\right)$$

When $m = 6$, $x = \frac{5(6)^2}{\sqrt{6^2+13}} = \frac{180}{7}$.

$$\frac{dR}{dm} = \frac{5(6)^3+130(6)}{(6^2+13)^{3/2}}\left(10\left(\frac{180}{7}\right)-0.1\left(\frac{180}{7}\right)^2+\frac{5(6)^2\left(10-0.2\cdot\frac{180}{7}\right)}{\sqrt{6^2+13}}\right) \approx 1713.14$$

21. The number of lots ordered per year is $\frac{N}{x}$. Let C represent the inventory cost.

$$C = \frac{N}{x}(F+Bx)+\frac{x}{2}A = \frac{FN}{x}+BN+\frac{Ax}{2}$$

$$\frac{dC}{dx} = -\frac{FN}{x^2}+\frac{A}{2};\quad -\frac{FN}{x^2}+\frac{A}{2}=0,\quad x=\sqrt{\frac{2FN}{A}}$$

Section 4.6

Concepts Review

1. $f(x)$; $-f(x)$

3. $x=-a, x=b, x=-\frac{d}{c}$; $y=\frac{1}{c}$

Problem Set 4.6

1. $f(x) = x^2 - ax$

$f(x) = (-x)^2 - a(-x) = x^2 + ax$; no symmetry unless $a = 0$.

$f(0) = 0$; $0 = x^2 - ax$; $0 = x(x-a)$; intercepts: $(0, 0)$, $(a, 0)$

$f'(x) = 2x - a$; $f'(x) = 0$ when $x = \frac{a}{2}$

Increasing: $\left[\frac{a}{2}, \infty\right)$; decreasing: $\left(-\infty, \frac{a}{2}\right]$

Minimum $f\left(\frac{a}{2}\right) = \frac{a^2}{4} - \frac{a^2}{2} = -\frac{a^2}{4}$

$f''(x) = 2$; concave up on $(-\infty, \infty)$

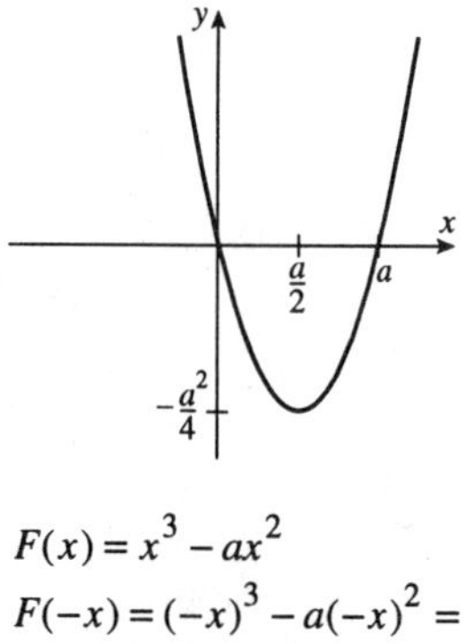

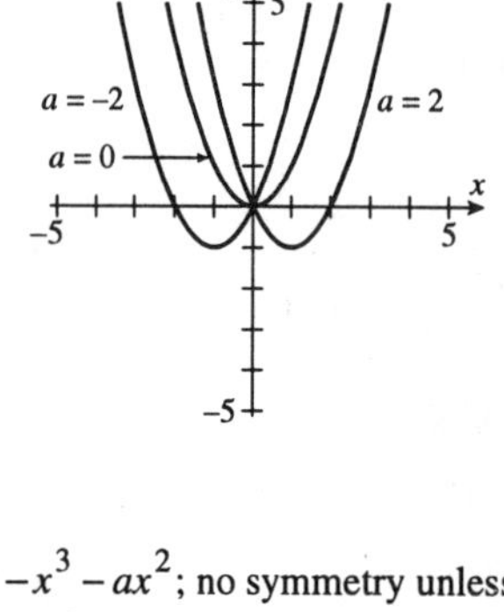

3. $F(x) = x^3 - ax^2$

$F(-x) = (-x)^3 - a(-x)^2 = -x^3 - ax^2$; no symmetry unless $a = 0$

$F(0) = 0$; $0 = x^3 - ax^2$; $0 = x^2(x-a)$; intercepts: $(0, 0)$, $(a, 0)$

$F'(x) = 3x^2 - 2ax$; $x(3x - 2a) = 0$; $x = 0, \frac{2a}{3}$

Increasing: $(-\infty, 0] \cup \left[\frac{2a}{3}, \infty\right)$ if $a > 0$; $\left(-\infty, \frac{2a}{3}\right] \cup [0, \infty)$ if $a < 0$

Decreasing: $\left[0, \frac{2a}{3}\right]$ if $a > 0$; $\left[\frac{2a}{3}, 0\right]$ if $a < 0$

Local maximum: $F(0) = 0$ if $a > 0$, $F\left(\frac{2a}{3}\right) = \frac{8a^3}{27} - \frac{4a^3}{9} = -\frac{4a^3}{27}$ if $a < 0$

Local minimum: $F\left(\frac{2a}{3}\right) = -\frac{4a^3}{27}$ if $a > 0$, $F(0) = 0$ if $a < 0$

$F''(x) = 6x - 2a$; $F''(x) = 0$ when $x = \frac{a}{3}$

Concave up: $\left(\frac{a}{3}, \infty\right)$; concave down: $\left(-\infty, \frac{a}{3}\right)$

Inflection point: $F\left(\frac{a}{3}\right) = -\frac{2a^3}{27}$

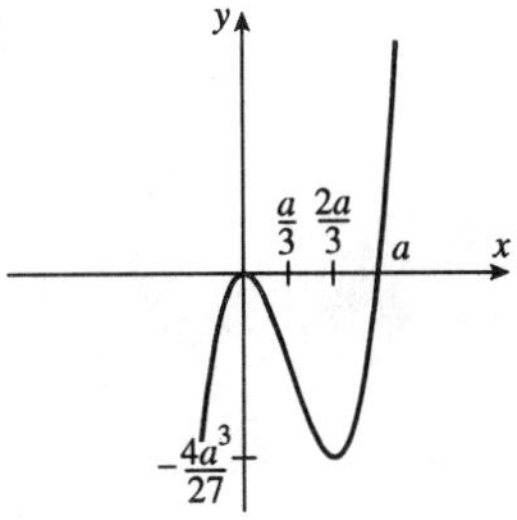

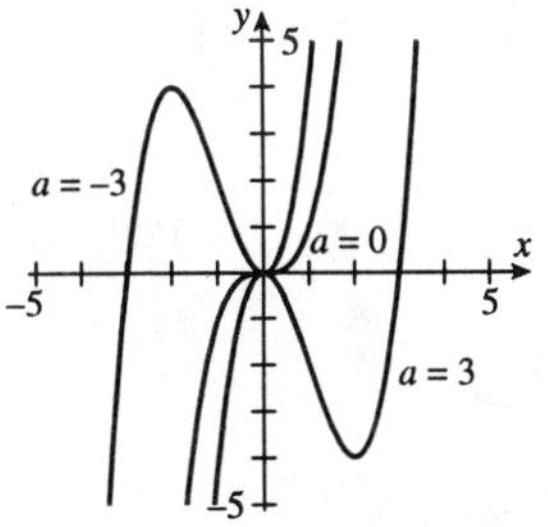

5. $f(x) = Ax^2e^{-kx}$, $A > 0$, $k > 0$

$f(-x) = A(-x)^2 e^{kx} = Ax^2 e^{kx}$; no symmetry

Range: $[0, \infty)$

$f'(x) = Ax^2(-ke^{-kx}) + (e^{-kx})(2Ax) = Ax(2 - kx)e^{-kx}$; $f'(x) = 0$ when $x = 0, \frac{2}{k}$

Increasing: $\left[0, \frac{2}{k}\right]$; decreasing: $(-\infty, 0] \cup \left[\frac{2}{k}, \infty\right)$

Local maximum: $f\left(\frac{2}{k}\right) = \frac{4A}{k^2}e^{-2}$; local minimum: $f(0) = 0$

$f''(x) = A(2 - 4kx + k^2x^2)e^{-kx}$

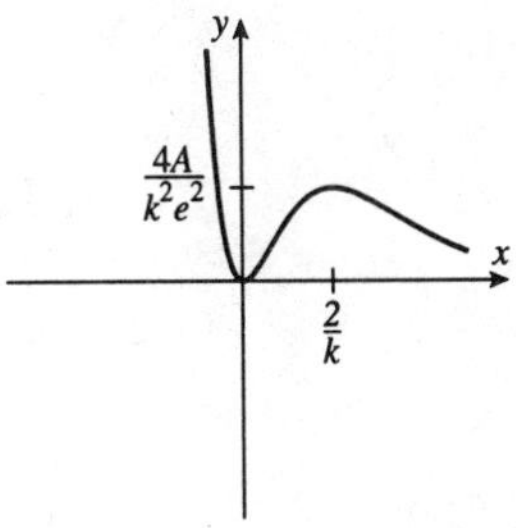

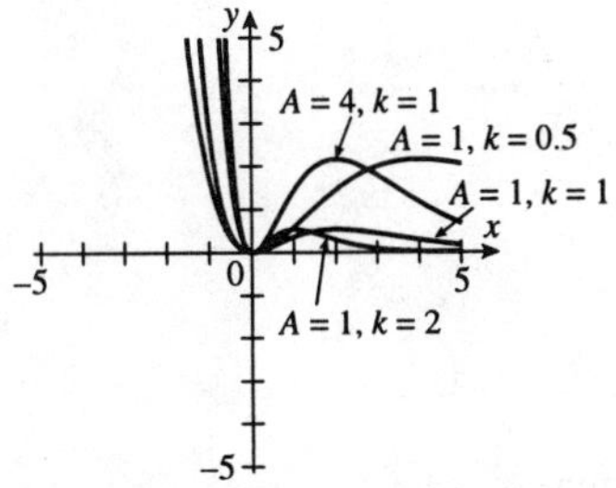

7. $h(x) = \dfrac{x}{x+a}$; domain $(-\infty, -a) \cup (-a, \infty)$; if $a = 0$, $h(x) = 1$; $x \neq 0$

$h(-x) = \dfrac{-x}{-x+a} = \dfrac{x}{x-a}$; no symmetry unless $a = 0$

Intercept (0, 0); asymptotes $x = -a$, $y = 1$

$h'(x) = \dfrac{(x+a)-x}{(x+a)^2} = \dfrac{a}{(x+a)^2}$

If $a > 0$, $h(x)$ is increasing on $(-\infty, -a) \cup (-a, \infty)$.
If $a < 0$, $h(x)$ is decreasing on $(-\infty, -a) \cup (-a, \infty)$.

$h''(x) = \dfrac{-2a}{(x+a)^3}$

If $a > 0$, $h(x)$ is concave up on $(-\infty, -a)$, concave down $(-a, \infty)$.
If $a < 0$, $h(x)$ is concave up on $(-a, \infty)$, concave down $(-\infty, -a)$.

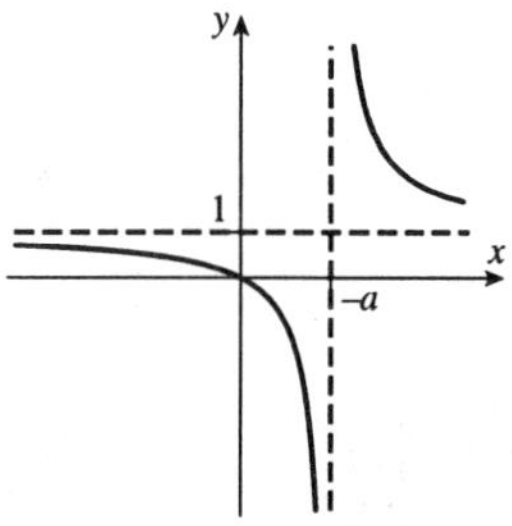

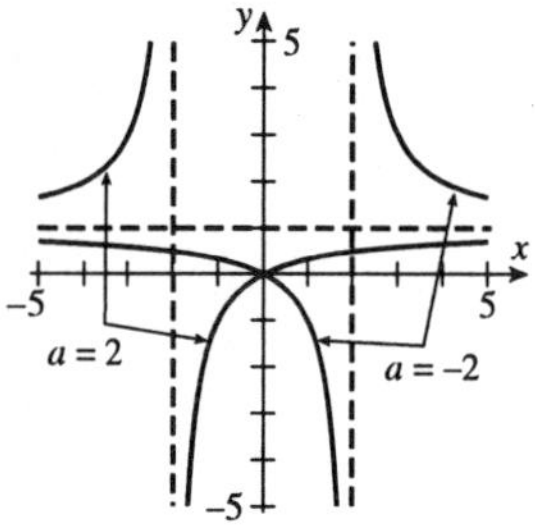

9. $f(x) = e^{-(x-a)^2}$; range $(0, \infty)$

$f(-x) = e^{-(-x-a)^2} = e^{-(x+a)^2}$; no symmetry unless $a = 0$

Intercept: $(0,\ e^{-a^2})$

$f'(x) = (-2x + 2a)e^{-(x-a)^2} = 2(a-x)e^{-(x-a)^2}$; $f'(x) = 0$ when $x = a$

Increasing: $(-\infty, a]$; decreasing: $[a, \infty)$

Local maximum: $f(a) = e^0 = 1$

$f''(x) = 2(2x^2 - 4ax + 2a^2 - 1)e^{-(x-a)^2}$; $f''(x) = 0$ when $x = a + \dfrac{\sqrt{2}}{2},\ a - \dfrac{\sqrt{2}}{2}$

Concave up: $\left(-\infty,\ a - \dfrac{\sqrt{2}}{2}\right) \cup \left(a + \dfrac{\sqrt{2}}{2},\ \infty\right)$

Concave down: $\left(a - \dfrac{\sqrt{2}}{2},\ a + \dfrac{\sqrt{2}}{2}\right)$

Inflection points: $x = a - \dfrac{\sqrt{2}}{2},\ a + \dfrac{\sqrt{2}}{2}$

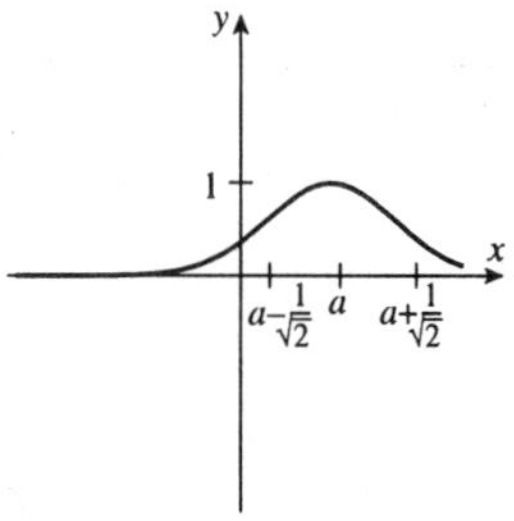

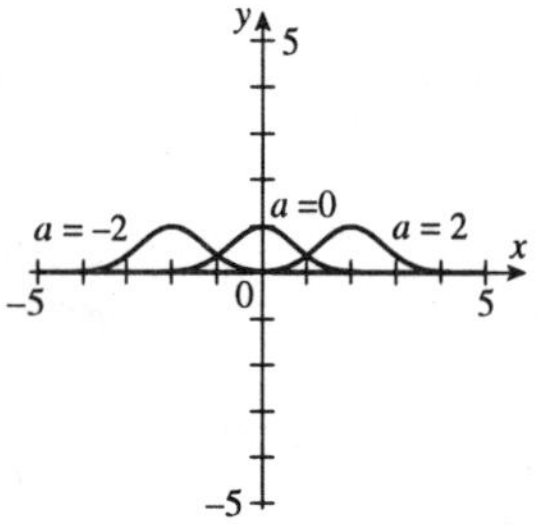

11. $g(x) = 2x\sqrt{x-a}$; domain $[a, \infty)$

$$g'(x) = \frac{2x}{2\sqrt{x-a}} + 2\sqrt{x-a} = \frac{3x-2a}{\sqrt{x-a}};\ g'(x) = 0 \text{ when } x = \frac{2a}{3}$$

$x = \frac{2a}{3}$ is only in the domain of $g(x)$ if $a < 0$.

Increasing: $\left[\frac{2a}{3}, \infty\right)$; decreasing $\left[a, \frac{2a}{3}\right]$

Local minimum: $g\left(\frac{2a}{3}\right) = \frac{4a}{3}\sqrt{-\frac{a}{3}},\ a < 0$

$$g''(x) = \frac{3x-4a}{2(x-a)^{3/2}};\ g''(x) = 0 \text{ when } x = \frac{4a}{3}$$

$x = \frac{4a}{3}$ is only in the domain of $g(x)$ if $a \ge 0$.

Concave down: $\left[a, \frac{4a}{3}\right)$; concave up: $\left(\frac{4a}{3}, \infty\right)$

Inflection point: $g\left(\frac{4a}{3}\right) = \frac{8a}{3}\sqrt{\frac{a}{3}}$, $a \ge 0$.

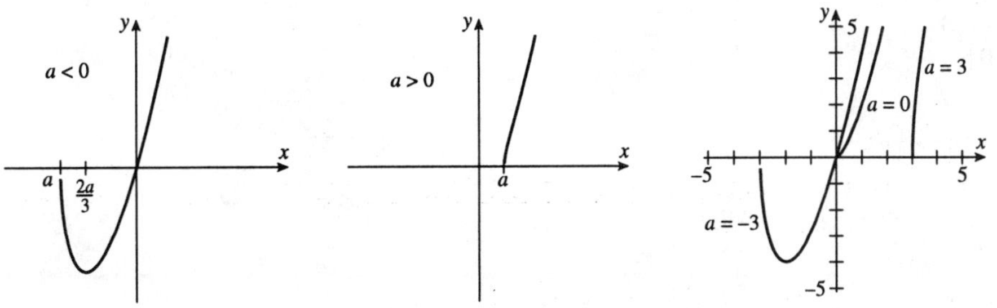

13.

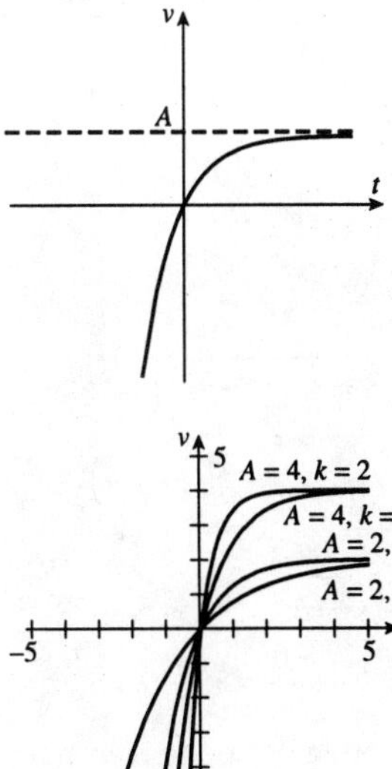

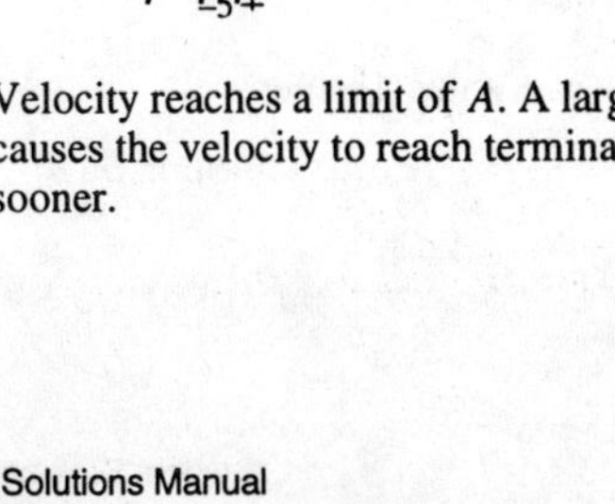

Velocity reaches a limit of A. A larger value of k causes the velocity to reach terminal velocity sooner.

15.

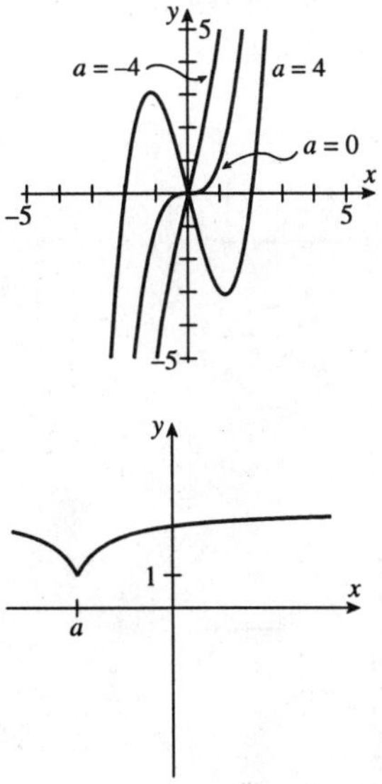

17.

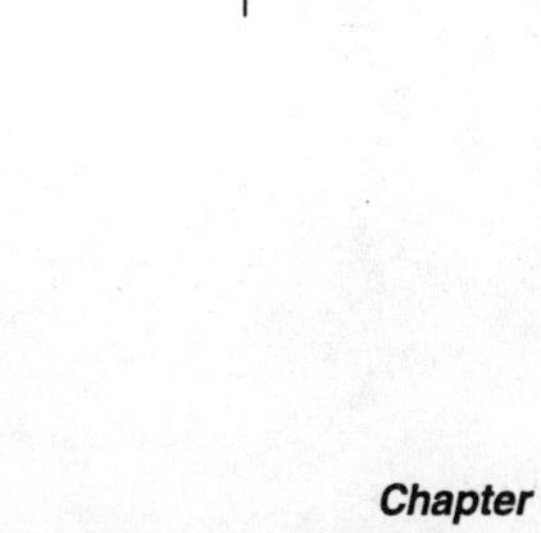

19.

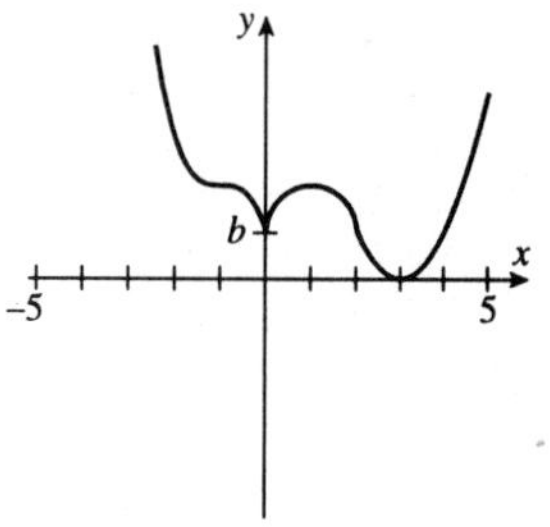

Section 4.7

Concepts Review

1. Continuous; (a, b); $f(b)-f(a)=f'(c)(b-a)$

3. $F(x)=G(x)+C$

Problem Set 4.7

1. $f'(x)=2x+2$
$$\frac{f(2)-f(-2)}{2-(-2)}=\frac{8-0}{4}=2$$
$2c+2=2;\ c=0$

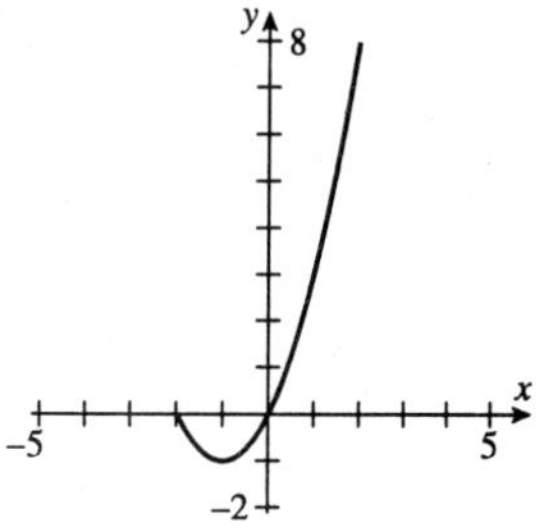

3. $g'(x)=x^2$
$$\frac{g(2)-g(-2)}{2-(-2)}=\frac{\frac{8}{3}-\left(-\frac{8}{3}\right)}{4}=\frac{4}{3}$$
$c^2=\frac{4}{3};\ c=-\frac{2\sqrt{3}}{3},\ \frac{2\sqrt{3}}{3}$

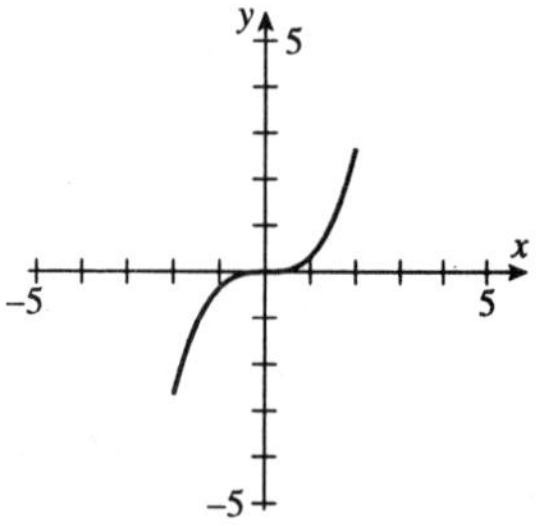

5. The Mean Value Theorem does not apply because $F(t)$ is not continuous on $[-1, 4]$.

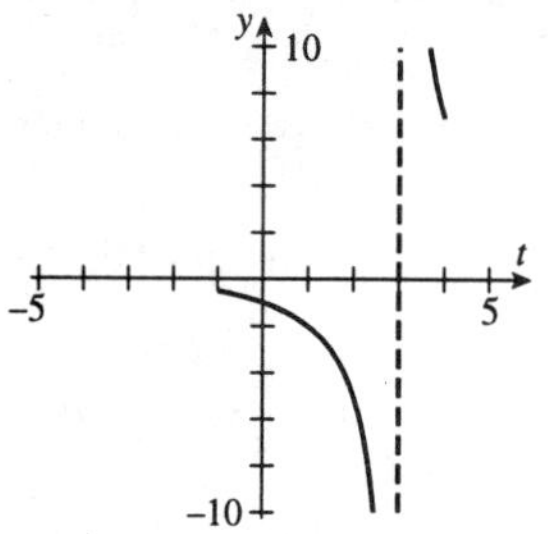

7. $h'(x)=\frac{2}{3}x^{-1/3}$
$$\frac{h(2)-h(0)}{2-0}=\frac{\sqrt[3]{4}-0}{2}=\frac{\sqrt[3]{4}}{2}$$
$\frac{2}{3}c^{-1/3}=\frac{4^{1/3}}{2};\ c=\frac{16}{27}$

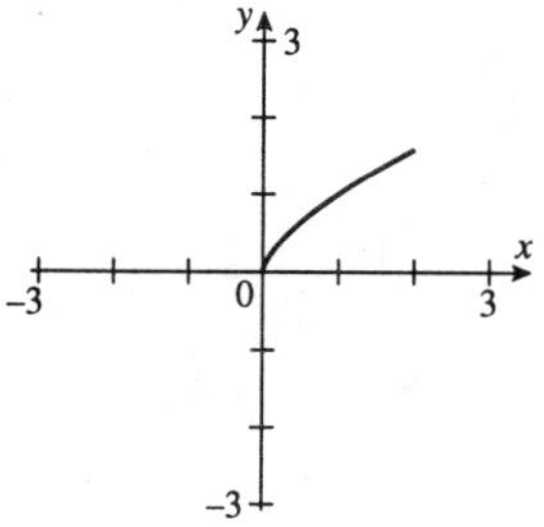

9. The Mean Value Theorem does not apply because $\phi(x)$ is not continuous on $\left[-1, \frac{1}{2}\right]$.

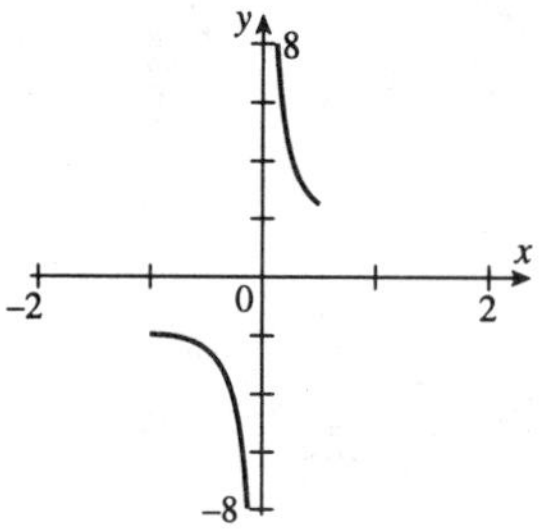

11. $f(t)=55t;\ f'(t)=55$
$$\frac{f(2)-f(0)}{2-0}=\frac{112-0}{2}=56$$
At some time during the trip, Johnny must have gone 56 miles per hour.

13. By the Mean Value Theorem
$\frac{f(b)-f(a)}{b-a}=f'(c)$ for some c in (a, b).
Since $f(b)=f(a)$, $\frac{0}{b-a}=f'(c);\ f'(c)=0$.

15. By the Monotonicity Theorem, f is increasing on the intervals (a, x_0) and (x_0, b).
To show that $f(x_0) > f(x)$ for x in (a, x_0), consider f on the interval $(a, x_0]$.
f satisfies the conditions of the Mean Value Theorem on the interval $[x, x_0]$ for x in (a, x_0).
So for some c in (x, x_0),
$f(x_0) - f(x) = f'(c)(x_0 - x)$.
Because $f'(c) > 0$ and $x_0 - x > 0$,
$f(x_0) - f(x) > 0$, so $f(x_0) > f(x)$.
Similar reasoning shows that $f(x) > f(x_0)$ for x in (x_0, b).
Therefore, f is increasing on (a, b).

17. $F'(x) = 0$ and $G(x) = 0$, $G'(x) = 0$.
By Theorem B,
$F(x) = G(x) + C$: $F(x) = 0 + C = C$.

19. Let $G(x) = Dx$; $F'(x) = D$ and $G'(x) = D$.
By Theorem B, $F(x) = G(x) + C$; $F(x) = Dx + C$.

21. Suppose there is more than one zero between successive distinct zeros of f'. That is, there are a and b such that $f(a) = f(b) = 0$ with a and b between successive distinct zeros of f'. Then by Rolle's Theorem, there is a c between a and b such that $f'(c) = 0$. This contradicts the supposition that a and b lie between successive distinct zeros.

23. By applying the Mean Value Theorem and taking the absolute value of both sides

$$\frac{|f(x_2) - f(x_1)|}{|x_2 - x_1|} = |f'(c)|, \text{ for some } c \text{ in } (x_1, x_2).$$

Since $|f'(x)| \le M$ for all x in (a, b),

$$\frac{|f(x_2) - f(x_1)|}{|x_2 - x_1|} \le M;$$

$$|f(x_2) - f(x_1)| \le M|x_2 - x_1|.$$

25. a.

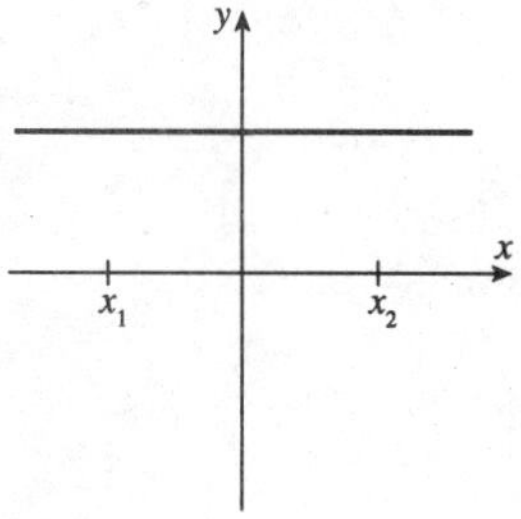

b.

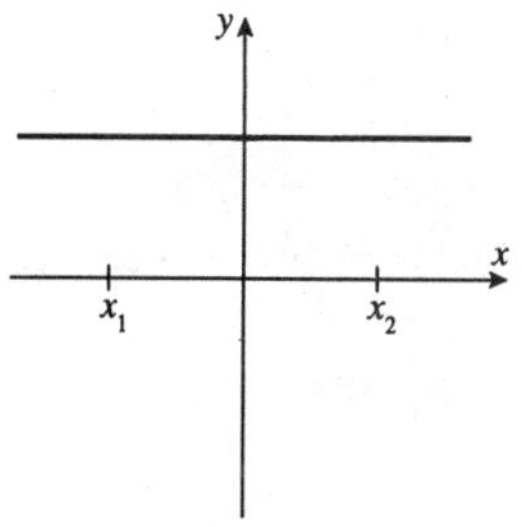

27. $[f(x)^2]' = 2f(x)f'(x)$
Because $f(x) \ge 0$ and $f'(x) \ge 0$ on I, $[f(x)^2]' \ge 0$ on I. As a consequence of the Mean Value Theorem, $f(x_2)^2 - f(x_1)^2 \ge 0$ for all $x_2 > x_1$ on I. Therefore f^2 is nondecreasing.

29. Let $f(x) = \sqrt{x}$ so $f'(x) = \frac{1}{2\sqrt{x}}$. Apply the Mean Value Theorem to f on the interval $[x, x+2]$ for $x > 0$. Thus $\sqrt{x+2} - \sqrt{x} = \frac{1}{2\sqrt{c}}(2) = \frac{1}{\sqrt{c}}$ for some c in $(x, x+2)$. Observe $\frac{1}{\sqrt{x+2}} < \frac{1}{\sqrt{c}} < \frac{1}{\sqrt{x}}$. Thus as $x \to \infty$, $\frac{1}{\sqrt{c}} \to 0$.
Therefore $\lim_{x\to\infty}\left(\sqrt{x+2} - \sqrt{x}\right) = \lim_{x\to\infty} \frac{1}{\sqrt{c}} = 0$.

31. Let d be the difference in distance between horse A and horse B as a function of time t.
Then d' is the difference in speeds.
Let t_0 and t_1 and be the start and finish times of the race.
$d(t_0) = d(t_1) = 0$
By the Mean Value Theorem,

$$\frac{d(t_1) - d(t_0)}{t_1 - t_0} = d'(c) \text{ for some } c \text{ in } (t_0, t_1).$$

Therefore $d'(c) = 0$ for some c in (t_0, t_1).

33. Suppose $x > c$. Then by the Mean Value Theorem, $f(x) - f(c) = f'(a)(x - c)$ for some a in (c, x).
Since f is concave up, $f'' > 0$ and by the Monotonicity Theorem f' is increasing.
Therefore $f'(a) > f'(c)$ and
$f(x) - f(c) = f'(a)(x - c) > f'(c)(x - c)$
$f(x) > f(c) + f'(c)(x - c)$, $x > c$
Suppose $x < c$. Then by the Mean Value Theorem, $f(c) - f(x) = f'(a)(c - x)$ for some a in (x, c).
Since f is concave up, $f'' > 0$, and by the Monotonicity Theorem f' is increasing.
Therefore, $f'(c) > f'(a)$ and
$f(c) - f(x) = f'(a)(c - x) < f'(c)(c - x)$.

$-f(x) < -f(c) + f'(c)(c - x)$

$f(x) > f(c) - f'(c)(c - x)$

$f(x) > f(c) + f'(c)(x - c), \ x < c$

Therefore $f(x) > f(c) + f'(c)(x - c), \ x \neq c$.

Section 4.8 Chapter Review

Concepts Test

1. True. Max-Min Existence Theorem

3. True. For example, let $f(x) = \sin x$.

5. True. $f'(x) = 18x^5 + 16x^3 + 4x$;

   $f''(x) = 90x^4 + 48x^2 + 4$, which is greater than zero for all x.

7. True. When $f'(x) > 0$, $f(x)$ is increasing.

9. True. $f(x) = ax^2 + bx + c$;

   $f'(x) = 2ax + b$; $f''(x) = 2a$

11. False. $f(x)$ is decreasing if $b < 0$.

13. True. The asymptote occurs at $y = \frac{a}{-1} = -a$.

15. True. The function is differentiable on (0, 2).

17. True. $\frac{dy}{dx} = \cos x$; $\frac{d^2y}{dx^2} = -\sin x$; $-\sin x = 0$ has infinitely many solutions.

19. True. $\frac{dy}{dx} = \cos x \, e^{\sin x}$;

   $\frac{d^2y}{dx^2} = \cos^2 x \, e^{\sin x} - \sin x \, e^{\sin x}$; $\cos^2 x = \sin x$ has infinitely many solutions.

21. True. By the Mean Value Theorem, the derivative must be zero between each pair of distinct x-intercepts.

23. False. Let $f(x) = g(x) = 2x$, $f'(x) > 0$ and $g'(x) > 0$ for all x, but $f(x)g(x) = 4x^2$ is decreasing on $(-\infty, 0)$.

25. True. If the function is nondecreasing, $f'(x)$ must be greater than or equal to zero, and if $f'(x) \geq 0$, f is nondecreasing. This can be seen using the Mean Value Theorem.

27. False. For example, let $f(x) = e^x$.

29. True. There are at most two stationary points, only one of which can be a maximum.

Sample Test Problems

1. $f'(x) = -\frac{2}{x^3}$; no stationary points

   Critical points: $x = -2, -\frac{1}{2}$

   Minimum: $f(-2) = \frac{1}{4}$, maximum: $f\left(-\frac{1}{2}\right) = 4$

3. $f'(x) = 12x^3 - 12x^2 = 12x^2(x - 1)$;

   $f'(x) = 0$ when $x = 0, 1$

   Critical points: $x = -2, 0, 1, 3$

   $f(-2) = 80, f(0) = 0, f(1) = -1, f(3) = 135$

   Minimum: $f(1) = -1$, maximum: $f(3) = 135$

5. $f'(x) = 10x^4 - 20x^3 + 7$;

   $f'(x) = 0$ when $x \approx 0.847, 1.90$

   Critical points: $x = -1, 0.847, 1.90, 3$

   $f(-1) = -14, f(0.847) \approx 4.23, f(1.90) \approx -2.34,$ $f(3) = 102$

   Minimum: $f(0) = -14$, maximum: $f(3) = 102$

7. $f'(x) = 3x^2 - 3$; $f'(x) = 0$ when $x = \pm 1$

   Increasing: $(-\infty, -1] \cup [1, \infty)$

   $f''(x) = 6x$

   Concave down: $(-\infty, 0)$

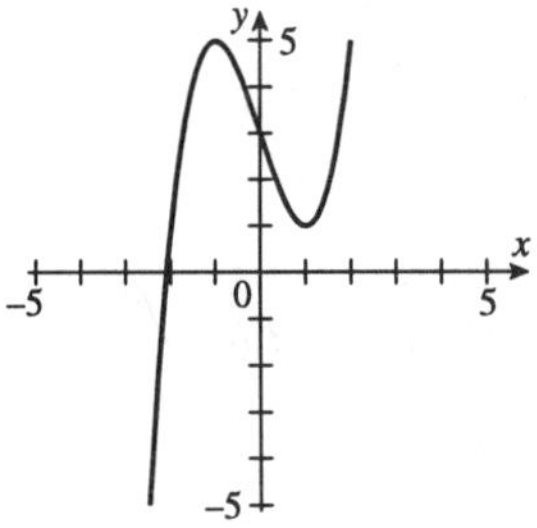

9. $f'(x) = 1 + 4x^3 - 20x^4$;

   $f'(x) = 0$ when $x \approx -0.43, 0.53$

   Increasing: $[-0.43, 0.53]$

   $f''(x) = 12x^2 - 80x^3$;

   $f''(x) = 0$ when $x = 0, 0.15$

   Concave down: $(0.15, \infty)$

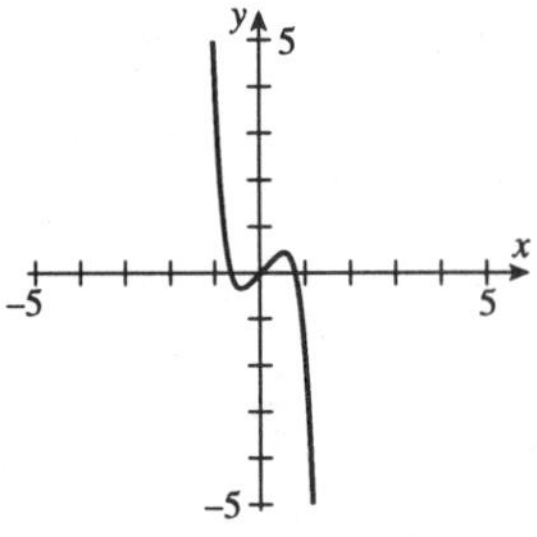

11. $f(x) = x^2(x-a),\ a > 0$

$f'(x) = x^2 + (x-a)(2x) = 3x^2 - 2ax;$

$f'(x) = 0$ when $x = 0,\ \dfrac{2a}{3}$

Increasing: $(-\infty,\ 0] \cup \left[\dfrac{2a}{3},\ \infty\right)$

Decreasing: $\left[0,\ \dfrac{2a}{3}\right]$; local maximum $f(0) = 0$,

local minimum $f\left(\dfrac{2a}{3}\right) = -\dfrac{4a^3}{27}$

$f''(x) = 6x - 2a;\ f''(x) = 0$ when $x = \dfrac{a}{3}$

Inflection point: $f\left(\dfrac{a}{3}\right) = -\dfrac{2a^3}{27}$

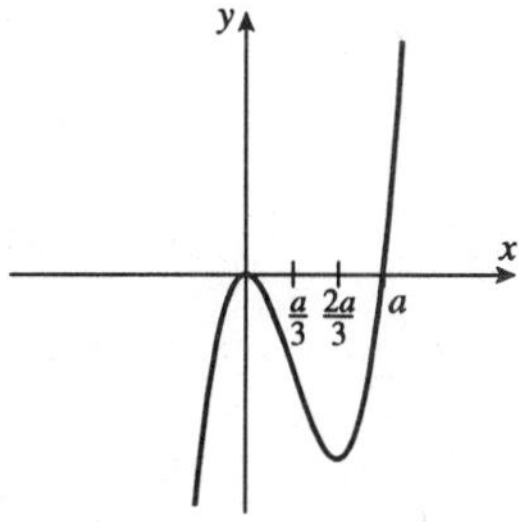

13. $f(x) = x^4 - 32x;\ f'(x) = 4x^3 - 32;$

$f'(x) = 0$ when $x = 2$

Local minimum $f(2) = -48$

$f''(x) = 12x^2;\ f''(x) > 0$ except at $x = 0$,

so there is no inflection point.

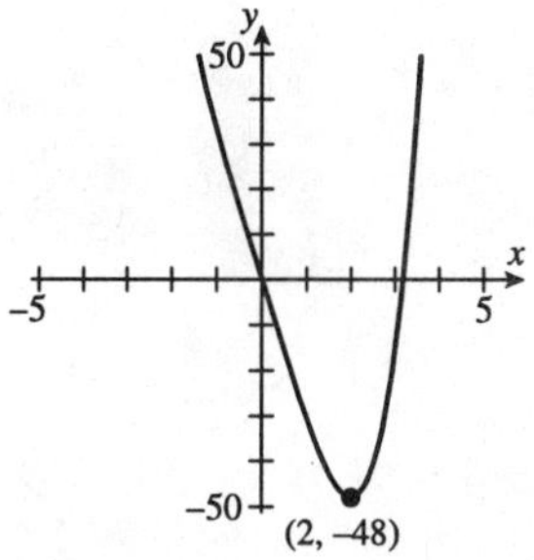

15. $f(x) = x \ln(x-3)$; domain $(3, \infty)$

$f'(x) = x \cdot \left(\dfrac{1}{x-3}\right) + \ln(x-3) = \dfrac{x}{x-3} + \ln(x-3)$

$f''(x) = \dfrac{x-6}{(x-3)^2};\ f''(x) = 0$ when $x = 6$

Inflection point: $f(6) \approx 6.6$

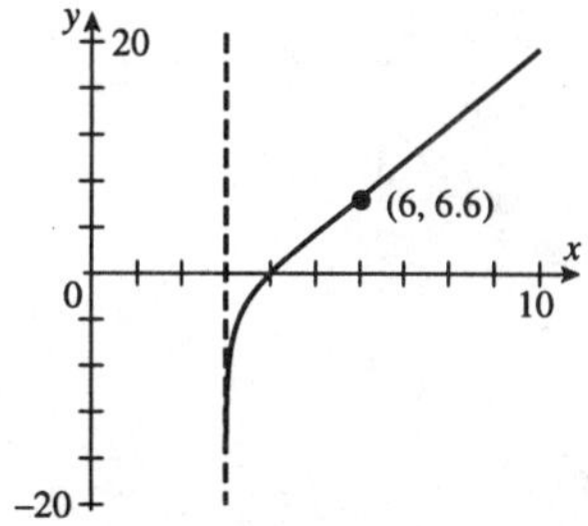

17. $f(x) = 3x^4 - 4x^3$

$f'(x) = 12x^3 - 12x^2;\ 12x^2(x-1) = 0;\ x = 0,\ 1$

$f''(x) = 36x^2 - 24x;\ f''(x) = 0$ when $x = 0,\ \dfrac{2}{3}$

$f''(0) = 0,\ f''(1) = 12;\ f(1) = -1$ is a local minimum.

Inflection points: $f(0) = 0,\ f\left(\dfrac{2}{3}\right) \approx -0.593$

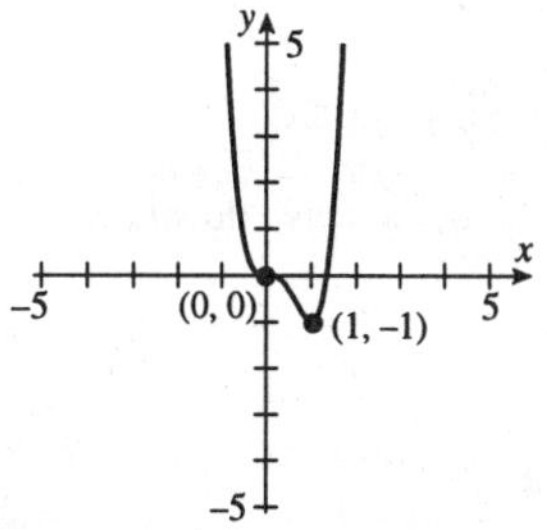

19.

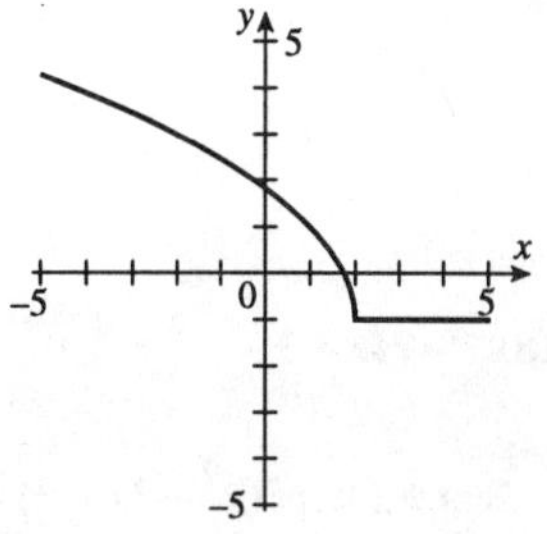

21. Let x be the length of a turned up side and let l be the (fixed) length of the sheet of metal.

$V = x(16-2x)l = 16xl - 2x^2 l$

$\dfrac{dV}{dx} = 16l - 4xl;\ V' = 0$ when $x = 4$

$\dfrac{d^2V}{dx^2} = -4l;\ x = 4$ is a maximum.

23. Let x be the width and y the height of a page. $A = xy$. Because of the margins,

$(y-4)(x-3) = 27$ or $y = \dfrac{27}{x-3} + 4$

$A = \dfrac{27x}{x-3} + 4x;$

$$\frac{dA}{dx}=\frac{(x-3)(27)-27x}{(x-3)^2}+4=-\frac{81}{(x-3)^2}+4$$

$\frac{dA}{dx}=0$ when $x=-\frac{3}{2}, \frac{15}{2}$

$x=\frac{15}{2}$; $y=10$

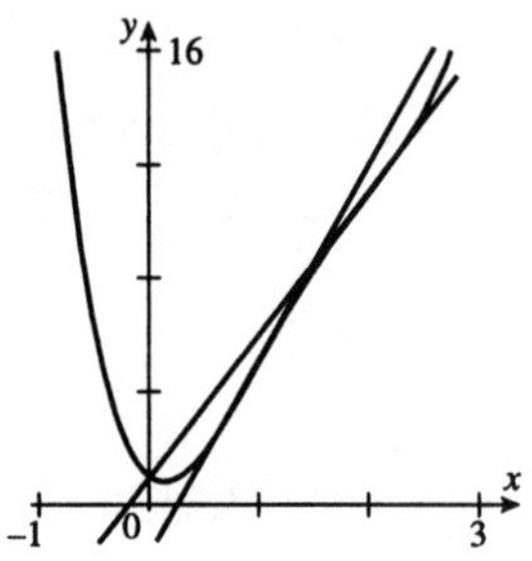

25. $f'(x)=\begin{cases}\frac{x}{2}+\frac{3}{2} & \text{if } -2\le x\le 0\\ -\frac{x+2}{3} & \text{if } 0\le x\le 2\end{cases}$

$\frac{x}{2}+\frac{3}{2}=0$; $x=-3$, which is not in the domain.

$-\frac{x+2}{3}=0$; $x=-2$, which is not in the domain.

Critical numbers: $x=-2, 0, 2$

$f(-2)=0, f(0)=2, f(2)=0$

Minima $f(-2)=0, f(2)=0$, maximum $f(0)=2$.

$f''(x)=\begin{cases}\frac{1}{2} & \text{if } -2\le x\le 0\\ -\frac{1}{3} & \text{if } 0\le x\le 2\end{cases}$

Concave up [–2, 0), concave down (0, 2]

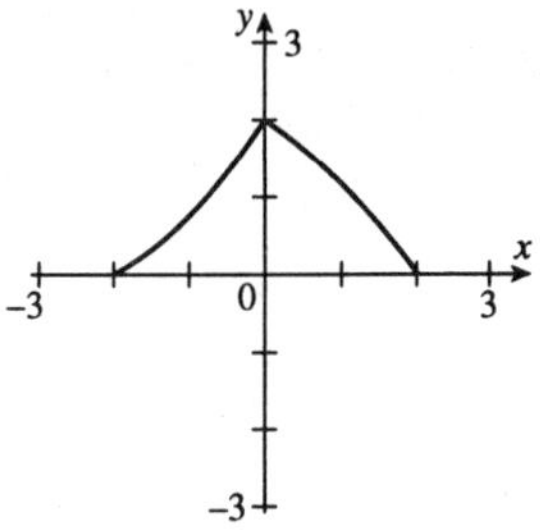

27. $\frac{dy}{dx}=4x^3-18x^2+24x-3$

$\frac{d^2y}{dx^2}=12x^2-36x+24$; $12(x^2-3x+2)=0$;

$x=1, 2$

Inflection points: $x=1, y=5$ and $x=2, y=11$

Slope at $x=1$: $\left.\frac{dy}{dx}\right|_{x=1}=7$

Tangent line: $y-5=7(x-1)$; $y=7x-2$

Slope at $x=2$: $\left.\frac{dy}{dx}\right|_{x=2}=5$

Tangent line: $y-11=5(x-2)$; $y=5x+1$

Chapter 5

Section 5.1

Concepts Review

1. rx^{r-1}; $\frac{x^{r+1}}{r+1}$, $r \neq -1$

3. $\frac{x^4+3x^2+1}{9}$

Problem Set 5.1

1. $\int 4\,dx = 4x + C$

3. $\int\left(3x^2+\sqrt{2}\right)dx = x^3+\sqrt{2}x+C$

5. $\int x^{2/3}dx = \frac{3}{5}x^{5/3}+C$

7. $\int(6x^2-6x+1)dx = 2x^3-3x^2+x+C$

9. $\int(3e^x+x^3)dx = 3e^x+\frac{1}{4}x^4+C$

11. $\int\left(\frac{4}{x^5}-\frac{3}{x}\right)dx = \int(4x^{-5}-3x^{-1})dx$
$= -x^{-4}-3\ln|x|+C$

13. $\int\frac{4x^6+3x^5-8}{x^5}dx = \int(4x+3-8x^{-5})dx$
$= 2x^2+3x+2x^{-4}+C$

15. $\int\left(x^3+\sqrt{x}\right)dx = \frac{1}{4}x^4+\frac{2}{3}x^{3/2}+C$

17. $\int\left(3e^y-\frac{2}{y^2}+\frac{2}{y}\right)dy = \int(3e^y-2y^{-2}+2y^{-1})dy$
$= 3e^y+2y^{-1}+2\ln|y|+C$

19. $\int\frac{e^x-2e^{2x}+1}{e^x}dx = \int(1-2e^x+e^{-x})dx$
$= x-2e^x-e^{-x}+C$

21. $\int(3\sin t-2\cos t)dt = -3\cos t-2\sin t+C$

23. Guess: $(3x+1)^5+C$
Check: $\frac{d}{dx}(3x+1)^5 = 15(3x+1)^4$
New guess: $\frac{1}{5}(3x+1)^5+C$
Check: $\frac{d}{dx}\frac{1}{5}(3x+1)^5 = 3(3x+1)^4$
$\int(3x+1)^5 3dx = \frac{1}{5}(3x+1)^5+C$

25. Guess: $(5x^3-18)^8+C$
Check: $\frac{d}{dx}(5x^3-18)^8 = 120x^2(5x^3-18)^7$
New guess: $\frac{1}{8}(5x^3-18)^8+C$
Check: $\frac{d}{dx}\frac{1}{8}(5x^3-18)^8 = 15x^2(5x^3-18)^7$
$\int(5x^3-18)^7 15x^2\,dx = \frac{1}{8}(5x^3-18)^8+C$

27. Guess: $e^{2x^5+9}+C$
Check: $\frac{d}{dx}e^{2x^5+9} = 10x^4e^{2x^5+9}$
New guess: $\frac{3}{10}e^{2x^5+9}+C$
Check: $\frac{d}{dx}\frac{3}{10}e^{2x^5+9} = 3x^4e^{2x^5+9}$
$\int 3x^4e^{2x^5+9}dx = \frac{3}{10}e^{2x^5+9}+C$

29. Guess: $\ln(5x^3+3x^2-8)+C$
Check: $\frac{d}{dx}\ln(5x^3+3x^2-8)$
$= \frac{1}{5x^3+3x^2-8}(15x^2+6x)$
$= \frac{15x^2+6x}{5x^3+3x^2-8}$
New guess: $\frac{1}{3}\ln(5x^3+3x^2-8)+C$
Check: $\frac{d}{dx}\frac{1}{3}\ln(5x^3+3x^2-8) = \frac{5x^2+2x}{5x^3+3x^2-8}$
$\int\frac{5x^2+2x}{5x^3+3x^2-8}dx = \frac{1}{3}\ln(5x^3+3x^2-8)+C$

31. $\sqrt[3]{2t^2-11}=(2t^2-11)^{1/3}$

Guess: $(2t^2-11)^{4/3}+C$

Check: $\frac{d}{dt}(2t^2-11)^{4/3}=\frac{4}{3}(2t^2-11)^{1/3}\cdot 4t$

$=\frac{16}{3}t\sqrt[3]{2t^2-11}$

New guess: $\frac{9}{16}(2t^2-11)^{4/3}+C$

Check: $\frac{d}{dt}\frac{9}{16}(2t^2-11)^{4/3}=3t\sqrt[3]{2t^2-11}$

$\int 3t\sqrt[3]{2t^2-11}\,dt=\frac{9}{16}(2t^2-11)^{4/3}+C$

33. Guess: $\sin^5 x+C$

Check: $\frac{d}{dx}\sin^5 x=5\sin^4 x\cos x$

$\int \sin^4 x\cos x\,dx=\frac{1}{5}\sin^5 x+C$

35. Guess: $\sin^6 x^2+C$

Check: $\frac{d}{dx}\sin^6 x^2=6\sin^5 x^2\cdot\cos x^2\cdot 2x$

$=12(\sin^5 x^2)(x\cos x^2)$

$\int(\sin^5 x^2)(x\cos x^2)dx=\frac{1}{12}\sin^6 x^2+C$

37. Guess: $(x^2+1)^4+C$

Check: $\frac{d}{dx}(x^2+1)^4=8x(x^2+1)^3$

This method does not work.

$\int(x^2+1)^3x^2dx=\int(x^8+3x^6+3x^4+x^2)dx$

$=\frac{1}{9}x^9+\frac{3}{7}x^7+\frac{3}{5}x^5+\frac{1}{3}x^3+C$

39. $f'(x)=\int(3x+1)dx=\frac{3}{2}x^2+x+C_1$

$f(x)=\int\left(\frac{3}{2}x^2+x+C_1\right)dx$

$=\frac{1}{2}x^3+\frac{1}{2}x^2+C_1x+C_2$

41. $f'(x)=\int x^{1/2}dx=\frac{2}{3}x^{3/2}+C_1$

$f(x)=\int\left(\frac{2}{3}x^{3/2}+C_1\right)dx$

$=\frac{4}{15}x^{5/2}+C_1x+C_2$

43. $f'(x)=\int x^{-2}\,dx=-x^{-1}+C_1$

$f(x)=\int(-x^{-1}+C_1)dx$

$=-\ln|x|+C_1x+C_2$

45. The Product Rule for derivatives says $\frac{d}{dx}[f(x)g(x)]=f(x)g'(x)+f'(x)g(x)$. Thus $\int[f(x)g'(x)+f'(x)g(x)]dx=f(x)g(x)+C$.

47. Let $f(x)=x^2,\ g(x)=\sqrt{x-1}$.

$f'(x)=2x,\ g'(x)=\frac{1}{2\sqrt{x-1}}$

$\int\left[\frac{x^2}{2\sqrt{x-1}}+2x\sqrt{x-1}\right]dx=x^2\sqrt{x-1}+C$

49. $\frac{d}{dx}\left[\frac{\sqrt{x^4-1}}{x}\right]=\frac{-\sqrt{x^4-1}}{x^2}+\frac{4x^3}{2x\sqrt{x^4-1}}$

$=\frac{-(x^4-1)+2x^4}{x^2\sqrt{x^4-1}}=\frac{x^4+1}{x^2\sqrt{x^4-1}}$

Thus, $\int\frac{x^4+1}{x^2\sqrt{x^4-1}}dx=\frac{\sqrt{x^4-1}}{x}+C.$

51. $\int f''(x)dx=\int\frac{d}{dx}f'(x)dx=f'(x)+C$

$f'(x)=\sqrt{x^3+1}+\frac{3x^3}{2\sqrt{x^3+1}}=\frac{5x^3+2}{2\sqrt{x^3+1}}$ so

$\int f''(x)dx=\frac{5x^3+2}{2\sqrt{x^3+1}}+C.$

53. The Product Rule for derivatives says that

$\frac{d}{dx}[f^m(x)g^n(x)]=f^m(x)[g^n(x)]'+[f^m(x)]'g^n(x)$

$=f^m(x)[ng^{n-1}(x)g'(x)]+[mf^{m-1}(x)f'(x)]g^n(x)$

$=f^{m-1}(x)g^{n-1}(x)[nf(x)g'(x)+mg(x)f'(x)]$. Thus

$\int f^{m-1}(x)g^{n-1}(x)[nf(x)g'(x)+mg(x)f'(x)]dx=f^m(x)g^n(x)+C.$

55. If $x > 0$, then $|x| = x$ and $\int |x| dx = \frac{1}{2}x^2 + C$.

If $x < 0$, then $|x| = -x$ and $\int |x| dx = -\frac{1}{2}x^2 + C$.

$$\int |x| dx = \begin{cases} \frac{1}{2}x^2 + C & \text{if } x > 0 \\ -\frac{1}{2}x^2 + C & \text{if } x < 0 \end{cases}$$

57. a. $F_1(x) = -x\cos x + \sin x$
$F_2(x) = -2\cos x - x\sin x$
$F_3(x) = x\cos x - 3\sin x$
$F_4(x) = 4\cos x + x\sin x$

b. $F_{16}(x) = 16\cos x + x\sin x$

Section 5.2

Concepts Review

1. Differential equation

3. Separate variables

Problem Set 5.2

1. $\frac{1}{2}(4 - x^2)^{-1/2}(-2x) + \frac{x}{(4 - x^2)^{1/2}} = 0$

3. $(-C_1 \sin x - C_2 \cos x) + (C_1 \sin x + C_2 \cos x) = 0$

5. $\int dy = \int (3x^2 + 1) dx$
$y = x^3 + x + C$
$C = 4 - (1^3 + 1) = 2$

7. $\int 2y\, dy = \int x\, dx$
$y^2 = \frac{1}{2}x^2 + C$
$C = 3^2 - \frac{1}{2}(2^2) = 7$

9. $\int \frac{1}{y} dy = \int 3dt$
$\ln|y| = 3t + C$
$C = \ln 2 - 3(0) = \ln 2$
$\ln y = 3t + \ln 2$
$y = 2e^{3t}$

11. $\int ds = \int (3t^2 + 4t - 1) dt$
$s = t^3 + 2t^2 - t + C$
$C = 5 - (2^3 + 2 \cdot 2^2 - 2) = -9$

13. $\int e^{-y} dy = \int (2x + 1)^4 dx$
$-e^{-y} = \frac{1}{10}(2x + 1)^5 + C$
$C = -e^{-6} - \frac{1}{10}(2 \cdot 0 + 1)^5 = -\left(e^{-6} + \frac{1}{10}\right)$
$y = -\ln\left[e^{-6} + \frac{1}{10} - \frac{1}{10}(2x + 1)^5\right]$

15. $\frac{dy}{dx} = 4x$
$\int dy = \int 4x\, dx$
$y = 2x^2 + C$
$C = 2 - 2(1^2) = 0$

17. $a = \frac{dv}{dt} = t$
$\int dv = \int t\, dt$
$v = \frac{ds}{dt} = \frac{1}{2}t^2 + v_0 = \frac{1}{2}t^2 + 2$
$\int ds = \int \left(\frac{1}{2}t^2 + 2\right) dt$
$s = \frac{1}{6}t^3 + 2t + s_0 = \frac{1}{6}t^3 + 2t$

19. $a = \frac{dv}{dt} = \sin 2t$
$\int dv = \int (\sin 2t) dt$
$v = \frac{ds}{dt} = -\frac{1}{2}\cos 2t + C = -\frac{1}{2}\cos 2t + \frac{1}{2} + v_0$
$= -\frac{1}{2}\cos 2t + \frac{1}{2}$
$\int ds = \int \left(-\frac{1}{2}\cos 2t + \frac{1}{2}\right) dt$
$s = -\frac{1}{4}\sin 2t + \frac{1}{2}t + s_0 = -\frac{1}{4}\sin 2t + \frac{1}{2}t + 10$

21. $v = -32t + 96$
$v = 0$ at $t = 3$
At $t = 3$, $s = -16(3^2) + 96(3) + 0 = 144$ ft.

23. $\frac{dv}{dt} = -5.28$
$\int dv = -\int 5.28 dt$
$v = \frac{ds}{dt} = -5.28t + v_0$
$\int ds = \int (-5.28t + v_0) dt$
$s = -2.64t^2 + v_0 t + s_0$
For $s_0 = 1000$, $v_0 = 56$, and $t = 4.5$,
$v = 32.24$ ft/s and $s = 1198.54$ ft

25. $\frac{dV}{dt} = -kS$

Since $V = \frac{4}{3}\pi r^3$ and $S = 4\pi r^2$,

$4\pi r^2 \frac{dr}{dt} = -k4\pi r^2$ and $\frac{dr}{dt} = -k.$

$\int dr = -\int k\,dt$

$r = -kt + C$

$2 = -k(0) + C$ and $0.5 = -k(10) + C.$

$C = 2,\ k = \frac{3}{20}$

27. $v_{esc} = \sqrt{2gR}$

For the Moon, $v_{esc} \approx \sqrt{2(0.165)(32)(1080 \cdot 5280)}$
≈ 7760 ft/s ≈ 1.470 mi/s.
For Venus, $v_{esc} \approx \sqrt{2(0.85)(32)(3800 \cdot 5280)}$
$\approx 33{,}038$ ft/s ≈ 6.257 mi/s.
For Jupiter, $v_{esc} \approx 194{,}369$ ft/s ≈ 36.812 mi/s.
For the Sun, $v_{esc} \approx 2{,}021{,}752$ ft/s ≈ 382.908 mi/s.

29. $a = \frac{dv}{dt} = \frac{\Delta v}{\Delta t} = \frac{60-45}{10} = 1.5$ mi/h/s

31. For the first 10 s, $a = \frac{dv}{dt} = 6t,\ v = 3t^2$, and $s = t^3$. So $v(10) = 300$ and $s(10) = 1000$. After 10 s, $a = \frac{dv}{dt} = -10$, $v = -10(t - 10) + 300$, and $s = -5(t-10)^2 + 300(t-10) + 1000$. $v = 0$ at $t = 40$, at which time $s = 5500$ m.

33. a.

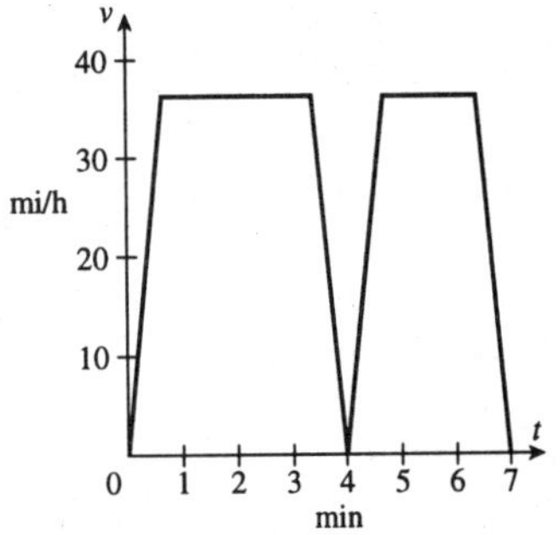

b. Since the trip that involves 1 min more travel time at v_m is 0.6 mi longer,

$v_m = 0.6$ mi/min
$= 36$ mi/h.

c. From part b, $v_m = 0.6$ mi/min. Note that the average speed during acceleration and decleration is $\frac{v_m}{2} = 0.3$ mi/min. Let t be the time spent between stop C and stop D at the constant speed v_m, so $0.6t + 0.3(4 - t) = 2$ miles. Therefore, $t = 2\frac{2}{3}$ min and the time spent accelerating is $\frac{4 - 2\frac{2}{3}}{2} = \frac{2}{3}$ min.

$a = \frac{0.6 - 0}{\frac{2}{3}} = 0.9 \text{ mi / min}^2.$

35. a. $\frac{dV}{dt} = C_1\sqrt{h}$ where h is the depth of the water. Here, $V = \pi r^2 h = 100h$, so $h = \frac{V}{100}$.

Hence $\frac{dV}{dt} = C_1\frac{\sqrt{V}}{10}$; $V(0) = 1600,\ V(40) = 0.$

b. $\int 10V^{-1/2}dV = \int C_1 dt;\ 20\sqrt{V} = C_1 t + C_2;$

$C_2 = 20 \cdot 40 = 800;\ C_1 = -\frac{800}{40} = -20$

$V = \frac{1}{400}(-20t + 800)^2$

c. $V(10) = \frac{1}{400}(-200 + 800)^2 = 900 \text{ cm}^3$

37. Initially, $v = -32t$ and $s = -16t^2 + 16$. $s = 0$ when $t = 1$. Later, the ball falls 9 ft in a time given by $0 = -16t^2 + 9$, or $\frac{3}{4}$ s, and on impact has a velocity of $-32\left(\frac{3}{4}\right) = -24$ ft/s. By symmetry, 24 ft/s must be the velocity right after the first bounce. So

a. $v(t) = \begin{cases} -32t & \text{for } 0 \le t \le 1 \\ -32(t-1) + 24 & \text{for } 1 < t \le 2.5 \end{cases}$

b. $9 = -16t^2 + 16 \Rightarrow t \approx 0.66$ s; s also equals 9 at the apex of the first rebound at $t = 1.75$ s.

Section 5.3

Concepts Review

1. $2 + 4 + 6 + 8 = 20$

3. Riemann sum

Problem Set 5.3

1. $2+5+8+11+14=40$

3. $\frac{2}{\frac{1}{5}+1}+\frac{2}{\frac{2}{5}+1}+\frac{2}{\frac{3}{5}+1}+\frac{2}{\frac{4}{5}+1}+\frac{2}{1+1}=\frac{1627}{252}$

5. $-1+2-4+8-16=-11$

7. $1+0-1+0+1+0=1$

9. $\sum_{i=1}^{98} i$

11. $\sum_{i=1}^{69} \frac{1}{i}$

13. $\sum_{i=1}^{n} a_i$

15. $\sum_{i=1}^{n} f(c_i)$

17. $\sum_{i=1}^{10} \frac{3i}{5}\cdot\frac{1}{5}=\frac{33}{5}$

19.

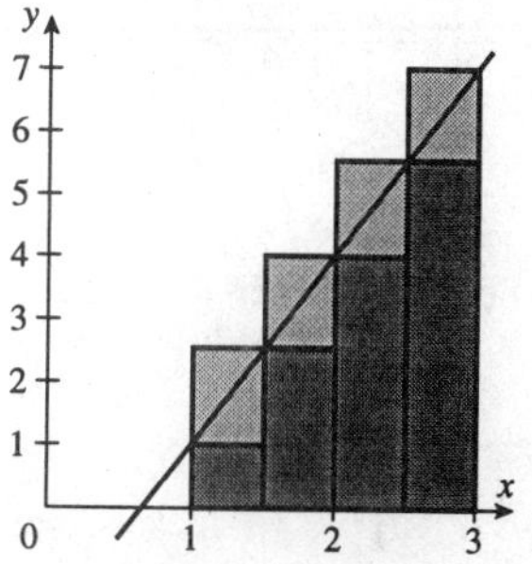

Using smaller rectangles,

$$A \approx \frac{1}{2}\left(1+\frac{5}{2}+4+\frac{11}{2}\right)=\frac{13}{2}.$$

Using larger rectangles,

$$A \approx \frac{1}{2}\left(\frac{5}{2}+4+\frac{11}{2}+7\right)=\frac{19}{2}.$$

$$A=2+\frac{1}{2}(2\cdot 6)=8$$

21.

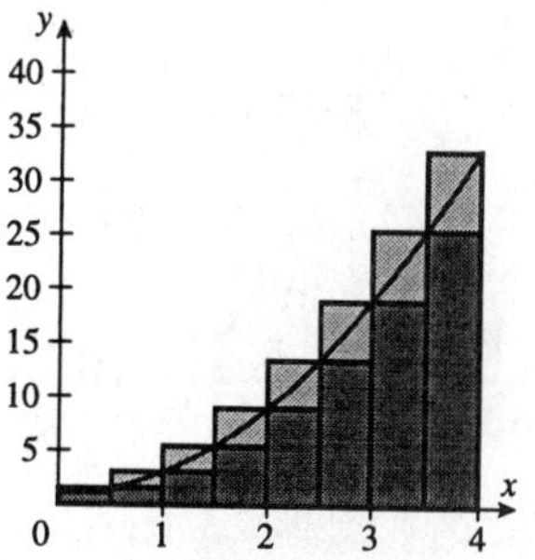

Using smaller rectangles,

$$A \approx \frac{1}{2}\left(1+\frac{3}{2}+3+\frac{11}{2}+9+\frac{27}{2}+19+\frac{51}{2}\right)=39.$$

Using larger rectangles,

$$A \approx \frac{1}{2}\left(\frac{3}{2}+3+\frac{11}{2}+9+\frac{27}{2}+19+\frac{51}{2}+33\right)=55.$$

23. $$\sum_{i=1}^{10}\left\{\frac{1}{2}\left[0+\frac{2-0}{10}(i-1)\right]^2+1\right\}\left(\frac{1}{5}\right)=\frac{157}{50}=3.14$$

$$\sum_{i=1}^{10}\left\{\frac{1}{2}\left[0+\frac{2-0}{10}i\right]^2+1\right\}\left(\frac{1}{5}\right)=\frac{177}{50}=3.54$$

$$\sum_{i=1}^{100}\left\{\frac{1}{2}\left[0+\frac{2-0}{100}(i-1)\right]^2+1\right\}\left(\frac{1}{50}\right)=3.3134$$

$$\sum_{i=1}^{100}\left\{\frac{1}{2}\left[0+\frac{2-0}{100}i\right]^2+1\right\}\left(\frac{1}{50}\right)=3.3534$$

$$\sum_{i=1}^{1000}\left\{\frac{1}{2}\left[0+\frac{2-0}{1000}(i-1)\right]^2+1\right\}\left(\frac{1}{500}\right)=3.331334$$

$$\sum_{i=1}^{1000}\left\{\frac{1}{2}\left[0+\frac{2-0}{1000}i\right]^2+1\right\}\left(\frac{1}{500}\right)=3.335334$$

The area to one decimal place is 3.3.

25. $$\sum_{i=1}^{10}\ln\left[1+\frac{4-1}{10}(i-1)\right]\left(\frac{3}{10}\right)\approx 2.3316$$

$$\sum_{i=1}^{10}\ln\left[1+\frac{4-1}{10}i\right]\left(\frac{3}{10}\right)\approx 2.7475$$

$$\sum_{i=1}^{100}\ln\left[1+\frac{4-1}{100}(i-1)\right]\left(\frac{3}{100}\right)\approx 2.5243$$

$$\sum_{i=1}^{100}\ln\left[1+\frac{4-1}{100}i\right]\left(\frac{3}{100}\right)\approx 2.5659$$

$$\sum_{i=1}^{1000}\ln\left[1+\frac{4-1}{1000}(i-1)\right]\left(\frac{3}{1000}\right)\approx 2.5431$$

$$\sum_{i=1}^{1000}\ln\left[1+\frac{4-1}{1000}i\right]\left(\frac{3}{1000}\right)\approx 2.5473$$

The area to one decimal place is 2.5.

27. $\sum_{i=1}^{10}\cos\left[0+\frac{\pi}{10}(i-1)\right]\left(\frac{\pi}{10}\right)\approx 0.3141593$

$\sum_{i=1}^{10}\cos\left[0+\frac{\pi}{10}i\right]\left(\frac{\pi}{10}\right)\approx -0.3141593$

$\sum_{i=1}^{100}\cos\left[0+\frac{\pi}{100}(i-1)\right]\left(\frac{\pi}{100}\right)\approx 0.0314159$

$\sum_{i=1}^{100}\cos\left[0+\frac{\pi}{100}i\right]\left(\frac{\pi}{100}\right)\approx -0.0314159$

$\sum_{i=1}^{1000}\cos\left[0+\frac{\pi}{1000}(i-1)\right]\left(\frac{\pi}{1000}\right)\approx 0.0031416$

$\sum_{i=1}^{1000}\cos\left[0+\frac{\pi}{1000}i\right]\left(\frac{\pi}{1000}\right)\approx -0.0031416$

The area to two decimal places is 0.

29. $\sum_{i=1}^{10}(i+4)2^i$

31. $\sum_{i=1}^{10} i\sin\left(\frac{\pi}{i}\right)$

33. a. $\sum_{k=0}^{10}\left(\frac{1}{2}\right)^k=\frac{1-\left(\frac{1}{2}\right)^{11}}{\frac{1}{2}}=2-\left(\frac{1}{2}\right)^{10}$, so

$\sum_{k=1}^{10}\left(\frac{1}{2}\right)^k=1-\left(\frac{1}{2}\right)^{10}.$

b. $\sum_{k=0}^{10}2^k=\frac{1-2^{11}}{-1}=2^{11}-1$, so

$\sum_{k=1}^{10}2^k=2^{11}-2.$

Section 5.4

Concepts Review

1. Definite integral; $\int_a^b f(x)dx$

3. $\frac{15}{2}$

Problem Set 5.4

1.

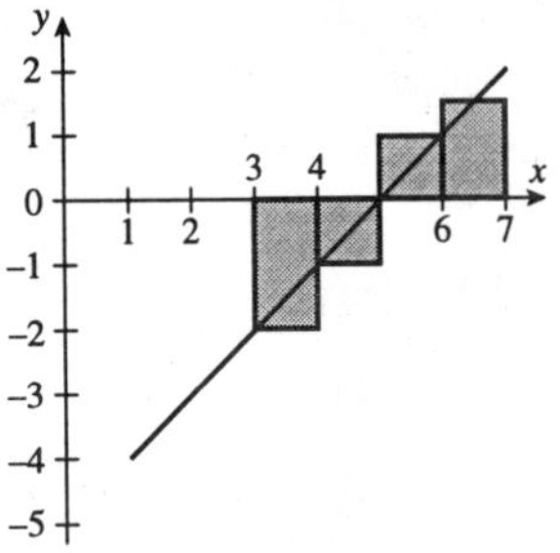

$\sum_{i=1}^{4}f(\bar{x}_i)\Delta x=(-2)+(-1)+(1)+\left(\frac{3}{2}\right)=-\frac{1}{2}$

This is the difference between the areas of the first two rectangles and those of the last two.

3.

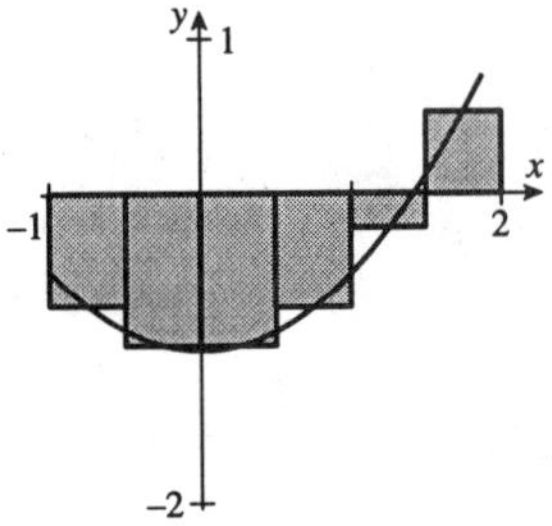

$\sum_{i=1}^{6}f(\bar{x}_i)\Delta x=\frac{1}{2}\left[\left(-\frac{23}{32}\right)+\left(-\frac{31}{32}\right)+\left(-\frac{31}{32}\right)+\left(-\frac{23}{32}\right)+\left(-\frac{7}{32}\right)+\left(\frac{17}{32}\right)\right]=-\frac{49}{16}$

This is the difference between the areas of the first five rectangles and that of the last one.

5. $\sum_{i=1}^{10}\left\{\left[-2+\frac{4}{10}\left(i-\frac{1}{2}\right)\right]^3+1\right\}\left(\frac{4}{10}\right)=4$

$\sum_{i=1}^{100}\left\{\left[-2+\frac{4}{100}\left(i-\frac{1}{2}\right)\right]^3+1\right\}\left(\frac{4}{100}\right)=4$

$\sum_{i=1}^{1000}\left\{\left[-2+\frac{4}{1000}\left(i-\frac{1}{2}\right)\right]^3+1\right\}\left(\frac{4}{1000}\right)$

$=4$

Perfectly accurate

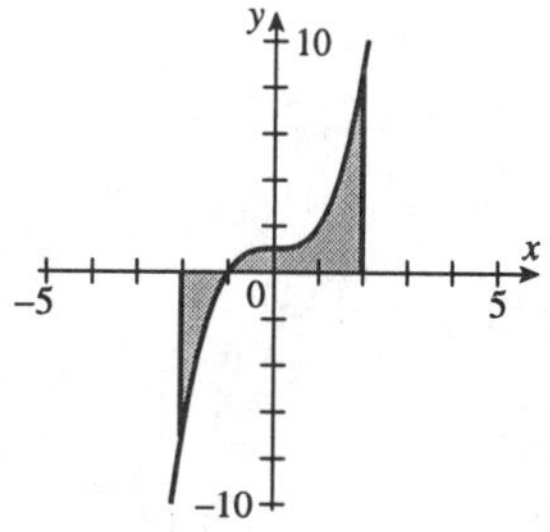

The integral is the shaded area above the x-axis minus that below.

7. $\sum_{i=1}^{10} \cos\left[\frac{2}{10}\left(i-\frac{1}{2}\right)\right]^2\left(\frac{2}{10}\right) \approx 0.456353$

$\sum_{i=1}^{100} \cos\left[\frac{2}{100}\left(i-\frac{1}{2}\right)\right]^2\left(\frac{2}{100}\right) = 0.461411$

$\sum_{i=1}^{1000} \cos\left[\frac{2}{1000}\left(i-\frac{1}{2}\right)\right]^2\left(\frac{2}{1000}\right) = 0.46146$

Accurate to three decimal places

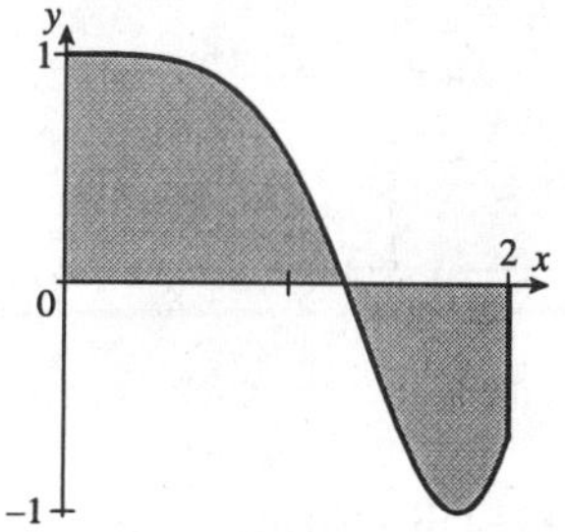

The integral is the shaded area above the x-axis minus that below.

9. $\sum_{i=1}^{10} \ln\left[1+\frac{2}{10}\left(i-\frac{1}{2}\right)\right]\left(\frac{2}{10}\right) \approx 1.296944280$

$\sum_{i=1}^{100} \ln\left[1+\frac{2}{100}\left(i-\frac{1}{2}\right)\right]\left(\frac{2}{100}\right) \approx 1.295847980$

$\sum_{i=1}^{1000} \ln\left[1+\frac{2}{1000}\left(i-\frac{1}{2}\right)\right]\left(\frac{2}{1000}\right) \approx 1.295836977$

Accurate to at least four decimal places

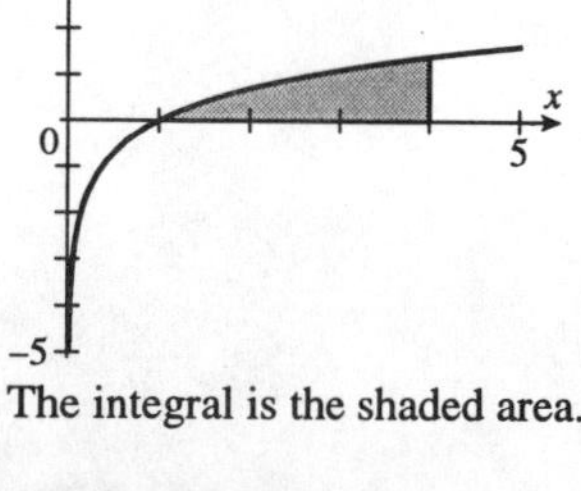

The integral is the shaded area.

11. $\sum_{i=1}^{10} \left\{\left[-1+\frac{3}{10}\left(i-\frac{1}{2}\right)\right]^2 - 1\right\}\left(\frac{3}{10}\right)$

≈ -0.0225000000

$\sum_{i=1}^{100} \left\{\left[-1+\frac{3}{100}\left(i-\frac{1}{2}\right)\right]^2 - 1\right\}\left(\frac{3}{100}\right)$

≈ -0.0002250000

$\sum_{i=1}^{1000} \left\{\left[-1+\frac{3}{1000}\left(i-\frac{1}{2}\right)\right]^2 - 1\right\}\left(\frac{3}{1000}\right)$

≈ -0.0000022500

Accurate to five decimal places

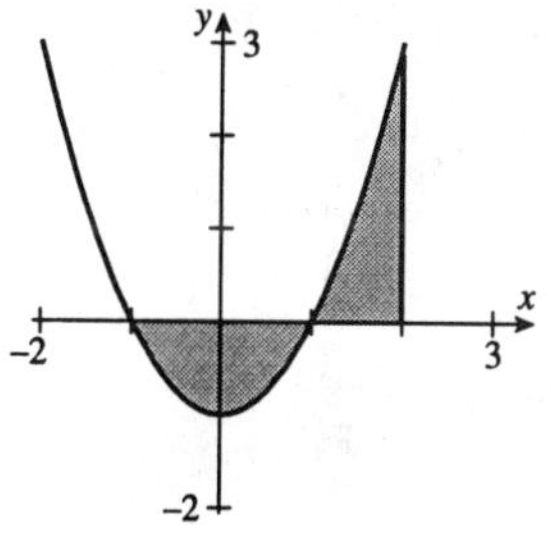

The integral is the shaded area above the curve minus that below.

13. $\int_{-2}^{2}(x^3+1)dx = 4$

15. $\int_{0}^{2}\cos x^2\, dx \approx 0.4614614624$

17. $\int_{1}^{3}\ln x\, dx = 3\ln 3 - 2 \approx 1.295836866$

19. $\int_{-1}^{2}(x^2-1)dx = 0$

21.

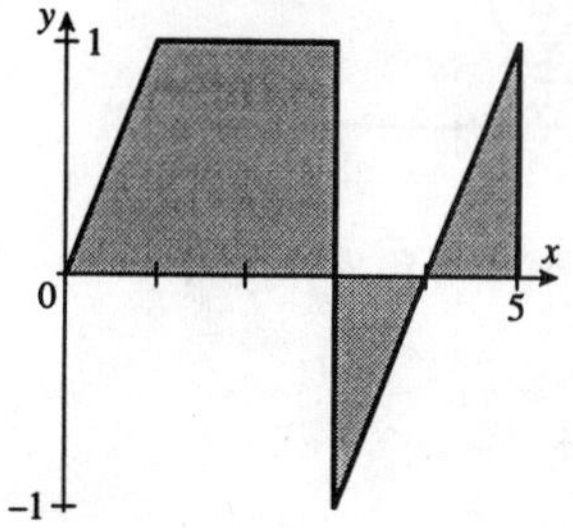

The second and third shaded regions cancel, so the integral is the area of the first shaded region:

$\int_0^5 f(x)\,dx = \frac{1}{2}(1)(1)+(2)(1) = 2\frac{1}{2}$

23.

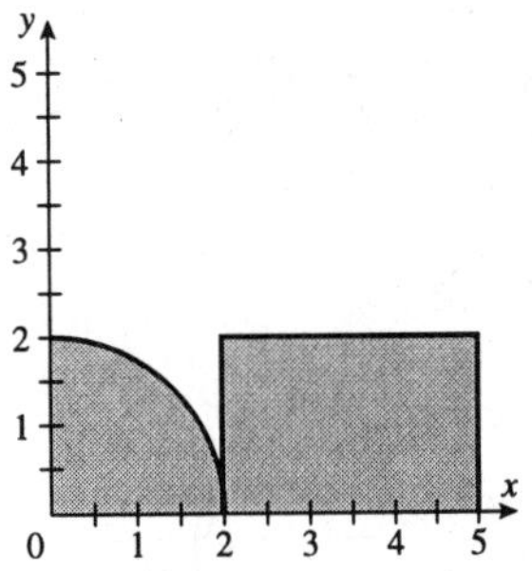

$$\int_0^5 f(x)\,dx = \frac{1}{4}\pi 2^2 + (3)(2) = \pi + 6$$

Section 5.5

Concepts Review

1. Antiderivative; $F(b) - F(a)$

3. $\dfrac{26}{3}$

Problem Set 5.5

1. $\left[x^3 - x^2 + 3x\right]_{-1}^{2} = 15$

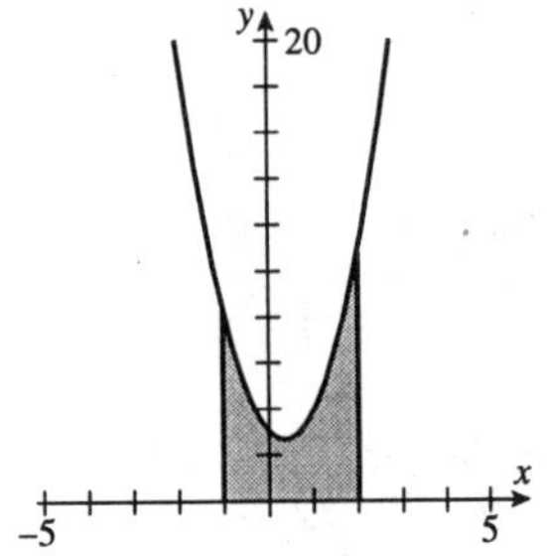

The integral is the area of the shaded region.

3. $\left[3e^x + 3x\right]_{-1}^{2} = 3(e^2 - e^{-1}) + 9$

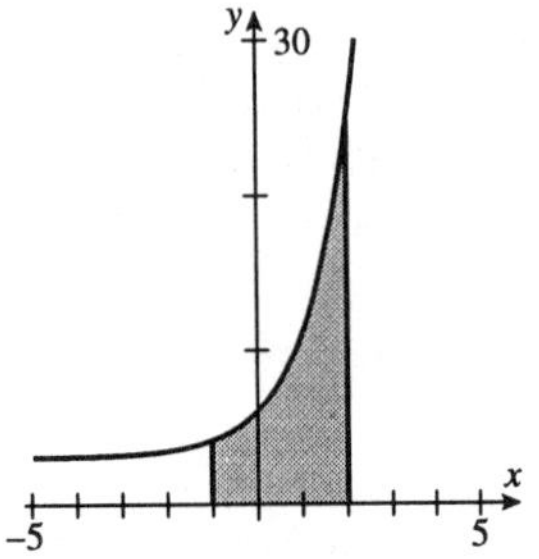

The integral is the area of the shaded region.

5. $\left[\frac{1}{3}y^3 - \frac{1}{2y^2}\right]_{-4}^{-2} = \frac{56}{3} - \frac{3}{32} = \frac{1783}{96}$

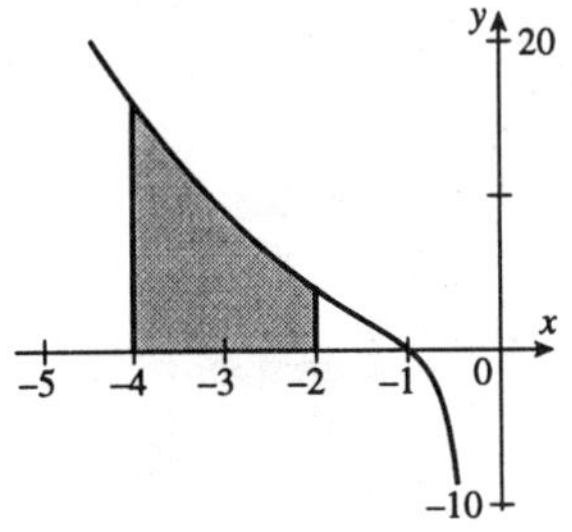

The integral is the area of the shaded region.

7. $\left[2e^{t/2}\right]_0^4 = 2(e^2 - 1)$

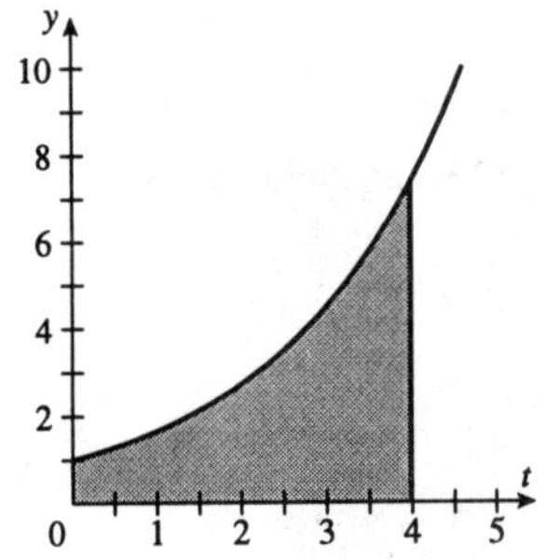

The integral is the area of the shaded region.

9. $\left[\frac{2}{3}y^{3/2} + \ln|y|\right]_2^4 = \frac{16 - 4\sqrt{2}}{3} + \ln 2$

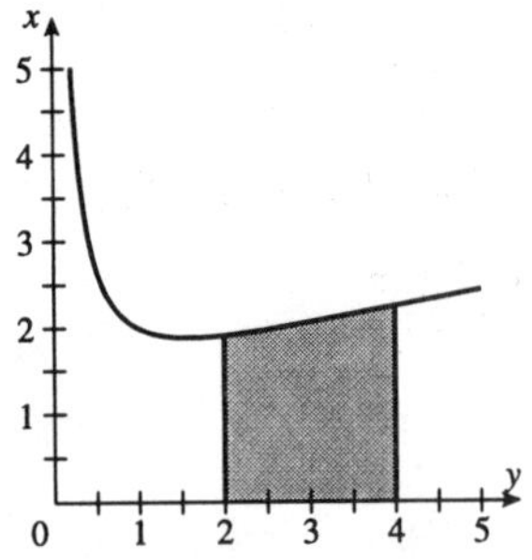

The integral is the area of the shaded region.

11. $[\sin x]_0^{\pi/2} = 1$

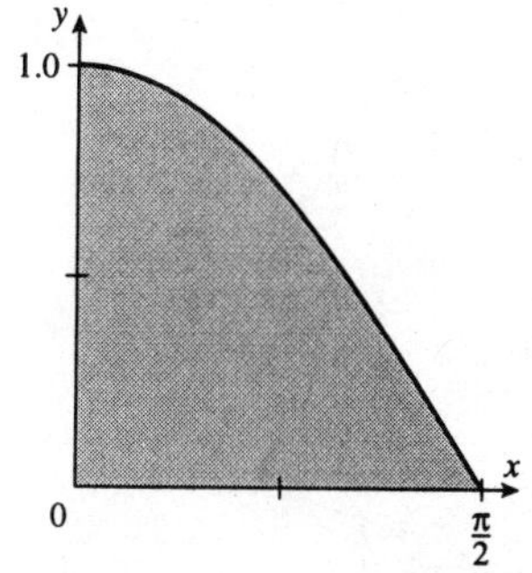

The integral is the area of the shaded region.

13. $\left[\frac{2}{5}x^5 - x^3 + 5x\right]_0^1 = 4\frac{2}{5}$

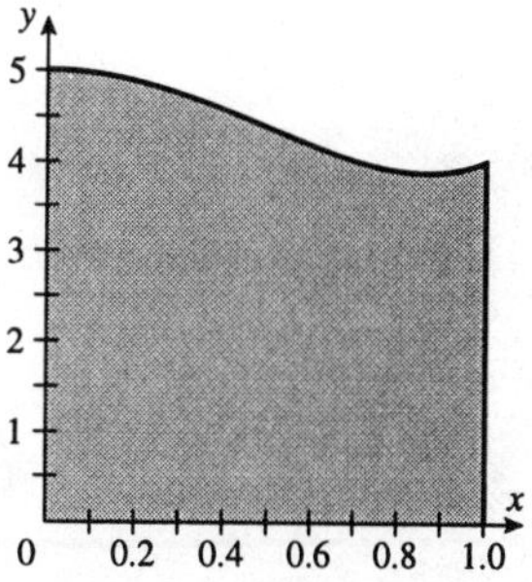

The integral is the area of the shaded region.

15. Guess: $(x^2+1)^{11}$

Check: $\frac{d}{dx}(x^2+1)^{11} = (x^2+1)^{10}(22x)$

New guess: $\frac{1}{11}(x^2+1)^{11}$

Check: $\frac{d}{dx}\frac{1}{11}(x^2+1)^{11} = (x^2+1)^{10}(2x)$

$\frac{1}{11}\left[(x^2+1)^{11}\right]_0^1 = \frac{2047}{11}$

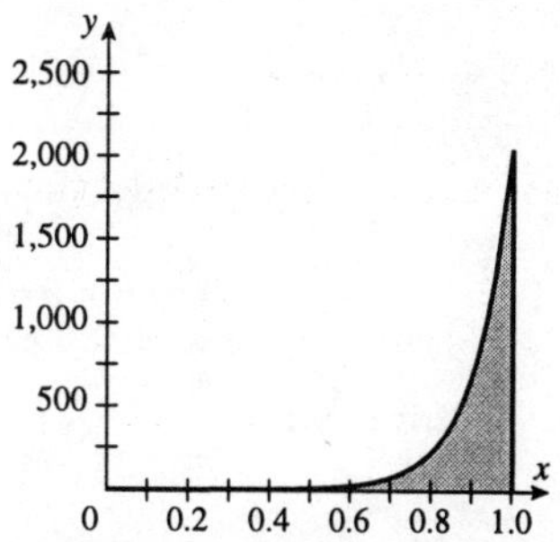

The integral is the area of the shaded region.

17. Guess: $-(t+2)^{-1}$

Check: $\frac{d}{dx}[-(t+2)^{-1}] = (t+2)^{-2}$

$\left[-(t+2)^{-1}\right]_{-1}^{3} = \frac{4}{5}$

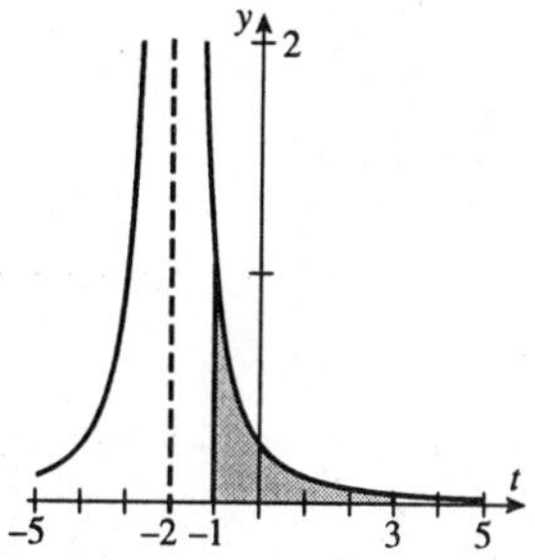

The integral is the area of the shaded region.

19. Guess: $\frac{1}{2}\ln(x^2+1)$

Check: $\frac{d}{dx}\left[\frac{1}{2}\ln(x^2+1)\right] = \frac{1}{x^2+1}\cdot x$

$\left[\frac{1}{2}\ln(x^2+1)\right]_0^5 = \frac{1}{2}\ln 26$

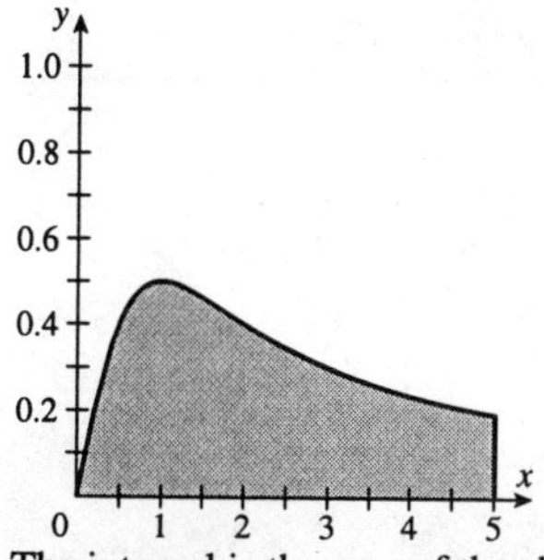

The integral is the area of the shaded region.

21. Guess: sin(ln x)

Check: $\cos(\ln x)\cdot\frac{1}{x}$

$[\sin(\ln x)]_1^{10} = \sin(\ln 10)$

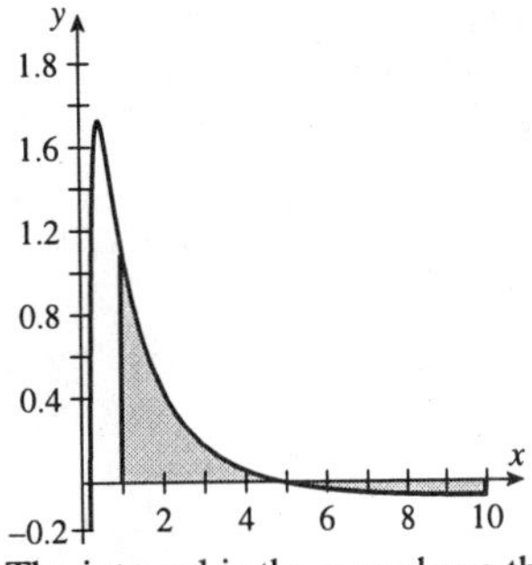

The integral is the area above the x-axis minus the area below.

23. Guess: $-\dfrac{1}{3}\cos^3 x$

Check: $\dfrac{d}{dx}\left(-\dfrac{1}{3}\cos^3 x\right) = (-\cos^2 x)(-\sin x)$

$$\left[-\frac{1}{3}\cos^3 x\right]_0^{\pi/2} = \frac{1}{3}$$

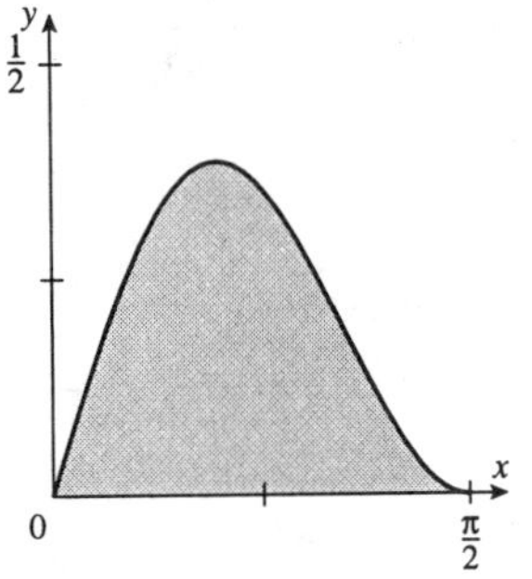

The integral is the area of the shaded region.

25. Guess: $x^2 - \cos x$

Check: Works

$$\left[x^2 - \cos x\right]_0^{\pi/2} = \frac{\pi^2}{4} + 1$$

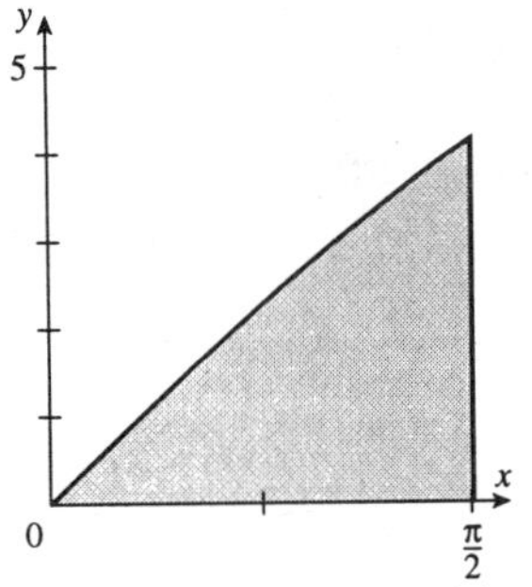

The integral is the area of the shaded region.

27. Guess: $\dfrac{2}{3}x^{3/2} + \dfrac{1}{3}(2x+1)^{3/2}$

Check: Works

$$\left[\frac{2}{3}x^{3/2} + \frac{1}{3}(2x+1)^{3/2}\right]_0^4 = \frac{16}{3} + 9 - \frac{1}{3} = 14$$

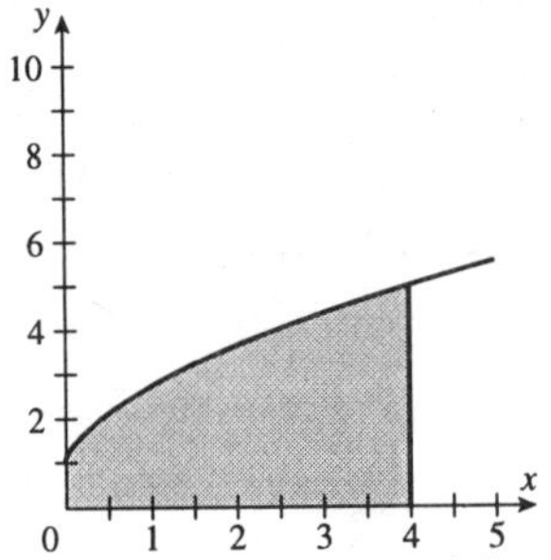

The integral is the area of the shaded region.

29. Guess: $\dfrac{a}{2\pi}\sin\left(\dfrac{2\pi}{a}x\right)$

Check: Works

$$\left[\frac{a}{2\pi}\sin\left(\frac{2\pi}{a}x\right)\right]_0^a = 0$$

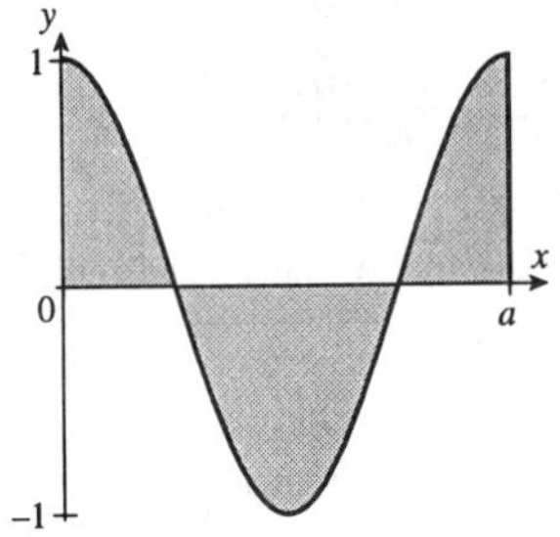

The area above the x-axis and that below cancel.

31. $\left[x^2 \sin x - 2\sin x + 2x\cos x\right]_0^1 = 2\cos(1) - \sin(1)$

33. $\left[\dfrac{1}{2}\sin x^2\right]_0^1 = \dfrac{1}{2}\sin(1)$

35. $\left[\dfrac{1}{3}\sin x^3\right]_0^1 = \dfrac{1}{3}\sin(1)$

37. The integral can be found exactly when $n = im - 1$ for some positive integer i. For $i = 1$, it is easy to see that the antiderivative is $-\dfrac{1}{m}\sin(x^m)$.

39. $\frac{d}{dx}\left(\frac{1}{2}x|x|\right) = x$ for $x \ge 0$ and $-x$ for $x < 0$, or $|x|$.

$$\left[\frac{1}{2}x|x|\right]_a^b = \frac{1}{2}(b|b| - a|a|)$$

Section 5.6

Concepts Review

1. Interval Additive; 1

3. $\sin^3 x$; $2x\sin^3(x^2)$

Problem Set 5.6

1. $\left[\frac{2}{3}x^{3/2}\right]_0^1 + \left[\frac{1}{3}x^3\right]_1^4 = \frac{65}{3}$

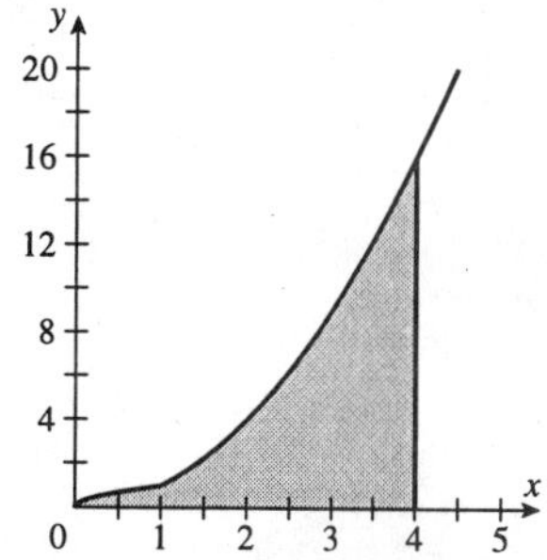

3. $\left[e^x\right]_0^2 + \left[xe^2\right]_2^4 = 3e^2 - 1$

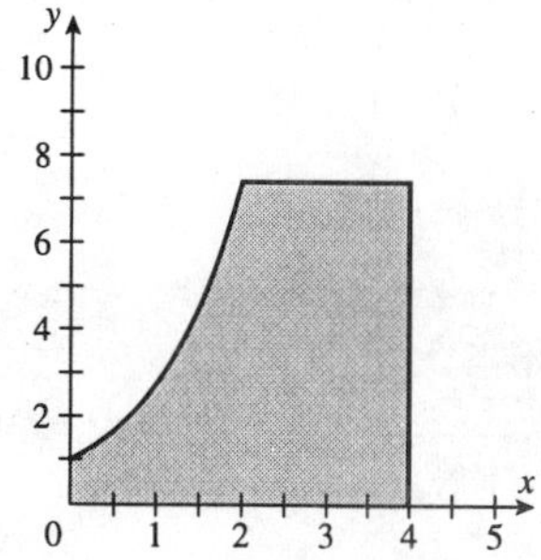

5. $[\sin x]_0^{\pi/2} + [-\sin x]_{\pi/2}^4 = 2 - \sin(4)$

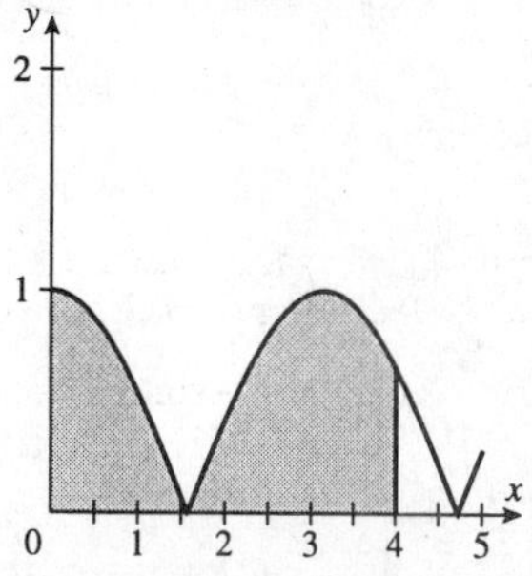

7. By Theorem 5.6.D, $G'(x) = 2x + 1$

9. By Theorem 5.6.D, $G'(x) = e^{-x^2}$

11. Switching limits and adding a minus sign, then applying Theorem 5.6.D yields $G'(x) = x\tan x$, $-\frac{\pi}{2} < x < \frac{\pi}{2}$.

13. $\frac{dG}{d(2+\sin x)} \cdot \frac{d(2+\sin x)}{dx}$

$$= \left[\frac{2}{2+\sin x} + \cos(2+\sin x)\right](\cos x)$$

15. Using the hint, switching limits and changing signs yields

$$G(x) = -\int_0^x \sqrt{1+s^4}\,ds + \int_0^{x^3} \sqrt{1+s^4}\,ds.$$

Then $G'(x) = -\sqrt{1+x^4} + \frac{dG}{d(x^3)} \cdot \frac{d(x^3)}{dx}$

$$= -\sqrt{1+x^4} + 3x^2\sqrt{1+x^{12}}.$$

17. $\frac{dy}{dx} = \frac{x}{\sqrt{a^2+x^2}}$

$$\frac{d^2y}{dx^2} = \frac{1}{\sqrt{a^2+x^2}} - \frac{x^2}{\sqrt{(a^2+x^2)^3}}$$

$$= \frac{a^2}{\sqrt{(a^2+x^2)^3}}$$, which is always positive.

19. $\int_0^1 1\,dx = 1$; $\int_0^1 1 + x^4\,dx = \left[x + \frac{x^5}{5}\right]_0^1 = \frac{6}{5}$

21. $\frac{1}{3-1}\int_1^3 4x^3\,dx = \frac{1}{2}\left[x^4\right]_1^3 = 40$

23. $\frac{1}{3-(-1)}\left[2x + \frac{1}{2}x|x|\right]_{-1}^3 = \frac{13}{4}$

(See Problem 39 of Section 5.5.)

25. $\sum_{i=1}^{100} \sin\left(\cos\left(\frac{i-\frac{1}{2}}{10}\right)\right)\left(\frac{1}{10}\right) \approx -0.046609$

27. 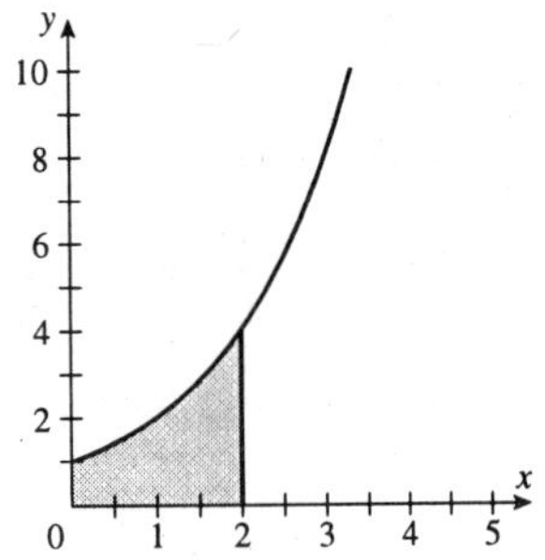

Using $c = 1$ yields $2(2-0) = 4$.

29. 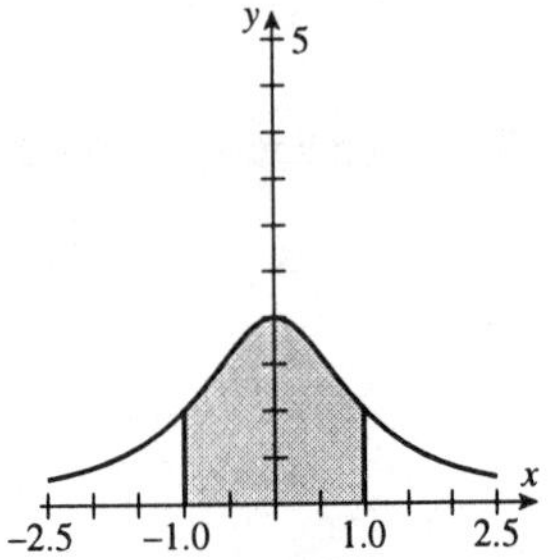

Using $c = \frac{1}{2}$ yields $\frac{8}{5}(1+1) = \frac{16}{5}$.

31. True. The slope of the antiderivative $F(x)$ will be nonnegative and $F(b)$ will be greater than or equal to $F(a)$.

33. False. $a = -1,\ b = 1, f(x) = x$ is a counterexample.

35. True. $\int_a^b f(x)dx - \int_a^b g(x)dx$
$= \int_a^b [f(x) - g(x)]dx$

37. $\frac{1}{b-a}\int_a^b v(t)dt = \frac{1}{b-a}\int_a^b \frac{ds}{dt}\cdot dt$
$= \frac{1}{b-a}\int_a^b ds = \frac{s(b)-s(a)}{b-a}$

Section 5.7

Concepts Review

1. $\int_1^2 \frac{1}{3}u^4 du$

3. Unhelpful

Problem Set 5.7

1. $u = 3x+2,\ du = 3dx$
$$\int \sqrt{u}\cdot\frac{1}{3}du = \frac{2}{9}u^{3/2} = \frac{2}{9}(3x+2)^{3/2} + C$$

3. $u = 3x+2,\ du = 3dx$
$$\int \cos(u)\cdot\frac{1}{3}du = \frac{1}{3}\sin u = \frac{1}{3}\sin(3x+2) + C$$

5. $u = x^2+4,\ du = 2x\,dx$
$$\int e^u\cdot\frac{1}{2}du = \frac{1}{2}e^u = \frac{1}{2}e^{x^2+4} + C$$

7. $u = x^2+4,\ du = 2x\,dx$
$$\int \sin(u)\cdot\frac{1}{2}du = -\frac{1}{2}\cos u = -\frac{1}{2}\cos(x^2+4) + C$$

9. $u = x^2+4,\ du = 2x\,dx$
$$\int \frac{1}{u}\cdot\frac{1}{2}du = \frac{1}{2}\ln|u| = \frac{1}{2}\ln(x^2+4) + C$$

11. $u = \ln x,\ du = \frac{1}{x}dx$
$$\int u\,du = \frac{1}{2}u^2 = \frac{1}{2}\ln^2 x + C$$

13. $u = \sqrt{t}+4,\ du = \frac{1}{2\sqrt{t}}dt$
$$\int 2u^3 du = \frac{1}{2}u^4 = \frac{1}{2}\left(\sqrt{t}+4\right)^4 + C$$

15. $u = 3x+1,\ du = 3dx$
$$\int_1^4 u^3\cdot\frac{1}{3}du = \left[\frac{1}{12}u^4\right]_1^4 = \frac{85}{4}$$

17. $u = t^2+9,\ du = 2t\,dt$
$$\int_9^{13}\frac{1}{u}\cdot\frac{1}{2}du = \left[\frac{1}{2}\ln u\right]_9^{13} = \frac{1}{2}(\ln 13 - \ln 9)$$

19. $$\sum_{i=1}^{100}\frac{\frac{\left(i-\frac{1}{2}\right)}{100}}{\left\{\left[\frac{\left(i-\frac{1}{2}\right)}{100}\right]^3 + 4\frac{\left(i-\frac{1}{2}\right)}{100} + 1\right\}^2}\cdot\frac{1}{100} \approx 0.45679$$

$\left(\frac{d}{dx}(x^3+4x+1)\text{ is not a constant times }x.\right)$

21. $u = \sin\theta,\ du = \cos\theta\,d\theta$
$$\int_0^{1/2} u^3 du = \left[\frac{1}{4}u^4\right]_0^{1/2} = \frac{1}{64}$$

23. $u = e^x+1,\ du = e^x dx$
$$\int_2^{e+1} u^{10}du = \left[\frac{1}{11}u^{11}\right]_2^{e+1} = \frac{1}{11}\left[(e+1)^{11} - 2048\right]$$

25. $u = \pi x^2,\ du = 2\pi x\,dx$

$$\int_0^{\pi} \sin(u)\cdot\frac{1}{2\pi}du = \left[-\frac{1}{2\pi}\cos u\right]_0^{\pi} = \frac{1}{\pi}$$

27. $\displaystyle\sum_{i=1}^{100}\left(\frac{i-\frac{1}{2}}{100}\right)^2 \sin\left[\pi\left(\frac{i-\frac{1}{2}}{100}\right)^2\right]\cdot\frac{1}{100} \approx 0.2187$

29. $u = \sqrt{t}+1,\ du = \dfrac{1}{2\sqrt{t}}dt$

$$\int_2^3 \frac{2}{u^3}du = \left[-\frac{1}{u^2}\right]_2^3 = \frac{5}{36}$$

31. $u = -x,\ du = -dx$

$\int_{-a}^{-b} f(u)\cdot -du = \int_{-b}^{-a} f(u)du$

But u is a dummy variable that might as well be x.

33. $\displaystyle\frac{1}{12}\int_6^{18} T(t)\,dt = \left[70t - \frac{96}{\pi}\cos\left[\frac{\pi}{12}(t-9)\right]\right]_6^{18}$

$$= 2\frac{35\pi + 4\sqrt{2}}{\pi} \approx 73.6$$

Section 5.8 Chapter Review

Concepts Test

1. True. Theorem 5.1.D

3. False. $\sin^2 x \neq 1 - \cos x$

5. True. At any given height, speed on the downward trip is the negative of speed on the upward.

7. True. $\displaystyle\sum_{i=1}^{10}(a_i+1)^2 = \sum_{i=1}^{10}a_i^2 + 2\sum_{i=1}^{10}a_i + \sum_{i=1}^{10}1$

9. False. $\int_0^1 e^{x^2}dx$ is a counterexample.

11. False. $\int_{-1}^1 x\,dx$ is a counterexample.

13. False. For any n, the midpoint sum tends to be closer to the true value of the integral.

15. True. Theorem 5.6.A

17. True. Theorem 5.6.A plus the definition of absolute value.

19. False. $|\sin x| \neq 4\sin x$, in general

21. True. Functions with the same derivative differ at most by a constant, or apply the Fundamental Theorem of Calculus.

23. True. Odd-exponent terms cancel themselves out over the interval.

25. True. Both sides will be nonnegative, but some canceling of terms may occur on the left side.

27. True. The left side is the limit of the appropriate right-endpoint Riemann sum.

Sample Test Problems

1. $\left[\frac{1}{4}x^4 - x^3 + 2x^{3/2}\right]_0^1 = \frac{5}{4}$

3. $\left[\frac{1}{4}\ln(t^4+9)\right]_0^2 = \frac{1}{2}(\ln 5 - \ln 3)$

5. Numerical answer ≈ 16.4526; e^{x^2} is not formally integrable.

7. $\int dy = \int (x+1)^{1/2}dx;\ y = \frac{2}{3}(x+1)^{3/2} + \frac{38}{3}$

9. $\int \frac{1}{y^4}dy = \int e^t dt;\ -\frac{1}{3y^3} = e^t - \left(\frac{1}{3}+e\right)$

$$y = -\sqrt[3]{\frac{1}{3\left[e^t - \left(\frac{1}{3}+e\right)\right]}}$$

11. $\frac{dy}{dx} = \frac{x^2}{2};\ \int dy = \int \frac{x^2}{2}dx;\ y = \frac{x^3}{6}+1$

13. $s = -16t^2 + 48t + 448$; $s = 0$ at $t = 7$;
then $v = -32(7) + 48 = -176$ ft/s

15. $\displaystyle\sum_{i=1}^{4}\left[\left(\frac{i}{2}\right)^2 - 1\right]\left(\frac{1}{2}\right) = \frac{7}{4}$

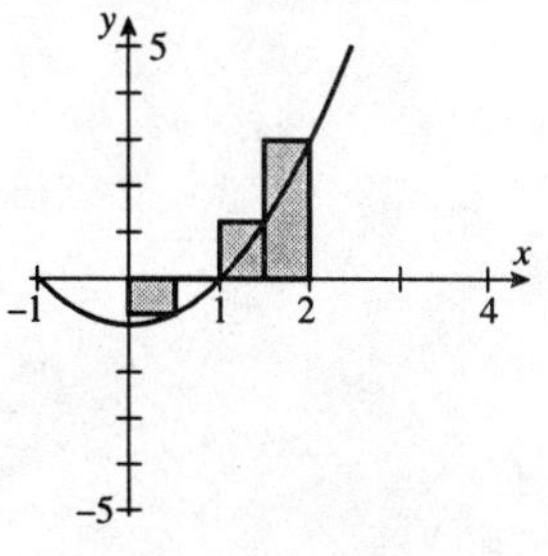

17. $\int_0^3 (2-\sqrt{x+1})^2\,dx$

$= \int_0^3 x+5-4\sqrt{x+1}\,dx$

$= \left[\frac{1}{2}x^2+5x-\frac{8}{3}(x+1)^{3/2}\right]_0^3 = \frac{5}{6}$

19. $\int_2^4 \frac{5x^2-1}{x^2}\,dx = \int_2^4 5-\frac{1}{x^2}\,dx = \left[5x+\frac{1}{x}\right]_2^4 = \frac{39}{4}$

21. a. $\sum_{n=2}^{78} \frac{1}{n}$

b. $\sum_{n=1}^{50} nx^{2n}$

23. a. $\int_0^1 (1-x)dx + \int_1^4 (x-1)dx$

$= \left[x-\frac{1}{2}x^2\right]_0^1 + \left[\frac{1}{2}x^2-x\right]_1^4 = 5$

b. $\int_0^1 (x^2-3x+2)dx + \int_1^2 (-x^2+3x-2)dx$

$+\int_2^4 (x^2-3x+2)dx$

$= \left[\frac{1}{3}x^3-\frac{3}{2}x^2+2x\right]_0^1 + \left[-\frac{1}{3}x^3+\frac{3}{2}x^2-2x\right]_1^2$

$+\left[\frac{1}{3}x^3-\frac{3}{2}x^2+2x\right]_2^4 = \frac{17}{3}$

25. $\frac{1}{(-1)-(-4)}\int_{-4}^{-1} 3x^2\,dx = \frac{1}{3}\left[x^3\right]_{-4}^{-1} = 21 = 3c^2$

$c = \sqrt{7}$

27. a. $\int_0^4 \sqrt{x}\,dx = \frac{2}{3}\left[x^{3/2}\right]_0^4 = \frac{16}{3}$

b. $\int_1^3 x^2\,dx = \frac{1}{3}\left[x^3\right]_1^3 = \frac{26}{3}$

Chapter 6

Section 6.1

Concepts Review

1. $\int_a^b f(x)dx;\ -\int_a^b f(x)dx$

3. $\int_a^b (g(x)-f(x))\,dx;\ f(x)=g(x)$

Problem Set 6.1

1. $A=\int_{-1}^{2}(x^2+1)dx=\left[\frac{1}{3}x^3+x\right]_{-1}^{2}=6$

3. $A=\int_{-1}^{1}\sqrt{1-x^4}\,dx\approx 1.74804$
(Numerical methods were used to evaluate this integral.)

5. To find the intersection points, solve $2-x^2=x$; $x=-2, 1$.
$A=\int_{-2}^{1}[(2-x^2)-x]dx=\int_{-2}^{1}(-x^2-x+2)dx$
$\left[-\frac{x^3}{3}-\frac{x^2}{2}+2x\right]_{-2}^{1}=\frac{9}{2}$

7. $\int_{0.158594}^{3.14619}(\ln x+2-x)dx\approx 1.94909$
(Numerical methods were used to find the intersection points and to evaluate the integral.)

9. To find the intersection points, solve
$y+1=3-y^2$;
$y=-2, 1$.
$A=\int_{-2}^{1}[(3-y^2)-(y+1)]dy=\int_{-2}^{1}(-y^2-y+2)dy$
$=\left[-\frac{y^3}{3}-\frac{y^2}{2}+2y\right]_{-2}^{1}=\frac{9}{2}$

11. 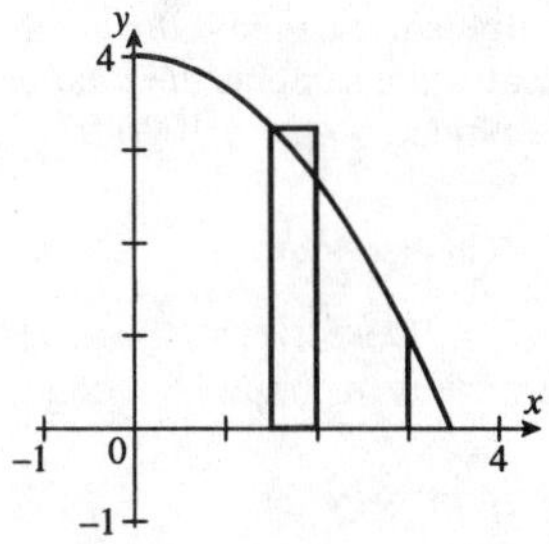

$\int_0^3\left(4-\frac{1}{3}x^2-0\right)dx=\left[4x-\frac{1}{9}x^3\right]_0^3=9$

13. 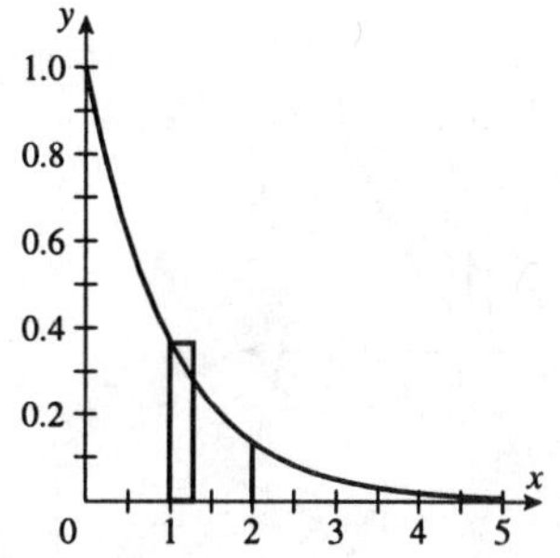

$\int_0^2(e^{-x}-0)=\left[-e^{-x}\right]_0^2=-e^{-2}+1$
≈ 0.864665

15. 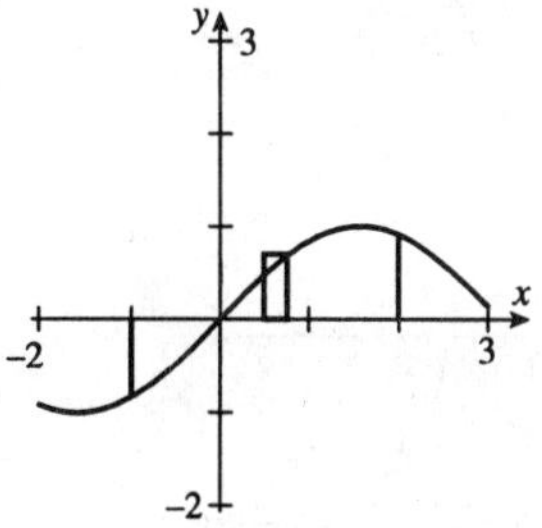

$-\int_{-1}^{0}\sin x\,dx+\int_0^2\sin x\,dx$
$=[\cos x]_{-1}^{0}+[-\cos x]_0^2=1-\cos(-1)-(\cos 2-1)$
≈ 1.87584

17. 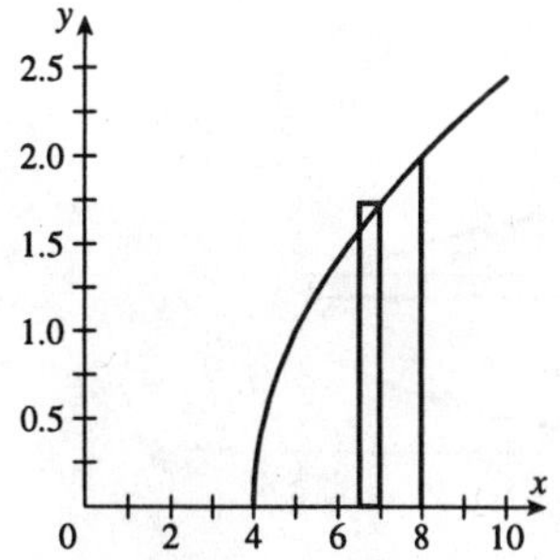

$\int_4^8\sqrt{x-4}\,dx=\left[\frac{2}{3}(x-4)^{3/2}\right]_4^8=\frac{16}{3}$

19.

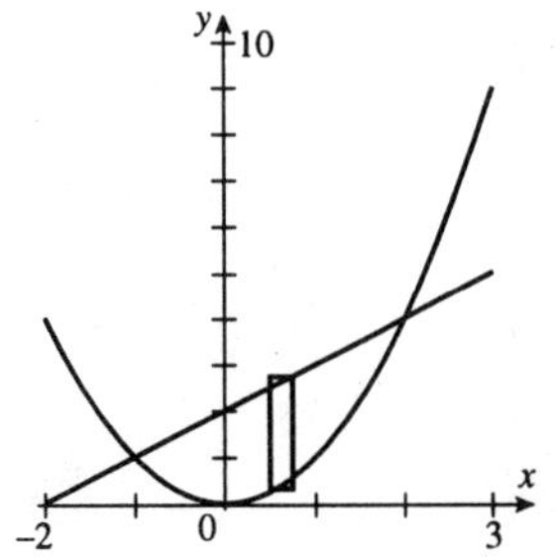

$$\int_{-1}^{2}(x+2-x^2)dx=\left[\frac{1}{2}x^2+2x-\frac{1}{3}x^3\right]_{-1}^{2}=\frac{9}{2}$$

21.

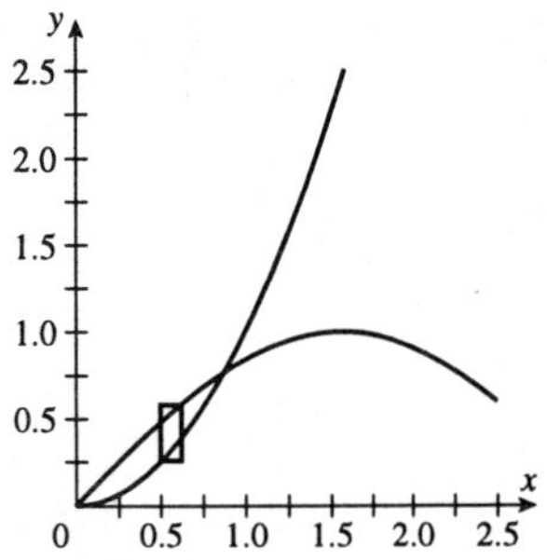

$$\int_{0}^{0.876726}(\sin x-x^2)dx=\left[-\cos x-\frac{1}{3}x^3\right]_{0}^{0.876726}$$

≈ 0.135698

(Numerical methods were used to find the intersection points and to evaluate the integral.)

23.

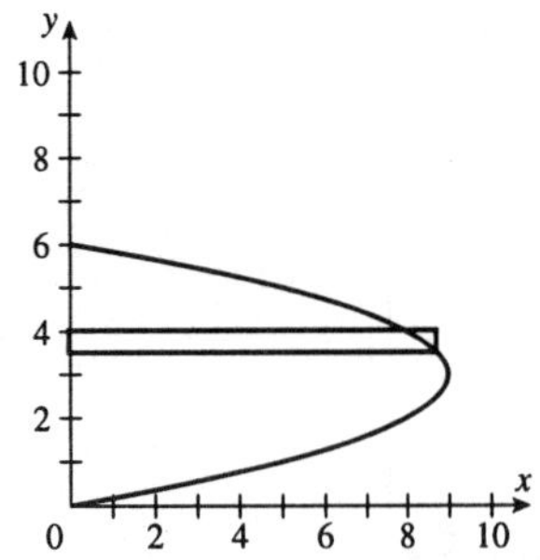

$$\int_{0}^{6}(6y-y^2-0)dy=\left[3y^2-\frac{1}{3}y^3\right]_{0}^{6}=36$$

25.

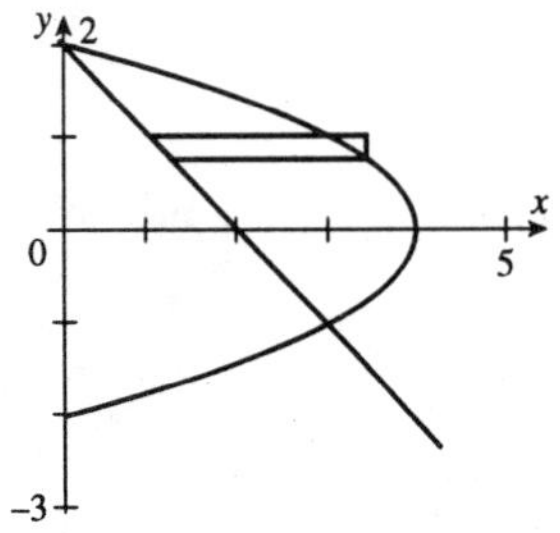

$$\int_{-1}^{2}[4-y^2-(2-y)]dy=\int_{-1}^{2}(2-y^2+y)dy$$

$$=\left[2y-\frac{1}{3}y^3+\frac{1}{2}y^2\right]_{-1}^{2}=\frac{9}{2}$$

27.

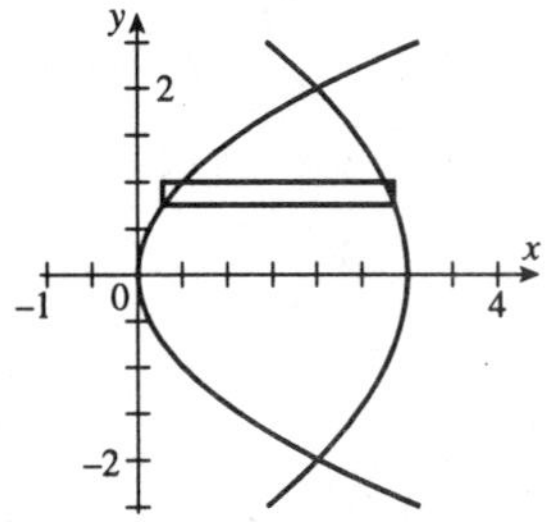

$$\int_{-2}^{2}\left(3-\frac{1}{4}y^2-\frac{1}{2}y^2\right)dy=\int_{-2}^{2}\left(3-\frac{3}{4}y^2\right)dy$$

$$=\left[3y-\frac{1}{4}y^3\right]_{-2}^{2}=8$$

29.

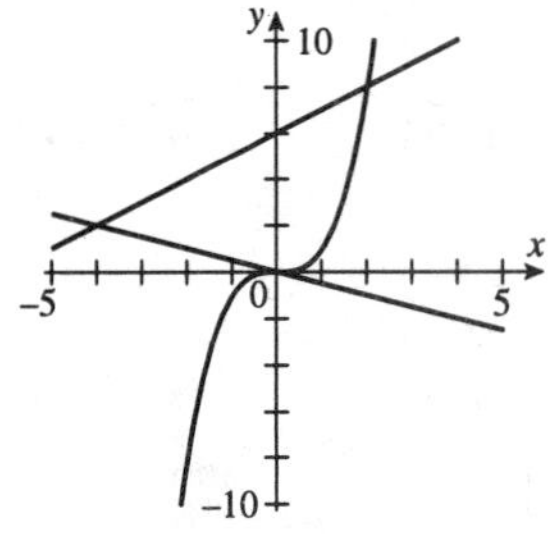

$$\int_{-4}^{0}\left(x+6+\frac{1}{2}x\right)dx+\int_{0}^{2}(x+6-x^3)dx$$

$$=\left[\frac{1}{2}x^2+6x+\frac{1}{4}x^2\right]_{-4}^{0}+\left[\frac{1}{2}x^2+6x-\frac{1}{4}x^4\right]_{0}^{2}$$

$=12+10=22$

31. For $-1\le t\le 9$, $v(t)=3t^2+36\ge 0$ so

$$\int_{-1}^{9}|v(t)|dt=\int_{-1}^{9}(3t^2+36)dt$$

$=[t^3+36t]_{-1}^{9}=1090.$

Since the velocity is not negative for $-1\le t\le 9$, the displacement is the same as the total distance traveled by the object; both are 1090 ft.

33. $s(t)=\int v(t)dt=\int(2t-4)dt=t^2-4t+C$

Since $s(0)=0$, $C=0$ and $s(t)=t^2-4t$. $s=12$ when $t=6$, so it takes the object 6 seconds to get $s=12$.

$$|2t-4|=\begin{cases}4-2t & 0\le t<2\\ 2t-4 & 2\le t\end{cases}$$

$\int_{0}^{2}|2t-4|dt=[-t^2+4t]_0^2=4$, so the object travels a distance of 4 cm in the first two seconds.

$\int_2^x |2t-4|dt = [t^2-4t]_2^x = x^2-4x+4$

$x^2-4x+4=8$ when $x=2+2\sqrt{2}$, so the object takes $2+2\sqrt{2}\approx 4.83$ seconds to travel a total distance of 12 centimeters.

35. Equation of line through (–2, 4) and (3, 9): $y=x+6$

Equation of line through (2, 4) and (–3, 9): $y=-x+6$

$$A(A)=\int_{-3}^{0}[9-(-x+6)]dx+\int_{0}^{3}[9-(x+6)]dx$$

$$=\int_{-3}^{0}(3+x)dx+\int_{0}^{3}(3-x)dx$$

$$=\left[3x+\frac{1}{2}x^2\right]_{-3}^{0}+\left[3x-\frac{1}{2}x^2\right]_{0}^{3}=9$$

$$A(B)=\int_{-3}^{-2}[(-x+6)-x^2]dx+\int_{-2}^{0}[(-x+6)-(x+6)]dx$$

$$=\int_{-3}^{-2}(-x^2-x+6)dx+\int_{-2}^{0}(-2x)dx$$

$$=\left[-\frac{1}{3}x^3-\frac{1}{2}x^2+6x\right]_{-3}^{-2}+\left[-x^2\right]_{-2}^{0}=\frac{37}{6}$$

$$A(C)=A(B)=\frac{37}{6}\text{ (by symmetry)}$$

$$A(D)=\int_{-2}^{0}[(x+6)-x^2]dx+\int_{0}^{2}[(-x+6)-x^2]dx$$

$$=\left[-\frac{1}{3}x^3+\frac{1}{2}x^2+6x\right]_{-2}^{0}+\left[-\frac{1}{3}x^3-\frac{1}{2}x^2+6x\right]_{0}^{2}$$

$$=\frac{44}{3}$$

$A(A)+A(B)+A(C)+A(D)=36$

$$A(A+B+C+D)=\int_{-3}^{3}(9-x^2)dx=\left[9x-\frac{1}{3}x^3\right]_{-3}^{3}$$

$$=36$$

37. $$\int_{\pi/6}^{5\pi/6}\left(\sin x-\frac{1}{2}\right)dx+\int_{5\pi/6}^{13\pi/6}\left(\frac{1}{2}-\sin x\right)dx+\int_{13\pi/6}^{17\pi/6}\left(\sin x-\frac{1}{2}\right)dx$$

$$=\left[-\cos x-\frac{1}{2}x\right]_{\pi/6}^{5\pi/6}+\left[\frac{1}{2}x+\cos x\right]_{5\pi/6}^{13\pi/6}+\left[-\cos x-\frac{1}{2}x\right]_{13\pi/6}^{17\pi/6}=3\sqrt{3}$$

Section 6.2

Concepts Review

1. $\pi r^2 h$

3. $\pi x^4 \Delta x$

Problem Set 6.2

1. $$V=\pi\int_0^2(x^2+1)^2dx=\pi\int_0^2(x^4+2x^2+1)dx$$

$$=\pi\left[\frac{1}{5}x^5+\frac{2}{3}x^3+x\right]_0^2=\frac{206\pi}{15}$$

3. a. $$V=\pi\int_0^2(4-x^2)^2dx=\pi\int_0^2(16-8x^2+x^4)dx$$

$$=\pi\left[16x-\frac{8}{3}x^3+\frac{1}{5}x^5\right]_0^2=\frac{256\pi}{15}$$

b. $x=\sqrt{4-y}$

$$V=\pi\int_0^4(4-y)dy=\pi\left[4y-\frac{1}{2}y^2\right]_0^4=8\pi$$

5. 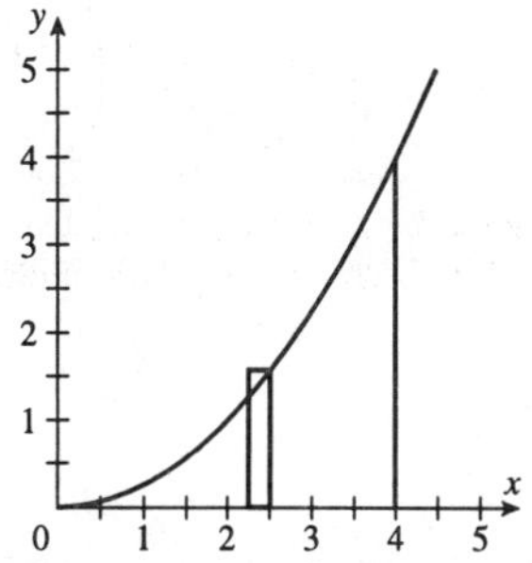

$$V=\pi\int_0^4\frac{x^4}{16}dx=\frac{\pi}{16}\left[\frac{x^5}{5}\right]_0^4=\frac{64\pi}{5}\approx 40.21$$

7. 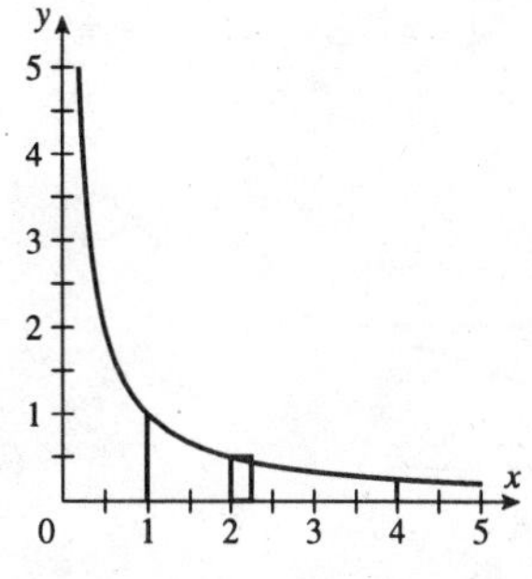

$$V=\pi\int_1^4\frac{1}{x^2}dx=\pi\left[\frac{-1}{x}\right]_1^4=\frac{3\pi}{4}\approx 2.36$$

9.

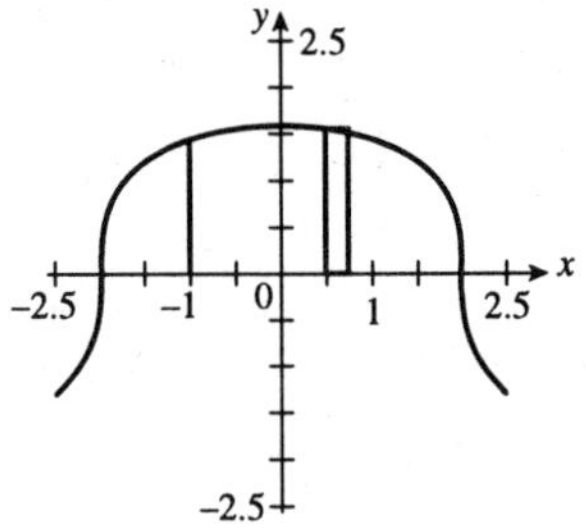

$$V = \pi\int_{-1}^{2}(4 - x^2)^{2/3}\,dx \approx 19.1676$$

(Numerical methods were used to evaluate the integral.)

11.

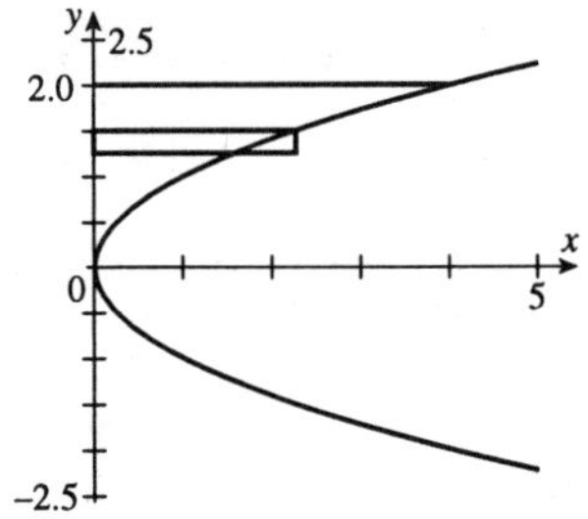

$$V = \pi\int_0^2 y^4\,dy = \pi\left[\frac{y^5}{5}\right]_0^2 = \frac{32\pi}{5} \approx 20.1062$$

13.

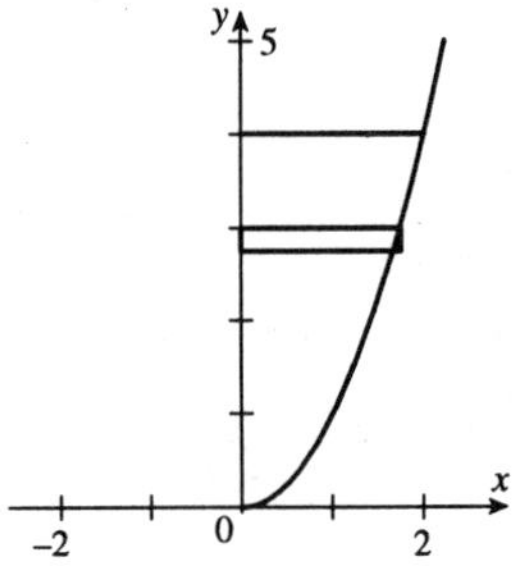

$$V = \pi\int_0^4 y\,dy = \pi\left[\frac{y^2}{2}\right]_0^4 = 8\pi \approx 25.1327$$

15.

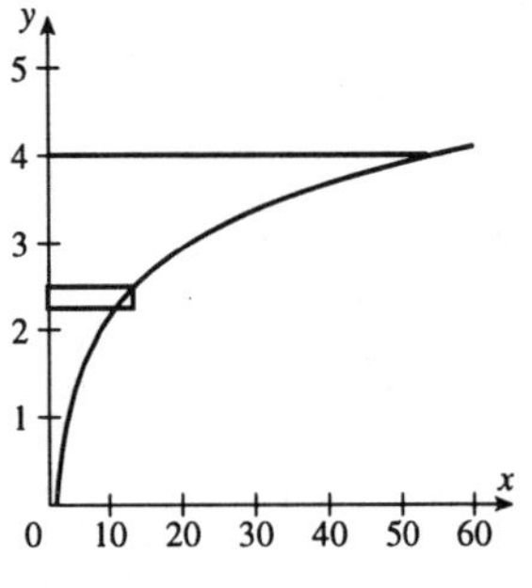

$$V = \pi\int_0^4 e^{2y}\,dy = \pi\left[\frac{e^{2y}}{2}\right]_0^4 = \frac{\pi}{2}(e^8 - 1)$$

$$\approx 4680.91$$

17. The equation of the upper half of the ellipse is

$y = b\sqrt{1 - \frac{x^2}{a^2}}$ or $y = \frac{b}{a}\sqrt{a^2 - x^2}$.

$$V = \pi\int_{-a}^{a}\frac{b^2}{a^2}(a^2 - x^2)\,dx$$

$$= \frac{b^2\pi}{a^2}\left[a^2x - \frac{x^3}{3}\right]_{-a}^{a}$$

$$= \frac{b^2\pi}{a^2}\left[2a^3 - \frac{2a^3}{3}\right] = \frac{4}{3}ab^2\pi$$

19. Solve $2x + 4 = e^x$. Find the intersection points; $x \approx -1.92722$, $x \approx 2.10547$. (Numerical methods were used to find the intersection points.)

$$V = \pi\int_{-1.92722}^{2.10547}\left[(2x+4)^2 - (e^x)^2\right]dx$$

$$= \pi\int_{-1.92722}^{2.10547}(4x^2 + 16x + 16 - e^{2x})\,dx$$

$$V = \pi\left[\frac{4}{3}x^3 + 8x^2 + 16x - \frac{1}{2}e^{2x}\right]_{-1.92722}^{2.10547}$$

$$\approx 183.981$$

21. $$V = \pi\int_0^4\left[\left(\frac{\sqrt{y}}{2}\right)^2 - \left(\frac{y}{4}\right)^2\right]^2 dy$$

$$= \pi\int_0^4\left(\frac{y}{4} - \frac{y^2}{16}\right)dy$$

$$= \pi\left[\frac{y^2}{8} - \frac{y^3}{48}\right]_0^4$$

$$= \frac{2\pi}{3} \approx 2.0944$$

23.

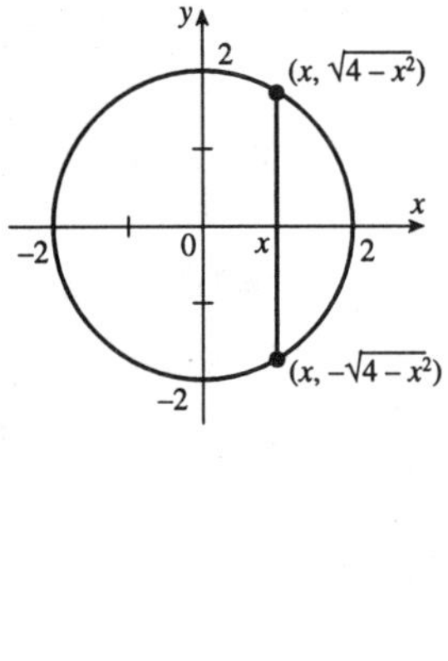

The square at x has sides of length $2\sqrt{4-x^2}$, as shown.

$V = \int_{-2}^{2} 4(4-x^2)dx$

$= 4\left[4x - \frac{x^3}{3}\right]_{-2}^{2} = \frac{128}{3}$

25. The square at x has sides of length $\sqrt{\cos x}$.

$V = \int_{-\pi/2}^{\pi/2} \cos x dx = [\sin x]_{-\pi/2}^{\pi/2} = 2$

27. The square at x has sides of length $\sqrt{1-x^2}$.

$V = \int_0^1 (1-x^2)dx = \left[x - \frac{x^3}{3}\right]_0^1 = \frac{2}{3}$

29. a. Revolving about the line $x = 4$, the radius of the disk at y is $4-\sqrt[3]{y^2} = 4-y^{2/3}$.

$V = \pi\int_0^8 (16-8y^{2/3}+y^{4/3})dy$

$= \pi\left[16y - \frac{24y^{5/3}}{5} + \frac{3y^{7/3}}{7}\right]_0^8$

$= \frac{1024\pi}{35} \approx 91.914$

b. Revolving about the line $y = 8$, the inner radius of the disk at x is $8-\sqrt{x^3} = 8-x^{3/2}$.

$V = \pi\int_0^4 \left[8^2 - (8-x^{3/2})^2\right]dx$

$= \pi\int_0^4 (16x^{3/2} - x^3)dx$

$= \pi\left[\frac{32x^{5/2}}{5} - \frac{x^4}{4}\right]_0^4$

$= \frac{704\pi}{5} \approx 442.336$

31. Let the x-axis lie along the diameter at the base perpendicular to the water level and slice perpendicular to the x-axis. Let $x = 0$ be at the center. The slice has base length $2\sqrt{r^2-x^2}$ and height $\frac{hx}{r}$.

$V = \frac{2h}{r}\int_0^r x\sqrt{r^2-x^2}\,dx$

$= \frac{2h}{r}\left[-\frac{1}{3}(r^2-x^2)^{3/2}\right]_0^r = \frac{2h}{r}\left(\frac{1}{3}r^3\right) = \frac{2}{3}r^2h$

33. a. If the depth of the tank is h, the volume is

$V = \pi\int_0^h \sqrt{\frac{y}{k}}dy = \frac{\pi}{\sqrt{k}}\left[\frac{2}{3}y^{3/2}\right]_0^h$

$= \frac{2\pi}{3\sqrt{k}}h^{3/2}$.

The volume as a function of the depth of the tank is $V(y) = \frac{2\pi}{3\sqrt{k}}y^{3/2}$.

b. It is given that $\frac{dV}{dt} = -m\sqrt{y}$.

From part a, $\frac{dV}{dt} = \frac{\pi}{\sqrt{k}}y^{1/2}\frac{dy}{dt}$.

Thus, $\frac{\pi}{\sqrt{k}}\sqrt{y}\frac{dy}{dt} = -m\sqrt{y}$ and $\frac{dy}{dt} = \frac{-m\sqrt{k}}{\pi}$ which is constant.

35. If two solids have the same cross sectional area at every x in $[a, b]$, then they have the same volume.

Section 6.3

Concepts Review

1. Circle

3. $\int_a^b \{[f'(t)]^2 + [g'(t)]^2\}^{1/2} dt$

Problem Set 6.3

1. $y' = 3$,

$L = \int_1^4 \sqrt{1+9}dx = \sqrt{10}[x]_1^4 = 3\sqrt{10}$

When $x = 1, y = 8$ and when $x = 4, y = 17$. By the distance formula,

$L = \sqrt{(4-1)^2 + (17-8)^2} = 3\sqrt{10}$.

3.

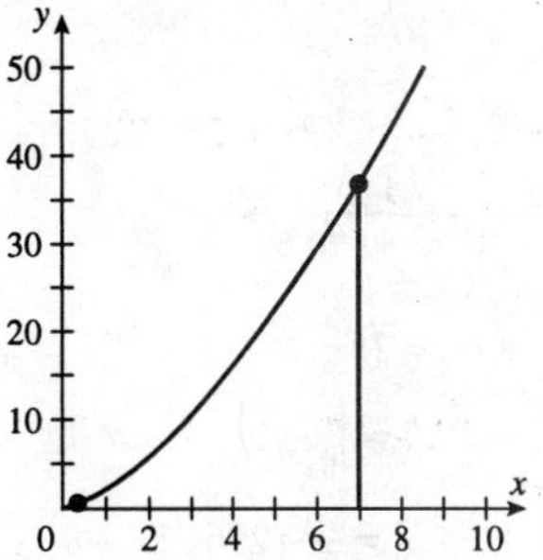

$y' = 3\sqrt{x}$, $L = \int_{1/3}^{7} \sqrt{1+9x}dx$

$= \frac{2}{27}\left[(1+9x)^{3/2}\right]_{1/3}^{7} = \frac{112}{3}$

5.

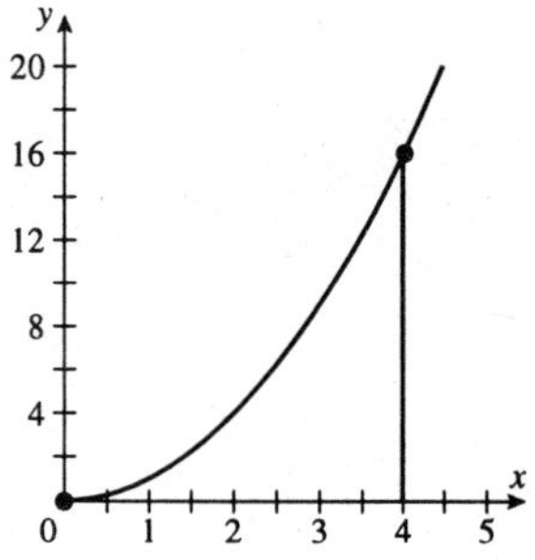

$y' = 2x, \;\; L = \int_0^4 \sqrt{1+4x^2}\,dx \approx 16.8186$

(Numerical methods were used to evaluate the integral.)

7.

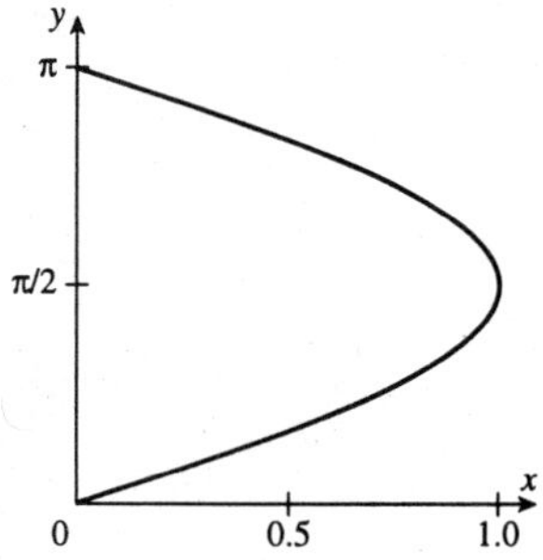

$\dfrac{dx}{dy} = \cos y$

$L = \int_0^{\pi} \sqrt{1+\cos^2 y}\,dy \approx 3.8202$

(Numerical methods were used to evaluate the integral.)

9.

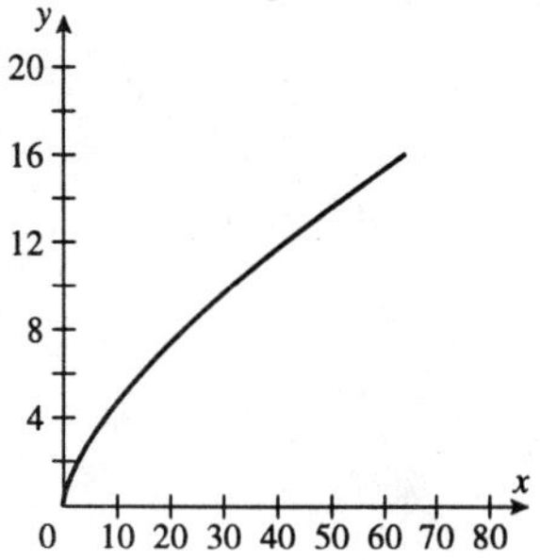

$f'(t) = 3t^2, \;\; g'(t) = 2t$

$L = \int_0^4 \sqrt{9t^4 + 4t^2}\,dt = \int_0^4 t\sqrt{9t^2+4}\,dt$

$= \dfrac{1}{27}[(9t^2+4)^{3/2}]_0^4 = \dfrac{1}{27}\left(296\sqrt{37} - 8\right)$

≈ 66.3888

11.

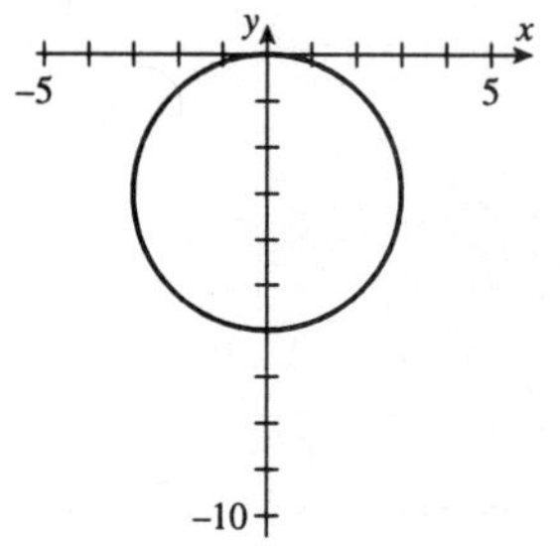

$f'(t) = 3\cos t, \;\; g'(t) = -3\sin t$

$L = \int_0^{2\pi} \sqrt{9\cos^2 t + 9\sin^2 t}\,dt$

$= \int_0^{2\pi} 3\,dt = [3t]_0^{2\pi} = 6\pi \approx 18.8496$

13.

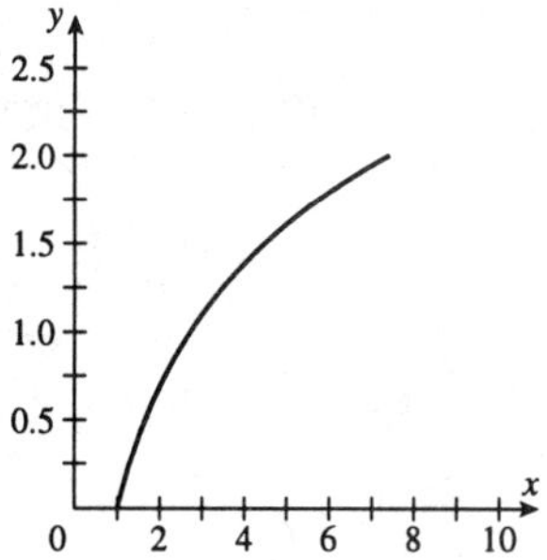

$f'(t) = 1, \;\; g'(t) = \dfrac{1}{t}$

$L = \int_1^{e^2} \sqrt{1 + \dfrac{1}{t^2}}\,dt = \int_1^{e^2} \dfrac{1}{t}\sqrt{t^2+1}\,dt$

≈ 6.78865

(Numerical methods were used to evaluate the integral.)

The function in Problem 8 can be written as $y = \ln x$ between $x = 1$ and $x = e^2$ which can be parametrized as $x = t, y = \ln t$; $1 \le t \le e^2$. The curves are the same.

15. a.

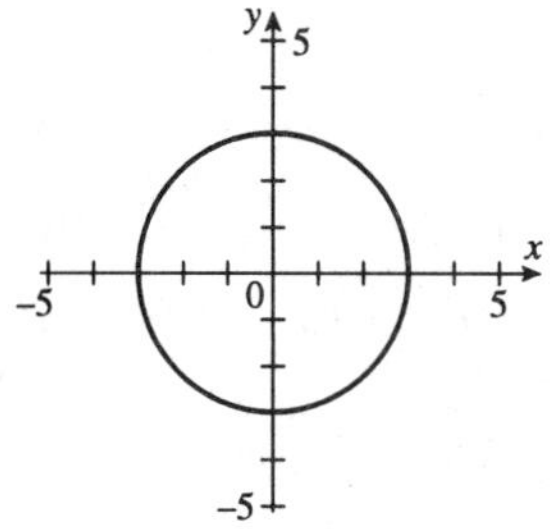

b.

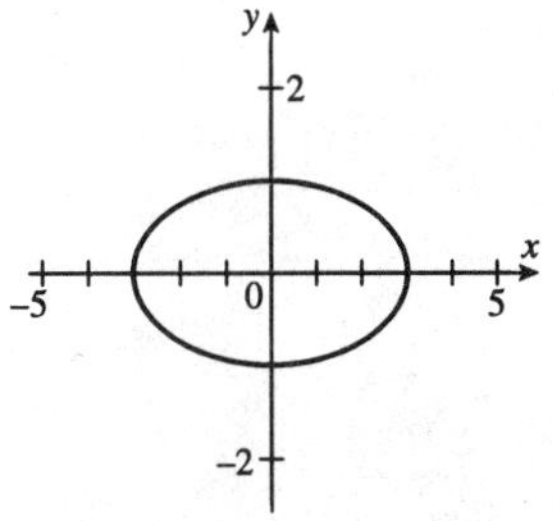

c.

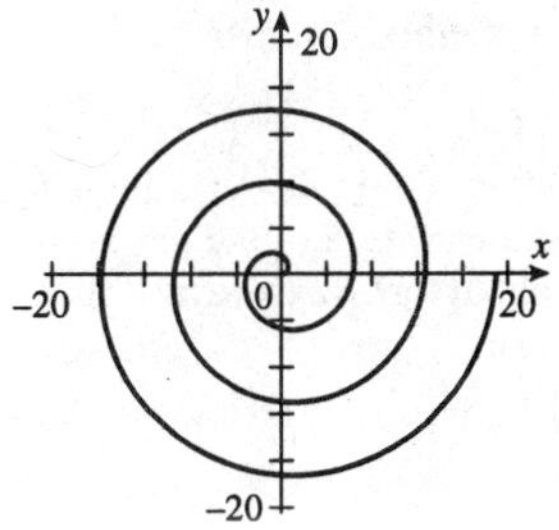

d.

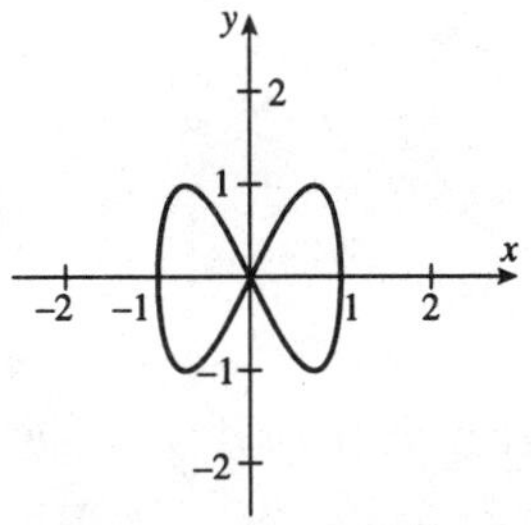

e.

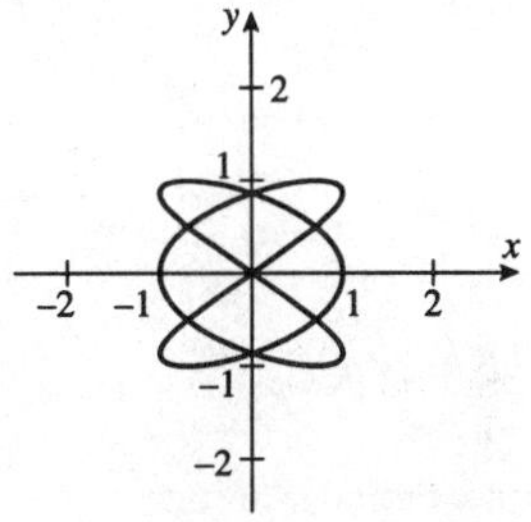

f.

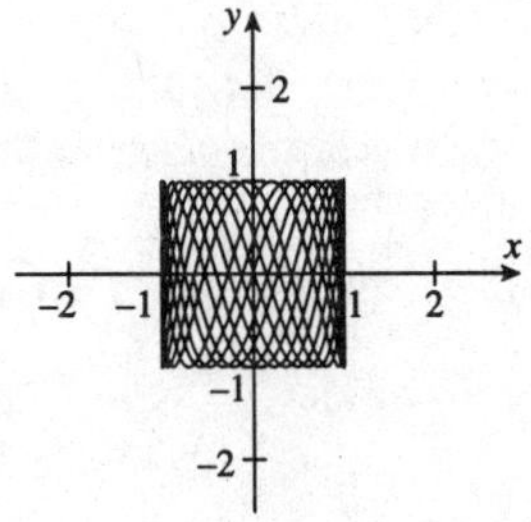

17.

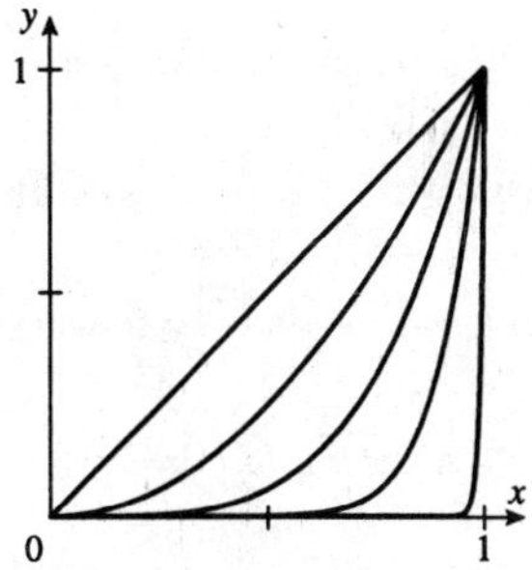

$y = x,\ y' = 1,$

$L = \int_0^1 \sqrt{2}\,dx = \left[\sqrt{2}\,x\right]_0^1 = \sqrt{2} \approx 1.41421$

$y = x^2,\ y' = 2x,\ L = \int_0^1 \sqrt{1+4x^2}\,dx \approx 1.47894$

$y = x^4,\ y' = 4x^3,\ L = \int_0^1 \sqrt{1+16x^6}\,dx \approx 1.60023$

$y = x^{10},\ y' = 10x^9,$

$L = \int_0^1 \sqrt{1+100^{18}}\,dx \approx 1.75441$

$y = x^{100},\ y' = 100x^{99},$

$L = \int_0^1 \sqrt{1+10{,}000x^{198}}\,dx \approx 1.95167$

Numerical methods were used to evaluate the integrals.
When $n = 10{,}000$ the length will be close to 2.

19. a. $\overline{OT} = \text{length}(\widehat{PT}) = a\theta$

b. Written response.

c. $x = \overline{OT} - \overline{PQ} = a\theta - a\sin\theta = a(\theta - \sin\theta)$
$y = \overline{CT} - \overline{CQ} = a - a\cos\theta = a(1-\cos\theta)$

21. a. Using $\theta = \omega t$, the point P is at $x = a\omega t - a\sin(\omega t),\ \ y = a - a\cos(\omega t)$ at time t.

$x'(t) = a\omega - a\omega\cos(\omega t) = a\omega(1-\cos(\omega t))$

$y'(t) = a\omega\sin(\omega t)$

$\frac{ds}{dt} = \sqrt{[x'(t)]^2 + [y'(t)]^2}$

$= \sqrt{2a^2\omega^2 - 2a^2\omega^2\cos(\omega t)}$

$= 2a\omega\sqrt{\frac{1}{2}(1-\cos(\omega t))}$

$= 2a\omega\left|\sin\frac{\omega t}{2}\right|$

(Use a half-angle formula.)

b. The speed is a maximum when $\left|\sin\frac{\omega t}{2}\right| = 1$, which occurs when $t = \frac{\pi}{\omega}(2k+1)$. The speed is a minimum when $\left|\sin\frac{\omega t}{2}\right| = 0$, which occurs when $t = \frac{2k\pi}{\omega}$.

c. From Problem 19a, the distance traveled by the wheel is $a\theta$, so at time t, the wheel has gone $a\theta = a\omega t$ miles. Since the car is going 60 miles per hour, the wheel has gone $60t$ miles at time t. Thus, $a\omega = 60$ and the maximum speed of the bug on the wheel is $2a\omega = 2(60) = 120$ miles per hour.

23. a. From Section 5.6

$\frac{dy}{dx} = \sqrt{64\sin^2 x\cos^4 x - 1}$

$L = \int_{\pi/6}^{\pi/3}\sqrt{1 + 64\sin^2\cos^4 x - 1}dx$

$= \int_{\pi/6}^{\pi/3} 8\sin x\cos^2 x dx$

$= \left[-\frac{8}{3}\cos^3 x\right]_{\pi/6}^{\pi/3} = \sqrt{3} - \frac{1}{3} \approx 1.39872$

b. $f'(t) = at\cos t$

$g'(t) = at\sin t$

$L = \int_{-1}^{1}\sqrt{a^2t^2\cos^2 t + a^2t^2\sin^2 t}dt$

$= \int_{-1}^{1}|at|dt = \int_0^1 at\,dt - \int_{-1}^{0} at\,dt$

$= \left[\frac{at^2}{2}\right]_0^1 - \left[\frac{at^2}{2}\right]_{-1}^{0}$

$= a$

25. $f'(x) = -\frac{x}{\sqrt{25 - x^2}}$

$A = 2\pi\int_{-2}^{3}\sqrt{25 - x^2}\sqrt{1 + \frac{x^2}{25 - x^2}}dx$

$= 2\pi\int_{-2}^{3} 5dx = 10\pi[x]_{-2}^{3} = 50\pi$

27. $f'(x) = 3x^2$

$A = 2\pi\int_0^1 x^3\sqrt{1 + 9x^4}dx$

$= 2\pi\left[\frac{1}{54}(1 + 9x^4)^{3/2}\right]_0^1 = \frac{\pi}{27}(10^{3/2} - 1)$

≈ 3.56312

The integral in Problem 27 is easier to find exactly.

29. $f'(x) = \cos x$

$A = 2\pi\int_0^{\pi}\sin x\sqrt{1 + \cos^2 x}dx$

≈ 14.4236

(Numerical methods were used to evaluate the integral.)

31. $f'(t) = -2t;\ g'(t) = e^t$

$A = 2\pi\int_0^1 e^t\sqrt{4t^2 + e^{2t}}dt \approx 23.8581$

(Numerical methods were used to evaluate the integral.)

33. Put the center of a circle of radius a at $(a, 0)$. Revolving the portion of the circle from $x = b$ to $x = b + h$ about the x-axis results in the surface in question. (See figure.)

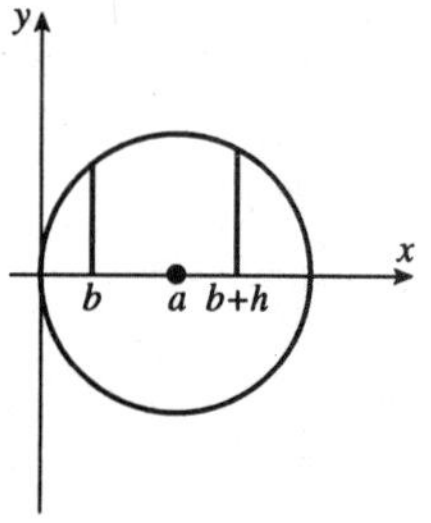

The equation of the top half of the circle is

$y = \sqrt{a^2 - (x - a)^2}$.

$f'(x) = \frac{a - x}{\sqrt{a^2 - (x - a)^2}}$

$A = 2\pi\int_b^{b+h}\sqrt{a^2 - (x - a)^2}\sqrt{1 + \frac{(a - x)^2}{a^2 - (x - a)^2}}dx$

$= 2\pi\int_b^{b+h} adx = 2\pi a[x]_b^{b+h}$

$= 2\pi ah$

A right circular cylinder of radius a and height h has surface area $2\pi ah$.

35. $f'(t) = -a\sin t,\ g'(t) = a\cos t$

Since the circle is being revolved about the line $x = b$, the surface area is

$A = 2\pi\int_0^{2\pi}(b - a\cos t)\sqrt{a^2\sin^2 t + a^2\cos^2 t}dt$

$= 2\pi a\int_0^{2\pi}(b - a\cos t)dt$

$= 2\pi a[bt - a\sin t]_0^{2\pi}$

$= 4\pi^2 ab$

Section 6.4

Concepts Review

1. $F \cdot (b-a)$; $\int_a^b F(x)dx$

3. 300

Problem Set 6.4

1. By Hooke's Law, $F\left(\frac{1}{2}\right) = 8 = k\left(\frac{1}{2}\right)$, so $k = 16$.

$W = \int_0^{1/2} 16x dx = [8x^2]_0^{1/2} = 2$ foot-pounds

3. By Hooke's Law, $F(2) = 200 = k(2)$, so $k = 100$.
$W = \int_0^4 100x dx = [50x^2]_0^4 = 800$ ergs

5. $W = \int_0^d kx dx = \left[\frac{1}{2}kx^2\right]_0^d$
$= \frac{1}{2}k(d^2 - 0) = \frac{1}{2}kd^2$

7. a.

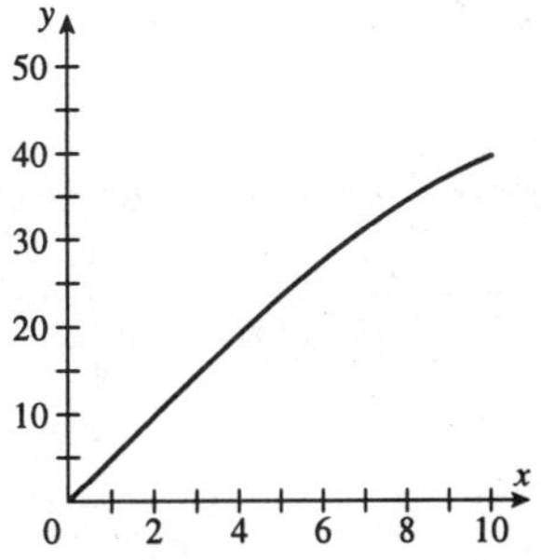

It appears that the function is approximately linear for $0 \le x \le 6$. $F(6) \approx 28 = k(6)$, $k \approx 4.67$.

b. To stretch the spring 2 inches:
$W = \int_0^2 (5x - 0.01x^3)dx$
$= \left[\frac{5}{2}x^2 - 0.0025x^4\right]_0^2$
$= 9.96$ inch-pounds
To stretch the spring 6 inches:
$W = \int_0^6 (5x - 0.01x^3)dx$
$= \left[\frac{5}{2}x^2 - 0.0025x^4\right]_0^6$
$= 86.76$ inch pounds

c. $F = 30$ when $x \approx 6.57$.
$W \approx \int_0^{6.57} (5x - 0.01x^3)dx$
$= \left[\frac{5}{2}x^2 - 0.0025x^4\right]_0^{6.57} \approx 103.254$
The spring is not approximately linear at this point. Written response.

9. $\Delta W \approx \delta \cdot 10 \cdot \left(4 - \frac{4y}{5}\right)\Delta y(10-y)$
$W = \int_0^5 8\delta(y^2 - 15y + 50)dy$
$= 8(62.4)\left[\frac{y^3}{3} - \frac{15y^2}{2} + 50y\right]_0^5$
$= 52{,}000$ foot-pounds

11. $\Delta W \approx \delta \cdot 10 \cdot \left(\frac{3}{4}y + 3\right)\Delta y(9-y)$
$W = \int_0^4 \frac{15}{2}\delta(36 + 5y - y^2)dy$
$= \frac{15}{2}(62.4)\left[36y + \frac{5y^2}{2} - \frac{y^3}{3}\right]_0^4$
$\approx 76{,}128$ foot-pounds

13. Let $y = 0$ be at the bottom of the tank. A disk of thickness Δy at height y has weight of about $25\pi\delta\Delta y$, and must be lifted $10 - y$ feet.
$W = \int_0^{10} 25\pi\delta(10-y)dy$
$= 1250\pi\int_0^{10}(10-y)dy$
$= 1250\pi\left[10y - \frac{1}{2}y^2\right]_0^{10} = 62{,}500\pi$
$\approx 196{,}350$ foot-pounds

15. The total force on the face of the piston is $A \cdot f(x)$ if the piston is x inches from the cylinder head. The work done by moving the piston from x_1 to x_2 is $W = \int_{x_1}^{x_2} A \cdot f(x)dx = A\int_{x_1}^{x_2} f(x)dx$. This is the work done by the gas in moving the piston. The work done by the piston to compress the gas is the opposite of this or $A\int_{x_2}^{x_1} f(x)dx$.

17. Since $pv^{1.4} = c$ and $v = 16$ when $p = 40$, $c \approx 1940.12$ and $p(v) = cv^{-1.4}$. If the area of the face of the piston is 2 square inches, then the piston is 8 inches from the cylinder head when the volume is 16 cubic inches, and 1 inch from the cylinder head when the volume is 2 cubic inches.

$v = 2x$

$W = 2\int_1^8 c(2x)^{-1.4}\,dx$

$\approx 3880.24(2^{-1.4})\left[\frac{1}{-0.4x^{0.4}}\right]_1^8$

≈ 2075.84 foot-pounds

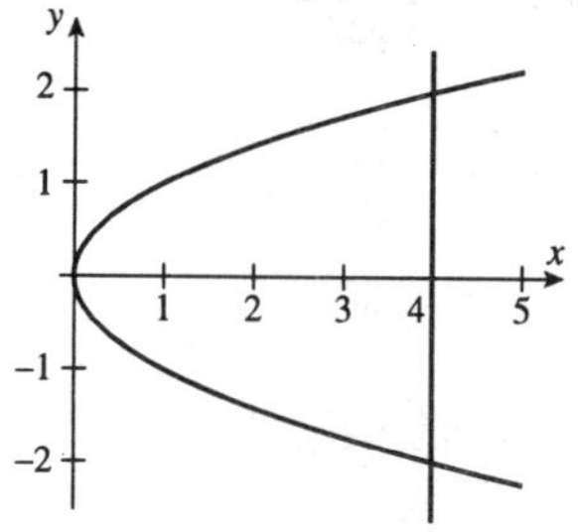

19. The total work is equal to the work W_1 to haul the load by itself and the work W_2 to haul the rope by itself.

$W_1 = 200 \cdot 500 = 100{,}000$ foot-pounds

Let $y = 0$ be the bottom of the shaft. When the rope is at y, $\Delta W_2 \approx 2\Delta y(500 - y)$.

$$W_2 = \int_0^{500} 2(500-y)dy = 2\left[500y - \frac{y^2}{2}\right]_0^{500}$$

$= 250{,}000$ foot-pounds

$W = W_1 + W_2 = 100{,}000 + 250{,}000$

$= 350{,}000$ foot-pounds

21. $f(x) = \frac{k}{x^2}$; $f(4000) = 5000$ so $5000 = \frac{k}{4000^2}$;

so $k = 80{,}000{,}000{,}000$

$$W = \int_{4000}^{4200} \frac{80{,}000{,}000{,}000}{x^2}dx$$

$$= 80{,}000{,}000{,}000\left[-\frac{1}{x}\right]_{4000}^{4200}$$

$$= \frac{20{,}000{,}000}{21} \approx 952{,}381 \text{ mile-pounds}$$

23. The relationship between the height of the bucket and time is $y = 2t$, so $t = \frac{1}{2}y$. When the bucket is a height y, the sand has been leaking out of the bucket for $\frac{1}{2}y$ seconds. The weight of the bucket and sand is $100 + 500 - 3\left(\frac{1}{2}y\right) = 600 - \frac{3}{2}y$. The work to lift the bucket of sand from y to $y + \Delta y$ is $\left(600 - \frac{3}{2}y\right)\Delta y$.

$$W = \int_0^{80}\left(600 - \frac{3}{2}y\right)dy$$

$$= \left[600y - \frac{3}{4}y^2\right]_0^{80}$$

$= 43{,}200$ foot-pounds

25. The total work is equal to the work W_1 needed to fill pipe plus the work W_2 needed to fill the tank. Let $\delta = 62.4$.

$$\Delta W_1 = \delta\pi\left(\frac{1}{2}\right)^2 \Delta y(y)$$

$$W_1 = \int_0^{30} \frac{\delta\pi y}{4}dy = \frac{(62.4)\pi}{4}\left[\frac{1}{2}y^2\right]_0^{30}$$

$\approx 22{,}054$ foot-pounds

The cross sectional area at height y feet ($30 \le y \le 50$) is πr^2 where

$r = \sqrt{10^2 - (40-y)^2} = \sqrt{-y^2 + 80y - 1500}$.

$$\Delta W_2 = \delta\pi r^2 \Delta y(y) = \delta\pi(-y^3 + 80y^2 - 1500y)\Delta y$$

$$W_2 = \int_{30}^{50} \delta\pi(-y^3 + 80y^2 - 1500y)dy$$

$$= (62.4)\pi\left[-\frac{y^4}{4} + \frac{80y^3}{3} - 750y^2\right]_{30}^{50}$$

$\approx 10{,}455{,}220$ foot-pounds

$W = W_1 + W_2 \approx 10{,}477{,}274$ foot-pounds

27. Since $\delta\left(\frac{1}{3}\pi a^2\right)(8) = 300$, $a = \sqrt{\frac{225}{2\pi\delta}}$.

When the buoy is at z feet ($0 \le z \le 2$), the radius r at the water level is

$$r = \left(\frac{8+z}{8}\right)a = \sqrt{\frac{225}{2\pi\delta}}\left(\frac{8+z}{8}\right).$$

$$F = \delta\left(\frac{1}{3}\pi r^2\right)(8+z) - 300$$

$$= \frac{75}{128}(8+z)^3 - 300$$

$$W = \int_0^2\left[\frac{75}{128}(8+z)^3 - 300\right]dz$$

$$= \left[\frac{75}{512}(8+z)^4 - 300z\right]_0^2$$

$$= \frac{8475}{32} \approx 265 \text{ foot-pounds}$$

Section 6.5

Concepts Review

1. Right; $\frac{4 \cdot 1 + 6 \cdot 3}{10} = 2.2$

3. 1; 3

Problem Set 6.5

1. $\bar{x} = \frac{4\cdot 2 + 6(-2) + 9\cdot 1}{19} = \frac{5}{19}$

3. $\bar{x} = \frac{\int_0^9 x\sqrt{x}dx}{\int_0^9 \sqrt{x}dx} = \frac{\int_0^9 x^{3/2}dx}{\int_0^9 x^{1/2}dx}$

$= \frac{\left[\frac{2}{5}x^{5/2}\right]_0^9}{\left[\frac{2}{3}x^{3/2}\right]_0^9} = \frac{\frac{486}{5}}{18} = 5.4$

5. $M_y = 3\cdot 1 + 2\cdot 7 + 4(-2) + 6(-1) + 2\cdot 4$
$= 11$
$M_x = 3\cdot 1 + 2\cdot 1 + 4\cdot(-5) + 6\cdot 0 + 2\cdot 6$
$= -3$
$m = 3 + 2 + 4 + 6 + 2 = 17$
$\bar{x} = \frac{11}{17};\ \bar{y} = -\frac{3}{17}$

7. Let region 1 be the region bounded by $x = -2$, $x = 2$, $y = 0$, and $y = 1$, so $m_1 = 4\cdot 1 = 4$.
By symmetry, $\bar{x}_1 = 0$ and $\bar{y}_1 = \frac{1}{2}$. Therefore $M_{1y} = \bar{x}_1 \cdot m_1 = 0$ and $M_{1x} = \bar{y}_1 \cdot m_1 = 2$.
Let region 2 be the region bounded by $x = -2$, $x = 1$, $y = -1$, and $y = 0$, so $m_2 = 3\cdot 1 = 3$.
By symmetry, $\bar{x}_2 = -\frac{1}{2}$ and $\bar{y}_2 = -\frac{1}{2}$. Therefore $M_{2y} = \bar{x}_2 \cdot m_2 = -\frac{3}{2}$ and $M_{2x} = \bar{y}_2 \cdot m_2 = -\frac{3}{2}$.
$\bar{x} = \frac{M_{1y} + M_{2y}}{m_1 + m_2} = \frac{-\frac{3}{2}}{7} = -\frac{3}{14}$
$\bar{y} = \frac{M_{1x} + M_{2y}}{m_1 + m_2} = \frac{\frac{1}{2}}{7} = \frac{1}{14}$

9. Let region 1 be the region bounded by $x = -2$, $x = 2$, $y = 2$, and $y = 4$, so $m_1 = 4\cdot 2 = 8$. By symmetry, $\bar{x}_1 = 0$ and $\bar{y}_1 = 3$. Therefore, $M_{1y} = \bar{x}_1 \cdot m_1 = 0$ and $M_{1x} = \bar{y}_1 \cdot m_1 = 24$. Let region 2 to be the region bounded by $x = -1$, $x = 2$, $y = 0$, and $y = 2$, so $m_2 = 3\cdot 2 = 6$. By symmetry, $\bar{x}_2 = \frac{1}{2}$ and $\bar{y}_2 = 1$. Therefore, $M_{2y} = \bar{x}_2 \cdot m_2 = 3$ and $M_{2x} = \bar{y}_2 \cdot m_2 = 6$. Let region 3 be the region bounded by $x = 2$, $x = 4$, $y = 0$, and $y = 1$, so $m_3 = 2\cdot 1 = 2$. By symmetry, $\bar{x}_3 = 3$ and $\bar{y}_3 = \frac{1}{2}$. Therefore, $M_{3y} = \bar{x}3 \cdot m_3 = 6$ and $M_{3x} = \bar{y}_3 \cdot m_3 = 1$.

$\bar{x} = \frac{M_{1y} + M_{2y} + M_{3y}}{m_1 + m_2 + m_3} = \frac{9}{16}$
$\bar{y} = \frac{M_{1x} + M_{2x} + M_{3x}}{m_1 + m_2 + m_3} = \frac{31}{16}$

11.

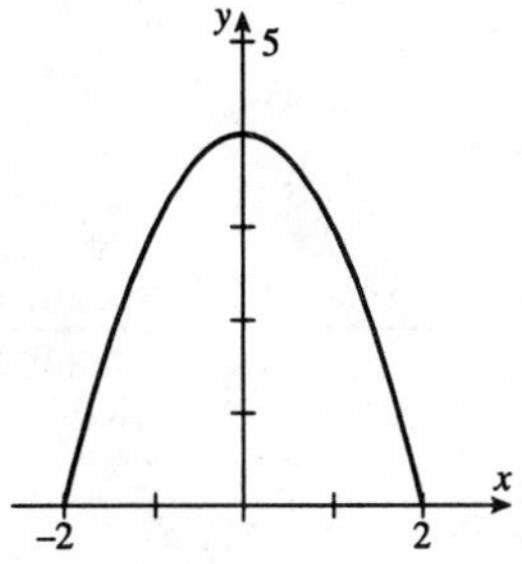

By symmetry $\bar{x} = 0$.

$\bar{y} = \frac{\frac{1}{2}\int_{-2}^{2}(4 - x^2)^2 dx}{\int_{-2}^{2}(4 - x^2)dx}$

$= \frac{\frac{1}{2}\int_{-2}^{2}(16 - 8x^2 + x^4)dx}{\int_{-2}^{2}(4 - x^2)dx}$

$= \frac{\frac{1}{2}\left[16x - \frac{8}{3}x^3 + \frac{1}{5}x^5\right]_{-2}^{2}}{\left[4x - \frac{1}{3}x^3\right]_{-2}^{2}} = \frac{\frac{256}{15}}{\frac{32}{3}} = \frac{8}{5}$

13.

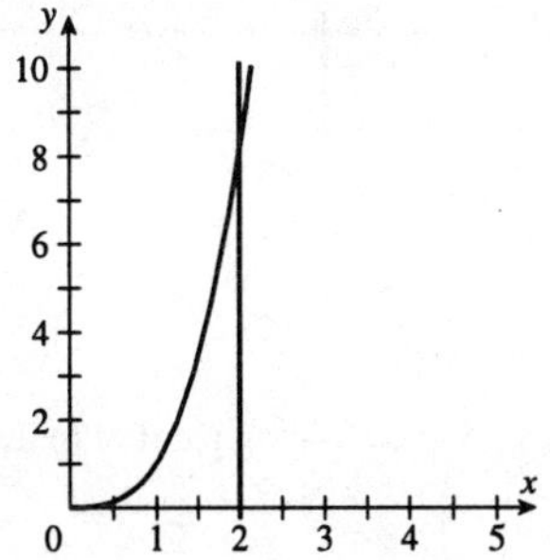

$\bar{x} = \frac{\int_0^2 x(x^3)dx}{\int_0^2 x^3 dx} = \frac{\left[\frac{1}{5}x^5\right]_0^2}{\left[\frac{1}{4}x^4\right]_0^2} = \frac{8}{5}$

$\bar{y} = \frac{\frac{1}{2}\int_0^2 x^6 dx}{\int_0^2 x^3 dx} = \frac{\frac{1}{2}\left[\frac{1}{7}x^7\right]_0^2}{\left[\frac{1}{4}x^4\right]_0^2} = \frac{16}{7}$

15.

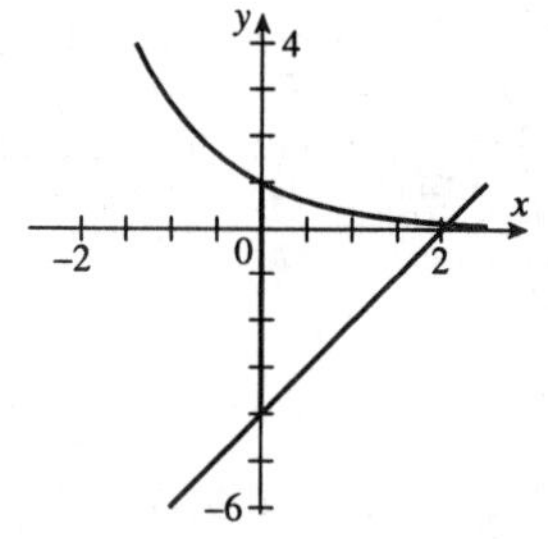

$$\bar{x} = \frac{\int_0^{2.0635} x[e^{-x} - (2x-4)]dx}{\int_0^{2.0635} (e^{-x} - 2x + 4)dx} \approx \frac{3.26934}{4.86896}$$

≈ 0.67

$$\bar{y} = \frac{\frac{1}{2}\int_0^{2.0635} [e^{-2x} - (2x-4)^2]dx}{\int_0^{2.0635} (e^{-x} - 2x + 4)dx}$$

$$= \frac{\frac{1}{2}\left[-\frac{1}{2}e^{-2x} - 16x + 8x^2 - \frac{4}{3}x^3\right]_0^{2.0635}}{[-e^{-x} - x^2 + 4x]_0^{2.0635}}$$

$$\approx -\frac{5.08755}{4.86896} \approx -1.04$$

(Numerical methods were used to find the intersection point of the curves and to evaluate the integral in the numerator of $\bar{x}$.)

17.

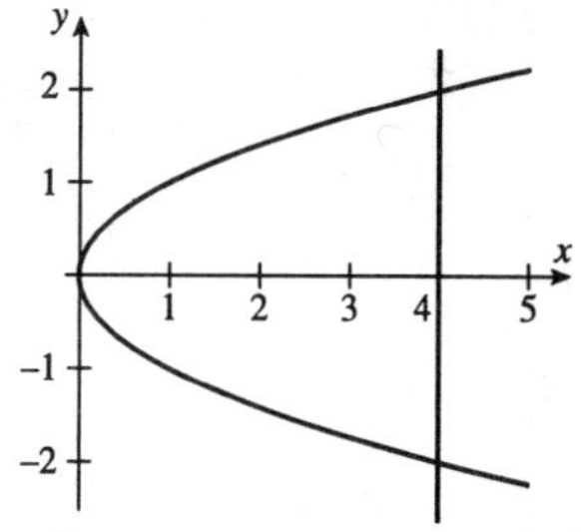

By symmetry, $\bar{y} = 0$. Slicing parallel to the $\bar{x}$-axis,

$$\bar{x} = \frac{\frac{1}{2}\int_{-2}^{2}(16 - y^4)dy}{\int_{-2}^{2}(4 - y^2)dy} = \frac{\frac{1}{2}\left[16y - \frac{1}{5}y^5\right]_{-2}^{2}}{\left[4y - \frac{1}{3}y^3\right]_{-2}^{2}}$$

$$= \frac{\frac{128}{5}}{\frac{32}{3}} = 2.4.$$

19. From Problem 13, the centroid is $\left(\frac{8}{5}, \frac{16}{7}\right)$. Thus, the distance traveled by the centroid is

$2\pi\left(\frac{8}{5}\right) = \frac{16\pi}{5}.$

$$A = \int_0^2 x^3 dx = \left[\frac{1}{4}x^4\right]_0^2 = 4$$

By Pappus's Theorem, the volume is

$4\left(\frac{16\pi}{5}\right) \approx 40.2124$. Using disks, the volume is

$$V = \pi\int_0^8 (4 - y^{2/3})dy = \pi\left[4y - \frac{3}{5}y^{5/3}\right]_0^8 = \frac{64}{5}\pi$$

which agrees with the result of Pappus's Theorem.

21. The volume of a sphere of radius a is $\frac{4}{3}\pi a^3$. If the semicircle $y = \sqrt{a^2 - x^2}$ is revolved about the x-axis is the result of a sphere of radius a. The centroid of the region travels a distance of $2\pi\bar{y}$.

The area of the region is $\frac{1}{2}\pi a^2$. Pappus's Theorem says that

$$(2\pi\bar{y})\left(\frac{1}{2}\pi a^2\right) = \pi^2 a^2 \bar{y} = \frac{4}{3}\pi a^3.$$

$\bar{y} = \frac{4a}{3\pi}$, $\bar{x} = 0$ (by symmetry)

23. a. If a slice at y is rotated about the line $y = e$, $\Delta V \approx 2\pi(e - y)w(y)\Delta y$.

Therefore, $V = 2\pi\int_c^d (e - y)dy$.

b. $M_x = \int_c^d yw(y)dy, \ m = \int_c^d w(y)dy$

$$\bar{y} = \frac{\int_c^d yw(y)dy}{\int_c^d w(y)dy}$$

The distance traveled by the centroid is

$$2\pi(e - \bar{y}) = 2\pi\left(e - \frac{\int_c^d yw(y)dy}{\int_c^d w(y)dy}\right)$$

$$= 2\pi\left(\frac{\int_c^d (e - y)w(y)dy}{\int_c^d w(y)dy}\right).$$

$A = \int_c^d w(y)dy$

By Pappus's Theorem,

$V = 2\pi(e - \bar{y})A = 2\pi\int_c^d (e - y)w(y)dy.$

This gives the same result as part a.

25. a. The area of a regular polygon P of $2n$ sides is $2r^2 n\sin\frac{\pi}{2n}\cos\frac{\pi}{2n}$. (To find this consider the isosceles triangles with one vertex at the center of the polygon and the other vertices on adjacent corners of the polygon. Each such triangle has base of length $2r\sin\frac{\pi}{2n}$ and height $r\cos\frac{\pi}{2n}$.) Since P is a regular polygon the centroid is at its center. The distance from the centroid to any side is $r\cos\frac{\pi}{2n}$, so the centroid travels a distance of

$2\pi r\cos\frac{\pi}{2n}$.

Thus, by Pappus's Theorem, the volume of the resulting solid is

$$\left(2\pi r\cos\frac{\pi}{2n}\right)\left(2r^2 n\sin\frac{\pi}{2n}\cos\frac{\pi}{2n}\right)$$

$$= 4\pi r^3 n\sin\frac{\pi}{2n}\cos^2\frac{\pi}{2n}.$$

b. $$\lim_{n\to\infty} 4\pi r^3 n\sin\frac{\pi}{2n}\cos^2\frac{\pi}{2n}$$

$$\lim_{n\to\infty}\frac{\sin\frac{\pi}{2n}}{\frac{\pi}{2n}}2\pi^2 r^3\cos^2\frac{\pi}{2n}$$

$$= 2\pi^2 r^3$$

As $n\to\infty$, the regular polygon approaches a circle. Using Pappus's Theorem on the circle of area πr^2 whose centroid (= center) travels a distance of $2\pi r$, the volume of the solid is $(\pi r^2)(2\pi r) = 2\pi^2 r^3$ which agrees with the results from the polygon.

Section 6.6 Chapter Review

Concepts Test

1. False. $\int_0^\pi \cos x\,dx = 0$ because half of the area lies above the x-axis and half below the x-axis.

3. False. The statement would be true if either $f(x) \ge g(x)$ or $g(x) \ge f(x)$ for $a \le x \le b$. Consider Problem 1 with $f(x) = \cos x$ and $g(x) = 0$.

5. True. Since the cross sections in all planes parallel to the bases have the same area, the integrals used to compute the volumes will be equal.

7. False. The area is $\int_0^1 \sqrt{x}\,dx = \left[\frac{2}{3}x^{3/2}\right]_0^1 = \frac{2}{3}$.

9. False. A spiral with no space between successive curves would have infinite length.

11. False. If the cone-shaped tank is placed with the point downward, then the amount of water that needs to be pumped from near the bottom of the tank is much less than the amount that needs to be pumped from near the bottom of the cylindrical tank.

13. True. This is the definition of the center of mass.

15. True. By symmetry, the centroid is on the line $x = \frac{\pi}{2}$, so the centroid travels a distance of $2\pi\left(\frac{\pi}{2}\right) = \pi^2$.

17. True. Since the density is proportional to the square of the distance from the midpoint, equal masses are on either side of the midpoint.

Sample Test Problems

1. $A = \int_0^1 (x - x^2)dx = \left[\frac{x}{2} - \frac{x^3}{3}\right]_0^1 = \frac{1}{6}$

3. $y = x - x^2;\ x^2 - x + y = 0$

$$x = \frac{1 \pm \sqrt{1-4y}}{2}$$

$$V_2 = \pi\int_0^{1/4}\left[\left(\frac{1+\sqrt{1-4y}}{2}\right)^2 - \left(\frac{1-\sqrt{1-4y}}{2}\right)^2\right]dy$$

$$= \pi\int_0^{1/4}\sqrt{1-4y}\,dy$$

$$= \pi\left[-\frac{1}{6}(1-4y)^{3/2}\right]_0^{1/4} = \frac{\pi}{6}$$

5. $$V_4 = \pi\int_0^{1/4}\left[\left(2 - \frac{1-\sqrt{1-4y}}{2}\right)^2 - \left(2 - \frac{1+\sqrt{1-4y}}{2}\right)^2\right]dy$$

$$= \pi\int_0^{1/4} 3\sqrt{1-4y}\,dy = \pi\left[-\frac{1}{2}(1-4y)^{3/2}\right]_0^{1/4} = \frac{\pi}{2}$$

7. From Problem 1, $A = \frac{1}{6}$.

From Problem 6, $\bar{x} = \frac{1}{2}$ and $\bar{y} = \frac{1}{10}$.

$$V_1 = 2\pi\left(\frac{1}{10}\right)\left(\frac{1}{6}\right) = \frac{\pi}{30}$$

$$V_2 = 2\pi\left(\frac{1}{2}\right)\left(\frac{1}{6}\right) = \frac{\pi}{6}$$

$$V_3 = 2\pi\left(\frac{1}{10}+1\right)\left(\frac{1}{6}\right) = \frac{11\pi}{30}$$
$$V_4 = 2\pi\left(2-\frac{1}{2}\right)\left(\frac{1}{6}\right) = \frac{\pi}{2}$$

9. $W = \int_0^8 (62.4)(5^2)\pi(10-y)\,dy$
$= 1560\pi\int_0^8 (10-y)\,dy$
$= 1560\pi\left[10y - \frac{y^2}{2}\right]_0^8 = 74,880\pi$
$\approx$ 235,242 foot-pounds

11. a. $A = \int_0^3 (3x - x^2)\,dx = \left[\frac{3x^2}{2} - \frac{x^3}{3}\right]_0^3 = 4.5$

b. $A = \int_0^9 \left(\sqrt{y} - \frac{y}{3}\right)dy = \left[\frac{2}{3}y^{3/2} - \frac{y^2}{6}\right]_0^9 = 4.5$

13. $V = \pi\int_0^3 (9x^2 - x^4)\,dx = \pi\left[3x^3 - \frac{x^5}{5}\right]_0^3$
$= 32.4\pi \approx 101.8$
The centroid of the region is at
$$\bar{x} = \frac{\int_0^3 (3x^2 - x^3)\,dx}{\int_0^3 (3x - x^2)\,dx} = \frac{\left[x^3 - \frac{x^4}{4}\right]_0^3}{\left[\frac{3x^2}{2} - \frac{x^3}{3}\right]_0^3} = 1.5$$
$$\bar{y} = \frac{\frac{1}{2}\int_0^3 (9x^2 - x^4)\,dx}{\int_0^3 (3x - x^2)\,dx} = 3.6,$$ so the centroid moves a distance of 7.2π. The volume is $(4.5)(7.2\pi) = 32.4\pi \approx 101.8$

15. $L = \int_0^\pi \sqrt{1+\cos^2 x}\,dx \approx 3.82$
(Numerical methods were used to evaluate the integral.)

17. $V = \int_{-2}^2 \left(\sqrt{4-x^2}\right)^2 dx = \left[4x - \frac{x^3}{3}\right]_{-2}^2 = \frac{32}{3}$

19. $V = \pi\int_a^b \left[f^2(x) - g^2(x)\right]dx$

21. $M_y = \delta\int_a^b x[f(x) - g(x)]\,dx$
$M_x = \frac{\delta}{2}\int_a^b \left[f^2(x) - g^2(x)\right]dx$

23. $A_1 = 2\pi\int_a^b f(x)\sqrt{1+[f'(x)]^2}\,dx$
$A_2 = 2\pi\int_a^b g(x)\sqrt{1+[g'(x)]^2}\,dx$
$A_3 = \pi\left[f^2(a) - g^2(a)\right]$
$A_4 = \pi\left[f^2(b) - g^2(b)\right]$
Total surface area $= A_1 + A_2 + A_3 + A_4$

Chapter 7

Section 7.1

Concepts Review

1. $f(x_1) \neq f(x_2)$

3. Monotonic; increasing; decreasing

Problem Set 7.1

1. $f(x)$ is one-to-one and has an inverse. Since $f(4) = 2,\ f^{-1}(2) = 4$.

3. $f(x)$ is not one-to-one and so it does not have an inverse.

5. $f(x)$ is one-to-one and has an inverse. Since $f(-1) = 2,\ f^{-1}(2) = -1$.

7. $f'(x) = -15x^4 - 1 < 0$ for all x. Therefore $f(x)$ is strictly decreasing and has an inverse.

9. $f'(x) = \sec^2 x > 0$ for $-\frac{\pi}{2} < x < \frac{\pi}{2}$. Therefore $f(x)$ is strictly increasing and has an inverse.

11. $f'(x) = 6e^{2x} > 0$ for all x. Therefore $f(x)$ is strictly increasing and has an inverse.

13. $f'(x) = \sqrt{x^2 + 2} > 0$ for all x. Therefore $f(x)$ is strictly increasing and has an inverse.

15. $y = 3x - 1$
$y + 1 = 3x$
$\frac{y+1}{3} = x$
$f^{-1}(y) = \frac{y+1}{3}$
$f^{-1}(x) = \frac{1}{3}(x+1)$
$f^{-1}(f(x)) = f^{-1}(3x-1) = \frac{1}{3}(3x-1+1)$
$= x$
$f(f^{-1}(x)) = f\left(\frac{1}{3}(x+1)\right) = 3\left(\frac{1}{3}\right)(x+1) - 1$
$= x$

17. $y = \sqrt{2x+5}$
$y^2 = 2x + 5$
$\frac{y^2 - 5}{2} = x$
$f^{-1}(y) = \frac{y^2 - 5}{2}$
$f^{-1}(x) = \frac{x^2 - 5}{2}$
$f^{-1}(f(x)) = f^{-1}\left(\sqrt{2x+5}\right)$
$= \frac{\left(\sqrt{2x+5}\right)^2 - 5}{2}$
$= \frac{2x+5-5}{2}$
$= x$
$f(f^{-1}(x)) = \sqrt{2\left(\frac{x^2-5}{2}\right) + 5}$
$= \sqrt{x^2 - 5 + 5}$
$= x$

19. $y = \frac{1}{x-5}$
$x - 5 = \frac{1}{y}$
$x = \frac{1}{y} + 5$
$f^{-1}(y) = \frac{1}{y} + 5$
$f^{-1}(x) = \frac{1}{x} + 5$
$f^{-1}(f(x)) = f^{-1}\left(\frac{1}{x-5}\right)$
$= \frac{1}{\left(\frac{1}{x-5}\right)} + 5$
$= x - 5 + 5 = x$
$f(f^{-1}(x)) = f\left(\frac{1}{x} + 5\right)$
$= \frac{1}{\frac{1}{x} + 5 - 5}$
$= x$

21. $y = x^2,\ x \le 0$
$x = -\sqrt{y}$
$f^{-1}(y) = -\sqrt{y}$
$f^{-1}(x) = -\sqrt{x},\ x \ge 0$
$f(f^{-1}(x)) = f\left(-\sqrt{x}\right) = \left(-\sqrt{x}\right)^2 = x,\ x \ge 0$
$f^{-1}(f(x)) = f^{-1}(x^2) = -\sqrt{x^2} = -|x| = x,\ x \le 0$

Note that the domain of f is $x \le 0$, while the domain of f^{-1} is $x \ge 0$.

23. $y = (x-4)^3$

$y^{1/3} + 4 = x$

$f^{-1}(y) = y^{1/3} + 4$

$f^{-1}(x) = x^{1/3} + 4$

$f^{-1}(f(x)) = f^{-1}((x-4)^3)$

$= [(x-4)^3]^{1/3} + 4$

$= x - 4 + 4 = x$

$f(f^{-1}(x)) = f(x^{1/3} + 4)$

$= (x^{1/3} + 4 - 4)^3$

$= x$

25. $y = 3e^{5x}$

$\frac{y}{3} = e^{5x}$

$\ln\left(\frac{y}{3}\right) = 5x$

$x = \frac{1}{5}\ln\left(\frac{y}{3}\right)$

$f^{-1}(y) = \frac{1}{5}\ln\left(\frac{y}{3}\right)$

$f^{-1}(x) = \frac{1}{5}\ln\left(\frac{x}{3}\right)$

$f^{-1}(f(x)) = f^{-1}(3e^{5x})$

$= \frac{1}{5}\ln\left(\frac{3e^{5x}}{3}\right)$

$= \frac{1}{5}\ln(e^{5x})$

$= \frac{1}{5}(5x) = x$

$f(f^{-1}(x)) = f\left(\frac{1}{5}\ln\left(\frac{x}{3}\right)\right)$

$= 3e^{5\left(\frac{1}{5}\ln\left(\frac{x}{3}\right)\right)}$

$= 3\left(\frac{x}{3}\right) = x$

27. $y = \frac{x^3+1}{x^3+2}$

$y(x^3+2) = x^3+1$

$yx^3 + 2y = x^3 + 1$

$yx^3 - x^3 = 1 - 2y$

$x^3(y-1) = 1 - 2y$

$x^3 = \frac{1-2y}{y-1}$

$x = \sqrt[3]{\frac{1-2y}{y-1}}$

$f^{-1}(y) = \sqrt[3]{\frac{1-2y}{y-1}}$

$f^{-1}(x) = \sqrt[3]{\frac{1-2x}{x-1}}$

$f^{-1}(f(x)) = f^{-1}\left(\frac{x^3+1}{x^3+2}\right)$

$= \left[\frac{1-2\left(\frac{x^3+1}{x^3+2}\right)}{\left(\frac{x^3+1}{x^3+2}\right)-1}\right]^{1/3}$

$= \left[\frac{x^3+2-2x^3-2}{x^3+2} \cdot \frac{x^3+2}{x^3+1-x^3-2}\right]^{1/3}$

$= \left[\frac{-x^3}{-1}\right]^{1/3} = x$

$f(f^{-1}(x)) = f\left(\sqrt[3]{\frac{1-2x}{x-1}}\right)$

$= \frac{\frac{1-2x}{x-1}+1}{\frac{1-2x}{x-1}+2}$

$= \frac{1-2x+x-1}{x-1} \cdot \frac{x-1}{1-2x+2x-2}$

$= \frac{-x}{-1} = x$

29. The graph is decreasing on $(-\infty, -0.25]$ and increasing on $[-0.25, \infty)$. Restrict the domain to $(-\infty, -0.25]$ or restrict it to $[-0.25, \infty)$.

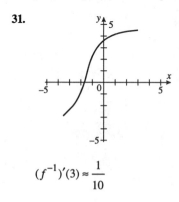

31.

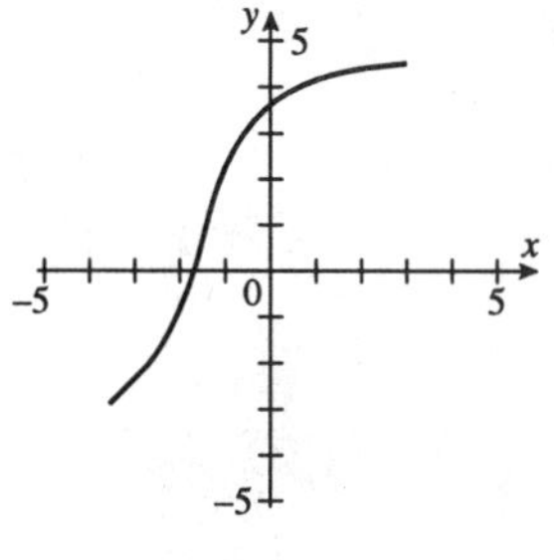

$(f^{-1})'(3) \approx \frac{1}{10}$

33. 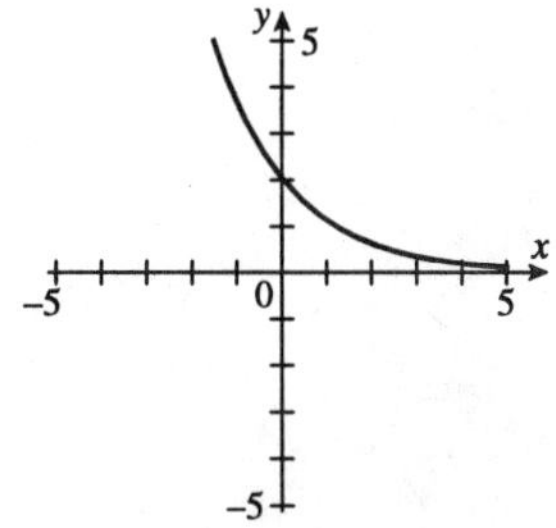

$(f^{-1})'(3) \approx -\frac{1}{3}$

35. $f'(x) = 15x^4 + 1$ and $y = 2$ corresponds to $x = 1$, so $(f^{-1})'(2) = \frac{1}{f'(1)} = \frac{1}{15+1} = \frac{1}{16}$.

37. $f'(x) = 2\sec^2 x$ and $y = 2$ corresponds to $x = \frac{\pi}{4}$, so $(f^{-1})'(2) = \frac{1}{f'\left(\frac{\pi}{4}\right)} = \frac{1}{2\sec^2\left(\frac{\pi}{4}\right)} = \frac{1}{2}\cos^2\left(\frac{\pi}{4}\right)$

$= \frac{1}{4}$.

39. $g^{-1}(f^{-1}(h(x))) = g^{-1}(f^{-1}(f(g(x))))$
$= g^{-1}(g(x)) = x$
$h(g^{-1}(f^{-1}(x))) = f(g(g^{-1}(f^{-1}(x))))$
$= f(f^{-1}(x)) = x$

41. f has an inverse because it is monotonic (increasing):
$\frac{df}{dx} = \sqrt{1+\cos^2 x} > 0$

a. $(f^{-1})'(A) = \frac{1}{f'\left(\frac{\pi}{2}\right)} = \frac{1}{\sqrt{1+\cos^2\left(\frac{\pi}{2}\right)}} = 1$

b. $(f^{-1})'(B) = \frac{1}{f'\left(\frac{5\pi}{6}\right)} = \frac{1}{\sqrt{1+\cos^2\left(\frac{5\pi}{6}\right)}} = \frac{1}{\sqrt{\frac{7}{4}}}$

$= \frac{2}{\sqrt{7}}$

c. $(f^{-1})'(0) = \frac{1}{f'(0)} = \frac{1}{\sqrt{1+\cos^2(0)}} = \frac{1}{\sqrt{2}}$

43. 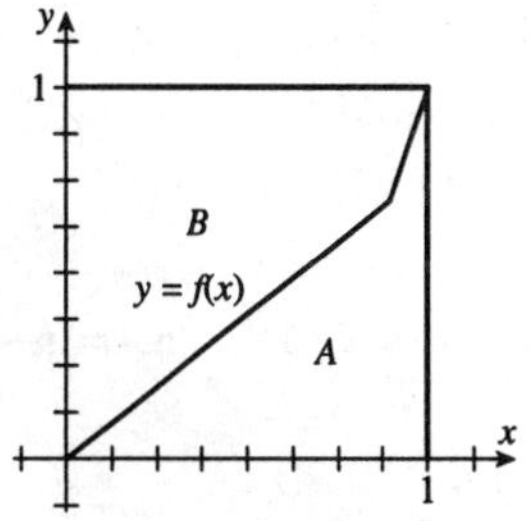

$\int_0^1 f^{-1}(y)dy =$ (Area of region B)
$= 1 -$ (Area of region A)
$= 1 - \int_0^1 f(x)dx = 1 - \frac{2}{5} = \frac{3}{5}$

Section 7.2

Concepts Review

1. $\int_1^x \left(\frac{1}{t}\right)dt$; $(0, \infty)$; $(-\infty, \infty)$

3. $\ln e = 1$; 2.72

Problem Set 7.2

1. a. Numerical methods were used to evaluate the integral $\int_1^6 \frac{1}{t}dt \approx 1.7918$.

b. Numerical methods were used to evaluate the integral $\int_1^{1.5} \frac{1}{t}dt \approx 0.4055$.

c. Numerical methods were used to evaluate the integral $\int_1^{81} \frac{1}{t}dt \approx 4.3944$.

d. Numerical methods were used to evaluate the integral $\int_1^{\sqrt{2}} \frac{1}{t}dt \approx 0.3466$

e. Numerical methods were used to evaluate the integral $\int_1^{1/36} \frac{1}{t}dt \approx -3.5835$

f. Numerical methods were used to evaluate the integral $\int_1^{48} \frac{1}{t}dt \approx 3.8712$

3. a. $e^{3\ln 2} = e^{\ln(2^3)} = e^{\ln 8} = 8$

b. $e^{\frac{(\ln 64)}{2}} = e^{\ln(64^{1/2})} = e^{\ln 8} = 8$

Written response.

5. $\ln e^{-x+2} = -x+2$

7. $e^{2\ln x} = e^{\ln(x^2)} = x^2 \quad (x > 0)$

9. $2\ln(x+1) - \ln x = \ln(x+1)^2 - \ln x = \ln\dfrac{(x+1)^2}{x}$

11. $\ln(x-2) - \ln(x+2) + 2\ln x = \ln\dfrac{x-2}{x+2} + \ln x^2 = \ln\dfrac{x^2(x-2)}{x+2} \quad (x > 2)$

13. $\dfrac{d}{dx}\ln(x-5)^4 = \dfrac{1}{(x-5)^4}\dfrac{d}{dx}(x-5)^4 = \dfrac{4(x-5)^3}{(x-5)^4} = \dfrac{4}{x-5}$

15. $\dfrac{dy}{dx} = \dfrac{d}{dx}[\ln x^3 + (\ln x)^3] = \dfrac{1}{x^3}\dfrac{d}{dx}x^3 + 3(\ln x)^2\dfrac{d}{dx}\ln x$

$= \dfrac{3x^2}{x^3} + \dfrac{3(\ln x)^2}{x} = \dfrac{3}{x}\left[1 + (\ln x)^2\right]$

17. $f'(x) = \dfrac{d}{dx}\ln\left(x + \sqrt{x^2-1}\right) = \dfrac{1}{x+\sqrt{x^2-1}}\dfrac{d}{dx}\left(x+\sqrt{x^2-1}\right)$

$= \dfrac{1}{x+\sqrt{x^2-1}}\left(1 + \dfrac{2x}{2\sqrt{x^2-1}}\right) = \dfrac{1}{x+\sqrt{x^2-1}}\left(\dfrac{\sqrt{x^2-1}+x}{\sqrt{x^2-1}}\right) = \dfrac{1}{\sqrt{x^2-1}}$

19. $f'(x) = \dfrac{d}{dx}\ln\sqrt[3]{x} = \dfrac{d}{dx}\dfrac{1}{3}\ln x = \dfrac{1}{3x}$

$f'(100) = \dfrac{1}{3(100)} = \dfrac{1}{300}$

21. $\dfrac{dy}{dx} = \dfrac{d}{dx}e^{\sqrt{x+1}} = e^{\sqrt{x+1}}\dfrac{d}{dx}\sqrt{x+1} = \dfrac{e^{\sqrt{x+1}}}{2\sqrt{x+1}}$

23. $\dfrac{dy}{dx} = \dfrac{d}{dx}(x^2e^x) = (2x)(e^x) + (x^2)(e^x) = (x^2+2x)e^x$

25. $e^{xy} + y = 2$

$\dfrac{d}{dx}e^{xy} + \dfrac{dy}{dx} = \dfrac{d}{dx}2$

$e^{xy}\dfrac{d}{dx}(xy) + \dfrac{dy}{dx} = 0$

$e^{xy}\left(y + x\dfrac{dy}{dx}\right) + \dfrac{dy}{dx} = 0$

$(xe^{xy}+1)\dfrac{dy}{dx} = -e^{xy}y$

$\dfrac{dy}{dx} = -\dfrac{ye^{xy}}{xe^{xy}+1}$

27. Let $u = \ln x$, so $du = \frac{1}{x}dx$.

$$\int \frac{\ln x}{x}dx = \int \ln x\left(\frac{1}{x}dx\right) = \int u\,du = \frac{u^2}{2} + C = \frac{(\ln x)^2}{2} + C$$

29. Let $u = x^4 + 1$, so $du = 4x^3 dx$.

$$\int \frac{x^3}{x^4+1}dx = \frac{1}{4}\int \frac{1}{x^4+1}(4x^3 dx) = \frac{1}{4}\int \frac{1}{u}du$$

$$= \frac{1}{4}\ln|u| + C = \frac{1}{4}\ln(x^4+1) + C$$

$$\int_0^3 \frac{x^3}{x^4+1} = \left[\frac{1}{4}\ln(x^4+1)\right]_0^3 = \frac{1}{4}(\ln 82 - \ln 1) = \frac{1}{4}\ln 82$$

31. Numerical methods were used to evaluate the integral $\int_{0.1}^{20} \cos(\ln x)dx \approx -8.3698$.

33. Let $u = 3x + 1$, so $du = 3dx$.

$$\int e^{3x+1}dx = \frac{1}{3}\int e^{3x+1}(3dx) = \frac{1}{3}\int e^u du = \frac{1}{3}e^u + C = \frac{1}{3}e^{3x+1} + C$$

35. Let $u = e^x - 1$, so $du = e^x dx$.

$$\int \frac{e^x}{e^x - 1}dx = \int \frac{1}{u}du = \ln|u| + C = \ln\left|e^x - 1\right| + C$$

$$\int_1^2 \frac{e^x}{e^x-1}dx = \left[\ln\left|e^x - 1\right|\right]_1^2 = \ln(e^2 - 1) - \ln(e - 1) = \ln\frac{e^2-1}{e-1} = \ln(e+1)$$

37. Numerical methods were used to evaluate the integral $\int_{-3}^{3} \exp\left(-\frac{1}{x^2}\right)dx = 2\int_0^3 \exp\left(-\frac{1}{x^2}\right)dx$.

$$2\lim_{E\to 0^+}\int_E^3 \exp\left(-\frac{1}{x^2}\right)dx \approx 3.1097$$

$\left(\text{Note that } \exp\left(-\frac{1}{0^2}\right) \text{ is undefined, but } \lim_{x\to 0}\exp\left(-\frac{1}{x^2}\right) = 0.\right)$

39. Let $u = \cos x$, then $du = -\sin x\, dx$. For $0 \le x \le \frac{\pi}{2}$,

$$\int \tan x\,dx = -\int \frac{-\sin x}{\cos x}dx = -\int \frac{1}{u}du = -\ln|u| + C = -\ln|\cos x| + C.$$

$$\int_0^{\pi/3} \tan x\,dx = [-\ln(\cos x)]_0^{\pi/3} = -\ln\left(\cos\frac{\pi}{3}\right) + \ln(\cos 0)$$

$$= -\ln\left(\frac{1}{2}\right) + \ln 1 = \ln 2 \approx 0.693$$

41. We need to find $\int_0^1 2\pi x e^{-x^2}dx$.

Let $u = -x^2$, then $du = -2x\,dx$.

$$\int 2\pi x e^{-x^2}dx = -\pi\int e^{-x^2}(-2x)dx = -\pi\int e^u du$$

$= -\pi e^u + C = -\pi e^{-x^2} + C$

$\int_0^1 2\pi x e^{-x^2}\,dx = -\pi\left[e^{-x^2}\right]_0^1 = -\pi(e^{-1} - e^0) = \pi(1 - e^{-1})$

43. $y = \frac{x^2}{4} - \ln\sqrt{x} = \frac{x^2}{4} - \frac{1}{2}\ln x$, so $\frac{dy}{dx} = \frac{x}{2} - \frac{1}{2x}$.

The length of the curve is $\int_1^2 \sqrt{1 + \left(\frac{dy}{dx}\right)^2}\,dx$.

Note that $1 + \left(\frac{dy}{dx}\right)^2 = 1 + \frac{1}{4}\left(x - \frac{1}{x}\right)^2 = \frac{1}{4}\left(x + \frac{1}{x}\right)^2 = \left[\frac{1}{2}\left(x + \frac{1}{x}\right)\right]^2$.

$\int_1^2 \sqrt{1 + \left(\frac{dy}{dx}\right)^2}\,dx = \int_1^2 \frac{1}{2}\left(x + \frac{1}{x}\right)dx = \frac{1}{2}\left[\frac{x^2}{2} + \ln x\right]_1^2 = \frac{1}{2}\left(\frac{4}{2} + \ln 2 - \frac{1}{2} - \ln 1\right)$

$= \frac{3}{4} + \frac{\ln 2}{2} \approx 1.097$

45. **a.** $f(x) = \ln(1.5 + \sin x)$

$f'(x) = \frac{1}{1.5 + \sin x}\frac{d}{dx}(1.5 + \sin x) = \frac{\cos x}{1.5 + \sin x}$

Extreme points occur when $f'(x) = 0$: $x = \frac{\pi}{2}, \frac{3\pi}{2}, \frac{5\pi}{2}$.

Absolute minimum: $\left(\frac{3\pi}{2}, \ln 0.5\right)$

Absolute maxima: $\left(\frac{\pi}{2}, \ln 2.5\right), \left(\frac{3\pi}{2}, \ln 2.5\right)$

b. $f''(x) = \frac{(1.5 + \sin x)(-\sin x) - (\cos x)(\cos x)}{(1.5 + \sin x)^2} = \frac{-1.5\sin x - 1}{(1.5 + \sin x)^2}$

Inflection points occur when $f''(x) = 0$: $x = \sin^{-1}\left(-\frac{2}{3}\right) + 2n\pi = -\sin^{-1}\left(\frac{2}{3}\right) + 2n\pi$ or

$x = \pi + \sin^{-1}\left(\frac{2}{3}\right) + 2n\pi$; on $[0, 3\pi]$ these values are $x = \pi + \sin^{-1}\left(\frac{2}{3}\right)$, $2\pi - \sin^{-1}\left(\frac{2}{3}\right)$.

Inflection points: $\left(\pi + \sin^{-1}\left(\frac{2}{3}\right), \ln\left(\frac{5}{6}\right)\right), \left(2\pi - \sin^{-1}\left(\frac{2}{3}\right), \ln\left(\frac{5}{6}\right)\right)$

47.

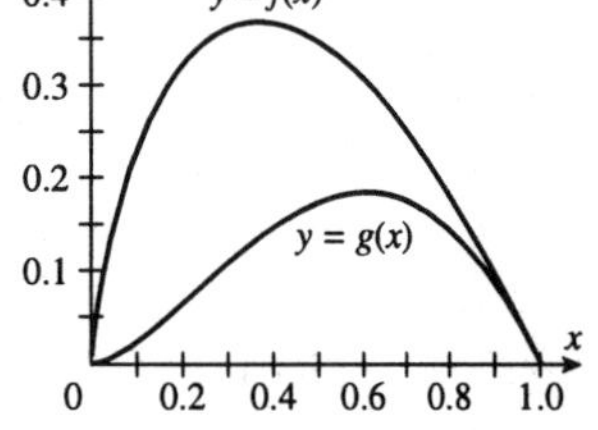

a. Numerical methods were used to evaluate the integral $\int_0^1 [f(x) - g(x)]dx = \int_0^1 (x - x^2)\ln\left(\frac{1}{x}\right)dx \approx 1.3889$.

b. $y = |f(x) - g(x)| = (x - x^2)\ln\left(\frac{1}{x}\right) = (x^2 - x)\ln x$ on (0, 1].

$\frac{dy}{dx} = (2x-1)(\ln x) + (x^2 - x)\left(\frac{1}{x}\right) = (2x-1)\ln x + x - 1$

Numerical methods were used to solve the equation $(2x - 1)\ln x + x - 1 = 0$ on (0, 1]: $x \approx 0.2356$ or $x = 1$.
The maximum value of $|f(x) - g(x)|$ on (0, 1] is approximately $f(0.2356) - g(0.2356) \approx 0.2603$.

49. $f(x) = e^{-x^2}$

$f'(x) = e^{-x^2} \cdot \frac{d}{dx}(-x^2) = -2xe^{-x^2}$

$f''(x) = -2e^{-x^2} - 2x(-2xe^{-x^2}) = 2(2x^2 - 1)e^{-x^2}$

Note that $f(x) \le f''(x)$ on $\left(-\infty, -\frac{\sqrt{3}}{2}\right] \cup \left[\frac{\sqrt{3}}{2}, \infty\right)$ and $f(x) \ge f''(x)$ on $\left[-\frac{\sqrt{3}}{2}, \frac{\sqrt{3}}{2}\right]$.

The graphs of $y = f(x)$ and $y = f''(x)$ are symmetric about the y-axis, so

$\int_{-3}^{3} |f(x) - f''(x)| = 2\int_0^3 |f(x) - f''(x)|$

$= 2\int_0^{\sqrt{3}/2} (f(x) - f''(x))dx + 2\int_{\sqrt{3}/2}^{3} (f''(x) - f(x))dx$

$= 2\int_0^{\sqrt{3}/2} (3 - 4x^2)e^{-x^2}\,dx + 2\int_{\sqrt{3}/2}^{3} (4x^2 - 3)e^{-x^2}\,dx$

≈ 4.2614

51. By l'Hôpital's Rule, $\lim\limits_{x\to 0} \frac{\sin x}{x} = \lim\limits_{x\to 0} \frac{\cos x}{1} = 1$.

The natural logarithm function is continuous at $x = 1$, $\lim\limits_{x\to 0} \ln\left(\frac{\sin x}{x}\right) = \ln 1 = 0$.

53. If $m > 0$, $\ln 4 > 1$ implies that $m \ln 4 > m$, or $\ln 4^m > m$. Therefore, $\ln x$ can be made as large as desired by choosing x sufficiently large (i.e., to obtain $\ln x > m$, choose $x = 4^m$). Since $y = \ln x$ is an increasing function, this implies $\lim\limits_{x\to\infty} \ln x = \infty$.

55. $\lim\limits_{n\to\infty}\left[\frac{1}{n+1} + \frac{1}{n+2} + \cdots + \frac{1}{2n}\right]$

$= \lim\limits_{n\to\infty}\left[\frac{1}{1+\frac{1}{n}} + \frac{1}{1+\frac{2}{n}} + \cdots + \frac{1}{1+\frac{n}{n}}\right]\frac{1}{n}$

$= \int_0^1 \frac{1}{1+x}\,dx$

$= \int_1^2 \frac{1}{u}\,du = \ln 2 \approx 0.693$

57. $\frac{1,000,000}{\ln 1,000,000} \approx 72,382$

There are about 72,400 primes less than 1,000,000.

59. $f(x) = e^x$ is an increasing function because $f'(x) = e^x > 0$. Therefore, $a < b$ (or $-a > -b$) implies $e^{-a} > e^{-b}$.

61. Use $x = 30$, $n = 8$, and $k = 0.25$.

$P_n(x) = \frac{(kx)^n e^{-kx}}{n!} = \frac{(0.25 \cdot 30)^8 e^{-0.25\cdot 30}}{8!} \approx 0.14$

63. Written response (graph)

a. $\lim_{x\to\infty} x^p e^{-x} = 0$

b. On $[0, \infty)$, the maximum occurs at $x = p$.

65. Using the first part of Theorem B, $e^b e^{a-b} = e^{b+(a-b)} = e^a$.

Dividing by e^b gives $\frac{e^a}{e^b} = e^{a-b}$.

Section 7.3

Concepts Review

1. $e^{\sqrt{3}\ln\pi}$; $e^{x\ln a}$

3. $\frac{\ln x}{\ln a}$

Problem Set 7.3

1. $\log_3 9 = x$

$3^x = 9$

$x = 2$

3. $\log_9 x = \frac{3}{2}$

$x = 9^{3/2} = \left(\sqrt{9}\right)^3 = 3^3 = 27$

5. $2\log_{10}\left(\frac{x}{3}\right) = 1$

$\log_{10}\left(\frac{x}{3}\right) = \frac{1}{2}$

$\frac{x}{3} = 10^{1/2}$

$x = 3\sqrt{10}$

7. $\log_2(x+1) - \log_2 x = 2$

$\log_2\left(\frac{x+1}{x}\right) = 2$

$\frac{x+1}{x} = 2^2$

$x + 1 = 4x$

$x = \frac{1}{3}$

9. $\log_5 13 = \frac{\ln 13}{\ln 5} \approx 1.59$

11. $\log_{11}(8.16)^{1/5} = \frac{1}{5}\log_{11} 8.16 = \frac{\ln 8.16}{5\ln 11} \approx 0.1751$

13. $2^x = 19$

$x\ln 2 = \ln 19$

$x = \frac{\ln 19}{\ln 2} \approx 4.2479$

15. $4^{3x-1} = 5$

$(3x - 1)\ln 4 = \ln 5$

$x = \frac{\ln 5 + \ln 4}{3} \approx 0.9986$

17. $\frac{d}{dx}(5^{x^2}) = 5^{x^2}\ln 5\frac{d}{dx}(x^2) = 5^{x^2}2x\ln 5$

19. $\frac{d}{dx}\log_2 e^x = \frac{d}{dx}x\log_2 e = \log_2 e$

$= \frac{\ln e}{\ln 2} = \frac{1}{\ln 2} \approx 1.4427$

21. $\frac{d}{dx}[2^x\ln(x+5)] = (2^x\ln 2)\ln(x+5) + (2^x)\frac{1}{x+5}$

$= 2^x\left[\ln 2\ln(x+5) + \frac{1}{x+5}\right]$

23. Let $u = x^2$, then $du = 2x\,dx$.

$\int x2^{x^2}dx = \frac{1}{2}\int 2x2^{x^2}dx = \frac{1}{2}\int 2^u du$

$= \frac{1}{2\ln 2}2^u + C = \frac{2^{x^2}}{2\ln 2} + C$

25. Let $u = \sqrt{x}$, then $du = \frac{1}{2\sqrt{x}}dx$.

$\int \frac{5^{\sqrt{x}}}{\sqrt{x}}dx = 2\int \frac{5^{\sqrt{x}}}{2\sqrt{x}}dx = 2\int 5^u du = \frac{2}{\ln 5}5^u + C$

$= \frac{2\cdot 5^{\sqrt{x}}}{\ln 5} + C$

$\int_1^4 \frac{5^{\sqrt{x}}}{\sqrt{x}}dx = \left[\frac{2\cdot 5^{\sqrt{x}}}{\ln 5}\right]_1^4 = \frac{2}{\ln 5}\left(5^{\sqrt{4}} - 5^{\sqrt{1}}\right)$

$= \frac{40}{\ln 5} \approx 24.8534$

27. $\frac{d}{dx}10^{(x^2)} = 10^{x^2}\ln 10\frac{d}{dx}x^2 = 10^{(x^2)}2x\ln 10$

$\frac{d}{dx}(x^2)^{10} = \frac{d}{dx}x^{20} = 20x^{19}$

$\frac{dy}{dx} = \frac{d}{dx}[10^{(x^2)} + (x^2)^{10}]$

$= 10^{(x^2)}2x\ln 10 + 20x^{19}$

29. $\frac{d}{dx}x^{\pi+1} = (\pi+1)x^{\pi}$

$\frac{d}{dx}(\pi+1)^x = (\pi+1)^x \ln(\pi+1)$

$\frac{dy}{dx} = \frac{d}{dx}[x^{\pi+1} + (\pi+1)^x]$

$= (\pi+1)x^{\pi} + (\pi+1)^x \ln(\pi+1)$

31. $y = (x^2+1)^{\ln x} = e^{(\ln x)\ln(x^2+1)}$

$\frac{dy}{dx} = e^{(\ln x)\ln(x^2+1)} \frac{d}{dx}[(\ln x)\ln(x^2+1)]$

$= e^{(\ln x)\ln(x^2+1)}\left[\frac{1}{x}\ln(x^2+1) + \ln x \frac{2x}{x^2+1}\right]$

$= (x^2+1)^{\ln x}\left(\frac{\ln(x^2+1)}{x} + \frac{2x\ln x}{x^2+1}\right)$

33. $f(x) = x^{\sin x} = e^{\sin x \ln x}$

$f'(x) = e^{\sin x \ln x}\frac{d}{dx}(\sin x \ln x)$

$= e^{\sin x \ln x}\left[(\sin x)\left(\frac{1}{x}\right) + (\cos x)(\ln x)\right]$

$= x^{\sin x}\left(\frac{\sin x}{x} + \cos x \ln x\right)$

$f'(1) = 1^{\sin 1}\left(\frac{\sin 1}{1} + \cos 1 \ln 1\right) = \sin 1 \approx 0.8415$

35. $\log_{1/2} x = \frac{\ln x}{\ln \frac{1}{2}} = \frac{\ln x}{-\ln 2} = -\log_2 x$

37. $M = 0.67\log_{10}(0.37E) + 1.46$

$\log_{10}(0.37E) = \frac{M-1.46}{0.67}$

$E = \frac{\left(10^{\frac{M-1.46}{0.67}}\right)}{0.37}$

Evaluating this expression for $M = 7$ and $M = 8$ gives $E \approx 5.017\times10^8$ kw-h and $E \approx 1.560\times10^{10}$ kw-h, respectively.

39. $r = 2^{1/12} \approx 1.0595$

Frequency of

$\overline{\text{C}} = 440(2^{1/12})^3 = 440\sqrt[4]{2} \approx 523.25$

41. WRONG 1:

$y = f(x)^{g(x)}$

$y' = g(x)f(x)^{g(x)-1}f'(x)$

WRONG 2:

$y = f(x)^{g(x)}$

$y' = f(x)^{g(x)}(\ln f(x))\cdot g'(x)$

$= f(x)^{g(x)}g'(x)\ln f(x)$

RIGHT:

$y = f(x)^{g(x)} = e^{g(x)\ln f(x)}$

$y' = e^{g(x)\ln f(x)}\frac{d}{dx}[g(x)\ln f(x)]$

$= f(x)^{g(x)}\left[g'(x)\ln f(x) + g(x)\frac{1}{f(x)}f'(x)\right]$

$= f(x)^{g(x)}g'(x)\ln f(x) + f(x)^{g(x)-1}g(x)f'(x)$

Note that RIGHT = WRONG 2 + WRONG 1.

43. $f(x) = \frac{a^x - 1}{a^x + 1}$

$f'(x) = \frac{(a^x+1)a^x\ln a - (a^x-1)a^x\ln a}{(a^x+1)^2}$

$= \frac{2a^x\ln a}{(a^x+1)^2}$

Since a is positive, a^x is always positive. $(a^x+1)^2$ is also always positive, thus $f'(x) > 0$ if $\ln a > 0$ and $f'(x) < 0$ if $\ln a < 0$. $f(x)$ is either always increasing or always decreasing, depending on a, so $f(x)$ has an inverse.

$y = \frac{a^x-1}{a^x+1}$

$y(a^x+1) = a^x - 1$

$a^x(y-1) = -1-y$

$a^x = \frac{1+y}{1-y}$

$x\ln a = \ln\frac{1+y}{1-y}$

$x = \frac{\ln\frac{1+y}{1-y}}{\ln a} = \log_a\frac{1+y}{1-y}$

$f^{-1}(y) = \log_a\frac{1+y}{1-y}$

$f^{-1}(x) = \log_a\frac{1+x}{1-x}$

45. a. $f_u(x) = x^u e^{-x}$

$f_u'(x) = ux^{u-1}e^{-x} - x^u e^{-x} = (u-x)x^{u-1}e^{-x}$

Since $f_u'(x) > 0$ on $[0, u)$ and $f_u'(x) < 0$ on (u, ∞), $f_u(x)$ attains its maximum at $x = u$.

b. $f_u(u) > f_u(u+1)$ means
$u^u e^{-u} > (u+1)^u e^{-(u+1)}$.

Multiplying by $\frac{e^{u+1}}{u^u}$ gives $e > \left(\frac{u+1}{u}\right)^u$.

$f_{u+1}(u+1) > f_{u+1}(u)$ means
$(u+1)^{u+1} e^{-(u+1)} > u^{u+1} e^{-u}$.

Multiplying by $\frac{e^{u+1}}{u^{u+1}}$ gives $\left(\frac{u+1}{u}\right)^{u+1} > e$.

Combining the two inequalities,

$$\left(\frac{u+1}{u}\right)^u < e < \left(\frac{u+1}{u}\right)^{u+1}.$$

c. From part (b), $e < \left(\frac{u+1}{u}\right)^{u+1}$.

Multiplying by $\frac{u}{u+1}$ gives

$$\frac{u}{u+1} e < \left(\frac{u+1}{u}\right)^u.$$

We showed $\left(\frac{u+1}{u}\right)^u < e$ in part (b), so

$$\frac{u}{u+1} e < \left(\frac{u+1}{u}\right)^u < e.$$

Since $\lim_{u\to\infty} \frac{u}{u+1} e = e$, this implies that

$$\lim_{u\to\infty} \left(\frac{u+1}{u}\right)^u = e, \text{ i.e., } \lim_{u\to\infty} \left(1+\frac{1}{u}\right)^u = e.$$

47.

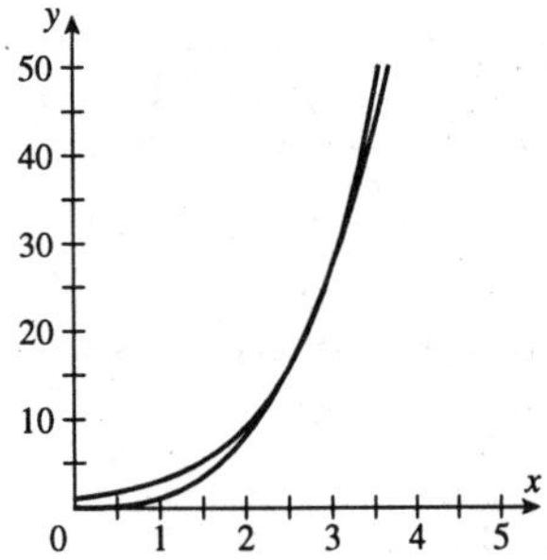

(2.4781, 15.2171), (3, 27)

49. (i) $a^x a^y = e^{x\ln a} e^{y\ln a} = e^{(x+y)\ln a} = a^{x+y}$

(iv) $(ab)^x = e^{x\ln ab} = e^{x\ln a + x\ln b} = e^{x\ln a} e^{x\ln b}$
$= a^x b^x$

Section 7.4

Concepts Review

1. ky; $\frac{ky(L-y)}{L}$

3. Half-life

Problem Set 7.4

1. $\frac{dy}{dt} = -5y$

$\int \frac{dy}{y} = \int -5dt$

$\ln y = -5t + C$

$y = e^{-5t+C} = y_0 e^{-5t}$

Since $y(0) = 4$, $y = 4e^{-5t}$.

3. $\frac{dy}{dt} = 0.006y$

$\int \frac{dy}{y} = \int 0.006dt$

$\ln y = 0.006t + C$

$y = e^{0.006t+C} = y_0 e^{0.006t}$

Since $y(10) = 2$, $y = 2e^{0.006(t-10)}$.

5. $y = y_0 e^{kt}$

$24,000 = 10,000e^{10k}$

$10k = \ln 2.4$

$k = \frac{\ln 2.4}{10} \approx 0.08755$

$y = 10,000e^{0.08755t}$

$y(25) = 10,000e^{0.08755\cdot 25} \approx 89,200$

7. $30,000 = 10,000e^{0.08755t}$

$t = \frac{\ln 3}{0.08755} \approx 12.55$ days

9. 1 year: (4.5 million)(1.032) ≈ 4.64 million

2 years: (4.5 million)$(1.032)^2 \approx 4.79$ million

10 years: (4.5 million)$(1.032)^{10} \approx 6.17$ million

100 years: (4.5 million)$(1.032)^{100} \approx 105$ million

11. $\frac{1}{2} = e^{810k}$

$k = \frac{\ln\left(\frac{1}{2}\right)}{810} \approx -8.557\times 10^{-4}$

$y = 10e^{(-8.557\times 10^{-4})t}$

$y(300) = 10e^{(-8.557\times10^{-4})(300)} \approx 7.736$
About 7.736 grams will remain.

13. $\frac{1}{2} = e^{5730k}$

$k = \frac{\ln\left(\frac{1}{2}\right)}{5730} \approx -1.210\times10^{-4}$

$0.7y_0 = y_0 e^{(-1.210\times10^{-4})t}$

$t = \frac{\ln 0.7}{-1.210\times10^{-4}} \approx 2950$

The fort burned down about 2950 years ago.

15. $\int \frac{dT}{T-75} = \int k\,dt$

$\ln(T-75) = kt + C$

$T = 75 + e^{kt+C} = 75 + Ae^{kt}$

Since $T(0) = 300$, $A = 225$.

Since $T(0.5) = 200$, $200 = 75 + 225e^{0.5k}$.

$e^{0.5k} = \frac{125}{225}$

$k = 2\ln\left(\frac{125}{225}\right) \approx -1.1756$

$T(3) = 75 + 225e^{(-1.1756)(3)} \approx 81.615$

The object will be about 81.6° F.
Written response.

17. **a.** $(\$375)(1.095)^2 \approx \$449.63$

b. $(\$375)\left(1+\frac{0.095}{12}\right)^{24} \approx \$453.13$

c. $(\$375)\left(1+\frac{0.095}{365}\right)^{730} \approx \$453.46$

d. $(\$375)e^{0.095\cdot 2} \approx \$453.47$

19. **a.** $\left(1+\frac{0.12}{12}\right)^{12t} = 2$

$1.01^{12t} = 2$

$12t = \log_{1.01} 2$

$t = \frac{1}{12}\log_{1.01} 2 = \frac{\ln 2}{12\ln 1.01} \approx 5.805$

It will take about 5.805 years or 5 years, 10 months.

b. $e^{0.12t} = 2$

$t = \frac{\ln 2}{0.12} \approx 5.776$

It will take about 5.776 years or 5 years, 9 months, and 9 days.

21. $(\$24)e^{0.06(1996-1626)} \approx \$1.0509\times10^{11}$
It would be worth about \$105 billion.

23. If t is the doubling time, then

$\left(1+\frac{p}{100}\right)^t = 2$

$t\ln\left(1+\frac{p}{100}\right) = \ln 2$

$t = \frac{\ln 2}{\ln\left(1+\frac{p}{100}\right)} \approx \frac{\ln 2}{\frac{p}{100}} = \frac{100\ln 2}{p} \approx \frac{70}{p}$

25.

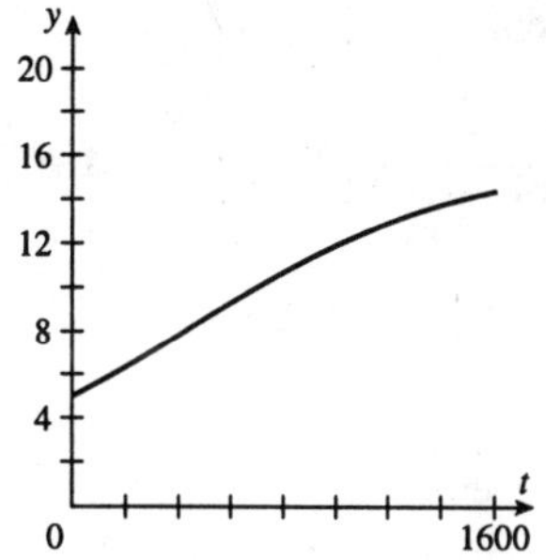

$y = \frac{16\cdot 5}{5+(16-5)e^{-0.00186t}} = \frac{80}{5+11e^{-0.00186t}}$

27. **a.** $\lim_{x\to0}(1-x)^{1/x} = \lim_{x\to0}\frac{1}{(1-x)^{\frac{1}{-x}}} = \frac{1}{e}$

b. $\lim_{x\to0}(1+3x)^{1/x} = \lim_{x\to0}\left[(1+3x)^{\frac{1}{3x}}\right]^3 = e^3$

c. $\lim_{n\to\infty}\left(\frac{n+2}{n}\right)^n = \lim_{n\to\infty}\left(1+\frac{2}{n}\right)^n$

$= \lim_{x\to0^+}(1+2x)^{1/x}$

$= \lim_{x\to0^+}\left[(1+2x)^{\frac{1}{2x}}\right]^2 = e^2$

d. $\lim_{n\to\infty}\left(\frac{n-1}{n}\right)^{2n} = \lim_{n\to\infty}\left(1-\frac{1}{n}\right)^{2n}$

$= \lim_{x\to0^+}(1-x)^{2/x}$

$= \lim_{x\to0^+}\left[(1-x)^{\frac{1}{-x}}\right]^{-2} = \frac{1}{e^2}$

29. $\frac{dy}{dt} = 0.012y + 60,000$

Using the result from Problem 28 with $a = 0.012$, $b = 60,000$, and $y_0 = 10,000,000$:

$$y = \left(10,000,000 + \frac{60,000}{0.012}\right)e^{0.012t} - \frac{60,000}{0.012}$$

$$y = 15,000,000e^{0.012t} - 5,000,000$$

$$y(25) = 15,000,000e^{0.012(25)} - 5,000,000 \approx 15,250,000$$

The population will be about 15.25 million.

31. Maximum population: $13,500,000 \text{ mi}^2 \cdot \frac{640 \text{ acres}}{1 \text{ mi}^2} \cdot \frac{1 \text{ person}}{\frac{1}{2} \text{ acre}} = 1.728 \times 10^{10}$ people

$$(5 \times 10^9)e^{0.019t} = 1.728 \times 10^{10}$$

$t = \dfrac{\ln\left(1.728 \times \frac{10}{5}\right)}{0.019} \approx 65.3$ years from 1987, or sometime in the year 2052.

Section 7.5

Concepts Review

1. Numerical; Euler's

3. Under

Problem Set 7.5

1. $\frac{dy}{dt} = -5y,\ y(0) = y_0 = 4$

$h = 0.2$ and $f(y, t) = -5y$, so use $y_{n+1} = (0.2)(-5y_n) + y_n = 0$. Exact solution is $y = 4e^{-5t}$.

n	t_n	Euler estimate $y_n = 0$	Exact solution $y = 4e^{-5t_n}$	Error
0	0.0	4	4	0
1	0.2	0	1.47	1.47
2	0.4	0	0.54	0.54
3	0.6	0	0.20	0.20
4	0.8	0	0.07	0.07
5	1.0	0	0.03	0.03

3. $\frac{dy}{dx} = 0.006y,\ y(10) = y_0 = 2$

$h = 0.2$ and $f(y, t) = 0.006y$, so use $y_{n+1} = (0.2)(0.006y_n) + y_n = 1.0012y_n$. Exact solution is $y = 2e^{0.006(t-10)}$.

n	t_n	Euler estimate $y_n = 1.0012y_{n-1}$	Exact solution $y = 2e^{0.006(t_n-10)}$	Error
0	10.0	2.0000000	2.0000000	
1	10.2	2.0024000	2.0024014	1.4×10^{-6}
2	10.4	2.0048029	2.0048058	2.9×10^{-6}
3	10.6	2.0072087	2.0072130	4.3×10^{-6}
4	10.8	2.0096174	2.0096231	5.7×10^{-6}
5	11.0	2.0120289	2.0120361	7.2×10^{-6}

5.

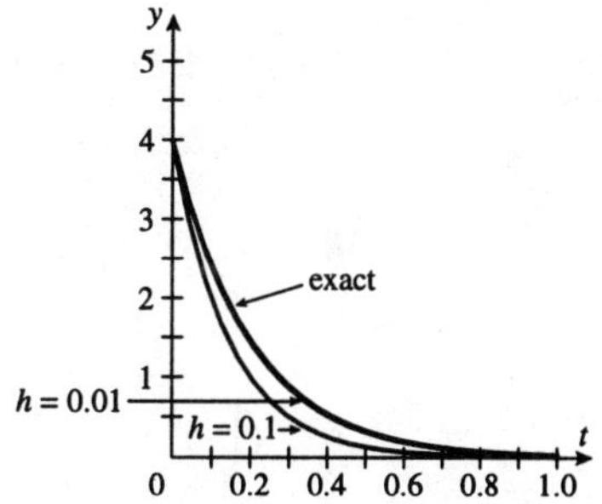

$\frac{dy}{dt} = -5y,\ y(0) = y_0 = 4$

For $h = 0.1$, use $y_{n+1} = (0.1)(-5y_n) + y_n = 0.5y_n$.

Right endpoint $y_{10} \approx 0.0039$ (error $\approx |0.027 - 0.0039| \approx 0.023$)

For $h = 0.01$, $y_{n+1} \approx (0.01)(-5y_n) + y_n = 0.95y_n$.

Right endpoint $y_{100} \approx 0.024$ (error $\approx |0.027 - 0.024| = 0.003$)

7.

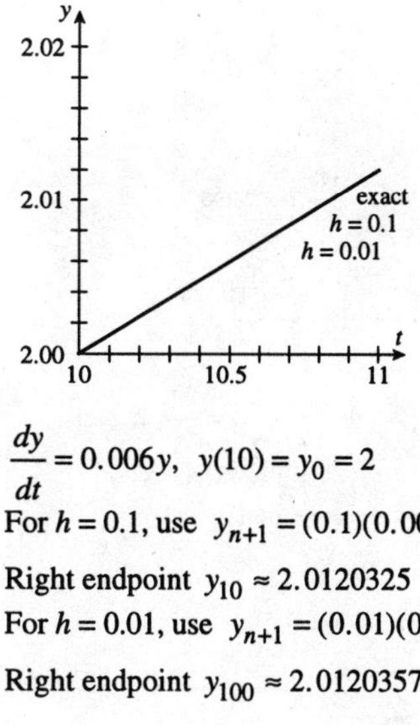

$\frac{dy}{dt} = 0.006y,\ y(10) = y_0 = 2$

For $h = 0.1$, use $y_{n+1} = (0.1)(0.006y_n) + y_n = 1.0006y_n$.

Right endpoint $y_{10} \approx 2.0120325$ (error $\approx |2.0120361 - 2.0120325| = 3.6\times10^{-6}$)

For $h = 0.01$, use $y_{n+1} = (0.01)(0.006y_n) + y_n = 1.00006y_n$.

Right endpoint $y_{100} \approx 2.0120357$ (error $\approx |2.0120361 - 2.0120357| = 4\times10^{-7}$)

9. $\frac{dy}{dt} = 0.10y\frac{500-y}{500} = \frac{500y-y^2}{5000}$

Use a loop based on the equations.

$t_{n+1} = t_n + h$

$y_{n+1} = \frac{500y_n - y_n^2}{5000}h + y_n$

with $t_0 = 0$, $y_0 = 20$.

h	Euler estimate of $y(100)$	Euler estimate of $y(200)$
10	500.00	500.00
1	499.55	500.00
0.1	499.47	500.00

The population reaches its limiting value in approximately 90 years. This is faster than in Example 2 because the growth rate k is larger. Written response.

11. $\frac{dy}{dt} = \frac{y^2}{5}$, $y(0) = 1$

Use the method of separation of variables because the variables can be separated.

$\int \frac{5}{y^2}dy = \int dt$

$-5y^{-1} = t + C$

$y = -\frac{5}{t+C}$

Since $y(0) = 1$, $C = -5$.

$y = \frac{5}{5-t}$

$y(1) = \frac{5}{5-1} = \frac{5}{4}$

13. $\frac{dy}{dt} = \frac{y^2}{5} + t$, $y(0) = 1$

Use Euler's method because the variables cannot be separated.

$y_{n+1} = \left(\frac{y_n^2}{5} + t_n\right)h + y_n$

If $h = 0.005$, we obtain the value $y(1) \approx y_{200} \approx 1.85$.

Section 7.6

Concepts Review

1. $[0, \pi]$; arccos

3. 1

Problem Set 7.6

1. $\sin^{-1}\left(\frac{\sqrt{3}}{2}\right) = \frac{\pi}{3}$ because

$\sin\frac{\pi}{3} = \frac{\sqrt{3}}{2}$, $\frac{\pi}{3} \in \left[-\frac{\pi}{2}, \frac{\pi}{2}\right]$.

3. $\arcsin\left(-\frac{\sqrt{2}}{2}\right) = -\frac{\pi}{4}$ because

$\sin\left(-\frac{\pi}{4}\right) = -\frac{\sqrt{2}}{2}$, $-\frac{\pi}{4} \in \left[-\frac{\pi}{2}, \frac{\pi}{2}\right]$.

5. $\tan^{-1}\left(-\sqrt{3}\right) = -\frac{\pi}{3}$ because

$\tan\left(-\frac{\pi}{3}\right) = -\sqrt{3}$, $-\frac{\pi}{3} \in \left(-\frac{\pi}{2}, \frac{\pi}{2}\right)$.

7. $\arccos\left(-\frac{1}{2}\right) = \frac{2\pi}{3}$ because

$\cos\left(\frac{2\pi}{3}\right) = -\frac{1}{2}$, $\frac{2\pi}{3} \in [0, \pi]$.

9. $\sec^{-1}(-2) = \cos^{-1}\left(-\frac{1}{2}\right) = \frac{2\pi}{3}$ because

$\cos\left(\frac{2\pi}{3}\right) = -\frac{1}{2}$, $\frac{2\pi}{3} \in [0, \pi]$.

11. $\sin(\sin^{-1} 0.541) = 0.541$

13. $\theta = \sin^{-1}\frac{x}{8}$

15. $\theta = \sin^{-1}\frac{5}{x}$

17. Let θ_1 be the angle opposite the side of length 3, and $\theta_2 = \theta_1 - \theta$. Then $\theta = \theta_1 - \theta_2$, $\tan\theta_1 = \frac{3}{x}$, and $\tan\theta_2 = \frac{1}{x}$.

$\theta = \tan^{-1}\frac{3}{x} - \tan^{-1}\frac{1}{x}$

19. $\frac{dy}{dx} = \frac{d}{dx}\cos^2(x-2) = 2\cos(x-2)\frac{d}{dx}\cos(x-2)$
$= 2\cos(x-2)[-\sin(x-2)]\frac{d}{dx}(x-2)$
$= -2\cos(x-2)\sin(x-2)$

21. $\frac{dy}{dx} = \frac{d}{dx}(\cot x \csc x)$
$= \left(\frac{d}{dx}\cot x\right)(\csc x) + (\cot x)\left(\frac{d}{dx}\csc x\right)$
$= (-\csc^2 x)(\csc x) + (\cot x)(-\csc x \cot x)$
$= (-\csc x)(\csc^2 x + \cot^2 x)$

23. $\frac{dy}{dx} = \frac{d}{dx}e^{\cot x} = e^{\cot x}\frac{d}{dx}\cot x = -e^{\cot x}\csc^2 x$

25. $\frac{dy}{dx} = \frac{d}{dx}[(\tan x)e^{\tan x}]$
$= (\sec^2 x)(e^{\tan x}) + (\tan x)(e^{\tan x}\sec^2 x)$
$= (1+\tan x)(e^{\tan x}\sec^2 x)$

27. $\frac{d}{dx}\sin^{-1}(x^2) = \frac{1}{\sqrt{1-(x^2)^2}}\frac{d}{dx}(x^2) = \frac{2x}{\sqrt{1-x^4}}$

29. $\frac{d}{dx}7\cos^{-1}\sqrt{2x} = \frac{-7}{\sqrt{1-\left(\sqrt{2x}\right)^2}}\frac{d}{dx}\sqrt{2x}$
$= \frac{-7}{\sqrt{1-2x}}\frac{2}{2\sqrt{2x}} = -\frac{7}{\sqrt{2x-4x^2}}$

31. Let $u = x^2$, then $du = 2x$.
$\int x\sin(x^2)dx = \frac{1}{2}\int \sin(x^2)\cdot 2x\,dx = \frac{1}{2}\int \sin u\,du$
$= -\frac{1}{2}\cos u + C = -\frac{1}{2}\cos(x^2) + C$

33. Let $u = \cos x$, then $du = -\sin x\,dx$.
$\int \tan x\,dx = \int \frac{\sin x}{\cos x}dx = -\int \frac{1}{\cos x}(-\sin x)dx$
$= -\int \frac{1}{u}du = -\ln|u| + C = -\ln|\cos x| + C$

35. Let $u = \tan x$, then $du = \sec^2 x\,dx$.
$\int \frac{\sec^2 x}{\tan x}dx = \int \frac{1}{u}du = \ln|u| + C = \ln|\tan x| + C$

37. Let $u = e^{2x}$, then $du = 2e^{2x}$.
$\int e^{2x}\cos(e^{2x})dx$
$= \frac{1}{2}\int \cos(e^{2x})(2e^{2x})dx = \frac{1}{2}\int \cos u\,du$
$= \frac{1}{2}\sin u + C = \frac{1}{2}\sin(e^{2x}) + C$

$\int_0^1 e^{2x}\cos(e^{2x})dx = \left[\frac{1}{2}\sin(e^{2x})\right]_0^1$
$= \left[\frac{1}{2}\sin(e^2) - \frac{1}{2}\sin(e^0)\right] = \frac{\sin e^2 - \sin 1}{2}$
≈ 0.0262

39. $\int_0^{\sqrt{2}/2}\frac{1}{\sqrt{1-x^2}}dx = \left[\sin^{-1}x\right]_0^{\sqrt{2}/2} = \frac{\pi}{4} - 0 = \frac{\pi}{4}$

41. $\int_{-1}^1 \frac{1}{1+x^2}dx = \left[\tan^{-1}x\right]_{-1}^1 = \tan^{-1}1 - \tan^{-1}(-1)$
$= \frac{\pi}{4} - \left(\frac{\pi}{4}\right) = \frac{\pi}{2}$

43. If $x^2y + y - 4 = 0$, then $y = \frac{4}{x^2+1}$.
$\text{Area} = \int_0^1 \frac{4}{x^2+1}dx = \left[4\tan^{-1}x\right]_0^1$
$= 4(\tan^{-1}1 - \tan^{-1}0) = 4\left(\frac{\pi}{4} - 0\right) = \pi$

45. $f(x) = 1.25e^x + 0.135\sin(12x)$
$f'(x) = 1.25e^x + 1.62\cos(12x)$
$f(x)$ is decreasing between $x \approx 0.2513$ and $x \approx 0.2585$ since $f'(x) < 0$ on this interval. (Numerical methods were used to solve $f'(x) = 0$.)

47. Let θ be the angle subtended for the viewer's eye.
$\theta = \tan^{-1}\left(\frac{12}{b}\right) - \tan^{-1}\left(\frac{2}{b}\right)$
$\frac{d\theta}{db} = \frac{1}{1+\left(\frac{12}{b}\right)^2}\left(-\frac{12}{b^2}\right) - \frac{1}{1+\left(\frac{2}{b}\right)^2}\left(-\frac{2}{b^2}\right)$
$= \frac{2}{b^2+4} - \frac{12}{b^2+144}$
$= \frac{10(24-b^2)}{(b^2+4)(b^2+144)}$

Since $\frac{d\theta}{db} > 0$ for $b \in \left[0, 2\sqrt{6}\right)$ and $\frac{d\theta}{db} < 0$ for $b \in \left(2\sqrt{6}, 0\right)$, the angle is maximized for $b = 2\sqrt{6} \approx 4.899$. The ideal distance is about 4.9 ft from the wall.

49.

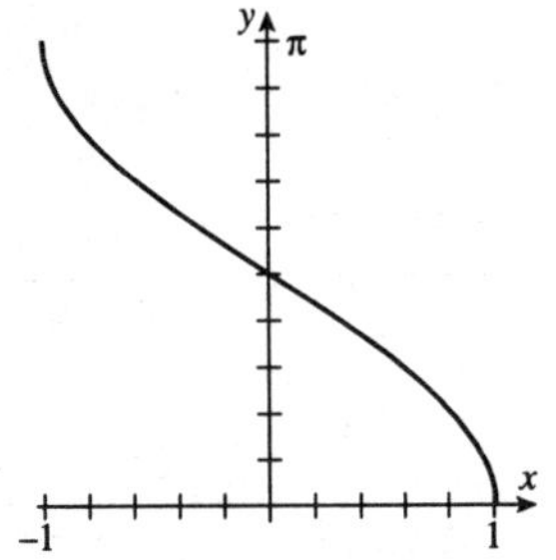

Conjecture:

$\frac{\pi}{2} - \arcsin x = \arccos x$

Proof:

$\cos\left(\frac{\pi}{2} - \arcsin x\right) = \sin(\arcsin x) = x$

Since $\left(\frac{\pi}{2} - \arcsin x\right) \in [0, \pi]$, this means

$\arccos x = \frac{\pi}{2} - \arcsin x$.

51. Let $u = \frac{x}{a}$, then $du = \frac{1}{a}dx$. Since $a > 0$,

$$\int \frac{dx}{\sqrt{a^2 - x^2}} = \int \frac{1}{\sqrt{1-\left(\frac{x}{a}\right)^2}} \frac{1}{a}dx = \int \frac{1}{\sqrt{1-u^2}} du$$

$$= \sin^{-1} u + C = \sin^{-1}\left(\frac{x}{a}\right) + C$$

53. Let $u = \frac{x}{a}$, then $du = \frac{1}{a}dx$.

$$\int \frac{dx}{a^2 + x^2} = \frac{1}{a}\int \frac{1}{1+\left(\frac{x}{a}\right)^2} \frac{1}{a}dx = \frac{1}{a}\int \frac{1}{1+u^2} du$$

$$= \frac{1}{a}\tan^{-1} u + C = \frac{1}{a}\tan^{-1}\left(\frac{x}{a}\right) + C$$

55. Recall that $\frac{d}{dx}\sin^{-1}\left(\frac{x}{a}\right) = \frac{1}{\sqrt{a^x - x^2}}$ (Problems 51–52).

$$\frac{d}{dx}\left[\frac{x}{2}\sqrt{a^2 - x^2} + \frac{a^2}{2}\sin^{-1}\frac{x}{a} + C\right]$$

$$= \frac{1}{2}\sqrt{a^2 - x^2} + \frac{x}{2}\frac{1}{2\sqrt{a^2 - x^2}}(-2x) + \frac{a^2}{2}\frac{1}{\sqrt{a^2 - x^2}} + 0$$

$$= \frac{1}{2}\sqrt{a^2 - x^2} + \frac{1}{2}\frac{-x^2 + a^2}{\sqrt{a^2 - x^2}} = \sqrt{a^2 - x^2}$$

57. a. $\theta = \cos^{-1}\left(\frac{x}{b}\right) - \cos^{-1}\left(\frac{x}{a}\right)$

$$\frac{d\theta}{dt} = \left(\frac{-1}{\sqrt{1-\left(\frac{x}{b}\right)^2}}\right)\left(\frac{1}{b}\right)\left(\frac{dx}{dt}\right) - \left(\frac{-1}{\sqrt{1-\left(\frac{x}{a}\right)^2}}\right)\left(\frac{1}{a}\right)\left(\frac{dx}{dt}\right)$$

$$= \left(\frac{1}{\sqrt{a^2 - x^2}} - \frac{1}{\sqrt{b^2 - x^2}}\right)\frac{dx}{dt}$$

b. $\theta = \tan^{-1}\left(\dfrac{a+x}{\sqrt{b^2-x^2}}\right) - \sin^{-1}\left(\dfrac{x}{b}\right)$

$$\frac{d\theta}{dt} = \left(\frac{1}{1+\left(\frac{(a+x)x}{\sqrt{b^2-x^2}}\right)^2}\right)\left(\frac{\sqrt{b^2-x^2}+\frac{(a+x)x}{\sqrt{b^2-x^2}}}{b^2-x^2}\right)\left(\frac{dx}{dt}\right) - \left(\frac{1}{\sqrt{1-\left(\frac{x}{b}\right)^2}}\right)\left(\frac{1}{b}\right)\left(\frac{dx}{dt}\right)$$

$$= \left[\left(\frac{b^2-x^2}{b^2-x^2+(a+x)^2}\right)\left(\frac{b^2+ax}{(b^2-x^2)^{3/2}}\right) - \frac{1}{\sqrt{b^2-x^2}}\right]\frac{dx}{dt}$$

$$= \left[\frac{b^2+ax}{(b^2+a^2+2ax)\sqrt{b^2-x^2}} - \frac{1}{\sqrt{b^2-x^2}}\right]\frac{dx}{dt}$$

$$= \left[-\frac{a^2+ax}{(b^2+a^2+2ax)\sqrt{b^2-x^2}}\right]\frac{dx}{dt}$$

59. Let $x(t)$ be the *horizontal* distance from the observer to the plane, in miles, t minutes *before* the plane is overhead. Let $t = 0$ when the distance to the plane is 3 miles. Then $x(0) = \sqrt{3^2 - 2^2} = \sqrt{5}$. The speed of the plane is 10 miles per minute, so $x(t) = \sqrt{5} - 10t$. The angle of elevation is $\theta(t) = \tan^{-1}\left(\dfrac{2}{x(t)}\right) = \tan^{-1}\left(\dfrac{2}{\sqrt{5}-10t}\right)$, so

$\dfrac{d\theta}{dt} = \dfrac{1}{1+\left(\frac{2}{\sqrt{5}-10t}\right)^2}\left(\dfrac{-2}{(\sqrt{5}-10t)^2}\right)(-10) = \dfrac{20}{(\sqrt{5}-10t)^2+4}$. When $t = 0$, $\dfrac{d\theta}{dt} = \dfrac{20}{9} \approx 2.22$ radians per minute.

61. Let x represent the length of the rope and let θ represent the angle of depression of the rope.
Then $\theta = \sin^{-1}\left(\dfrac{8}{x}\right)$, so $\dfrac{d\theta}{dt} = \dfrac{1}{\sqrt{1-\left(\frac{8}{x}\right)^2}}\dfrac{-8}{x^2}\dfrac{dx}{dt} = \dfrac{-8}{x\sqrt{x^2-64}}\dfrac{dx}{dt}$.

Substituting $x = 17$ and $\dfrac{dx}{dt} = -5$, we obtain $\dfrac{d\theta}{dt} = \dfrac{-8}{17\sqrt{17^2-64}}(-5) = \dfrac{8}{51}$.

The angle of depression is increasing at a rate of $\dfrac{8}{51} \approx 0.16$ radians per second.

63. $\dfrac{dy}{dt} = 0.2\cos^2(0.1y)$
$\int \sec^2(0.1y)dy = \int 0.2dt$
$10\tan(0.1y) = 0.2t + C$
$C = 0$ because $y = 0$ when $t = 0$.
$t = 50\tan(0.1y)$
Note that the function $\sec^2(0.1y)$ is undefined at $y = 5\pi$, so the equation is valid for y in $[0, 5\pi)$.
When $y = 2.5\pi$, $t = 50\tan(0.25\pi) = 50$.
The object reaches $y = 2.5\pi$ cm after 50 seconds.

The object approaches (but never reaches) 5π as $t \to \infty$, as we can see by writing $y = 10\tan^{-1}\left(\dfrac{t}{50}\right)$.

65. (ii) Recall that $\sin^2\theta + \cos^2\theta = 1$, so $\cos\theta = \pm\sqrt{1-\sin^2\theta}$.

If $\theta = \sin^{-1} x$, then $\theta \in \left[-\frac{\pi}{2}, \frac{\pi}{2}\right]$, so $\cos\theta \ge 0$.

Therefore, $\cos(\sin^{-1} x) = \sqrt{1-\sin^2(\sin^{-1} x)} = \sqrt{1-x^2}$.

(iii) Recall that $\tan^2\theta + 1 = \sec^2\theta$, so $\sec\theta = \pm\sqrt{1+\tan^2\theta}$.

If $\theta = \tan^{-1} x$, then $\theta \in \left(-\frac{\pi}{2}, \frac{\pi}{2}\right)$, so $\sec\theta > 0$.

Therefore, $\sec(\tan^{-1} x) = \sqrt{1+\tan^2(\tan^{-1} x)} = \sqrt{1+x^2}$.

(iv) Recall that $\tan^2\theta + 1 = \sec^2\theta$, so $\tan\theta = \pm\sqrt{\sec^2\theta - 1}$.

If $\theta = \sec^{-1} x$, we obtain $\tan(\sec^{-1} x) = \pm\sqrt{\sec^2(\sec^{-1} x) - 1} = \pm\sqrt{x^2 - 1}$.

If $x \ge 1$, then $\sec^{-1} x \in \left[0, \frac{\pi}{2}\right)$, so $\tan(\sec^{-1} x) \ge 0$.

If $x \le -1$, then $\sec^{-1} x \in \left(\frac{\pi}{2}, \pi\right]$, so $\tan(\sec^{-1} x) \le 0$.

Note that $\sec^{-1} x$ (and hence $\tan(\sec^{-1} x)$) is undefined when $-1 < x < 1$. Therefore,

$$\tan(\sec^{-1} x) = \begin{cases} \sqrt{x^2 - 1} \text{ if } x \ge 1 \\ -\sqrt{x^2 - 1} \text{ if } x \le 1 \end{cases}.$$

Section 7.7

Concepts Review

1. $\frac{e^x - e^{-x}}{2}$; $\frac{e^x + e^{-x}}{2}$

3. A hyperbola

Problem Set 7.7

1.

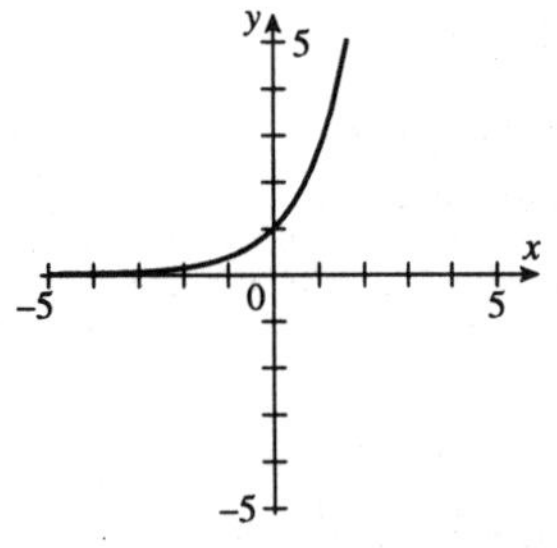

$$\cosh x + \sinh x = \frac{e^x + e^{-x}}{2} + \frac{e^x - e^{-x}}{2} = e^x$$

3. $\sinh x\cosh y+\cosh x\sinh y$

$$=\left(\frac{e^x-e^{-x}}{2}\right)\left(\frac{e^y+e^{-y}}{2}\right)+\left(\frac{e^x+e^{-x}}{2}\right)\left(\frac{e^y-e^{-y}}{2}\right)$$

$$=\frac{1}{4}(e^xe^y+e^xe^{-y}-e^{-x}e^y-e^{-x}e^{-y}+e^xe^y-e^xe^{-y}+e^{-x}e^y-e^{-x}e^{-y})$$

$$=\frac{1}{4}(2e^xe^y-2e^{-x}e^{-y})$$

$$=\frac{e^{x+y}-e^{-x-y}}{2}=\sinh(x+y)$$

5. Use the results of Problems 3 and 4.

$$\tanh(x+y)=\frac{\sinh(x+h)}{\cosh(x+y)}$$

$$=\frac{\sinh x\cosh y+\cosh x\sinh y}{\cosh x\cosh y+\sinh x\sinh y}\cdot\frac{\frac{1}{(\cosh x\cosh y)}}{\frac{1}{(\cosh x\cosh y)}}$$

$$=\frac{\frac{\sinh x}{\cosh x}+\frac{\sinh y}{\cosh y}}{1+\frac{\sinh x\sinh y}{\cosh x\cosh y}}$$

$$=\frac{\tanh x+\tanh y}{1+\tanh x\tanh y}$$

7. $\frac{dy}{dx}=\frac{d}{dx}\sinh^2 x$

$$=2\sinh x\frac{d}{dx}\sinh x$$

$$=2\sinh x\cosh x$$

9. $\frac{dy}{dx}=\frac{d}{dx}\cosh(x^2-1)$

$$=\sinh(x^2-1)\frac{d}{dx}(x^2-1)$$

$$=2x\sinh(x^2-1)$$

11. $\frac{dy}{dx}=\frac{d}{dx}(x^2\sinh x)=2x\sinh x+x^2\cosh x$

13. $y=\sinh 4x\cosh 2x$

$$=\left(\frac{e^{4x}-e^{-4x}}{2}\right)\left(\frac{e^{2x}+e^{-2x}}{2}\right)$$

$$=\frac{1}{4}(e^{6x}+e^{2x}-e^{-2x}-e^{-6x})$$

$$=\frac{1}{2}(\sinh 2x+\sinh 6x)$$

$$\frac{dy}{dx}=\frac{1}{2}\frac{d}{dx}(\sinh 2x+\sinh 6x)$$

$$=\frac{1}{2}(2\cosh 2x+6\cosh 6x)=\cosh 2x+3\cosh 6x$$

or (by using the Product Rule)

$$\frac{dy}{dx}=4\cosh 4x\cosh 2x+2\sinh 2x\sinh 4x$$

15. $\frac{dy}{dx}=\frac{d}{dx}\cosh^{-1}(x^3)=\frac{1}{\sqrt{(x^3)^2-1}}\frac{d}{dx}x^3$

$$=\frac{3x^2}{\sqrt{x^6-1}}$$

17. $\frac{dy}{dx}=\frac{d}{dx}[x\sinh^{-1}(-2x)]$

$$=\sinh^{-1}(-2x)+x\frac{1}{\sqrt{(-2x)^2+1}}(-2)$$

$$=-\sinh^{-1}(2x)-\frac{2x}{\sqrt{4x^2+1}}$$

19. $\frac{dy}{dx}=\frac{d}{dx}\sinh(\cos x)$

$$=\cosh(\cos x)\frac{d}{dx}\cos x=-\sin x\cosh(\cos x)$$

21. Let $u=x^2+3$, then $du=2x\,dx$.

$$\int x\cosh(x^2+3)dx=\frac{1}{2}\int[\cosh(x^2+3)]2x\,dx$$

$$=\frac{1}{2}\int\cosh u\,du=\frac{1}{2}\sinh u+C$$

$$=\frac{1}{2}\sinh(x^2+3)+C$$

23. Let $u=e^x$, then $du=e^x dx$.

$$\int e^x\sinh e^x dx=\int\sinh u\,du=\cosh u+C$$

$$=\cosh e^x+C$$

25. $\text{Area}=\int_0^{\ln 3}\cosh 2x\,dx=\left[\frac{1}{2}\sinh 2x\right]_0^{\ln 3}$

$$=\frac{e^{2\ln 3}-e^{-2\ln 3}}{4}-\frac{e^0-e^0}{4}$$

$$=\frac{1}{4}\left(9-\frac{1}{9}\right)=\frac{20}{9}$$

27. Volume $= \int_0^1 \pi\cosh^2 x\,dx = \frac{\pi}{2}\int_0^1 (1+\cosh 2x)dx$

$$= \frac{\pi}{2}\left[x + \frac{\sinh 2x}{2}\right]_0^1$$

$$= \frac{\pi}{2}\left(1 + \frac{\sinh 2}{2} - 0\right)$$

$$= \frac{\pi}{2} + \frac{\pi\sinh 2}{4} \approx 4.4193$$

29. $y = a\cosh\left(\frac{x}{a}\right) + C$

$$\frac{dy}{dx} = \sinh\left(\frac{x}{a}\right)$$

$$\frac{d^2y}{dx^2} = \frac{1}{a}\cosh\left(\frac{x}{a}\right)$$

We need to show that $\frac{d^2y}{dx^2} = \frac{1}{a}\sqrt{1+\left(\frac{dy}{dx}\right)^2}$.

Note that $1+\sinh^2\left(\frac{x}{a}\right) = \cosh^2\left(\frac{x}{a}\right)$ and $\cosh\left(\frac{x}{a}\right) > 0$.

Therefore, $\frac{1}{a}\sqrt{1+\left(\frac{dy}{dx}\right)^2} = \frac{1}{a}\sqrt{1+\sinh^2\left(\frac{x}{a}\right)} = \frac{1}{a}\sqrt{\cosh^2\left(\frac{x}{a}\right)}$

$$= \frac{1}{a}\cosh\left(\frac{x}{a}\right) = \frac{d^2y}{dx^2}$$

31. **a.**

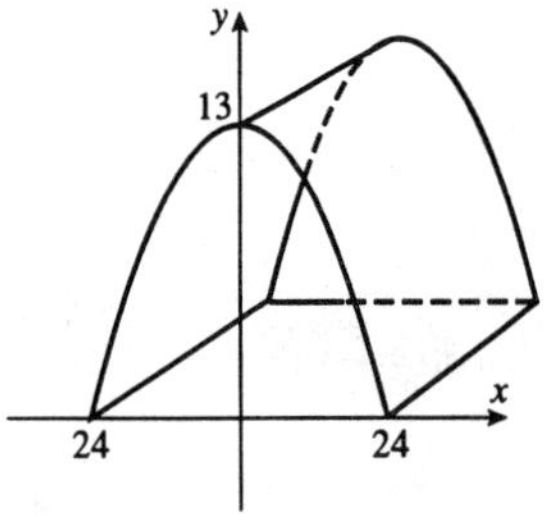

b. Area under the curve is

$$\int_{-24}^{24}\left[37 - 24\cosh\left(\frac{x}{24}\right)\right]dx = \left[37x - 576\sinh\left(\frac{x}{24}\right)\right]_{-24}^{24} \approx 422$$

Volume is about (422)(100) = 42,200 ft^3.

c. Length of the curve is

$$\int_{-24}^{24}\sqrt{1+\left(\frac{dy}{dx}\right)^2}\,dx = \int_{-24}^{24}\sqrt{1+\sinh^2\left(\frac{x}{24}\right)}dx = \int_{-24}^{24}\cosh\left(\frac{x}{24}\right)dx$$

$$= \left[24\sinh\left(\frac{x}{24}\right)\right]_{-24}^{24} = 48\sinh 1 \approx 56.4$$

Surface area $\approx (56.4)(100) = 5640$ ft^2

33. a. $(\sinh x+\cosh x)^r=\left(\frac{e^x-e^{-x}}{2}+\frac{e^x+e^{-x}}{2}\right)^r=(e^x)^r$

$=e^{rx}=\frac{e^{rx}-e^{-rx}}{2}+\frac{e^{rx}+e^{-rx}}{2}=\sinh rx+\cosh rx$

b. $(\cosh x-\sinh x)^r=\left(\frac{e^x+e^{-x}}{2}-\frac{e^x-e^{-x}}{2}\right)^r=(e^{-x})^r$

$=e^{-rx}=\frac{e^{rx}+e^{-rx}}{2}-\frac{e^{rx}-e^{-rx}}{2}=\cosh rx-\sinh rx$

35. $y(x)=b-a\cosh\left(\frac{x}{a}\right)$

$y(0) = 360$, so $360 = b - a$. That is, $b = a + 360$.

$y(315)=0$, so $a+360-a\cosh\left(\frac{315}{a}\right)=0$.

Numerical methods were used to obtain $a\approx 177.83$.

The equation is $y=537.83-177.83\cosh\left(\frac{x}{177.83}\right)$.

Section 7.8 Chapter Review

Concepts Test

1. False. ln 0 is undefined.

3. True. $\int_1^{e^3}\frac{1}{t}dt=\left[\ln|t|\right]_1^{e^3}=\ln e^3-\ln 1=3$

5. True. The range of $y = \ln x$ is the set of all real numbers.

7. False. $4\ln x=\ln(x^4)$

9. True. $f(g(x))=4+e^{\ln(x-4)}=4+(x-4)=x$
and
$g(f(x))=\ln(4+e^x-4)=\ln e^x=x$

11. True.

$\lim_{x\to 0^+}(\ln\sin x-\ln x)=\lim_{x\to 0^+}\ln\left(\frac{\sin x}{x}\right)=\ln 1=0$

13. False. ln π is a constant so $\frac{d}{dx}\ln\pi=0$.

15. True. $\frac{d}{dx}x^n=nx^{n-1}$ for any real number n.

17. False. $\frac{d}{dx}x^x=x^x(1+\ln x)$ by Section 7.3, Example 5.

19. True. The variables cannot be separated, but a solution can be approximated using Euler's method.

21. False. $\arcsin(\sin\ 2\pi)=\arcsin 0=0$

23. False. $\cosh(\ln 3)=\frac{e^{\ln 3}+e^{-\ln 3}}{2}=\frac{1}{2}\left(3+\frac{1}{3}\right)=\frac{5}{3}$

25. False. If $x\neq 0$ then $\cosh x>1$, so $\sin^{-1}(\cosh x)$ is undefined.

27. False. The logarithms are not defined for all real x.

29. False. $e^{0.11}<\left(1+\frac{0.12}{12}\right)^{12}$

Sample Test Problems

1. $\ln\frac{x^4}{2}=4\ln x-2$

$\frac{d}{dx}\ln\frac{x^4}{2}=\frac{d}{dx}(4\ln x-2)=\frac{4}{x}$

3. $\frac{d}{dx}e^{x^2-4x}=e^{x^2-4x}\frac{d}{dx}(x^2-4x)$

$=e^{x^2-4x}(2x-4)$

5. $\frac{d}{dx}\tan(\ln e^x)=\frac{d}{dx}\tan x=\sec^2 x$

7. $\dfrac{d}{dx}2\tanh\sqrt{x} = 2\text{sech}^2\sqrt{x}\dfrac{d}{dx}\sqrt{x} = \dfrac{\text{sech}^2\sqrt{x}}{\sqrt{x}}$

9. $\dfrac{d}{dx}\sinh^{-1}(\tan x) = \dfrac{1}{\sqrt{\tan^2 x+1}}\dfrac{d}{dx}\tan x$

$= \dfrac{\sec^2 x}{\sqrt{\tan^2 x+1}} = \dfrac{\sec^2 x}{\sqrt{\sec^2 x}} = |\sec x|$

11. $\dfrac{d}{dx}\sec^{-1} e^x = \dfrac{1}{|e^x|\sqrt{(e^x)^2-1}}\dfrac{d}{dx}e^x$

$= \dfrac{e^x}{e^x\sqrt{e^{2x}-1}} = \dfrac{1}{\sqrt{e^{2x}-1}}$

13. $\dfrac{d}{dx}3\ln(e^{5x}+1) = \dfrac{3}{e^{5x}+1}(5e^{5x}) = \dfrac{15e^{5x}}{e^{5x}+1}$

15. $\dfrac{d}{dx}\cos e^{\sqrt{x}} = -\sin e^{\sqrt{x}}\dfrac{d}{dx}e^{\sqrt{x}}$

$= (-\sin e^{\sqrt{x}})e^{\sqrt{x}}\dfrac{d}{dx}\sqrt{x} = -\dfrac{e^{\sqrt{x}}\sin e^{\sqrt{x}}}{2\sqrt{x}}$

17. $\dfrac{d}{dx}2\cos^{-1}\sqrt{x} = \dfrac{-2}{\sqrt{1-(\sqrt{x})^2}}\dfrac{d}{dx}\sqrt{x}$

$= \dfrac{-2}{\sqrt{1-x}}\dfrac{1}{2\sqrt{x}} = -\dfrac{1}{\sqrt{x-x^2}}$

19. $\dfrac{d}{dx}2\csc e^{\ln\sqrt{x}} = \dfrac{d}{dx}2\csc\sqrt{x}$

$= -2\csc\sqrt{x}\cot\sqrt{x}\dfrac{d}{dx}\sqrt{x} = -\dfrac{\csc\sqrt{x}\cot\sqrt{x}}{2\sqrt{x}}$

21. $\dfrac{d}{dx}4\tan 5x\sec 5x$

$= 20\sec^2 5x\sec 5x + 20\tan 5x\sec 5x\tan 5x$

$= 20\sec 5x(\sec^2 5x + \tan^2 5x)$

$= 20\sec 5x(2\sec^2 5x - 1)$

23. $\dfrac{d}{dx}x^{1+x} = \dfrac{d}{dx}e^{(1+x)\ln x}$

$= e^{(1+x)\ln x}\dfrac{d}{dx}[(1+x)\ln x]$

$= x^{1+x}\left[(1)(\ln x) + (1+x)\left(\dfrac{1}{x}\right)\right]$

$= x^{1+x}\left(\ln x + 1 + \dfrac{1}{x}\right)$

25. Let $u = 3x - 1$, then $du = 3\,dx$.

$\int e^{3x-1}dx = \dfrac{1}{3}\int e^{3x-1}3dx = \dfrac{1}{3}\int e^u du$

$= \dfrac{1}{3}e^u + C = \dfrac{1}{3}e^{3x-1} + C$

Check:

$\dfrac{d}{dx}\left(\dfrac{1}{3}e^{3x-1} + C\right) = \dfrac{1}{3}e^{3x-1}\dfrac{d}{dx}(3x-1) = e^{3x-1}$

27. Let $u = e^x$, then $du = e^x dx$.

$\int e^x \sin e^x dx = \int \sin u\, du = -\cos u + C = -\cos e^x + C$

Check:

$\dfrac{d}{dx}(-\cos e^x + C) = (\sin e^x)\dfrac{d}{dx}e^x = e^x \sin e^x$

29. Let $u = e^{x+3}+1$, then $du = e^{x+3}dx$.

$\int \dfrac{e^{x+2}}{e^{x+3}+1}dx = \dfrac{1}{e}\int \dfrac{1}{e^{x+3}+1}e^{x+3}dx = \dfrac{1}{e}\int \dfrac{1}{u}du$

$= \dfrac{1}{e}\ln|u| + C = \dfrac{\ln(e^{x+3}+1)}{e} + C$

Check:

$\dfrac{d}{dx}\left(\dfrac{\ln(e^{x+3}+1)}{e} + C\right) = \dfrac{1}{e}\dfrac{1}{e^{x+3}+1}\dfrac{d}{dx}(e^{x+3}+1)$

$= \dfrac{e^{x+3}e^{-1}}{e^{x+3}+1} = \dfrac{e^{x+2}}{e^{x+3}+1}$

31. Let $u = 2x$, then $du = 2\,dx$.

$\int \dfrac{4}{\sqrt{1-4x^2}}dx = 2\int \dfrac{1}{\sqrt{1-(2x)^2}}2dx$

$= 2\int \dfrac{1}{\sqrt{1-u^2}}du$

$= 2\sin^{-1} u + C = 2\sin^{-1} 2x + C$

Check:

$\dfrac{d}{dx}(2\sin^{-1} 2x + C) = 2\left(\dfrac{1}{\sqrt{1-(2x)^2}}\right)\dfrac{d}{dx}2x$

$= \dfrac{4}{\sqrt{1-4x^2}}$

33. Let $u = \ln x$, then $du = \dfrac{1}{x}dx$.

$\int \dfrac{-1}{x + x(\ln x)^2}dx = -\int \dfrac{1}{1+(\ln x)^2}\cdot\dfrac{1}{x}dx$

$= -\int \dfrac{1}{1+u^2}du = -\tan^{-1} u + C = -\tan^{-1}(\ln x) + C$

Check:
$$\frac{d}{dx}[-\tan^{-1}(\ln x)+C] = -\frac{1}{1+(\ln x)^2}\frac{d}{dx}\ln x = \frac{-1}{x+x(\ln x)^2}$$

35. $f(x) = x^2 - 4\sin x$

$f'(x) = 2x - 4\cos x$

Numerical methods were used to determine that $f'(x) < 0$ (and hence f is decreasing) on $(-\pi, 1.0299)$ and $f'(x) > 0$ (and hence f is increasing) on $(1.0299, \pi)$.

$f''(x) = 2 + 4\sin x$

The graph of f is concave upward on $\left[-\pi, -\frac{5\pi}{6}\right]$ and $\left[-\frac{\pi}{6}, \pi\right]$ because $f''(x) \geq 0$ on these intervals. The graph of f is concave downward on $\left[-\frac{5\pi}{6}, -\frac{\pi}{6}\right]$ because $f''(x) \leq 0$ on this interval. The absolute minimum value is $f(1.0299) \approx -2.3683$. The inflection points are $\left(-\frac{5\pi}{6}, \frac{25\pi^2}{36}+2\right)$ and $\left(-\frac{\pi}{6}, \frac{\pi^2}{36}+2\right)$.

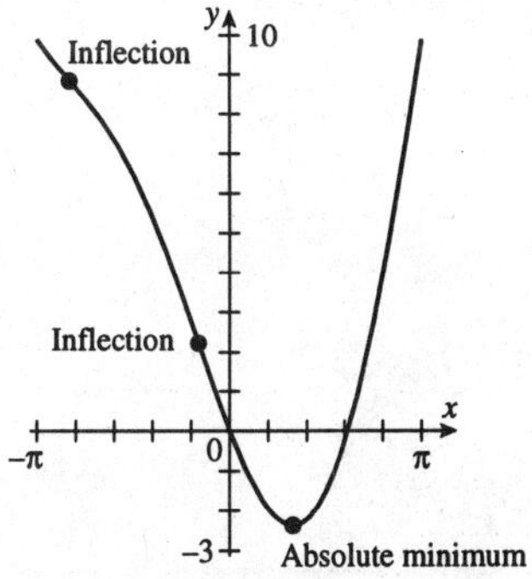

37. $f(x) = \frac{\ln(ax)}{x}$

$$f'(x) = \frac{x\left(\frac{1}{x}\right) - \ln(ax)(1)}{x^2} = \frac{1-\ln(ax)}{x^2}$$

f is increasing on $\left(0, \frac{e}{a}\right]$ because $f'(x) \geq 0$ on this interval.

f is decreasing on $\left[\frac{e}{a}, \infty\right)$ because $f'(x) \leq 0$ on this interval.

$$f''(x) = \frac{x^2\left(-\frac{1}{x}\right) - (1-\ln ax)(2x)}{x^4} = \frac{2\ln(ax) - 3}{x^3}$$

The graph of f is concave upward on $\left[\frac{e^{3/2}}{a}, \infty\right)$ because $f''(x) \geq 0$ on this interval.

The graph of f is concave downward on $\left(0, \frac{e^{3/2}}{a}\right]$ because $f''(x) \leq 0$ on this interval.

The absolute maximum value is $f\left(\frac{e}{a}\right) = \frac{a}{e}$.

The inflection point is $\left(\frac{e^{3/2}}{a}, \frac{3a}{2e^{3/2}}\right)$.

The graph shown illustrates the case $a = 4$.

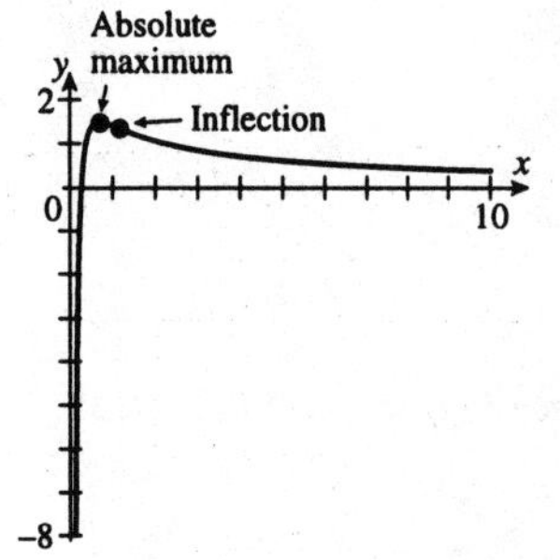

39. **a.** $(\$100)(1.12) = \$112$

b. $(\$100)\left(1+\dfrac{0.12}{12}\right)^{12} \approx \$112.68$

c. $(\$100)\left(1+\dfrac{0.12}{365}\right)^{365} \approx \$112.75$

d. $(\$100)e^{(0.12)(1)} \approx \$112.75$

41. $y = (\cos x)^{\sin x} = e^{\sin x \ln(\cos x)}$

$$\frac{dy}{dx} = e^{\sin x \ln(\cos x)} \frac{d}{dx}[\sin x \ln(\cos x)]$$

$$= e^{\sin x \ln(\cos x)}\left[\cos x \ln(\cos x) + (\sin x)\left(\frac{1}{\cos x}\right)(-\sin x)\right]$$

$$= (\cos x)^{\sin x}\left[\cos x \ln(\cos x) - \frac{\sin^2 x}{\cos x}\right]$$

At $x = 0$, $\dfrac{dy}{dx} = 1^0(1\ln 1 - 0) = 0.$

The tangent line has slope 0, so it is horizontal: $y = 1$.

Chapter 8

Section 8.1

Concepts Review

1. $\int u^5 du$

3. $\int_1^2 u^3 du$

Problem Set 8.1

1. Let $u = (x-1)$.
$du = dx$
$\int (x-1)^4 dx = \int u^4 du$
$= \frac{1}{5}u^5 + C$
$= \frac{1}{5}(x-1)^5 + C$

3. Let $u = x^2 + 1$.
$du = 2x\,dx$
$\frac{1}{2}du = x\,dx$
$\int x(x^2+1)^4 dx = \frac{1}{2}\int u^4 du$
$= \frac{1}{2}\cdot\frac{1}{5}u^5 + C$
$= \frac{1}{10}(x^2+1)^5 + C$

5. $\int \frac{dx}{x^2+1} = \tan^{-1} x + C$ from the integral tables

7. Let $u = x^2 + 1$.
$du = 2x\,dx$
$\frac{1}{2}du = x\,dx$
$\int \frac{x}{x^2+1}dx = \frac{1}{2}\int \frac{1}{u}du$
$= \frac{1}{2}\ln|u| + C$
$= \frac{1}{2}\ln\left|x^2+1\right| + C$

9. Let $u = 2 + t^2$
$du = 2t\,dt$
$\frac{1}{2}du = t\,dt$
$\int 3t\sqrt{2+t^2}\,dt = \frac{1}{2}\int 3\sqrt{u}\,du$
$= \frac{3}{2}\cdot\frac{2}{3}u^{3/2} + C$
$= (2+t^2)^{3/2} + C$

11. $\int \frac{\tan z}{\cos z}dz = \int \sec z \tan z\,dz$
$= \sec z + C$

13. Let $u = \sqrt{x}$.
$du = \frac{1}{2}x^{-1/2}dx$
$2du = x^{-1/2}dx$
$\int \frac{\cos\sqrt{x}}{\sqrt{x}}dx = 2\int \cos u\,du$
$= 2\sin u + C$
$= 2\sin\sqrt{x} + C$

15. Let $u = \sin x$.
$du = \cos x\,dx$
$\int_0^{\pi/2} \frac{\cos x}{1+\sin^2 x}dx = \int_0^1 \frac{1}{1+u^2}du = \left[\tan^{-1} u\right]_0^1$
$= \tan^{-1}(1) - \tan^{-1}(0)$
$= \frac{\pi}{4}$

17. By using long division,
$\frac{2x^2+x}{x+1} = 2x - 1 + \frac{1}{x+1}$
$\int \frac{2x^2+x}{x+1}dx = \int 2x - 1 + \frac{1}{x+1}dx$
$= x^2 - x + \int \frac{1}{x+1}dx$
Let $u = x + 1$.
$du = dx$
$\int \frac{1}{x+1}dx = \int \frac{1}{u}du = \ln|u| + C = \ln|x+1| + C$
$\int \frac{2x^2+x}{x+1}dx = x^2 - x + \ln|x+1| + C$

19. Let $u = \tan x$.
$du = \sec^2 x\,dx$
$\int \frac{\sqrt{\tan x}}{1-\sin^2 x}dx = \int \frac{\sqrt{\tan x}}{\cos^2 x}dx$
$= \int \sqrt{u}\,du$
$= \frac{2}{3}u^{3/2} + C$
$= \frac{2}{3}(\tan x)^{3/2} + C$

21. Let $u = \ln 4x^2$.

$du = \frac{1}{4x^2} \cdot 8x\,dx = \frac{2}{x}dx$

$\frac{1}{2}du = \frac{1}{x}dx$

$\int \frac{\cos(\ln 4x^2)dx}{x} = \frac{1}{2}\int \cos u\,du$

$= \frac{1}{2}\sin u + C$

$= \frac{1}{2}\sin(\ln 4x^2) + C$

23. Let $u = e^x$.

$du = e^x dx$

$\int \frac{5e^x}{\sqrt{1-e^{2x}}}dx = 5\int \frac{1}{\sqrt{1-u^2}}du$

$= 5\sin^{-1} u + C$

$= 5\sin^{-1}(e^x) + C$

25. Let $u = 1 - e^{2x}$.

$du = -2e^{2x}dx$

$-\frac{1}{2}du = e^{2x}dx$

$\int \frac{5e^{2x}}{\sqrt{1-e^{2x}}}dx = -\frac{5}{2}\int u^{-1/2}du$

$= -\frac{5}{2} \cdot 2u^{1/2} + C$

$= -5(1-e^{2x}) + C$

27. Let $u = x^2$.

$du = 2x\,dx$

$\frac{1}{2}du = x\,dx$

$\int_0^1 x10^{x^2}dx = \frac{1}{2}\int_0^1 10^u du$

$= \left[\frac{1}{2} \cdot \frac{10^u}{\ln 10}\right]_0^1$

$= \frac{1}{2} \cdot \frac{10}{\ln 10} - \frac{1}{2} \cdot \frac{1}{\ln 10}$

$= \frac{9}{2\ln 10}$

29. $\int \frac{\sin x - \cos x}{\sin x}dx = \int 1 - \frac{\cos x}{\sin x}dx$

Let $u = \sin x$.

$du = \cos x\,dx$

$\int \frac{\sin x - \cos x}{\sin x}dx = x - \int \frac{du}{u}$

$= x - \ln|u| + C$

$= x - \ln|\sin x| + C$

31. Let $u = z^2 + 4z - 3$.

$du = (2z+4)dz$

$du = 2(z+2)dz$

$\frac{1}{2}du = (z+2)dz$

$\int \frac{z+2}{\cot(z^2+4z-3)}dz = \frac{1}{2}\int \frac{1}{\cot u}du$

$= \frac{1}{2}\int \tan u\,du = \frac{1}{2}\ln|\sec u| + C$

$= \frac{1}{2}\ln\left|\sec(z^2+4z-3)\right| + C$

Use Formula 12 for $\int \tan u\,du$.

33. Let $u = e^x$.

$du = e^x dx$

$\int e^x \sec e^x dx = \int \sec u\,du$

$= \ln|\sec u + \tan u| + C$

$= \ln\left|\sec(e^x) + \tan(e^x)\right| + C$

Use Formula 14 for $\int \sec u\,du$.

35. $\int \frac{\sec^3 x + e^{\sin x}}{\sec x}dx = \int \sec^2 x + e^{\sin x}\cos x\,dx$

Using the simple substitution of $u = \sin x$ and simple integration formulas,

$\int \frac{\sec^3 x + e^{\sin x}}{\sec x} = \tan x + e^{\sin x} + C.$

37. Let $u = t^3 - 2$.

$du = 3t^2 dt$

$\frac{1}{3}du = t^2 dt$

$\int \frac{t^2\cos(t^3-2)}{\sin^2(t^3-2)}dt = \frac{1}{3}\int \frac{\cos u}{\sin^2 u}du$

Let $v = \sin u$.

$dv = \cos u\,du$

$\frac{1}{3}\int v^{-2}dv = -\frac{1}{3}v^{-1} + C$

$= -\frac{1}{3\sin(t^3-2)} + C.$

39. Let $u = t^3 - 2$.
$du = 3t^2 dt$
$\frac{1}{3} du = t^2 dt$
$\int \frac{t^2 \cos^2(t^3-2)}{\sin^2(t^3-2)} dt = \frac{1}{3} \int \frac{\cos^2 u}{\sin^2 u} du$
$= \frac{1}{3} \int \cot^2 u \, du$
$= \frac{1}{3}[-\cot u - u + C]$
$= \frac{1}{3}[-\cot(t^3-2) - (t^3-2) + C]$
$= \frac{1}{3}[-\cot(t^3-2) - t^3] + C$
Use Formula 36 with $n = 2$ for $\int \cot^2 u \, du$.

41. Let $u = \tan^{-1} 2t$.
$du = \frac{1}{1+4t^2} \cdot 2dt$
$\frac{1}{2} du = \frac{1}{1+4t^2} dt$
$\int \frac{e^{\tan^{-1} 2t}}{1+4t^2} dt = \frac{1}{2} \int e^u du$
$= \frac{1}{2} e^{\tan^{-1} 2t} + C$

43. Let $u = 3y^2$.
$du = 6y\, dy$
$\frac{1}{6} du = y\, dy$
$\int \frac{y}{\sqrt{16 - 9y^4}} dy = \frac{1}{6} \int \frac{1}{\sqrt{4^2 - u^2}} du$
$= \frac{1}{6} \sin^{-1}\left(\frac{u}{4}\right) + C$
$= \frac{1}{6} \sin^{-1}\left(\frac{3y^2}{4}\right) + C$

45. Let $u = \sec x$.
$du = \sec x \tan x \, dx$
$\int \frac{\sec x \tan x}{1 + \sec^2 x} dx = \int \frac{1}{1+u^2} du$
$= \tan^{-1} u + C$
$= \tan^{-1}(\sec x) + C$

47. Let $u = e^{3t}$.
$du = 3e^{3t} dt$
$\frac{1}{3} du = e^{3t} dt$
$\int \frac{e^{3t}}{\sqrt{4 - e^{6t}}} dt = \frac{1}{3} \int \frac{1}{\sqrt{2^2 - u^2}} du$
$= \frac{1}{3} \sin^{-1}\left(\frac{u}{2}\right) + C$
$= \frac{1}{3} \sin^{-1}\left(\frac{e^{3t}}{2}\right) + C$

49. Let $u = \cos x$.
$du = -\sin x \, dx$
$-du = \sin x \, dx$
$\int_0^{\pi/2} \frac{\sin x}{16 + \cos^2 x} dx = -\int_1^0 \frac{1}{16 + u^2} du$
$= -\left[\frac{1}{4} \tan^{-1}\left(\frac{u}{4}\right)\right]_1^0$
$= -\left[\frac{1}{4} \tan^{-1} 0 - \frac{1}{4} \tan^{-1}\left(\frac{1}{4}\right)\right]$
$= \frac{1}{4} \tan^{-1}\left(\frac{1}{4}\right)$

51. Let $u = \sqrt{2}\, t$.
$du = \sqrt{2}\, dt$
$\int \frac{dt}{t\sqrt{2t^2 - 9}} = 2 \int \frac{du}{u\sqrt{u^2 - 9}}$
$= 2 \cdot \frac{1}{3} \sec^{-1}\left|\frac{u}{3}\right| + C$
$= \frac{2}{3} \sec^{-1}\left|\frac{\sqrt{2}\, t}{3}\right| + C$
Use Formula 19 with $a = 3$ for $\int \frac{du}{u\sqrt{u^2 - 9}}$.

53. Let $u = 3x + 2$.
$x = \frac{u-2}{3}$
$du = 3\, dx$
$\frac{1}{3} du = dx$
$\int x\sqrt{3x + 2}\, dx = \int \frac{u-2}{3} \sqrt{u} \cdot \frac{1}{3} du$
$= \frac{1}{9} \int (u-2)\sqrt{u}\, du$
$= \frac{1}{9} \int u^{3/2} - 2u^{1/2}\, du$
$= \frac{1}{9}\left[\frac{2}{5} u^{5/2} - 2 \cdot \frac{2}{3} u^{3/2}\right] + C$
$= \frac{2}{45}(3x+2)^{5/2} - \frac{4}{27}(3x+2)^{3/2} + C$

55. Let $u = 4x$.

$du = 4\,dx$

$\frac{1}{4}du = dx$

$$\int \frac{dx}{9-16x^2} = \frac{1}{4}\int \frac{du}{3^2-u^2}du$$

$$= \frac{1}{4}\left[\frac{1}{2(3)}\ln\left|\frac{u+3}{u-3}\right|\right]+C$$

$$= \left[\frac{1}{24}\ln\left|\frac{4x+3}{4x-3}\right|\right]+C$$

Use Formula 18 with $a = 3$ for $\int \frac{du}{3^2-u^2}$.

57. $$\int x^2\sqrt{9-2x^2}\,dx = \int \sqrt{2}x^2\sqrt{\frac{9-2x^2}{2}}\,dx$$

$$= \sqrt{2}\int x^2\sqrt{\frac{9}{2}-x^2}\,dx$$

$$= \sqrt{2}\left[\frac{x}{8}\left(2x^2-\frac{9}{2}\right)\sqrt{\frac{9}{2}-x^2}+\frac{\left(\frac{81}{4}\right)}{8}\sin^{-1}\left(\frac{x}{\frac{3}{\sqrt{2}}}\right)+C\right]$$

$$= \frac{x}{16}(4x^2-9)\sqrt{9-2x^2}+\frac{81\sqrt{2}}{32}\sin^{-1}\left(\frac{\sqrt{2}\,x}{3}\right)+C$$

Use Formula 57 with $a = \frac{3}{\sqrt{2}}$ and $u = x$ for $\int x^2\sqrt{\frac{9}{2}-x^2}\,dx$.

59. $$\int \frac{dx}{\sqrt{5+3x^2}} = \frac{1}{\sqrt{3}}\int \frac{\sqrt{3}\,dx}{\sqrt{\left(\sqrt{5}\right)^2+\left(\sqrt{3}x\right)^2}}$$

$$= \frac{1}{\sqrt{3}}\ln\left|\sqrt{3}\,x+\sqrt{3x^2+5}\right|+C$$

Use Formula 45 with $a = \sqrt{5}$ and $u = \sqrt{3}\,x$ for $\int \frac{\sqrt{3}\,dx}{\sqrt{\left(\sqrt{5}\right)^2+\left(\sqrt{3}\,x\right)^2}}$.

61. Let $u = x + 1$.

$du = dx$

$x^2+2x-3 = x^2+2x+1-4$

$= (x+1)^2-4 = u^2-4$

$$\int \frac{\sqrt{x^2+2x-3}}{x+1}dx = \int \frac{\sqrt{u^2-4}}{u}du$$

$$= \sqrt{u^2-4}-2\sec^{-1}\frac{u}{2}+C$$

$$= \sqrt{(x+1)^2-4}-2\sec^{-1}\left(\frac{x+1}{2}\right)+C$$

Use Formula 47 with $a = 2$ for $\int \frac{\sqrt{u^2-4}}{u}du$.

63. Let $u = \sqrt{2}x$.

$du = \sqrt{2}dx$

$\frac{1}{\sqrt{2}}du = dx$

$$\int \frac{dx}{\sqrt{2x^2+1}} = \frac{1}{\sqrt{2}}\int \frac{du}{\sqrt{u^2+1}} = \ln\left|u+\sqrt{u^2+1}\right|+C$$

$$= \ln\left|\sqrt{2}\,x+\sqrt{2x^2+1}\right|+C$$

65. $\frac{\sqrt{2}\ln\left(\sqrt{2x^2+1}+\sqrt{2}x\right)}{2}-\frac{1}{2}\ln\left(x+\sqrt{x^2+\frac{1}{2}}\right)$

$=\frac{1}{\sqrt{2}}\ln\left(\frac{\sqrt{2x^2+1}+\sqrt{2}x}{x+\sqrt{x^2+\frac{1}{2}}}\right)$

$=\frac{1}{\sqrt{2}}\ln\left(\frac{\sqrt{2x^2+1}+\sqrt{2}x}{x+\frac{1}{\sqrt{2}}\sqrt{2x^2+1}}\right)$

$=\frac{1}{\sqrt{2}}\ln\left(\frac{\sqrt{2x^2+1}+\sqrt{2}x}{\frac{1}{\sqrt{2}}\left(\sqrt{2}x+\sqrt{2x^2+1}\right)}\right)$

$=\frac{1}{\sqrt{2}}\ln\left(\sqrt{2}\right)$

$=\frac{\sqrt{2}}{2}\ln 2^{1/2}=\frac{\sqrt{2}}{4}\ln 2$

Section 8.2

Concepts Review

1. $uv-\int v\,du$

3. 1

Problem Set 8.2

1. Let $u=x$.
$du=dx$
$dv=e^x dx$
$v=e^x$
$\int xe^x dx=xe^x-\int e^x dx$
$=xe^x-e^x+C$

3. Let $u=x$.
$du=dx$
$dv=\sin 3x\,dx$
$v=-\frac{1}{3}\cos 3x$
$\int x\sin 3x\,dx=-\frac{1}{3}x\cos 3x+\frac{1}{3}\int\cos 3x\,dx$
$=-\frac{1}{3}x\cos 3x+\frac{1}{9}\sin 3x+C$

5. Let $u=\arctan x$.
$du=\frac{1}{1+x^2}dx$
$dv=dx$
$v=x$
$\int\arctan x=x\arctan x-\int\frac{x}{1+x^2}dx$
Let $w=1+x^2$.
$dw=2x\,dx$
$\frac{1}{2}dw=x\,dx$
$x\arctan x-\frac{1}{2}\int\frac{1}{w}dw$
$=x\arctan x-\frac{1}{2}\ln w+C$
$=x\arctan x-\frac{1}{2}\ln(1+x^2)+C$

7. Let $u=t$.
$du=dt$
$dv=\sec^2 5t\,dt$
$v=\frac{1}{5}\tan 5t$
$\int t\sec^2 5t\,dt=\frac{t}{5}\tan 5t-\frac{1}{5}\int\tan 5t\,dt$
$=\frac{t}{5}\tan 5t+\frac{1}{5}\cdot\frac{1}{5}\ln|\cos 5t|+C$
$=\frac{t}{5}\tan 5t+\frac{1}{25}\ln|\cos 5t|+C$

9. Let $u=\ln x$.
$du=\frac{1}{x}dx$
$dv=x^{1/2}dx$
$v=\frac{2}{3}x^{3/2}$
$\int\sqrt{x}\ln x\,dx=\frac{2}{3}x^{3/2}\ln x-\frac{2}{3}\int x^{3/2}\cdot\frac{1}{x}dx$
$=\frac{2}{3}x^{3/2}\ln x-\frac{2}{3}\int x^{1/2}dx$
$=\frac{2}{3}x^{3/2}\ln x-\frac{4}{9}x^{3/2}+C$

11. Let $u=\arctan x$.
$du=\frac{1}{1+x^2}dx$
$dv=x\,dx$
$v=\frac{1}{2}x^2$
$\int x\arctan x=\frac{1}{2}x^2\arctan x-\frac{1}{2}\int x^2\cdot\frac{1}{1+x^2}dx$
$=\frac{1}{2}x^2\arctan x-\frac{1}{2}\int 1-\frac{1}{1+x^2}dx$
$=\frac{1}{2}x^2\arctan x-\frac{1}{2}x+\frac{1}{2}\arctan x+C$

13. Let $u = \ln w$.

$du = \frac{1}{w} dw$

$dv = w\,dw$

$v = \frac{1}{2}w^2$

$\int w \ln w\,dw = \frac{1}{2}w^2 \ln w - \int \frac{1}{2}w^2 \cdot \frac{1}{w} dw$

$= \frac{1}{2}w^2 \ln w - \frac{1}{2}\int w\,dw$

$= \frac{1}{2}w^2 \ln w - \frac{1}{4}w^2 + C$

15. Let $u = x$.

$du = dx$

$dv = \csc^2 x\,dx$

$v = -\cot x$

$\int_{\pi/6}^{\pi/2} x \csc^2 x\,dx = [-x \cot x]_{\pi/6}^{\pi/2} + \int_{\pi/6}^{\pi/2} \cot x\,dx$

$= \frac{\pi}{2\sqrt{3}} + [\ln|\sin x|]_{\pi/6}^{\pi/2}$

$= \frac{\pi}{2\sqrt{3}} + \ln 2$

17. Let $u = x$.

$du = dx$

$dv = a^x dx$

$v = \frac{a^x}{\ln a}$

$\int x a^x dx = \frac{x a^x}{\ln a} - \int \frac{a^x}{\ln a} dx$

$= \frac{x a^x}{\ln a} - \frac{1}{\ln a}\int a^x dx$

$= \frac{x a^x}{\ln a} - \frac{1}{\ln a}\left[\frac{a^x}{\ln a}\right] + C$

$= \frac{x a^x}{\ln a} - \frac{a^x}{(\ln a)^2} + C$

19. Let $u = x^2$.

$du = 2x\,dx$

$dv = e^x dx$

$v = e^x$

$\int x^2 e^x dx = x^2 e^x - \int 2x e^x dx$

Perform integration by parts a second time.

$u = x$

$du = dx$

$dv = e^x dx$

$v = e^x$

We have $x^2 e^x - 2x e^x + 2\int e^x dx$

$= x^2 e^x - 2x e^x + 2e^x + C$

21. Let $u = e^t$.

$du = e^t dt$

$dv = \cos t\,dt$

$v = \sin t$

$\int e^t \cos t\,dt = e^t \sin t - \int \sin t\, e^t dt$

Let $u = e^t$.

$du = e^t dt$

$dv = \sin t\,dt$

$v = -\cos t$

$\int e^t \cos t\,dt = e^t \sin t - \left[-e^t \cos t - \int -e^t \cos t\right]$

$\int e^t \cos t\,dt = e^t \sin t + e^t \cos t - \int e^t \cos t\,dt$

$2\int e^t \cos t\,dt = e^t \sin t + e^t \cos t + C$

$\int e^t \cos t\,dt = \frac{1}{2}e^t(\sin t + \cos t) + C$

23. Let $u = \sin(\ln x)$.

$du = \cos(\ln x) \cdot \frac{1}{x} dx$

$dv = dx$

$v = x$

$\int \sin(\ln x) dx = x \sin(\ln x) - \int x \cdot \frac{1}{x} \cos(\ln x) dx$

Let $u = \cos(\ln x)$.

$du = -\sin(\ln x) \cdot \frac{1}{x} dx$

$dv = dx$

$v = x$

$$\int \sin(\ln x)dx = x\sin(\ln x) - \left[x\cos(\ln x) - \int -x\sin(\ln x)\frac{1}{x}dx \right]$$

$$\int \sin(\ln x)dx = x\sin(\ln x) - x\cos(\ln x) - \int \sin(\ln x)dx$$

$$2\int \sin(\ln x)dx = x\sin(\ln x) - x\cos(\ln x) + C$$

$$\int \sin(\ln x)dx = \frac{1}{2}x[\sin(\ln x) - \cos(\ln x)] + C$$

25. First make a sketch.

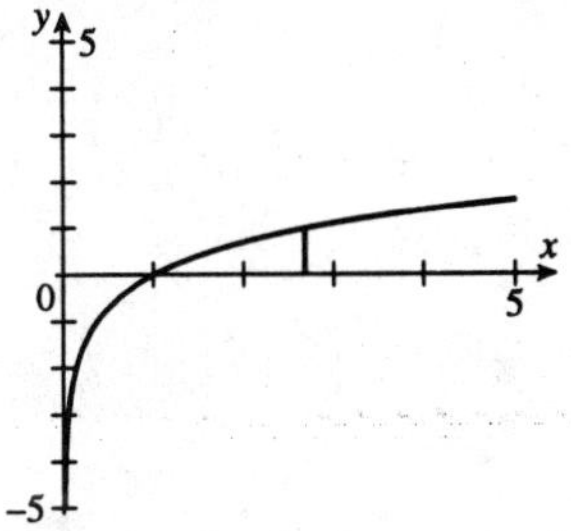

From the sketch, the area is given by

$\int_1^e \ln x\,dx$.

Let $u = \ln x$.

$du = \frac{1}{x}dx$

$dv = dx$

$v = x$

$$\int_1^e \ln x\,dx = [x\ln x]_1^e - \int_1^e x \cdot \frac{1}{x}dx$$

$= [x\ln x - x]_1^e$

$= (e - e) - (1(0) - 1)$

$= 1$

27. First make a sketch.

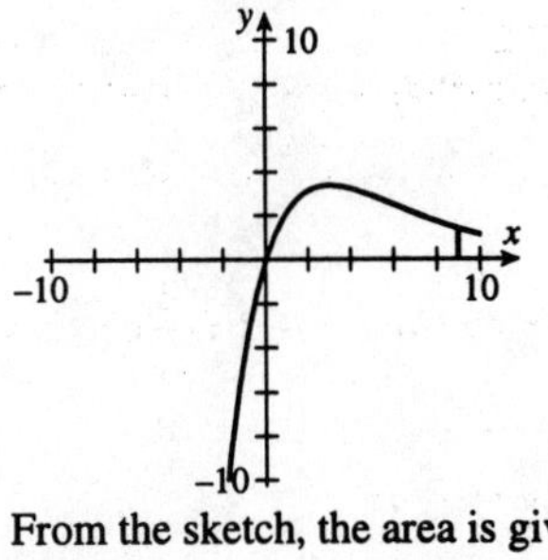

From the sketch, the area is given by

$\int_0^9 3xe^{-x/3}dx$.

Let $u = 3x$.

$du = 3\,dx$

$dv = e^{-x/3}dx$

$v = -3e^{-x/3}$

$$\int_0^9 3xe^{-x/3}dx = \left[-9xe^{-x/3}\right]_0^9 + \int_0^9 9e^{-x/3}dx$$

$= -81e^{-3} - [27e^{-x/3}]_0^9 = -81e^{-3} - 27e^{-3} + 27$

$= -108e^{-3} + 27$

29. Let $u = \cos^{n-1} x$.

$du = (n-1)\cos^{n-2} x \cdot (-\sin x)dx$

$dv = \cos x\, dx$

$v = \sin x$

$\int \cos^n x\, dx = \sin x \cos^{n-1} x - \int \sin x (n-1)\cos^{n-2} x(-\sin x)dx$

$= \sin x \cos^{n-1} x + (n-1)\int (1-\cos^2 x)\cos^{n-2} x\, dx$

$= \sin x \cos^{n-1} x + (n-1)\int \cos^{n-2} x - \cos^n x\, dx$

$= \sin x \cos^{n-1} x + (n-1)\int \cos^{n-2} x\, dx - (n-1)\int \cos^n x\, dx$

$n\int \cos^n x\, dx = \sin x \cos^{n-1} + (n-1)\int \cos^{n-2} x\, dx$

$$\int \cos^n x\, dx = \frac{\cos^{n-1} x \sin x}{n} + \frac{n-1}{n}\int \cos^{n-2} x\, dx$$

31. As in Example 7,

$$\int \sin^n x\, dx = \left[\frac{-\sin^{n-1} x \cos x}{n}\right]_0^{\pi/2} + \frac{n-1}{n}\int_0^{\pi/2} \sin^{n-2} x\, dx$$

$$= 0 + \frac{n-1}{n}\int_0^{\pi/2} \sin^{n-2} x\, dx.$$

Thus $\int_0^{\pi/2} \sin^7 x\, dx = \frac{6}{7}\int_0^{\pi/2} \sin^5 x\, dx$

$$= \frac{6}{7}\cdot\frac{4}{5}\cdot\frac{2}{3}\int_0^{\pi/2} \sin x\, dx$$

$$= \frac{16}{35}[-\cos x]_0^{\pi/2}$$

$$= \frac{16}{35}(0+1)$$

$$= \frac{16}{35}.$$

33. Let $u = \tan^{n-2} x$.

$du = (n-2)\tan^{n-3} x \cdot \sec^2 x$

$dv = \sec^2 x\, dx$

$v = \tan x$

$\int \sec^2 x \tan^{n-2} x\, dx = \tan^{n-2} x \tan x - \int (n-2)\tan^{n-2} x \sec^2 x\, dx$

$= \tan^{n-1} x - (n-2)\int \tan^{n-2} x \sec^2 x\, dx$

$(n-1)\int \sec^2 x \tan^{n-2} x\, dx = \tan^{n-1} x$

$$\int \sec^2 x \tan^{n-2} x\, dx = \frac{1}{n-1}\tan^{n-1} x$$

Thus,

$\int \tan^n x\, dx = \int \sec^2 x \tan^{n-2} x\, dx - \int \tan^{n-2} x\, dx$

$$= \frac{\tan^{n-1} x}{n-1} - \int \tan^{n-2} x\, dx$$

which is the desired result.

35. Let $u = x^n$.

$du = nx^{n-1}$

$dv = e^x dx$

$v = e^x$

$\int x^n e^x dx = x^n e^x - \int nx^{n-1} e^x dx$

$= x^n e^x - n\int x^{n-1} e^x dx$

$\int x^3 e^x dx = x^3 e^x - 3\int x^2 e^x dx$

$= x^3 e^x - 3[x^2 e^x - 2\int xe^x dx]$

$= x^3 e^x - 3x^2 e^x + 6\int xe^x dx$

$= x^3 e^x - 3x^2 e^x + 6\left[xe^x - \int e^x dx\right]$

$= x^3 e^x - 3x^2 e^x + 6xe^x - 6e^x + C$

37.

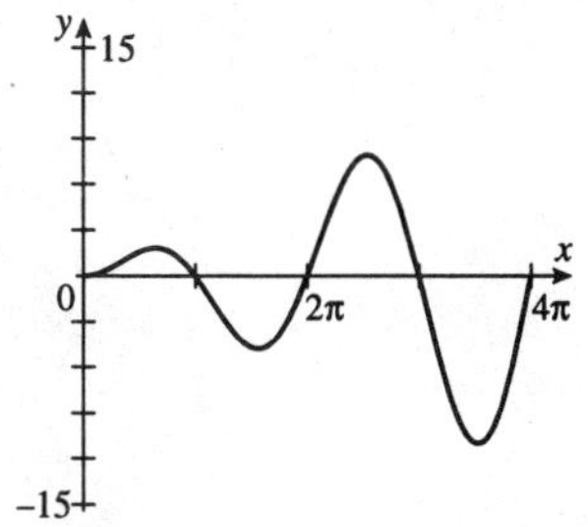

a. The area of the nth hump can be expressed by the integral

$\int_{\pi(n-1)}^{\pi n} x \sin x\, dx$ if n is odd and $-\int_{\pi(n-1)}^{\pi n} x \sin x\, dx$ if n is even.

Let $u = x$.

$du = dx$

$dv = \sin x\, dx$

$v = -\cos x$

$\int_{\pi(n-1)}^{\pi n} x \sin x\, dx = -[x \cos x]_{\pi(n-1)}^{\pi n} + \int_{\pi(n-1)}^{\pi n} \cos x\, dx$

$= [-x \cos x + \sin x]_{\pi(n-1)}^{\pi n}$

If n is odd,

$\int_{\pi(n-1)}^{\pi n} x \sin x\, dx = -\pi n(-1) + \pi(n-1)(1)$

$= \pi n + \pi n - \pi = 2\pi n - \pi$

If n is even,

$-\int_{\pi(n-1)}^{\pi n} x \sin x\, dx = -[-\pi n(1) + \pi(n-1)(-1)]$

$= -[-\pi n - \pi n + \pi]$

$= 2\pi n - \pi$

b. Slice at x. $\Delta V \approx (-x \sin x) 2\pi x \cdot \Delta x$.

$V = 2\pi \int_{\pi}^{2\pi} -x^2 \sin x\, dx$

Let $u = -x^2$.

$du = -2x$

$dv = \sin x\, dx$

$v = -\cos x$

$V = 2\pi\left(\left[x^2 \cos x\right]_{\pi}^{2\pi} - \int_{\pi}^{2\pi} 2x \cos x\, dx\right)$

Let $u = 2x$.
$du = 2\,dx$
$dv = \cos x\,dx$
$v = \sin x$

$$V = 2\pi\left(\left[x^2 \cos x\right]_{\pi}^{2\pi} - [2x\sin x]_{\pi}^{2\pi} + \int_{\pi}^{2\pi} 2\sin x\,dx\right)$$

$$= 2\pi\left(\left[x^2 \cos x\right]_{\pi}^{2\pi} - [2x\sin x]_{\pi}^{2\pi} + [-2\cos x]_{\pi}^{2\pi}\right)$$

$$= 2\pi(5\pi^2 - 4)$$

39. When computing an indefinite integral, a constant needs to be added to the result. If this had been done, we would have

$$\int \frac{1}{t}\,dt = 1 - \int -\frac{1}{t}\,dt$$

or $0 = 1 + C$, $C = -1$.
This is clearly a trivial solution, but it is correct.

Section 8.3

Concepts Review

1. $\int \frac{1+\cos 2x}{2}\,dx$

3. $2\tan t$

Problem Set 8.3

1. $\int \cos^2 x\,dx = \int \frac{1+\cos 2x}{2}\,dx$

$$= \int \frac{1}{2} + \frac{1}{2}\cos 2x\,dx$$

$$= \frac{1}{2}x + \frac{1}{4}\sin 2x + C$$

3. $\int \cos^3 x\,dx = \int \cos x \cos^2 x\,dx$

$$= \int \cos x(1 - \sin^2 x)dx$$

$$= \int \cos x - \cos x \sin^2 x\,dx$$

$$= \int \cos x\,dx - \int \cos x \sin^2 x\,dx$$

$$= \sin x - \frac{1}{3}\sin^3 x + C$$

5. Using the reduction formula from Example 7 of Section 8.2, $\int_0^{\pi/2} \sin^5 t\,dt = \frac{4}{5}\int_0^{\pi/2} \sin^3 t\,dt$

$$= \frac{4}{5}\cdot\frac{2}{3}\int_0^{\pi/2} \sin t\,dt$$

$$= \frac{8}{15}[-\cos t]_0^{\pi/2}$$

$$= \frac{8}{15}$$

7. $\int \sin^7 3x \cos^2 3x\,dx = \int \sin 3x(1-\cos^2 3x)^3 \cos^2 3x\,dx$

$= \int \left(\cos^2 3x - 3\cos^4 3x + 3\cos^6 3x - \cos^8 3x\right)\sin 3x\,dx$

Let $u = \cos 3x$.

$du = -3 \sin x\,dx$

$-\frac{1}{3}du = \sin x\,dx$

$\int \sin^7 3x \cos^2 3x\,dx = -\frac{1}{3}\int (u^8 - 3u^6 + 3u^4 - u^2)du = -\frac{1}{3}\left(\frac{1}{9}u^9 - \frac{3}{7}u^7 + \frac{3}{5}u^5 - \frac{1}{3}u^3\right) + C$

$= -\frac{1}{3}\cos^3 3x\left(\frac{1}{9}\cos^6 3x - \frac{3}{7}\cos^4 3x + \frac{3}{5}\cos^2 3x - \frac{1}{3}\right) + C$

9. $\int \cos^3\theta \sin^{-2}\theta\,d\theta = \int \cos\theta(1-\sin^2\theta)\sin^{-2}\theta\,d\theta = \int \left(\frac{\cos\theta}{\sin^2\theta} - \cos\theta\right)d\theta$

$= -\frac{1}{\sin\theta} - \sin\theta + C = -\csc\theta - \sin\theta + C$

11. $\int \sin^4 2t \cos^2 2t\,dt = \int \left(\frac{1-\cos 4t}{2}\right)^2\left(\frac{1+\cos 4t}{2}\right)dt$

$= \frac{1}{8}\int (1 - \cos 4t - \cos^2 4t + \cos^3 4t)dt$

$= \frac{1}{8}\int \left[1 - \cos 4t - \frac{1+\cos 8t}{2} + \cos 4t(1-\sin^2 4t)\right]dt$

$= \frac{1}{8}\int \left(\frac{1}{2} - \cos 4t \sin^2 4t - \frac{1}{2}\cos 8t\right)dt$

$= \frac{1}{8}\left(\frac{1}{2}t - \frac{1}{12}\sin^3 4t - \frac{1}{16}\sin 8t\right) + C$

13. $\int \sin 4y \cos 5y\,dy = \frac{1}{2}\int [\sin 9y + \sin(-y)]dy = \frac{1}{2}\int (\sin 9y - \sin y)dy$

$= \frac{1}{2}\left(-\frac{1}{9}\cos 9y + \cos y\right) + C = \frac{1}{2}\cos y - \frac{1}{18}\cos 9y + C$

15. $\int \sin 3t \sin t\,dt = -\frac{1}{2}\int \cos 4t - \cos 2t\,dt$

$= -\frac{1}{2}\cdot\frac{1}{4}\sin 4t + \frac{1}{4}\sin 2t + C$

$= -\frac{1}{8}\sin 4t + \frac{1}{4}\sin 2t + C$

17. Let $u = x + 4$.

$du = dx$

$\int \sqrt{x+4}\,dx = \int u^{1/2}du$

$= \frac{2}{3}u^{3/2} + C$

$= \frac{2}{3}(x+4)^{3/2} + C$

19. Let $x = 2\tan t$.

$dx = 2\sec^2 t\,dt$

$\sqrt{x^2+4} = 2\sec t$

$\int \sqrt{x^2+4}\,dx = \int 2\sec t \cdot 2\sec^2 t\,dt = 4\int \sec^3 t\,dt$

$= 4\left(\frac{1}{2}\sec t\tan t + \frac{1}{2}\int \sec t\,dt\right) = 2\sec t\tan t + 2\ln|\sec t + \tan t| + C$

$= 2\cdot\frac{\sqrt{x^2+4}}{2}\cdot\frac{x}{2} + 2\ln\left|\frac{\sqrt{x^2+4}}{2} + \frac{x}{2}\right| + C = \frac{1}{2}x\sqrt{x^2+4} + 2\ln\left|\sqrt{x^2+4} + x\right| + C$

Use the reduction formula for $\int \sec^3 t\,dt$ from Problem 34 of Section 8.2 and the formula for $\int \sec t\,dt$ from Example 4 of Section 8.1. Note that the $\frac{1}{2}$ within the natural logarithm was absorbed into the constant.

21. Let $u = \sqrt{x} + 2$.

$x = (u-2)^2$

$dx = 2(u-2)\,du$

$\int_1^4 \frac{dx}{\sqrt{x}+2} = \int_3^4 \frac{2(u-2)}{u}\,du = 2\int_3^4\left(1 - \frac{2}{u}\right)du$

$= 2[u - 2\ln u]_3^4$

$= 2\left(1 - 2\ln\frac{3}{4}\right)$

23. Make a sketch.

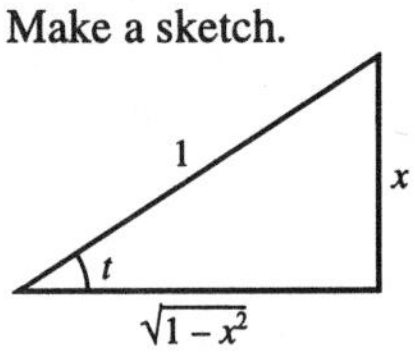

Let $x = \sin t$.

$dx = \cos t\,dt$

$\int \frac{\sqrt{1-x^2}}{x}\,dx = \int \frac{\cos t}{\sin t}\cdot\cos t\,dt = \int \frac{1-\sin^2 t}{\sin t}\,dt$

$= \int \csc t - \sin t\,dt$

$= -\ln|\csc t + \cot t| + \cos t + C$

$= -\ln\left|\frac{1}{x} + \frac{\sqrt{1-x^2}}{x}\right| + \sqrt{1-x^2} + C$

25. Make a sketch.

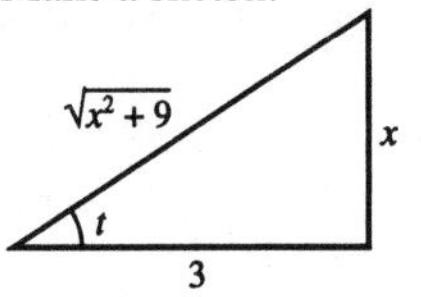

Let $x = 3\tan t$.

$dx = 3\sec^2 t\,dt$

$\int \frac{dx}{x\sqrt{x^2+9}} = \int \frac{1}{3\tan t\;3\sec t}\cdot 3\sec^2 t\,dt$

$= \frac{1}{3}\int \frac{\sec t}{\tan t}\,dt$

$= \frac{1}{3}\int \frac{1}{\cos t}\cdot\frac{\cos t}{\sin t}\,dt$

$= \frac{1}{3}\int \csc t\,dt$

$= -\frac{1}{3}\ln|\csc t + \cot t| + C$

$= -\frac{1}{3}\ln\left|\frac{\sqrt{x^2+9}}{x} + \frac{3}{x}\right| + C$

27. Make a sketch.

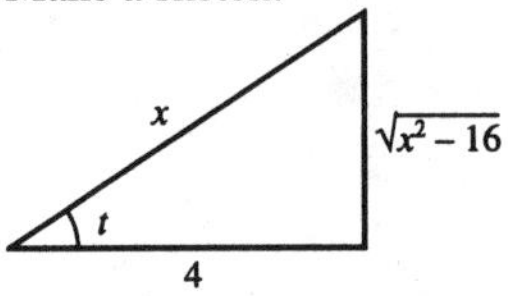

Let $x = 4\sec t$.

$dx = 4\sec t\tan t\,dt$

$\int_5^8 \frac{dx}{\sqrt{x^2-16}} = \int_{x=5}^{x=8} \frac{4\sec t\tan t\,dt}{4\tan t}$

$= \int_{x=5}^{x=8} \sec t\,dt$

$= \left[\ln|\sec t + \tan t|\right]_{x=5}^{x=8}$

$$= \left[\ln\left| \frac{x}{4} + \frac{\sqrt{x^2 - 16}}{4} \right| \right]_5^8$$

$$= \ln\left(2 + \sqrt{3}\right) - \ln(2)$$

29. $\int_0^{\pi} \pi(x + \sin x)^2\, dx$

$$= \pi \int_0^{\pi} x^2 + 2x \sin x + \sin^2 x\, dx$$

Using the methods of integration developed in Sections 8.1 to 8.3,

$$= \frac{5\pi^2}{2} + \frac{\pi^4}{3}$$

31. a. The volume can be represented by

$$\pi \int_{\arcsin k}^{\pi - \arcsin k} (\sin x - k)^2\, dx = 2\pi \int_{\arcsin k}^{\pi/2} \sin^2 x - 2k \sin x + k^2 dx$$

(by symmetry).

$$V = 2\pi \left[\frac{x}{2} - \frac{\sin 2x}{4} + 2k \cos x + k^2 x \right]_{\arcsin k}^{\pi/2}$$

$$= 2\pi \left(\frac{\pi}{4} + \frac{k^2 \pi}{2} - \frac{1}{2} \arcsin k + \frac{1}{2} \sin(\arcsin k) \cos(\arcsin k) - 2k \cos(\arcsin k) - k^2 \arcsin k \right)$$

$$= \frac{\pi}{2} \left[(\pi - 2 \arcsin k)(1 + 2k^2) - 6k\sqrt{1 - k^2} \right]$$

Use a graphing utility to graph the volume as a function of k $(0 \le k \le 1)$. The graph agrees with common sense: the maximum volume is $\frac{\pi^2}{2}$ when $k = 0$ and the minimum volume is 0 when $k = 1$.

33. a. Let $u = \sqrt{4 - x^2}$.

$$x = \sqrt{4 - u^2}$$

$$du = -\frac{x}{\sqrt{4 - x^2}} dx$$

$$dx = \frac{\sqrt{4 - x^2}}{-x} du = \frac{u}{-\sqrt{4 - u^2}} du$$

$$\int \frac{\sqrt{4 - x^2}}{x} dx = \int \frac{u}{\sqrt{4 - u^2}} \cdot \left(-\frac{u}{\sqrt{4 - u^2}} \right) du$$

$$= \int \frac{u^2}{u^2 - 4} du$$

Long division gives $\int 1 + \frac{4}{u^2 - 4} du$

$$= u - \ln\left| \frac{u + 2}{u - 2} \right| + C = \sqrt{4 - x^2} - \ln\left| \frac{\sqrt{4 - x^2} + 2}{\sqrt{4 - x^2} - 2} \right| + C$$

$$= \sqrt{4-x^2} - \ln\left|\frac{\left(\sqrt{4-x^2}+2\right)^2}{x^2}\right| + C$$

$$= \sqrt{4-x^2} - 2\ln\left|\frac{\sqrt{4-x^2}+2}{x}\right| + C$$

b. Let $x = 2\sin t$.
$dx = 2\cos t\,dt$

$2\cos t = \sqrt{4-x^2}$

$$\int \frac{\sqrt{4-x^2}}{x}dx = \int \frac{2\cos t}{2\sin t}\cdot 2\cos t\,dt$$

$$= 2\int \frac{\cos^2 t}{\sin t}dt$$

$$= 2\int \frac{(1-\sin^2 t)}{\sin t}dt$$

$$= 2\int \csc t - \sin t\,dt$$

$$= -2\ln|\csc t + \cot t| + 2\cos t$$

$$= -2\ln\left|\frac{2}{x} + \frac{\sqrt{4-x^2}}{x}\right| + \sqrt{4-x^2}$$

$$= -2\ln\left|\frac{2+\sqrt{4-x^2}}{x}\right| + \sqrt{4-x^2} + C$$

Section 8.4

Concepts Review

1. Proper

3. 2; 3; –1

Problem Set 8.4

1. $\dfrac{2}{x^2+2x} = \dfrac{2}{x(x+2)} = \dfrac{A}{x} + \dfrac{B}{x+2}$

$\dfrac{A(x+2)+Bx}{x(x+2)} = \dfrac{2}{x(x+2)}$

$A(x+2) + Bx = 2$
$Ax + Bx + 2A = 2$
$x(A+B) + 2A = 2$
$A + B = 0$ and $2A = 2$
$A = 1$
$B = -1$

$$\int \frac{2}{x^2+2x}dx = \int \frac{1}{x} - \frac{1}{x+2}dx$$
$$= \ln|x| - \ln|x+2| + C$$

3. $\dfrac{5x+3}{x^2-9} = \dfrac{5x+3}{(x+3)(x-3)} = \dfrac{A}{x+3} + \dfrac{B}{x-3}$

Solving for A and B yields $A = 2$ and $B = 3$.

$$\int \frac{5x+3}{x^2-9}dx = \int \frac{2}{x+3} + \frac{3}{x-3}dx$$
$$= 2\ln|x+3| + 3\ln|x-3| + C$$

5. $\dfrac{x-11}{x^2+3x-4} = \dfrac{x-11}{(x+4)(x-1)} = \dfrac{A}{x+4} + \dfrac{B}{x-1}$

Solving for A and B yields $A = 3$ and $B = -2$.

$$\int \frac{x-11}{x^2+3x-4}dx = \int \frac{3}{x+4} - \frac{2}{x-1}dx$$
$$= 3\ln|x+4| - 2\ln|x-1| + C$$

7. $\dfrac{2x^2+x-4}{x^3-x^2-2x} = \dfrac{2x^2+x-4}{x(x+1)(x-2)} = \dfrac{A}{x} + \dfrac{B}{x+1}$
$= \dfrac{C}{x-2}$

Solving for A, B, and C yields $A = 2$, $B = -1$, and $C = 1$.

$$\int \frac{2x^2+x-4}{x^3-x^2-2x}dx = \int \frac{2}{x} - \frac{1}{x+1} + \frac{1}{x-2}dx$$
$$= 2\ln|x| - \ln|x+1| + \ln|x-2| + C$$

9. Since the degree of the numerator is greater than the degree of the denominator, first use long division

$$\begin{array}{r} 3x - 3 + \frac{9x-6}{x^2+x-2} \\ x^2+x-2\,\overline{)\,3x^3} \\ \underline{3x^3+3x^2-6x} \\ -3x^2+6x \\ \underline{-3x^2-3x+6} \\ 9x-6 \end{array}$$

$$\frac{3x^3}{x^2+x-2} = 3x - 3 + \frac{9x-6}{(x-1)(x+2)}$$
$$= 3x - 3 + \frac{A}{x-1} + \frac{B}{x+2}$$

Solving for A and B yields $A = 1$, $B = 8$.

$$\int \frac{3x^3}{x^2+x-2}dx = \int 3x - 3 + \frac{1}{x-1} + \frac{8}{x+2}dx$$
$$= \frac{3}{2}x^2 - 3x + \ln|x-1| + 8\ln|x+2| + C$$

11. $\dfrac{x+1}{(x-3)^2}=\dfrac{A}{x-3}+\dfrac{B}{(x-3)^2}$

Solving for A and B yields $A = 1$ and $B = 4$.

$\int \dfrac{x+1}{(x-3)^2}dx=\int \dfrac{1}{x-3}+\dfrac{4}{(x-3)^2}dx$

$=\ln|x-3|-\dfrac{4}{x-3}+C$

13. $\dfrac{1}{(x^2+1)(x-1)}=\dfrac{Ax+B}{x^2+1}+\dfrac{C}{x-1}$

Solving for A, B, and C yields

$A=-\dfrac{1}{2},\ B=-\dfrac{1}{2},\text{ and } C=\dfrac{1}{2}.$

$\int \dfrac{1}{(x^2+1)(x-1)}dx=\int \dfrac{1}{2(x-1)}-\dfrac{x+1}{2(x^2+1)}dx$

$=\dfrac{1}{2}\int \dfrac{1}{x-1}dx-\dfrac{1}{2}\int \dfrac{x}{x^2+1}dx-\dfrac{1}{2}\int \dfrac{1}{x^2+1}dx$

$=\dfrac{1}{2}\ln|x-1|-\dfrac{1}{4}\ln(x^2+1)-\dfrac{1}{2}\arctan x+C$

15. a. $\dfrac{dy}{dt}=ky\dfrac{16-y}{16}$

$\dfrac{16dy}{y(16-y)}=k\,dt$

Integrating both sides yields

$\ln|y|-\ln|y-16|=kt+C$

$\dfrac{y}{y-16}=Ce^{kt}$

$y=\dfrac{-16Ce^{kt}}{1-Ce^{kt}},\ y(0)=2,\ y(50)=4$

so $C=-\dfrac{1}{7},\ k\approx 0.016946$

b. Let $t = 90$.
Then $y \approx 6.3$ billion.

c. Solve the equation $-\dfrac{16Ce^{kt}}{1-Ce^{kt}}=9$ for t using the values $C=-\dfrac{1}{7},\ k\approx 0.016946.\ t\approx 130$

Therefore, this will occur around the year 2055.

17. a. Separating variables, we obtain

$\dfrac{dx}{(a-x)(b-x)}=k\,dt$

$\dfrac{\ln|a-x|-\ln|b-x|}{a-b}=kt+C$

$\dfrac{1}{a-b}\ln\left|\dfrac{a-x}{b-x}\right|=kt+C$

$\dfrac{a-x}{b-x}=Ce^{(a-b)kt}$

Since $x = 0$ when $t = 0$, $C=\dfrac{a}{b}$, and we obtain

$a-x=(b-x)\dfrac{a}{b}e^{(a-b)kt}.$

$x(t)=\dfrac{a(1-e^{(a-b)kt})}{1-\frac{a}{b}e^{(a-b)kt}}$

$=\dfrac{ab(1-e^{(a-b)kt})}{b-ae^{(a-b)kt}}$

b. Since $b > a$ and $k > 0$, $e^{(a-b)kt}\to 0$ as $t\to\infty$. Thus,

$x\to\dfrac{ab(1)}{b-0}=a.$

c. $x(t)=\dfrac{8(1-e^{-2kt})}{4-2e^{-2kt}}$

$x(20) = 1$, so $4-2e^{-40k}=8-8e^{-40k}$

$6e^{-40k}=4$

$k=-\dfrac{1}{40}\ln\dfrac{2}{3}$

$e^{-2kt}=e^{t/20\ln 2/3}=e^{\ln(2/3)^{t/20}}=\left(\dfrac{2}{3}\right)^{t/20}$

$x(t)=\dfrac{4\left(1-\left(\frac{2}{3}\right)^{t/20}\right)}{2-\left(\frac{2}{3}\right)^{t/20}}$

$x(60)=\dfrac{4\left(1-\left(\frac{2}{3}\right)^{3}\right)}{2-\left(\frac{2}{3}\right)^{3}}=\dfrac{38}{23}\approx 1.65$ grams

19. Separating variables, we obtain

$\dfrac{dy}{(A-y)(B+y)}=k\,dt$

$\dfrac{-\ln(A-y)+\ln(B+y)}{A+B}=kt+C$

$\dfrac{1}{A+B}\ln\left|\dfrac{B+y}{A-y}\right|=kt+C$

$\dfrac{B+y}{A-y}=Ce^{(A+B)kt}$

$y(t)=\dfrac{ACe^{(A+B)kt}-B}{1+Ce^{(A+B)kt}}$

Section 8.5

Concepts Review

1. $\lim_{x\to a} f(x);\ \lim_{x\to a} g(x)$

3. $\frac{\sec^2 x}{1};\ 1;\ \lim_{x\to 0}\cos x \neq 0$

Problem Set 8.5

1. $\lim_{x\to 0}\frac{\sin x - 2x}{x} = \lim_{x\to 0}\frac{\cos x - 2}{1} = -1$

3. $\lim_{x\to 0}\frac{x - 2\sin x}{\tan x} = \lim_{x\to 0}\frac{1-2\cos x}{\sec^2 x} = \frac{1-2(1)}{1} = -1$

5. $\lim_{x\to\infty}\frac{\ln(x^{100})}{x} = \lim_{x\to\infty}\frac{1}{x^{100}}\cdot 100x^{99} = 0$

7. $\lim_{x\to\infty}\frac{x^{10}}{e^x} = \lim_{x\to\infty}\frac{10x^9}{e^x}$

Repeating the process eventually gives

$\lim_{x\to\infty}\frac{C}{e^x} = 0$ where C is 10!

9. $\lim_{x\to 1^+}\frac{x^2-2x+2}{x^2-1}$ is not an indeterminate form.

Plotting the function $y = \frac{x^2-2x+2}{x^2-1}$, we see that

$\lim_{x\to 1^+}\frac{x^2-2x+2}{x^2-1} = \infty.$

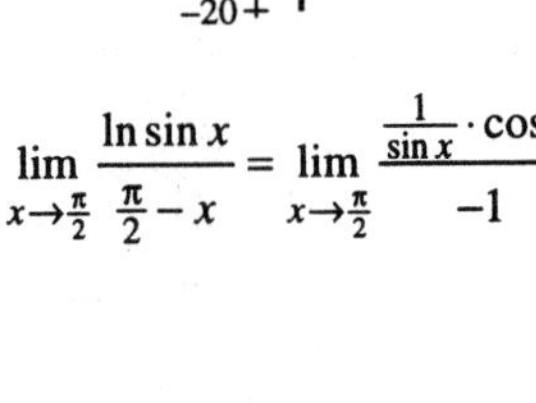

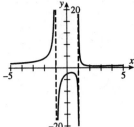

11. $\lim_{x\to\frac{\pi}{2}}\frac{\ln\sin x}{\frac{\pi}{2}-x} = \lim_{x\to\frac{\pi}{2}}\frac{\frac{1}{\sin x}\cdot\cos x}{-1} = \lim_{x\to\frac{\pi}{2}} -\cot x = 0$

13. $\lim_{x\to\frac{\pi}{2}}\frac{\sec x+1}{\tan x} = \lim_{x\to\frac{\pi}{2}}\frac{\sec x\tan x}{\sec^2 x} = \lim_{x\to\frac{\pi}{2}}\frac{\tan x}{\sec x}$

$= \lim_{x\to\frac{\pi}{2}}\frac{\sec^2 x}{\sec x\tan x} = \lim_{x\to\frac{\pi}{2}}\frac{\sec x}{\tan x}$

$= \lim_{x\to\frac{\pi}{2}}\frac{1}{\cos x}\cdot\frac{\cos x}{\sin x} = \lim_{x\to\frac{\pi}{2}}\frac{1}{\sin x} = 1$

15. $\lim_{x\to\infty}\frac{\ln(\ln x)}{\ln x} = \lim_{x\to\infty}\frac{\frac{1}{\ln x}\cdot\frac{1}{x}}{\frac{1}{x}} = \lim_{x\to\infty}\frac{1}{\ln x} = 0$

17. $\lim_{t\to 1}\frac{\sqrt{t}-1}{\ln t} = \lim_{t\to 1}\frac{\frac{1}{2}t^{-1/2}-1}{\frac{1}{t}}$

$= \frac{\frac{1}{2}-1}{1}$

$= -\frac{1}{2}$

19. Let $y = (2x)^{x^2}$.

$\ln y = x^2\ln(2x)$

$\lim_{x\to 0^+}\ln y = \lim_{x\to 0^+}\frac{\ln(2x)}{\frac{1}{x^2}} = \lim_{x\to 0^+}\frac{\frac{1}{x}}{-\frac{2}{x^2}}$

$= \lim_{x\to 0^+} -\frac{x^2}{2} = 0$

$\lim_{x\to 0^+} y = \exp\left(\lim_{x\to 0^+}\ln y\right) = 1$

21. $\lim_{x\to\frac{\pi}{2}^-}(\cos x)^{\tan x} = 0$ since $\cos x \to 0$ and $\tan x \to \infty$.

23. $\lim_{x\to 0}\frac{\ln(\cos 3x)}{2x^2} = \lim_{x\to 0}\frac{\frac{1}{\cos 3x}\cdot(-\sin 3x)(3)}{4x}$

$= \lim_{x\to 0} -\frac{3\tan 3x}{4x}$

$= \lim_{x\to 0} -\frac{9\sec^2 3x}{4}$

$= -\frac{9}{4}$

25. Let $y = x^{e/x}$.

$\ln y = \frac{e}{x}\ln x$

$\lim_{x\to\infty}\ln y = \lim_{x\to\infty}\frac{e\ln x}{x} = \lim_{x\to\infty}\frac{e}{x} = 0$

$\lim_{x\to\infty} y = \exp\left(\lim_{x\to\infty}\ln y\right) = 1$

27. The limit is of the form $\frac{0}{0}$.

$$\lim_{x\to 0} \frac{\int_0^x \sqrt{1+\sin t}\, dt}{x} = \lim_{x\to 0} \sqrt{1+\sin x} = 1$$

29. $\sqrt{1+e^{-t}} > 1$ for all t, so $\int_1^x \sqrt{1+e^{-t}}\, dt > \int_1^x dt = x-1$. The limit is of the form $\frac{\infty}{\infty}$.

$$\lim_{x\to\infty} \frac{\int_1^x \sqrt{1+e^{-t}}\, dt}{x} = \lim_{x\to\infty} \frac{\sqrt{1+e^{-x}}}{1} = 1$$

31. It would not have helped us because we proved $\lim_{x\to 0} \frac{\sin x}{x} = 1$ in order to find the derivative of $\sin x$.

33. In order for the limit to be a finite constant, $\frac{ax^4+bx^3+1}{(x-1)\sin \pi x}$ must be an interdeterminate form. Therefore, $a + b = -1$.

Now since we have an indeterminate form $\frac{0}{0}$ we have

$$\lim_{x\to 1} \frac{ax^4+bx^3+1}{(x-1)\sin \pi x} = \lim_{x\to 1} \frac{4ax^3+3bx^2}{\sin \pi x + \pi(x-1)\cos \pi x}.$$

In order for this limit to be a finite constant, $\frac{4ax^3+3bx^2}{\sin \pi x + \pi(x-1)\cos \pi x}$ must be an indeterminate form. Therefore, $4a + 3b = 0$. Combining this with $a + b = -1$, we obtain $a = 3$, $b = -4$.

$$\lim_{x\to 1} = \frac{3x^4-4x^3+1}{(x-1)\sin \pi x} = \lim_{x\to 1} \frac{12x^3-12x^2}{\sin \pi x + \pi(x-1)\cos \pi x}$$

$$= \lim_{x\to 1} \frac{36x^2-24x}{2\pi \cos \pi x - \pi^2 (x-1)\sin \pi x} = \frac{6}{\pi}$$

$a = 3,\ b = -4,\ c = \frac{6}{\pi}$

35. a. Let $y = x^x$.

$$\lim_{x\to 0^+} \ln y = \lim_{x\to 0^+} x\ln x = \lim_{x\to 0^+} \frac{\ln x}{\frac{1}{x}} = \lim_{x\to 0^+} \frac{\frac{1}{x}}{-\frac{1}{x^2}} = 0$$

$$\lim_{x\to 0^+} y = \exp\left(\lim_{x\to 0^+} \ln y\right) = 1$$

b. Let $y = (x^x)^x$.

$$\lim_{x\to 0^+} \ln y = \lim_{x\to 0^+} x\ln(x^x) = 0$$

$$\lim_{x\to 0^+} y = \exp\left(\lim_{x\to 0^+} \ln y\right) = 1$$

c. Let $y = x^{(x^x)}$.

$$\lim_{x\to 0^+} \ln y = \lim_{x\to 0^+} x^x \ln x = -\infty$$

$$\lim_{x\to 0^+} y = \exp\left(\lim_{x\to 0^+} \ln y\right) = 0$$

d. $y = \left((x^x)^x\right)^x$.

$$\lim_{x\to 0^+} \ln y = \lim_{x\to 0^+} x \ln\left((x^x)^x\right) = 0$$

$$\lim_{x\to 0^+} y = \exp\left(\lim_{x\to 0^+} \ln y\right) = 1$$

e. Let $y = \lim_{x\to 0^+} x^{\left(x^{(x^x)}\right)}$.

$$\lim_{x\to 0^+} \ln y = \lim_{x\to 0^+} x^{(x^x)} \ln x = \lim_{x\to 0^+} \frac{\ln x}{\frac{1}{x^{(x^x)}}} = \lim_{x\to 0^+} \frac{\frac{1}{x}}{\frac{-x^{(x^x)}\left[x^x(\ln x+1)\ln x+\frac{x^x}{x}\right]}{\left(x^{(x^x)}\right)^2}}$$

$$= \lim_{x\to 0^+} \frac{-x^{(x^x)}}{x^x x(\ln x)^2 + x^x x\ln x + x^x} = \frac{0}{1\cdot 0+1\cdot 0+1} = 0$$

Note: $\lim_{x\to 0^+} x(\ln x)^2 = \lim_{x\to 0^+} \frac{(\ln x)^2}{\frac{1}{x}} = \lim_{x\to 0^+} \frac{2\ln x \frac{1}{x}}{-\frac{1}{x^2}} = \lim_{x\to 0^+} -2x\ln x = 0$

$$\lim_{x\to 0^+} y = \exp\left(\lim_{x\to 0^+} \ln y\right) = 1$$

37. a.

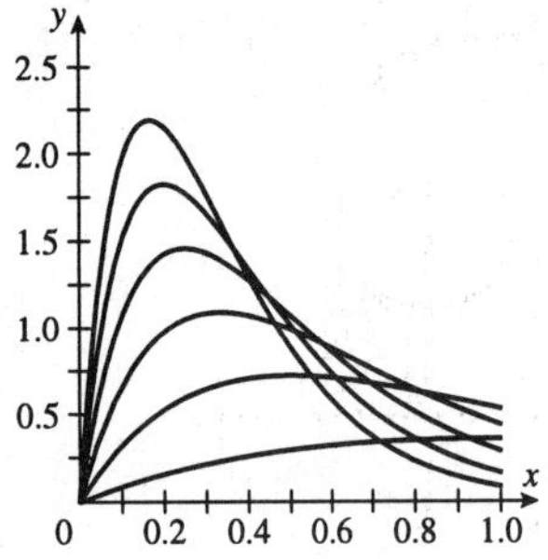

b. $n^2xe^{-nx} = \frac{n^2x}{e^{nx}}$, so the limit is of the form $\frac{\infty}{\infty}$.

$$\lim_{n\to\infty} \frac{n^2x}{e^{nx}} = \lim_{n\to\infty} \frac{2nx}{xe^{nx}}$$

This limit is of the form $\frac{\infty}{\infty}$.

$$\lim_{n\to\infty} \frac{2nx}{xe^{nx}} = \lim_{n\to\infty} \frac{2x}{x^2e^{nx}} = 0$$

c. $\int_0^1 xe^{-x}dx = \left[-xe^{-x} - e^{-x}\right]_0^1 = 1 - \frac{2}{e}$

$\int_0^1 4xe^{-2x}dx = \left[-2xe^{-2x} - e^{-2x}\right]_0^1 = 1 - \frac{3}{e^2}$

$\int_0^1 9xe^{-3x}dx = \left[-3xe^{-3x} - e^{-3x}\right]_0^1 = 1 - \frac{4}{e^3}$

$\int_0^1 16xe^{-4x}dx = \left[-4xe^{-4x} - e^{-4x}\right]_0^1 = 1 - \frac{5}{e^4}$

$\int_0^1 25xe^{-5x} = \left[-5xe^{-5x} - e^{-5x}\right]_0^1 = 1 - \frac{6}{e^5}$

$\int_0^1 36e^{-6x}dx = \left[-6xe^{-6x} - e^{-6x}\right]_0^1 = 1 - \frac{7}{e^6}$

d. Guess: $\lim_{n\to\infty} \int_0^1 n^2xe^{-nx}dx = 1$

$$\int_0^1 n^2xe^{-nx}dx = \left[-nxe^{-nx} - e^{-nx}\right]_0^1$$

$$= -(n+1)e^{-n} + 1 = 1 - \frac{n+1}{e^n}$$

$$\lim_{n\to\infty} \int_0^1 n^2xe^{-nx}dx = \lim_{n\to\infty}\left(1 - \frac{n+1}{e^n}\right)$$

$= 1 - \lim_{n\to\infty} \frac{n+1}{e^n}$ if this last limit exists. The limit is of the form $\frac{\infty}{\infty}$.

$$\lim_{n\to\infty} \frac{n+1}{e^n} = \lim_{n\to\infty} \frac{1}{e^n} = 0, \text{ so}$$

$$\lim_{n\to\infty} \int_0^1 n^2 x e^{-nx} dx = 1.$$

Section 8.6

Concepts Review

1. Converge

3. $p > 1$

Problem Set 8.6

In this section and the review, it is understood that $[g(x)]_a^{\infty}$ means $\lim_{b\to\infty} [g(x)]_a^b$ and likewise for similar expressions.

1. 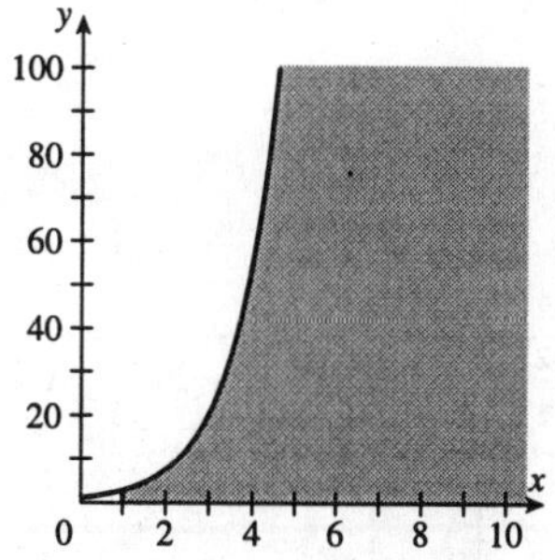

$$\int_1^{\infty} e^x dx = \left[e^x\right]_1^{\infty} = \infty$$

3. 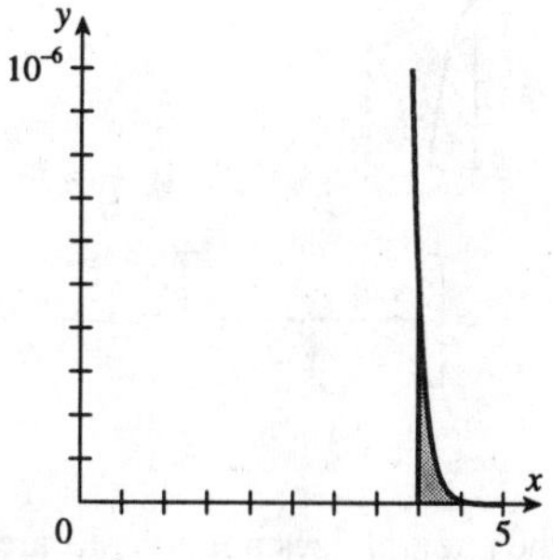

$$\int_4^{\infty} x e^{-x^2} dx = \left[-\frac{1}{2}e^{-x^2}\right]_4^{\infty} = 0 + \frac{1}{2}e^{-16} = \frac{1}{2}e^{-16}$$

5. 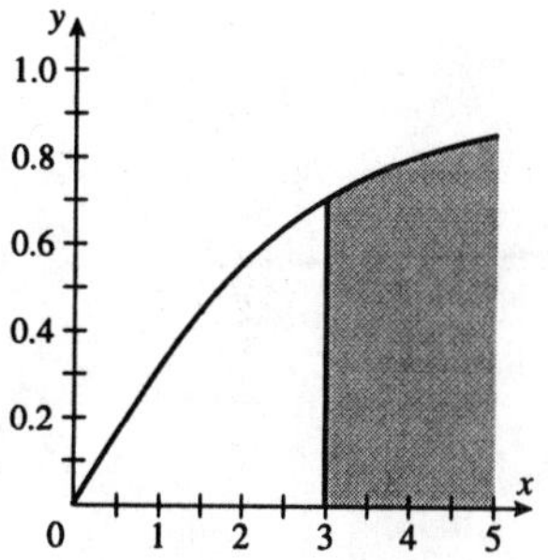

$$\int_3^{\infty} \frac{x\,dx}{\sqrt{9+x^2}} = \left[\sqrt{9+x^2}\right]_3^{\infty} = \infty$$

7. 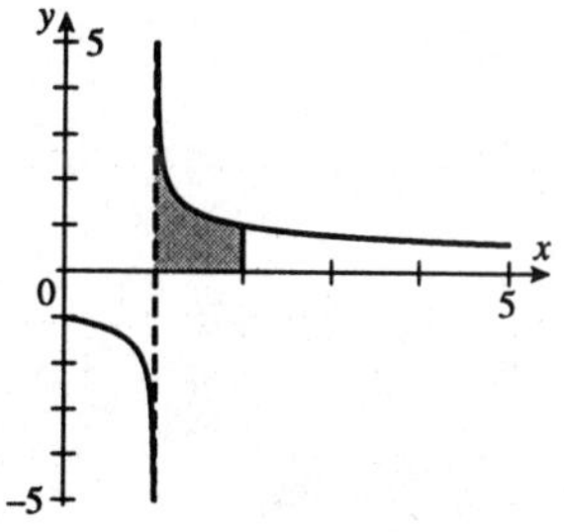

$$\int_1^2 \frac{dx}{(x-1)^{1/3}} = \left[\frac{3}{2}(x-1)^{2/3}\right]_1^2 = \frac{3}{2}(1) - 0 = \frac{3}{2}$$

9. 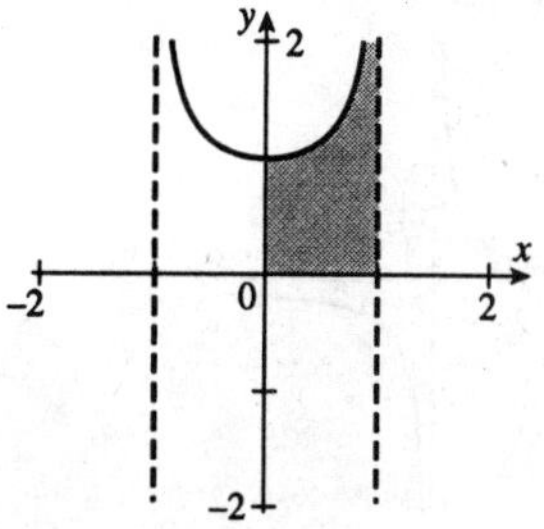

$$\int_0^1 \frac{dx}{\sqrt{1-x^2}} = [\arcsin x]_0^1 = \frac{\pi}{2}$$

11. 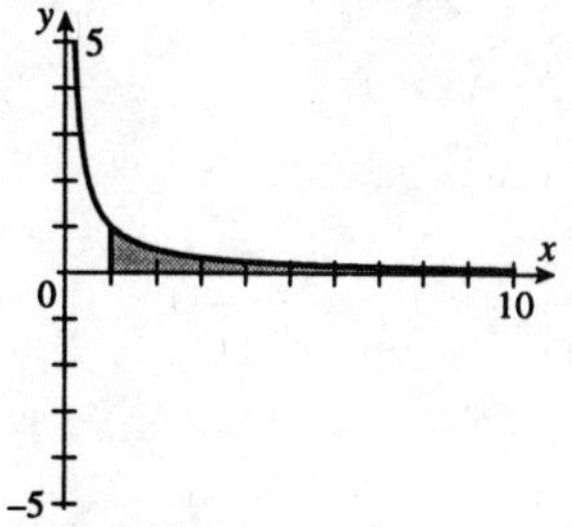

$$\int_1^{\infty} \frac{dx}{x^{1.01}} = \left[-\frac{100}{x^{0.01}}\right]_1^{\infty} = 0 + 100 = 100$$

13.

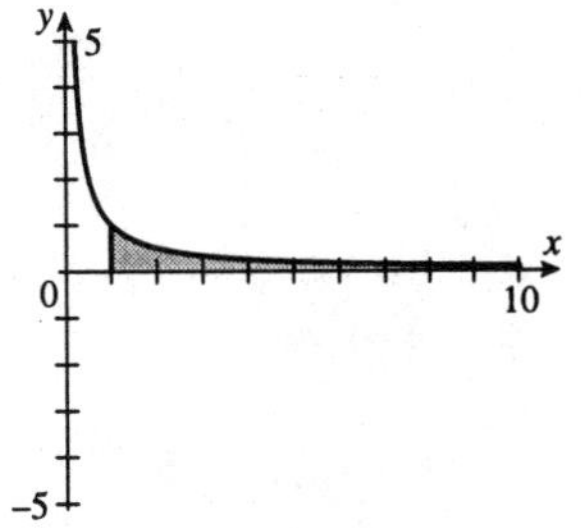

$$\int_1^{\infty}\frac{dx}{x^{0.99}}=\left[100x^{0.01}\right]_1^{\infty}=\infty$$

15.

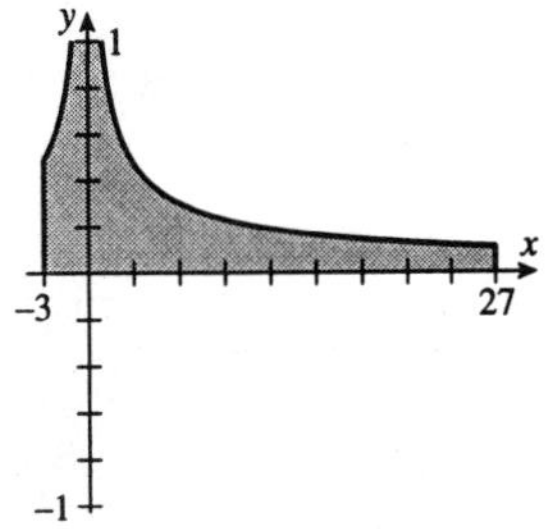

$$\int_{-1}^{27}x^{-2/3}dx=\int_{-1}^{0}x^{-2/3}dx+\int_0^{27}x^{-2/3}dx$$

$$=\left[3x^{1/3}\right]_{-1}^{0}+\left[3x^{1/3}\right]_0^{27}=3+9=12$$

17.

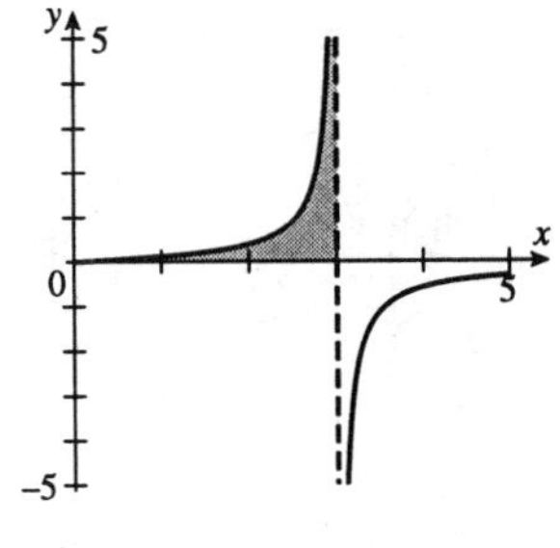

$$\int_0^3\frac{x}{9-x^2}dx=\left[-\frac{1}{2}\ln\left|9-x^2\right|\right]_0^3=\infty$$

19.

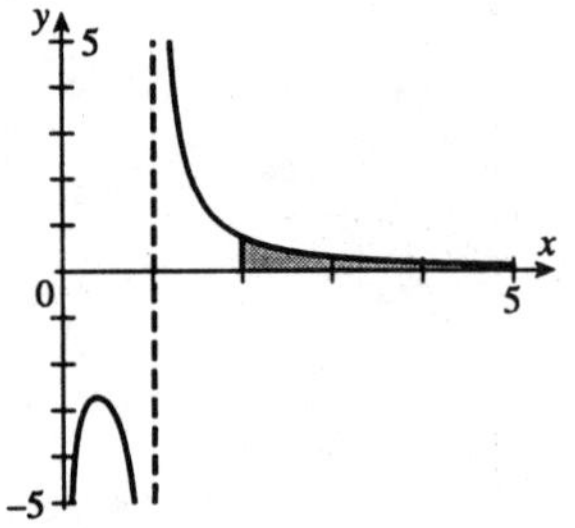

$$\int_2^{\infty}\frac{dx}{x\ln x}=\left[\ln|\ln x|\right]_2^{\infty}=\infty$$

21.

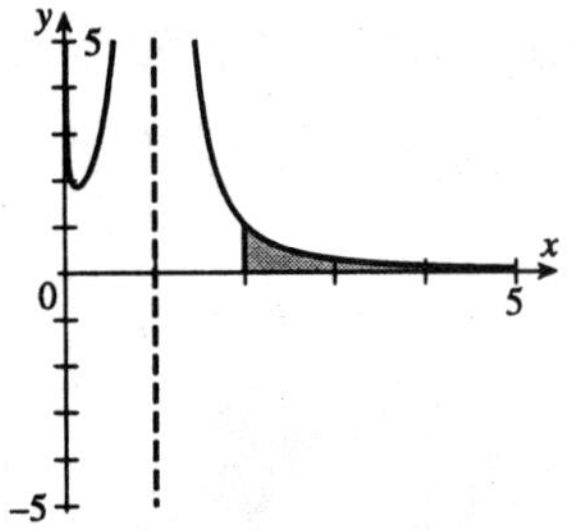

$$\int_2^{\infty}\frac{dx}{x(\ln x)^2}=\left[-\frac{1}{\ln x}\right]_2^{\infty}=0+\frac{1}{\ln 2}=\frac{1}{\ln 2}$$

23.

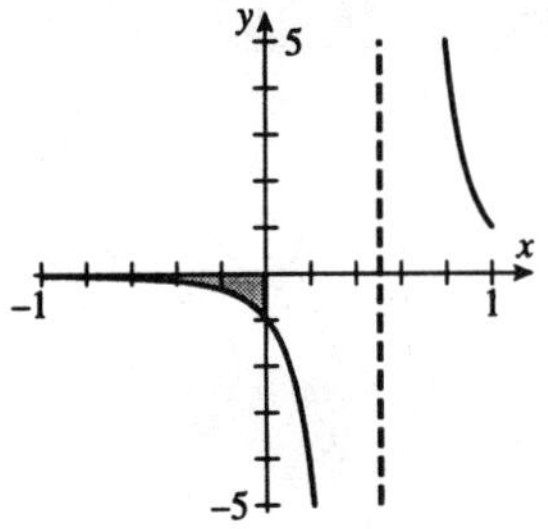

$$\int_{-\infty}^{0}\frac{dx}{(2x-1)^3}=\left[-\frac{1}{4(2x-1)^2}\right]_{-\infty}^{0}=-\frac{1}{4}-0$$

$$=-\frac{1}{4}$$

Note that the integral does not give the area of the shaded region. The area is $\int_{-\infty}^{0}\frac{dx}{(2x-1)^3}=\frac{1}{4}.$

25.

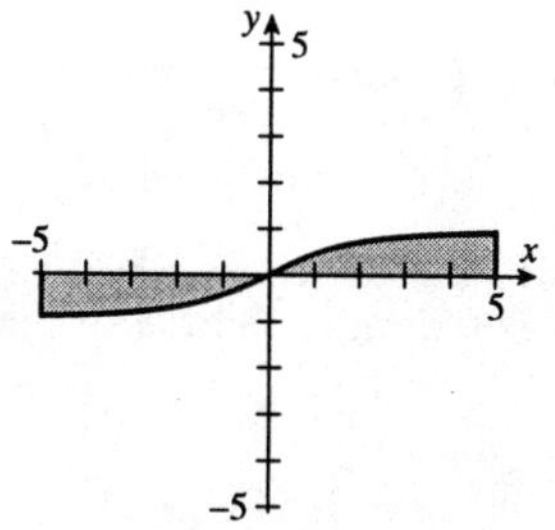

$$\int_{-\infty}^{\infty} \frac{x}{\sqrt{x^2+4}} = \int_{-\infty}^{0} \frac{x}{\sqrt{x^2+4}}dx + \int_{0}^{\infty} \frac{x}{\sqrt{x^2+4}}dx$$

$$= \left[\sqrt{x^2+4}\right]_{-\infty}^{0} + \left[\sqrt{x^2+4}\right]_{0}^{\infty} \text{ diverges.}$$

The interval diverges since both $\int_{-\infty}^{0} \frac{x}{\sqrt{x^2+4}}dx$ and $\int_{0}^{\infty} \frac{x}{\sqrt{x^2+4}}dx$ diverge.

Note that integral for the area is $\int_{0}^{\infty} \frac{x}{\sqrt{x^2+4}}dx - \int_{-\infty}^{0} \frac{x}{\sqrt{x^2+4}}dx$.

27.

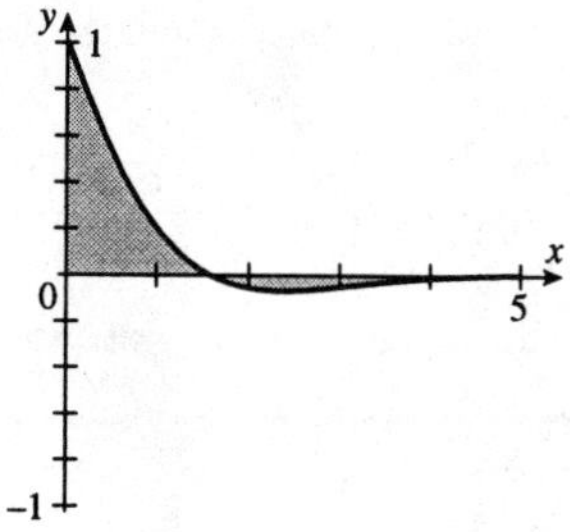

$$\int_0^{\infty} e^{-x} \cos x\, dx = \frac{1}{2e^x}(\sin x - \cos x)\Big|_0^{\infty}$$

$$= 0 - \frac{1}{2}(0-1)$$

$$= \frac{1}{2}$$

Note that the integral does not give the area of the shaded region since $e^{-x} \cos x$ has values above and below the x-axis in $[0, \infty)$.

29.

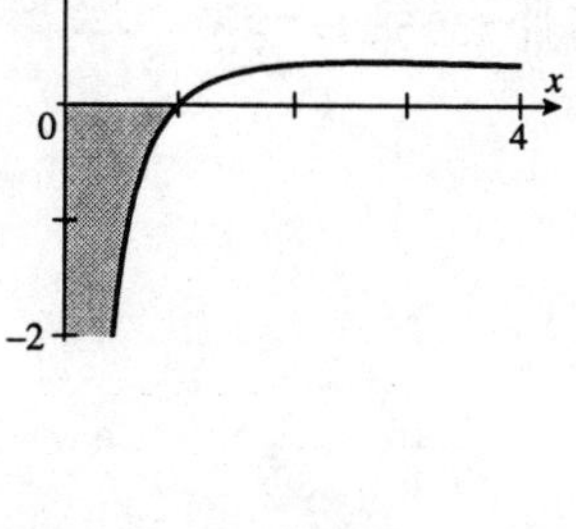

$$\int_0^1 \frac{\ln x}{x}dx = \left[\frac{(\ln x)^2}{2}\right]_0^1 = 0-(-\infty)^2 = -\infty$$

Note that the integral for the area is $-\int_0^1 \frac{\ln x}{x}dx$.

31. The area is given by

$$\int_1^{\infty} \frac{2}{4x^2-1}dx = \frac{1}{2}[\ln(2x-1)-\ln(2x+1)]_1^{\infty}$$

$$= \frac{1}{2}\left[\ln\left(\frac{2x-1}{2x+1}\right)\right]_1^{\infty}$$

$$= \frac{1}{2}\left(0-\ln\left(\frac{1}{3}\right)\right) = \frac{1}{2}\ln 3$$

Note:. $\lim_{x\to\infty} \ln\left(\frac{2x-1}{2x+1}\right) = 0$ since

$\lim_{x\to\infty}\left(\frac{2x-1}{2x+1}\right) = 1\cdot$

33. The integral would take the form

$$k\int_{3960}^{\infty} \frac{1}{x}dx = [k\ln x]_{3960}^{\infty} = \infty$$

which would make it impossible to send anything out of the earth's gravitational field.

35. $\int_0^{\infty} e^{-rt} f(t)\,dt = \int_0^{\infty} 100{,}000e^{-0.08t}$

$$= \left[-\frac{1}{0.08}100{,}000e^{-0.08t}\right]_0^{\infty}$$

$= 1{,}250{,}000$

37. a. $\int_0^{\infty} f(x)\,dx = \int_0^{\infty} \alpha e^{-\alpha x}dx$

$$= \left[-\frac{\alpha}{\alpha}e^{-\alpha x}\right]_0^{\infty}$$

$= 0 + 1 = 1$

b. $\int_0^{\infty} f(x)dx = \int_0^{\infty} \alpha x e^{-\alpha x}dx$

$u = \alpha x$

$du = \alpha dx$

$dv = e^{-\alpha x}dx$

$v = -\frac{1}{2}e^{-\alpha x}$

We have $\left[-xe^{-\alpha x}\right]_0^{\infty} + \int_0^{\infty} e^{-\alpha x}dx$

$$= \left[0+\left(-\frac{1}{\alpha}e^{-\alpha x}\right)\right]_0^{\infty}$$

$$= \frac{1}{\alpha}$$

39. a. $\int_{-\infty}^{\infty} \sin x\,dx = \int_{-\infty}^{0} \sin x\,dx + \int_{0}^{\infty} \sin x\,dx$

$= \lim_{a\to\infty}[-\cos x]_0^a + \lim_{a\to-\infty}[-\cos x]_a^0$

Both limits do not converge since $-\cos x$ is oscillating between -1 and 1, so the integral diverges.

b. $\lim_{a\to\infty}\int_{-a}^{a} \sin x\,dx = \lim_{a\to\infty}[-\cos x]_{-a}^{a}$

$= \lim_{a\to\infty}[-\cos a + \cos(-a)]$

$= \lim_{a\to\infty}[-\cos a + \cos a]$

Since $\cos x$ is an even function

$= \lim_{a\to\infty} 0 = 0.$

41. For example, the region under the curve $y = \frac{1}{x}$ to the right of $x = 1$.

Rotated about the x-axis the volume is $\pi\int_1^{\infty} \frac{1}{x^2}dx = \pi$, while rotated about the y-axis, the volume is $\pi\int_0^1\left(\frac{1}{y^2}-1\right)dy$ which diverges.

43. Let us first evaluate the indefinite integral

$$\int \frac{x}{\sqrt{9-x^2}}dx$$

$u = 9-x^2$

$du = -2x\,dx$

We have $-\frac{1}{2}\int u^{-1/2}du = -u^{1/2} + C$

$= -(9-x^2)^{1/2} + C.$

Since this is clearly defined at $x = 0$,

$$\int_{-3}^{3} \frac{x}{\sqrt{9-x^2}}dx = \int_{-3}^{0} \frac{x}{\sqrt{9-x^2}} + \int_0^3 \frac{x}{\sqrt{9-x^2}}dx$$

$$= \left[-(9-x^2)^{1/2}\right]_{-3}^{0} + \left[-(9-x^2)^{1/2}\right]_0^3$$

$= -3 + 0 + (-0 + 3) = 0$

45. $\int_0^{\infty} \frac{1}{x^p}dx = \int_0^1 \frac{1}{x^p}dx + \int_1^{\infty} \frac{1}{x^p}dx$

If $p > 1$, $\int_0^1 \frac{1}{x^p}dx = \left[\frac{1}{-p+1}x^{-p+1}\right]_0^1$ diverges

since $\lim_{x\to 0^+} x^{-p+1} = \infty$.

If $p < 1$ and $p \neq 0$, $\int_1^{\infty} \frac{1}{x^p}dx = \left[\frac{1}{-p+1}x^{-p+1}\right]_1^{\infty}$

diverges since $\lim_{x\to\infty} x^{-p+1} = \infty$.

If $p = 0$, $\int_0^{\infty} dx = \infty$.

If $p = 1$, both $\int_0^1 \frac{1}{x} dx$ and $\int_1^{\infty} \frac{1}{x} dx$ diverge.

47. $\int_0^8 (x-8)^{-2/3} dx = [3(x-8)^{1/3}]_0^8$
$= 3(0) - 3(-2)$
$= 6$

49. $\int_0^b \ln x\, dx = [x \ln x - x]_0^b = b \ln b - b$
$= b(\ln b - 1)$
So either $b = 0$ (reject) or $\ln b = 1$
$b = e$

51. a. $\Gamma(1) = \int_0^{\infty} x^0 e^{-x} dx$
$= [-e^{-x}]_0^{\infty} = 1$

b. $\Gamma(n+1) = \int_0^{\infty} x^n e^{-x} dx$
Let $u = x^n$, $dv = e^{-x} dx$,
$du = nx^{n-1} dx, v = -e^{-x}$.
$\Gamma(n+1) = [-x^n e^{-x}]_0^{\infty} + \int_0^{\infty} nx^{n-1} e^{-x} dx$
$= 0 + n\int_0^{\infty} x^{n-1} e^{-x} dx$
$= n\Gamma(n)$

c. From parts (a) and (b),
$\Gamma(1) = 1$, $\Gamma(2) = 1 \cdot \Gamma(1) = 1$,
$\Gamma(3) = 2 \cdot \Gamma(2) = 2 \cdot 1 = 2!$.
Suppose $\Gamma(n) = (n-1)!$, then by part (b),
$\Gamma(n+1) = n\Gamma(n) = n[(n-1)!] = n!$.

Section 8.7 Chapter Review

Concepts Test

1. True. $du = 2x\,dx$ will get rid of the $x\,dx$ in the integral.

3. True. The resulting integral will have the form of the derivative of an arctan function.

5. True. The resulting integrand will be of the form $\frac{1}{\sqrt{a^2 - x^2}}$.

7. True. Since $du = 2t - 1$, the resulting integrand will be of the form $\frac{1}{u}$.

9. True. See Section 8.3.

11. True. Let $x = 2 \tan \theta$.

13. True. Then expand and use the substitution $u = \sin x$.

15. True. Let $u = \ln x$.
$du = \frac{1}{x} dx$
$dv = x^2 dx$
$v = \frac{1}{3} x^3$

17. True. See Section 8.4. $A = \frac{1}{2}$, $B = -\frac{1}{2}$

19. False. It is a result of the Product Rule.

21. True. Use l'Hôpital's Rule.

23. False.
$\lim_{x \to \infty} xe^{-1/x} = \infty$ since $e^{-1/x} \to 1$ and $x \to \infty$ as $x \to \infty$.

25. False. See Example 7 of Section 8.5.

27. True. Use repeated applications of l'Hôpital's Rule.

29. False. $p > 1$. See Example 8 of Section 8.6.

31. False. Consider $\int_0^{\infty} \frac{1}{x+1} dx$.

33. False. $\int e^{-x^2} dx$ cannot be expressed in terms of elementary functions.

Sample Test Problems

1. $\int_0^4 \frac{t}{\sqrt{9+t^2}} dt = \left[\sqrt{9+t^2}\right]_0^4 = 5 - 3 = 2$

3. $\int_0^{\pi/2} e^{\cos x} \sin x\, dx = \left[-e^{\cos x}\right]_0^{\pi/2} = e - 1$

7. $\int \frac{y^3+y}{y+1}dy = \int y^2 - y + 2 - \frac{2}{1+y}dy$
$= \frac{1}{3}y^3 - \frac{1}{2}y^2 + 2y - 2\ln(1+y) + C$

9. $\int \frac{e^{2t}}{e^t - 2}dt = e^t + 2\ln(e^t - 2) + C$
(Use the substitution $u = e^t - 2$
$du = e^t dt$
which gives the integral $\int \frac{(u+2)}{u}du.$)

11. Let $u = \sqrt{3}y$.
$du = \sqrt{3}dy$
$\int_0^1 \frac{1}{\sqrt{2+3y^2}}dy = \frac{1}{\sqrt{3}}\int_0^{\sqrt{3}} \frac{1}{\sqrt{(\sqrt{2})^2 + u^2}}du$
$= \frac{1}{\sqrt{3}}\left[\ln\left|u + \sqrt{u^2+2}\right|\right]_0^{\sqrt{3}}$
$= \frac{1}{\sqrt{3}}\left[\ln\left|\sqrt{3}+\sqrt{5}\right| - \ln\left|\sqrt{2}\right|\right] = \frac{1}{\sqrt{3}}\ln\left(\frac{\sqrt{3}+\sqrt{5}}{\sqrt{2}}\right)$
Use Formula 45.

13. Using repeated integration by parts
$\int x^3 \sinh x\, dx = x(6+x^2)\cosh(x) - 3(2+x^2)\sinh(x) + C$

15. $u = \sqrt{x}$
$du = \frac{1}{2}x^{-1/2}dx$
$\int \frac{\sin\sqrt{x}}{\sqrt{x}}dx = 2\int \sin u\, du$
$= -2\cos\sqrt{x} + C$

17. Use integration by parts twice to obtain
$\int e^{t/3}\sin 3t\, dt = \frac{-3e^{t/3}(9\cos 3t - \sin 3t)}{82}.$

19. $\int_0^\infty \frac{dx}{(x+1)^2}$ is an improper integral which is equal
to $\left[-\frac{1}{x+1}\right]_0^\infty$
$= 0 + 1 = 1.$

21. Dividing, we get
$\frac{\sqrt{x}}{1+\sqrt{x}} = 1 - \frac{1}{1+\sqrt{x}}$
$= 1 - \frac{\sqrt{x}}{\sqrt{x}(1+\sqrt{x})}$
$= 1 - \frac{(1+\sqrt{x})-1}{\sqrt{x}(1+\sqrt{x})}$
$= 1 - \frac{(1+\sqrt{x})}{\sqrt{x}(1+\sqrt{x})} + \frac{1}{\sqrt{x}(1+\sqrt{x})}$
$= 1 - \frac{1}{\sqrt{x}} + \frac{1}{\sqrt{x}(1+\sqrt{x})}$
We have $\int \frac{\sqrt{x}}{1+\sqrt{x}}dx = \int 1 - \frac{1}{\sqrt{x}} + \frac{1}{\sqrt{x}(1+\sqrt{x})}dx$
$= x - 2\sqrt{x} + 2\ln(1+\sqrt{x}) + C.$

23. $\cos^3 x\sqrt{\sin x}\, dx = \cos x(1-\sin^2 x)\sqrt{\sin x}$
$= \cos x\left(\sqrt{\sin x} - \sin^{5/2} x\right)$
Let $u = \sin x$.
$du = \cos x\, dx$
$\int \cos^3 x\sqrt{\sin x}\, dx = \int u^{1/2} - u^{5/2}du$

23. $\cos^3 x\sqrt{\sin x}\,dx = \cos x(1-\sin^2 x)\sqrt{\sin x}$

$= \cos x\left(\sqrt{\sin x} - \sin^{5/2} x\right)$

Let $u = \sin x$.

$du = \cos x\,dx$

$\int \cos^3 x\sqrt{\sin x}\,dx = \int u^{1/2} - u^{5/2}du$

$= \frac{2}{3}u^{3/2} - \frac{2}{7}u^{7/2} + C$

$= \frac{2}{3}(\sin x)^{3/2} - \frac{2}{7}(\sin x)^{7/2} + C$

25. Let $u = 2x + 3$.

$du = 2\,dx$

$\int_{-2}^{0} \frac{dx}{2x+3} = \frac{1}{2}\int_{-1}^{3} \frac{du}{u}$

$= \frac{1}{2}\int_{-1}^{0} \frac{du}{u} + \frac{1}{2}\int_{0}^{3} \frac{du}{u}$

$= \frac{1}{2}[\ln u]_{-1}^{0} + \frac{1}{2}[\ln u]_{0}^{3}$

The integral diverges. This is an improper integral.

27. Draw a picture.

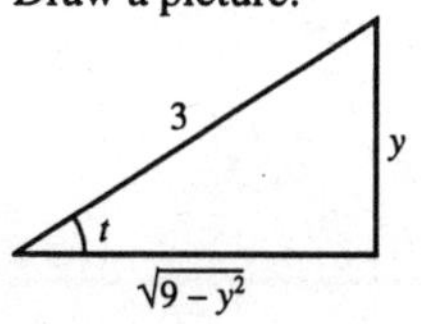

Let $y = 3 \sin t$.

$dy = 3 \cos t\,dt$

$\sqrt{9-y^2} = 3\cos t$

$\int \frac{\sqrt{9-y^2}}{y}dy = \int \frac{3\cos t}{3\sin t} \cdot 3\cos t\,dt$

$= 3\int \frac{1-\sin^2 t}{\sin t}$

$= 3\int \csc t - \sin t\,dt$

$= 3\left[-\ln|\csc t + \cot t| + \cos t\right] + C$

$= -3\ln\left|\frac{3}{y} + \frac{\sqrt{9-y^2}}{y}\right| + \sqrt{9-y^2} + C$

29. $u = \ln x$

$du = \frac{1}{x}dx$

$dv = x^3 dx$

$v = \frac{1}{4}x^4$

$\int x^3 \ln x = \frac{1}{4}x^4 \ln x - \frac{1}{4}\int x^4 \cdot \frac{1}{x}dx$

$= \frac{1}{4}x^4 \ln x - \frac{1}{4}\int x^3 dx$

$= \frac{1}{4}x^4 \ln x - \frac{1}{16}x^4 + C$

31. $\int_{-\infty}^{1} e^{2x}dx = \left[\frac{1}{2}e^{2x}\right]_{-\infty}^{1}$

$= \frac{1}{2}e^2$

(This is an improper integral.)

33. $u = x + 1$

$du = dx$

The improper integral becomes

$\int_0^{\infty} \frac{dx}{x+1} = \int_1^{\infty} \frac{du}{u}$

$= [\ln|u|]_1^{\infty}$

$= \infty$

35. $u = x^2 + 1$

$du = 2x\,dx$

The improper integral becomes

$\int_{-\infty}^{\infty} \frac{x}{x^2+1}dx$

$= \int_0^{\infty} \frac{x}{x^2+1}dx + \int_{-\infty}^{0} \frac{x}{x^2+1}dx$

$= \int_1^{\infty} \frac{du}{u} + \int_{\infty}^{1} \frac{du}{u}$

$= [\ln|u|]_1^{\infty} + [\ln|u|]_{\infty}^{1}$ which diverges.

The integral diverges since both $\int_0^{\infty} \frac{x}{x^2+1}dx$ and $\int_{-\infty}^{0} \frac{x}{x^2+1}dx$ diverge.

37. $\lim_{x\to 0} \frac{4x}{\tan x} = \lim_{x\to 0} \frac{4}{\sec^2 x} = 4$

39. $\lim_{x\to 0} \frac{\sin x - \tan x}{\frac{1}{3}x^2} = \lim_{x\to 0} \frac{\cos x - \sec^2 x}{\frac{2}{3}x}$

$= \lim_{x\to 0} \frac{-\sin x - 2\sec x(\sec x \tan x)}{\frac{2}{3}} = 0$

41. $\lim_{x\to 1^-} \frac{\ln(1-x)}{\cot \pi x} = \lim_{x\to 1^-} \frac{-\frac{1}{1-x}}{-\pi\csc^2 \pi x}$

$= \lim_{x\to 1^-} \frac{\sin^2 \pi x}{\pi(1-x)}$

$= \lim_{x\to 1^-} \frac{2\pi \sin \pi x \cos \pi x}{-\pi} = 0$

43. $\lim_{x\to\infty} \frac{2x^3}{\ln x} = \lim_{x\to\infty} \frac{6x^2}{\frac{1}{x}} = \lim_{x\to\infty} 6x^3 = \infty$

45. Let $y = x^x$. $\ln y = x \ln x$

$\lim_{x\to 0^+} \ln y = \lim_{x\to 0^+} \frac{\ln x}{\frac{1}{x}} = \lim_{x\to 0^+} \frac{\frac{1}{x}}{-\frac{1}{x^2}} = \lim_{x\to 0^+} -x = 0$

$\lim_{x\to 0^+} y = \exp\left(\lim_{x\to 0^+} \ln y\right) = 1$

47. $y = \lim_{t\to\infty} t^{1/t}$

$\ln y = \lim_{t\to\infty} \frac{1}{t} \ln t$

$= \lim_{t\to\infty} \frac{\frac{1}{t}}{1}$

$\ln y = 0$

$y = 1$

49. First plot the velocity from 0 to π (the point at which the person turns around).

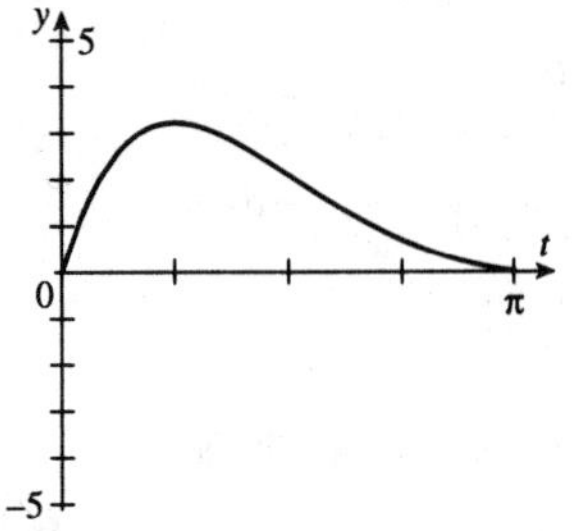

The distance walked is given by

$\int_0^{\pi} 10e^{-t} \sin t \, dt$.

Integrating by parts yields

$\left[-\frac{5\cos t}{e^t} - \frac{5\sin t}{e^t}\right]_0^{\pi} = 5 + \frac{5}{e^{\pi}} \approx 5.216$ feet.

Geometrically, the distance walked is the area under the curve from 0 to π.

51.

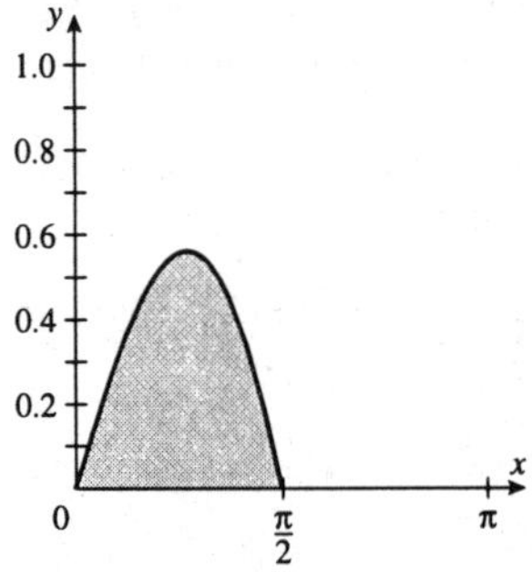

The volume is given by

$\pi \int [f(x)]^2 \, dx = \pi \int_0^{\pi/2} (x \cos x)^2 \, dx$

$= \pi \int_0^{\pi/2} x^2 \left(\frac{1 + \cos 2x}{2}\right) dx$

$= \pi \left[\frac{x^3}{6} + \frac{x}{4} \cos 2x + \frac{(2x^2 - 1)\sin 2x}{8}\right]_0^{\pi/2}$

$= \frac{\pi^4}{48} - \frac{\pi^2}{8}$

Chapter 9

Section 9.1

Concepts Review

1. A sequence

3. $p = f(p)$

Problem Set 9.1

1. $a_1 = 1;\ a_2 = \frac{2}{3};\ a_3 = \frac{3}{5};\ a_4 = \frac{4}{7};\ a_5 = \frac{5}{9}$

$\lim_{n\to\infty} \frac{n}{2n-1} = \lim_{n\to\infty} \frac{1}{2-\frac{1}{n}} = \frac{1}{2}$

3. $a_1 = \frac{5}{2};\ a_2 = \frac{17}{3};\ a_3 = \frac{37}{6};\ a_4 = \frac{65}{11};\ a_5 = \frac{101}{18}$

$\lim_{n\to\infty} \frac{4n^2+1}{n^2-2n+3} = \lim_{n\to\infty} \frac{4+\frac{1}{n^2}}{1-\frac{2}{n}+\frac{3}{n^2}} = 4$

5. $a_1 = -\frac{1}{2};\ a_2 = \frac{2}{3};\ a_3 = -\frac{3}{4};\ a_4 = \frac{4}{5};\ a_5 = -\frac{5}{6}$

$\lim_{n\to\infty} \frac{n}{n+1} = 1$, but the $(-1)^n$ causes the terms to alternate between positive and negative values, so the sequence does not converge.

7. $a_1 = \frac{\cos 1}{e} \approx 0.19877;\ a_2 = \frac{\cos 2}{e^2} \approx -0.05632;$

$a_3 = \frac{\cos 3}{e^3} \approx -0.04929;\ a_4 = \frac{\cos 4}{e^4} \approx -0.01197;$

$a_5 = \frac{\cos 5}{e^5} \approx 0.00191$

Numerical methods indicate that the sequence converges to 0.

9. $a_1 = \frac{-\pi}{4} \approx -0.78539;\ a_2 = \frac{\pi^2}{16} \approx 0.61685;$

$a_3 = \frac{-\pi^3}{64} \approx -0.48447;\ a_4 = \frac{\pi^4}{256} \approx 0.38050;$

$a_5 = \frac{-\pi^5}{1024} \approx -0.29885$

$\lim_{n\to\infty} \frac{(-\pi)^n}{4^n} = \lim_{n\to\infty} \left(-\frac{\pi}{4}\right)^n = 0$ since $-1 < -\frac{\pi}{4} < 1.$

11. $a_1 = 1.9;\ a_2 = 1.81;\ a_3 = 1.729;\ a_4 = 1.6561;$
$a_5 = 1.59049$

$\lim_{n\to\infty} (1+(0.9)^n) = 1 + \lim_{n\to\infty} 0.9^n = 1+0 = 1$ since $-1 < 0.9 < 1.$

13. $a_1 = \frac{\ln 1}{1} = 0;\ a_2 = \frac{\ln 2}{2} \approx 0.34657;$

$a_3 = \frac{\ln 3}{3} \approx 0.36620;\ a_4 = \frac{\ln 4}{4} \approx 0.34657;$

$a_5 = \frac{\ln 5}{5} \approx 0.32189$

Numerical methods indicate that $\lim_{n\to\infty} \frac{\ln n}{n} = 0.$

Apply l'Hôpital's Rule, $\lim_{x\to\infty} \frac{\ln x}{x} = \lim_{x\to\infty} \frac{1}{x} = 0;$

thus $\lim_{n\to\infty} \frac{\ln n}{n} = 0.$

15. $a_1 = 2;\ a_2 = 2.25;\ a_3 = \left(1+\frac{1}{3}\right)^3 \approx 2.3704;$

$a_4 = \left(1+\frac{1}{4}\right)^4 = 2.44140625;$

$a_5 = \left(1+\frac{1}{5}\right)^5 = 2.48832$

$\lim_{n\to\infty} \left(1+\frac{1}{n}\right)^n = e$

(Definition of the Natural Exponential Function, Section 1.4.)

17. $a_n = \frac{n}{n+1}$ or $a_n = 1 - \frac{1}{n+1}$; converges;

$\lim_{n\to\infty} \left(1-\frac{1}{n+1}\right) = 1 - \lim_{n\to\infty} \frac{1}{n+1} = 1$

19. $a_n = (-1)^n \frac{n}{2n-1};\ \lim_{n\to\infty} \frac{n}{2n-1}$

$= \lim_{n\to\infty} \frac{1}{2-\frac{1}{n}} = \frac{1}{2}$, but due to the $(-1)^n$, the terms of the sequence alternate between positive and negative, so the sequence diverges.

21. $a_n = \frac{n}{n^2-(n-1)^2} = \frac{n}{2n-1};$

$\lim_{n\to\infty} \frac{n}{2n-1} = \frac{1}{2}$; the sequence converges.

23. $a_1 = 1;\ a_2 = 1.5;\ a_3 = 1.75;\ a_4 = 1.875;$
$a_5 = 1.9375;\ a_6 = 1.96875$
It appears that $\lim_{n\to\infty} a_n = 2$. If $a_N = 2$, then
$a_{N+1} = 1 + \frac{1}{2}(2) = 2.$

25. $u_1 = \sqrt{3} \approx 1.732;\ u_2 = \sqrt{3+\sqrt{3}} \approx 2.175$
$u_3 = \sqrt{3+\sqrt{3+\sqrt{3}}} \approx 2.275;$
$u_4 \approx \sqrt{3+2.275} \approx 2.297;$
$u_5 \approx \sqrt{3+2.297} \approx 2.302;$
$u_6 \approx \sqrt{3+2.302} \approx 2.303;$
$u_7 \approx \sqrt{3+2.303} \approx 2.303$
Since $u_7 \approx u_6$, $\lim_{n\to\infty} u_n \approx 2.303.$
$\sqrt{3+2.303} = \sqrt{5.303} \approx 2.303$

27. $u_1 = 0;\ u_2 = 1;\ u_3 = 1.1;\ u_4 = 1.1^{1.1} \approx 1.105$
$u_5 \approx 1.1^{1.1105} \approx 1.1116;\ u_6 = 1.1^{1.1116} \approx 1.1118;$
$u_7 \approx 1.1^{1.1118} \approx 1.1118$
Since $u_7 \approx u_6$, $\lim_{n\to\infty} u_n \approx 1.1118.$

29. $x_1 = 2;\ x_2 = 3\tan^{-1} 2 \approx 3.3214;$
$x_3 \approx 3\tan^{-1} 3.3214 \approx 3.8351;$
$x_4 \approx 3\tan^{-1} 3.8351 \approx 3.9472;$
$x_5 \approx 3\tan^{-1} 3.9472 \approx 3.9680;$
$x_6 \approx 3\tan^{-1} 3.9680 \approx 3.9718;$
$x_7 \approx 3\tan^{-1} 3.9718 \approx 3.9724;$
$x_8 \approx 3\tan^{-1} 3.9724 \approx 3.9726;$
$x_9 \approx 3\tan^{-1} 3.9726 \approx 3.9726$
Since $x_9 \approx x_8$, $\lim_{n\to\infty} x_n \approx 3.9726.$

31. $g(x) = 1 + \frac{1}{2}x;\ x = 1 + \frac{1}{2}x;\ x = 2$ is the only fixed point of the system. $g'(2) = \frac{1}{2};\ \left|\frac{1}{2}\right| < 1$ so $x = 2$ is an attracting fixed point.

33. $g(x) = \sqrt{3+x};\ x = \sqrt{3+x};\ x^2 - x - 3 = 0;$
$x = \frac{1+\sqrt{13}}{2}$ is the fixed point of the system.
($x = \frac{1-\sqrt{13}}{2}$ is an extraneous solution of $x = \sqrt{3+x}$.)
$g'(x) = \frac{1}{2\sqrt{3+x}};\ \left|g'\left(\frac{1+\sqrt{13}}{2}\right)\right| \approx |0.217| < 1,$
so $\frac{1+\sqrt{13}}{2} \approx 2.303$ is an attracting fixed point of the system.

35. $g(x) = 1.1^x;\ x = 1.1^x;\ x \approx 1.1118$ and $x \approx 38.2287$ are the fixed points of the system.
$g'(x) = 1.1^x \cdot \ln 1.1;\ |g'(1.1118)| \approx |0.106| < 1$, but $|g'(38.2287)| \approx |3.644| > 1$, so only $x \approx 1.1118$ is an attracting fixed point of the system.

37. $a_1 = \frac{1}{2},\ a_2 = \frac{5}{4},\ a_3 = \frac{9}{8},\ a_4 = \frac{13}{16}$
a_n is positive for all n, and $a_{n+1} \le a_n$ for all $n \ge 2$, so $\{a_n\}$ converges to a limit $L \ge 0$.

39. $a_2 = \frac{3}{4};\ a_3 = \left(\frac{3}{4}\right)\left(\frac{8}{9}\right) = \frac{2}{3};$
$a_4 = \left(\frac{3}{4}\right)\left(\frac{8}{9}\right)\left(\frac{15}{16}\right) = \frac{5}{8};$
$a_5 = \left(\frac{3}{4}\right)\left(\frac{8}{9}\right)\left(\frac{15}{16}\right)\left(\frac{24}{25}\right) = \frac{3}{5}$
$a_n > 0$ for all n and $a_{n+1} < a_n$ since
$a_{n+1} = a_n\left(1 - \frac{1}{(n+1)^2}\right)$ and $1 - \frac{1}{(n+1)^2} < 1$, so $\{a_n\}$ converges to a limit $L \ge 0$.

41. $\left|\frac{n}{n+1} - 1\right| = \left|\frac{n-(n+1)}{n+1}\right| = \left|\frac{-1}{n+1}\right| = \frac{1}{n+1};$
$\frac{1}{n+1} < \varepsilon$ is the same as $\frac{1}{\varepsilon} < n+1$. For whatever ε is given, choose $N > \frac{1}{\varepsilon} - 1$ and
$n \ge N \Rightarrow \left|\frac{n}{n+1} - 1\right| < \varepsilon.$

43. Say $\lim_{n\to\infty} a_n = L$. Suppose that $\{a_n + b_n\}$ converges and that $\lim_{n\to\infty}(a_n + b_n) = M$. Then by Theorem A, $\lim_{n\to\infty}\left[(a_n + b_n) - a_n\right] = M - L$. But $a_n + b_n - a_n = b_n$, so the result would be that $\{b_n\}$ converges. Since $\{b_n\}$ diverges, $\{a_n + b_n\}$ must also diverge.

45. **a.** $f_3 = 2,\ f_4 = 3,\ f_5 = 5,\ f_6 = 8,$
$f_7 = 13,\ f_8 = 21,\ f_9 = 34,\ f_{10} = 55$

b. Using the formula,

$$f_1 = \frac{1}{\sqrt{5}}\left[\frac{1+\sqrt{5}}{2} - \frac{1-\sqrt{5}}{2}\right] = \frac{1}{\sqrt{5}}\left[\frac{2\sqrt{5}}{2}\right] = 1$$

$$f_2 = \frac{1}{\sqrt{5}}\left[\left(\frac{1+\sqrt{5}}{2}\right)^2 - \left(\frac{1-\sqrt{5}}{2}\right)^2\right] = \frac{1}{\sqrt{5}}\left[\frac{1+2\sqrt{5}+5-(1-2\sqrt{5}+5)}{4}\right] = \frac{1}{\sqrt{5}}\left[\frac{4\sqrt{5}}{4}\right] = 1.$$

$$\lim_{n\to\infty}\frac{f_{n+1}}{f_n} = \lim_{n\to\infty}\frac{\phi^{n+1} - (-1)^{n+1}\phi^{-n-1}}{\phi^n - (-1)^n\phi^{-n}} = \lim_{n\to\infty}\frac{\phi^{n+1} - \frac{(-1)^{n+1}}{\phi^{n+1}}}{\phi^n - \frac{(-1)^n}{\phi^n}} = \lim_{n\to\infty}\frac{\phi - \frac{(-1)^{n+1}}{\phi^{2n+1}}}{1 - \frac{(-1)^n}{\phi^{2n}}} = \phi$$

c. $\phi^2 - \phi - 1 = \left[\frac{1}{2}(1+\sqrt{5})\right]^2 - \frac{1}{2}(1+\sqrt{5}) - 1$

$= \left(\frac{3}{2} + \frac{\sqrt{5}}{2}\right) - \left(\frac{1}{2} - \frac{\sqrt{5}}{2}\right) - 1 = 0$

Therefore ϕ satisfies $x^2 - x - 1 = 0$.

Using the Quadratic Formula on $x^2 - x - 1 = 0$ yields

$$x = \frac{1 \pm \sqrt{1+4}}{2} = \frac{1 \pm \sqrt{5}}{2}.$$

$$\phi = \frac{1+\sqrt{5}}{2};\ -\frac{1}{\phi} = -\frac{2}{1+\sqrt{5}} = -\frac{2(1-\sqrt{5})}{1-5} = \frac{1-\sqrt{5}}{2}$$

47.

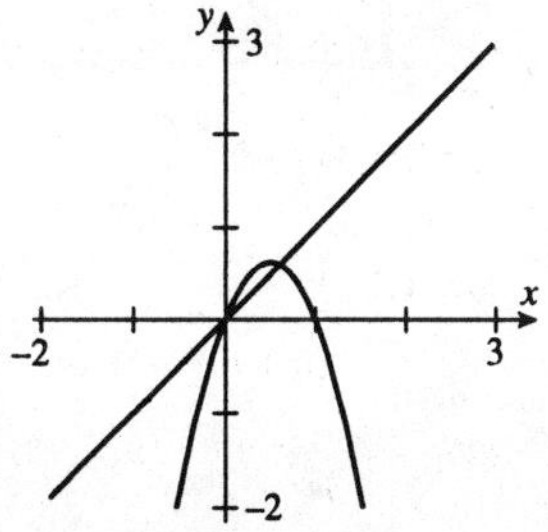

$x_0 = 1,\ x_1 = 0,\ x_2 = 0.\ x = 0$ is a fixed point of the system. $x_0 = -1,\ x_1 = -5,\ x_2 = -75,\ x_3 = -14{,}250$; does not converge. $x_0 = 0.5,\ x_1 = 0.625,\ x_2 \approx 0.5859,\ x_3 \approx 0.6065,\ x_4 \approx 0.5966,\ x_5 \approx 0.6017,\ x_6 \approx 0.5992$; converges to 0.6.

49.

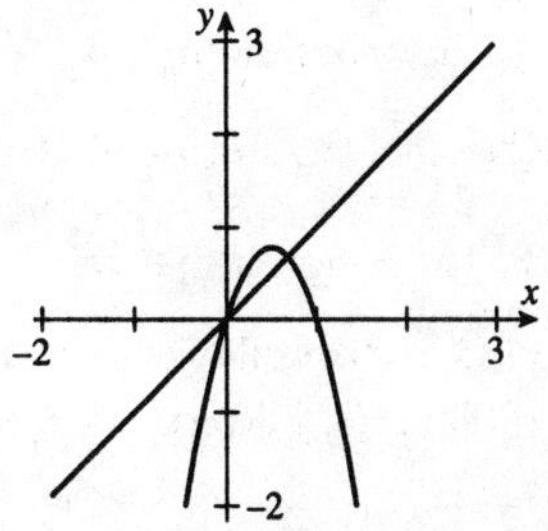

$x_0 = 1,\ x_1 = 0,\ x_2 = 0.\ x = 0$ is a fixed point of the system. $x_0 = 2,\ x_1 = -6.2,\ x_2 = -138.4$; does not converge. $x_0 = 0.5,\ x_1 = 0.775,\ x_2 \approx 0.54056,\ x_3 \approx 0.76990,\ x_4 = 0.54918,\ x_5 \approx 0.76750,\ x_6 \approx 0.55317$. Continued iteration yields $x_{29} \approx 0.76457,\ x_{30} \approx 0.55801,\ x_{31} \approx 0.76457,\ x_{32} \approx 0.55801.\ r_1 \approx 0.55801$ and $r_2 \approx 0.76457$

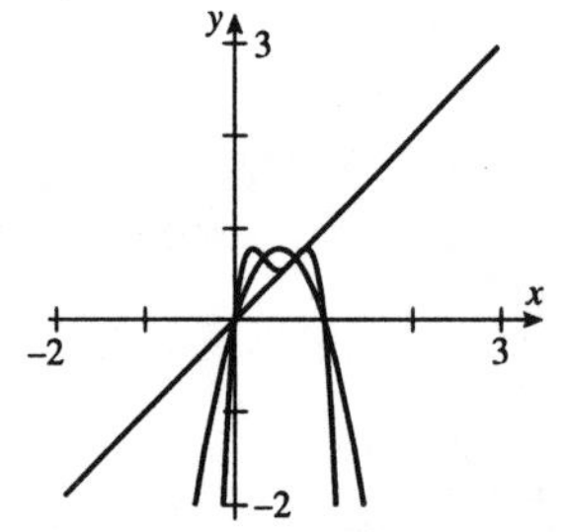

51. $f(x) = 3.5(x - x^2)$

$x_0 = 0.5,\ x_1 = 0.875,\ x_2 \approx 0.38281,\ x_3 \approx 0.82693,\ x_4 \approx 0.50090,$

$x_5 \approx 0.87500,\ x_6 \approx 0.38282,\ x_7 \approx 0.82694,\ x_8 \approx 0.50088,\ x_9 \approx 0.87500,\ x_{10} \approx 0.38282,\ x_{11} \approx 0.82694,$

$x_{12} \approx 0.50088$

$s_1 \approx 0.875,\ s_2 \approx 0.38282,\ s_3 \approx 0.82694,\ s_4 \approx 0.50088$

$g(x) = f(f(f(f(x))))$

53. Written response

Section 9.2

Concepts Review

1. An infinite series

3. $|r| < 1;\ \dfrac{a}{1-r}$

Problem Set 9.2

1. Geometric; converges; $\dfrac{1}{1-\frac{1}{5}} = \dfrac{5}{4}$

3. Sum of two geometric series; converges; $\dfrac{2}{1-\frac{1}{3}} + \dfrac{3}{1-\frac{1}{6}} = \dfrac{33}{5}$

5. $\lim\limits_{k\to\infty} \dfrac{k-3}{k} = 1 \neq 0$; diverges

7. Appears to converge to 0.5283627728

9. Appears to converge slowly to 2.404, accurate to 3 decimal places

11. Appears to converge to 0.1508916324

13. $\displaystyle\sum_{k=3}^{\infty} \frac{2}{(k-1)k} = \sum_{k=3}^{\infty}\left(\frac{2}{k-1} - \frac{2}{k}\right) = \left(\frac{2}{2} - \frac{2}{3}\right) + \left(\frac{2}{3} - \frac{2}{4}\right) + \left(\frac{2}{4} - \frac{2}{5}\right) + \ldots$

$\displaystyle = \frac{2}{2} + \left(-\frac{2}{3} + \frac{2}{3}\right) + \left(-\frac{2}{4} + \frac{2}{4}\right) + \ldots = 1$

15. $\displaystyle\sum_{k=1}^{\infty}\left[\frac{3}{(k+1)^2} - \frac{3}{k^2}\right] = \left(\frac{3}{4} - \frac{3}{1}\right) + \left(\frac{3}{9} - \frac{3}{4}\right) + \left(\frac{3}{16} - \frac{3}{9}\right) + \ldots = -\frac{3}{1} + \left(\frac{3}{4} - \frac{3}{4}\right) + \left(\frac{3}{9} - \frac{3}{9}\right) + \ldots = -3$

17. $0.013013013\ldots = \sum_{k=1}^{\infty} \frac{13}{1000}\left(\frac{1}{1000}\right)^{k-1} = \frac{\frac{13}{1000}}{1-\frac{1}{1000}} = \frac{13}{999}$

19. $0.4999\ldots = \frac{4}{10} + \sum_{k=1}^{\infty} \frac{9}{100}\left(\frac{1}{10}\right)^{k-1} = \frac{4}{10} + \frac{\frac{9}{100}}{1-\frac{1}{10}} = \frac{1}{2}$

21. Let $s = 1 - r$, so $r = 1 - s$. Since $0 < r < 2$, therefore $-1 < 1 - r < 1$, so

$|s| < 1$, so $\sum_{k=0}^{\infty} r(1-r)^k = \sum_{k=0}^{\infty}(1-s)s^k = \frac{1-s}{1-s} = 1$

23. The ball drops 100 feet, rebounds up $100\left(\frac{2}{3}\right)$ feet, drops $100\left(\frac{2}{3}\right)$ feet, rebounds up $100\left(\frac{2}{3}\right)^2$ feet, drops

$100\left(\frac{2}{3}\right)^2$, etc. The total distance it travels is

$$100 + 200\left(\frac{2}{3}\right) + 200\left(\frac{2}{3}\right)^2 + 200\left(\frac{2}{3}\right)^3 + \ldots$$

$$= -100 + 200 + 200\left(\frac{2}{3}\right) + 200\left(\frac{2}{3}\right)^2 + 200\left(\frac{2}{3}\right)^3 + \ldots$$

$$= -100 + \sum_{k=1}^{\infty} 200\left(\frac{2}{3}\right)^{k-1} = -100 + \frac{200}{1-\frac{2}{3}} = 500 \text{ feet}$$

25. \$1 billion + 75% of \$1 billion + 75% of 75% of \$1 billion + ... = $\sum_{k=0}^{\infty}(\$1 \text{ billion})0.75^k = \frac{\$1 \text{ billion}}{1-0.75} = \$4 \text{ billion}$

27. $\frac{1}{8}\cdot 1 + \frac{1}{8}\cdot\frac{1}{2} + \frac{1}{8}\cdot\frac{1}{4} + \ldots = \sum_{k=1}^{\infty}\frac{1}{8}\left(\frac{1}{2}\right)^{k-1} = \frac{\frac{1}{8}}{1-\frac{1}{2}} = \frac{1}{4}$ (NOTE: This can be seen intuitively because the square is covered by four congruent spirals.)

29. $\frac{3}{4} + \frac{3}{4}\left(\frac{1}{4}\cdot\frac{1}{4}\right) + \frac{3}{4}\left(\frac{1}{4}\cdot\frac{1}{4}\right)\left(\frac{1}{4}\cdot\frac{1}{4}\right) + \ldots = \sum_{k=1}^{\infty}\frac{3}{4}\left(\frac{1}{16}\right)^{k-1} = \frac{\frac{3}{4}}{1-\frac{1}{16}} = \frac{4}{5}$

The original does not need to be equilateral since each smaller triangle will have $\frac{1}{4}$ area of the previous larger triangle.

31. $100 + 10 + 1 + \frac{1}{10} + \frac{1}{100} + \ldots = \sum_{k=1}^{\infty} 100\left(\frac{1}{10}\right)^{k-1} = \frac{100}{1-\frac{1}{10}} = 111\frac{1}{9}$ yards

Also, one can see this by the following reasoning. In the time it takes the tortoise to run $\frac{d}{10}$ yards, Achilles will run d yards. Solve $d = 100 + \frac{d}{10}$. $d = \frac{1000}{9} = 111\frac{1}{9}$ yards

33. (Proof by contradiction) Assume $\sum_{k=1}^{\infty} ca_k$ converges, and $c \neq 0$. Then $\frac{1}{c}$ is defined, so $\sum_{k=1}^{\infty} a_k = \sum_{k=1}^{\infty}\frac{1}{c}ca_k = \frac{1}{c}\sum_{k=1}^{\infty} ca_k$ would also converge, by Theorem B (i).

35. a. The top block is supported *exactly* at its center of mass. The location of the center of mass of the top n blocks is the average of the locations of their individual centers of mass, so the nth block moves the center of mass left by $\frac{1}{n}$ of the location of its center of mass, that is, $\frac{1}{n}\cdot\frac{1}{2}$ or $\frac{1}{2n}$ to the left. But this is exactly how far the $(n+1)$th block underneath it is offset.

b. Since $\frac{1}{2}+\frac{1}{4}+\frac{1}{6}+\ldots=\frac{1}{2}\sum_{k=1}^{\infty}\frac{1}{k}$, which diverges, there is no limit to how far the top block can protrude.

37. (Proof by contradiction) Assume $\sum_{k=1}^{\infty}(a_k+b_k)$ converges. Since $\sum_{k=1}^{\infty}b_k$ converges, so would

$\sum_{k=1}^{\infty}a_k=\sum_{k=1}^{\infty}(a_k+b_k)+(-1)\sum_{k=1}^{\infty}b_k$, by Theorem B.

39. Taking vertical strips, the area is $1\cdot 1+1\cdot\frac{1}{2}+1\cdot\frac{1}{4}+1\cdot\frac{1}{8}+\cdots=\sum_{k=0}^{\infty}\left(\frac{1}{2}\right)^k$.

Taking horizontal strips, the area is $\frac{1}{2}\cdot 1+\frac{1}{4}\cdot 2+\frac{1}{8}\cdot 3+\frac{1}{16}\cdot 4+\cdots=\sum_{k=1}^{\infty}\frac{k}{2^k}$.

a. $\sum_{k=1}^{\infty}\frac{k}{2^k}=\sum_{k=0}^{\infty}\left(\frac{1}{2}\right)^k=\frac{1}{1-\frac{1}{2}}=2$

b. The moment about $x = 0$ is $\sum_{k=0}^{\infty}\left(\frac{1}{2}\right)^k\cdot(1)k=\sum_{k=1}^{\infty}\frac{k}{2^k}=2$.

$\bar{x}=\frac{\text{moment}}{\text{area}}=\frac{2}{2}=1$

Section 9.3

Concepts Review

1. $\lim_{n\to\infty}a_n=0$

3. The alternating harmonic series

Problem Set 9.3

1. $a_n=\frac{2}{3n+1}$; $\frac{2}{3n+1}>\frac{2}{3n+4}$, so $a_n>a_{n+1}$; $\lim_{n\to\infty}\frac{2}{3n+1}=0$. $S_9\approx 0.363$. The error made by using S_9 is not more than $a_{10}\approx 0.065$.
Using a computer algebra system, $S\approx 0.3287$.

3. $a_n=\frac{1}{\ln(n+1)}$; $\frac{1}{\ln(n+1)}>\frac{1}{\ln(n+2)}$, so $a_n>a_{n+1}$;
$\lim_{n\to\infty}\frac{1}{\ln(n+1)}=0$. $S_9\approx 1.137$. The error made by using S_9 is not more than $a_{10}\approx 0.417$.
Using a computer algebra system, $S\approx 0.9243$.

5. $a_n = \frac{\ln n}{n}$; $\frac{\ln n}{n} > \frac{\ln(n+1)}{n+1}$ is equivalent to

$\ln \frac{n^{n+1}}{(n+1)^n} > 0$ or $\frac{n^{n+1}}{(n+1)^n} > 1$ which is true for $n > 2$.

$S_9 \approx -0.041$; The error made by using S_9 is not more than $a_{10} \approx 0.230$.

Using a computer algebra system, $S \approx -0.1599$.

7. $\frac{|u_{n+1}|}{|u_n|} = \frac{\left|\left(-\frac{3}{4}\right)^{n+1}\right|}{\left|\left(-\frac{3}{4}\right)^n\right|} = \frac{3}{4} < 1$, so the series converges absolutely.

N	$\sum_{n=1}^{N}\left(-\frac{3}{4}\right)^n$
10	–0.40444
20	–0.42721
30	–0.42849
100	–0.42857
101	–0.42857

$\sum_{n=1}^{\infty}\left(-\frac{3}{4}\right)^n \approx -0.4286$

9. $\frac{|u_{n+1}|}{|u_n|} = \frac{\frac{n+1}{2^{n+1}}}{\frac{n}{2^n}} = \frac{n+1}{2n}$; $\lim_{n\to\infty} \frac{n+1}{2n} = \frac{1}{2} < 1$, so the series converges absolutely.

N	$\sum_{n=1}^{N}(-1)^{n+1}\frac{n}{2^n}$
10	0.21875
20	0.22222
30	0.22222
31	0.22222

$\sum_{n=1}^{\infty}(-1)^{n+1}\frac{n}{2^n} \approx 0.2222$

11. $\frac{|u_{n+1}|}{|u_n|} = \frac{\frac{1}{(n+1)(n+2)}}{\frac{1}{n(n+1)}} = \frac{n}{n+2}$; $\lim_{n\to\infty} \frac{n}{n+2} = 1$

The test is inconclusive.

N	$\sum_{n=1}^{N}(-1)^{n+1}\frac{1}{n(n+1)}$
10	0.38218
20	0.38516
30	0.38577
100	0.38625
150	0.38627
151	0.38632

$\sum_{n=1}^{\infty}(-1)^{n+1}\frac{1}{n(n+1)} \approx 0.3863$

13. $\sum_{n=1}^{\infty}(-1)^{n+1}\frac{1}{5n} = \frac{1}{5}\sum_{n=1}^{\infty}\frac{(-1)^{n+1}}{n}$ which converges since $\sum_{n=1}^{\infty}\frac{(-1)^{n+1}}{n}$ converges.

$\sum_{n=1}^{\infty}(-1)^{n+1}\frac{1}{5n} \approx \frac{1}{5}(0.69) = 0.138.$

15. $\lim_{n\to\infty}\frac{n}{10n+1} = \frac{1}{10} \neq 0$. Thus the sequence of partial sums does not converge; the series diverges.

17. $\lim_{n\to\infty}\frac{1}{n\ln n} = 0$; $\frac{1}{n\ln n} > \frac{1}{(n+1)\ln(n+1)}$ is equivalent to $(n+1)^{n+1} > n^n$ which is true for all $n > 0$. The series converges.

N	$\sum_{n=2}^{N}(-1)^{n+1}\frac{n}{n\ln n}$
10	–0.54658
20	–0.53448
30	–0.53121
100	–0.52749
200	–0.52688
300	–0.52670

$\sum_{n=2}^{\infty}(-1)^{n+1}\frac{1}{n\ln n} \approx -0.527$

19. $\lim_{n\to\infty} \dfrac{\frac{(n+1)^4}{2^{n+1}}}{\frac{n^4}{2^n}} = \lim_{n\to\infty} \dfrac{(n+1)^4}{2n^4} = \dfrac{1}{2} < 1$ so the series converges.

N	$\sum_{n=1}^{N} \frac{n^4}{2^n}$
10	127.74414
20	149.77052
30	149.99901
100	150

$\sum_{n=1}^{\infty} \frac{n^4}{2^n} \approx 150$

21. $\dfrac{|u_{n+1}|}{|u_n|} = \dfrac{1}{n}$; $\lim_{n\to\infty} \dfrac{1}{n} = 0 < 1$, so the series converges absolutely (hence converges).

N	$\sum_{n=1}^{N} (-1)^{n+1} \frac{n}{n!}$
10	0.36788
20	0.36788
30	0.36788

$\sum_{n=1}^{\infty} (-1)^{n+1} \frac{n}{n!} \approx 0.36788$

23. $\cos n\pi = (-1)^n = \dfrac{1}{(-1)}(-1)^{n+1}$ so the series is $(-1)\sum_{n=1}^{\infty} \dfrac{(-1)^{n+1}}{n}$, -1 times the alternating harmonic series.

$\sum_{n=1}^{\infty} \frac{\cos n\pi}{n} \approx -0.69$

25. $|\sin n| \le 1$, so $\left|\dfrac{\sin n}{n\sqrt{n}}\right| \le \dfrac{1}{n\sqrt{n}}$. $\sum_{n=1}^{\infty} \dfrac{1}{n\sqrt{n}} = \sum_{n=1}^{\infty} \dfrac{1}{n^{3/2}}$ converges since $\dfrac{3}{2} > 1$, thus $\sum_{n=1}^{\infty} \dfrac{\sin n}{n\sqrt{n}}$ converges absolutely (hence converges).

N	$\sum_{n=1}^{N} \frac{\sin n}{n\sqrt{n}}$
10	1.06799
20	1.05071
30	1.04702
50	1.04772

$\sum_{n=1}^{\infty} \frac{\sin n}{n\sqrt{n}} \approx 1.05$

27. $\dfrac{1}{\sqrt{n(n+1)}} > \dfrac{1}{\sqrt{(n+1)(n+2)}}$ and

$\lim_{n\to\infty} \dfrac{1}{\sqrt{n(n+1)}} = 0$ so the series converges.

N	$\sum_{n=1}^{N} (-1)^{n+1} \frac{1}{\sqrt{n(n+1)}}$
10	0.41804
50	0.45364
100	0.45850
500	0.46245
1000	0.46295

$\sum_{n=1}^{\infty} (-1)^{n+1} \frac{1}{\sqrt{n(n+1)}} \approx 0.46$

29. $\sum_{n=1}^{\infty} \dfrac{(-3)^{n+1}}{n^2} = \sum_{n=1}^{\infty} (-1)^{n+1} \dfrac{3^{n+1}}{n^2}$

$\dfrac{3^{n+1}}{n^2} > \dfrac{3^{n+2}}{(n+1)^2}$ only for $n = 1$ and

$\lim_{n\to\infty} \dfrac{3^{n+1}}{n^2} \ne 0$, so the series does not converge.

31. If $a_n > 0$, then $|a_n| = a_n$, so $\sum a_n = \sum |a_n|$. If some terms are negative, then by the triangle inequality, $\sum a_n < \sum |a_n|$ so $\sum |a_n|$ must diverge if $\sum a_n$ diverges.

33. The positive-term series is

$$1+\frac{1}{3}+\frac{1}{5}+\frac{1}{7}+\ldots=\sum_{n=1}^{\infty}\frac{1}{2n-1}$$

$$S_n=1+\frac{1}{3}+\frac{1}{5}+\ldots+\frac{1}{2n-1}$$

$$=1+\frac{1}{3}+\left(\frac{1}{5}+\frac{1}{7}+\frac{1}{9}\right)+\left(\frac{1}{11}+\frac{1}{13}+\frac{1}{15}+\frac{1}{17}+\frac{1}{19}+\frac{1}{21}+\frac{1}{23}+\frac{1}{25}+\frac{1}{27}\right)+\ \ldots\ +\frac{1}{2n-1}$$

$$<1+\frac{1}{3}+\frac{1}{3}+\frac{1}{3}+\ \ldots\ +\frac{1}{2n-1}$$

By taking n sufficiently large, any number of $\frac{1}{3}$'s can be introduced, so $\{S_n\}$ diverges.

The negative-term series is $-\frac{1}{2}-\frac{1}{4}-\frac{1}{6}-\frac{1}{8}-\ \ldots\ =-\frac{1}{2}\sum_{n=1}^{\infty}\frac{1}{n}$ which diverges, since the harmonic series diverges.

35. a. $1+\frac{1}{3}\approx 1.33$

b. $1+\frac{1}{3}-\frac{1}{2}\approx 0.833$

c. $1+\frac{1}{3}-\frac{1}{2}+\frac{1}{5}+\frac{1}{7}+\frac{1}{9}+\frac{1}{11}\approx 1.37$

$1+\frac{1}{3}-\frac{1}{2}+\frac{1}{5}+\frac{1}{7}+\frac{1}{9}+\frac{1}{11}-\frac{1}{4}\approx 1.13$

Written response

37. Written response

39. Consider $1-1+\frac{1}{2}-\frac{1}{4}+\frac{1}{3}-\frac{1}{9}+\ \ldots$

It is clear that $\lim_{n\to\infty} a_n=0$. Pairing successive terms, we obtain $\frac{1}{n}-\frac{1}{n^2}=\frac{n-1}{n^2}$.

N	$\sum_{n=1}^{N}\frac{n-1}{n^2}$
10	1.37920
20	2.00158
30	2.38284
100	3.55239
1000	5.84154

It appears that the series diverges.

$$\sum_{n=1}^{\infty}\left(\frac{1}{n}-\frac{1}{n^2}\right)=\sum_{n=1}^{\infty}\frac{1}{n}-\sum_{n=1}^{\infty}\frac{1}{n^2}\approx\sum_{n=1}^{\infty}\frac{1}{n}-1.6449$$

Since $\sum_{n=1}^{\infty}\frac{1}{n}$ diverges, $\sum_{n=1}^{\infty}\left(\frac{1}{n}-\frac{1}{n^2}\right)$ diverges.

41. $\frac{1}{n+1}+\frac{1}{n+2}+\cdots+\frac{1}{2n}=\left[\frac{1}{1+\frac{1}{n}}+\frac{1}{1+\frac{2}{n}}+\cdots+\frac{1}{1+\frac{n}{n}}\right]\left(\frac{1}{n}\right)$

This is a Riemann sum for the function $f(x)=\frac{1}{x}$ from $x = 1$ to 2 where $\Delta x=\frac{1}{n}$.

$$\lim_{n\to\infty}\sum_{k=1}^{n}\left[\frac{1}{1+\frac{k}{n}}\left(\frac{1}{n}\right)\right]=\int_1^2\frac{1}{x}dx=\ln 2$$

Section 9.4

Concepts Review

1. $f(1);\ f'(1);\ f''(1)$

3. $1+x+\frac{x^2}{2}+\frac{x^3}{6}+\frac{x^4}{24}$

Problem Set 9.4

1. $f(0)=1;\ f'(0)=2;\ f''(0)=4;\ f^{(3)}(0)=8;\ f^{(4)}(0)=16$

$P_4(x)=1+2x+2x^2+\frac{4}{3}x^3+\frac{2}{3}x^4$

$f(0.23)\approx 1+2(0.23)+2(0.23)^2+\frac{4}{3}(0.23)^3+\frac{2}{3}(0.23)^4\approx 1.5838882733$

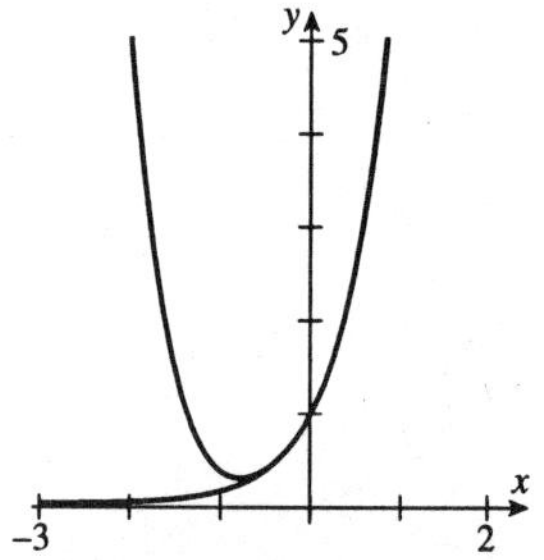

3. $f(0)=0;\ f'(0)=2;\ f''(0)=0;\ f^{(3)}(0)=-8;\ f^{(4)}(0)=0$

$P_4(x)=2x-\frac{4}{3}x^3$

$f(0.23)\approx 2(0.23)-\frac{4}{3}(0.23)^3\approx 0.4437773333$

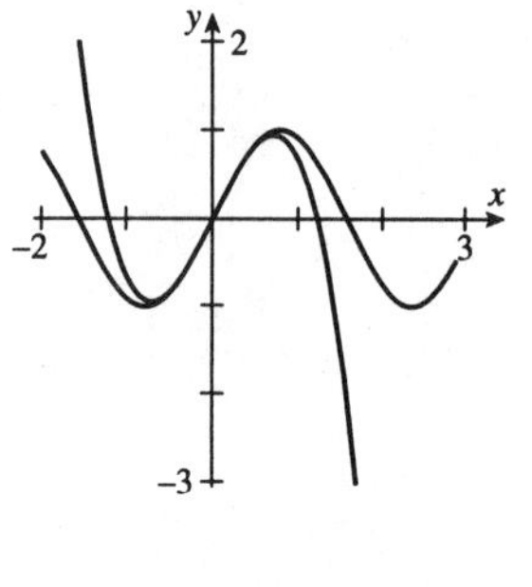

5. $f(0)=0$; $f'(0)=1$; $f''(0)=-1$; $f^{(3)}(0)=2$; $f^{(4)}(0)=-6$

$$P_4(x)=x-\frac{1}{2}x^2+\frac{1}{3}x^3-\frac{1}{4}x^4$$

$$f(0.23)\approx 0.23-\frac{1}{2}(0.23)^2+\frac{1}{3}(0.23)^3-\frac{1}{4}(0.23)^4=0.2069060642$$

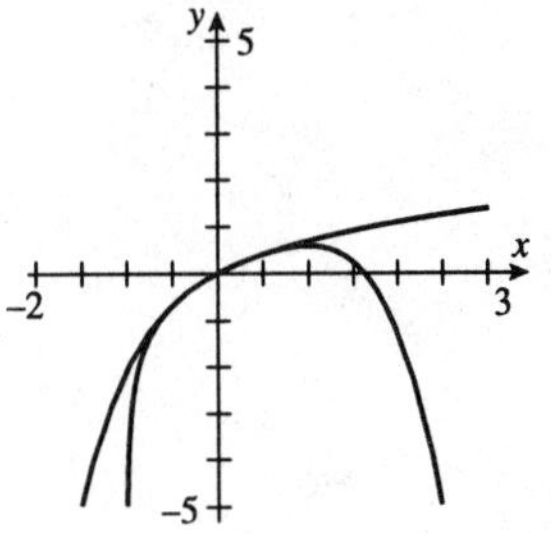

7. $f(0)=0$; $f'(0)=1$; $f''(0)=0$; $f^{(3)}(0)=-2$; $f^{(4)}(0)=0$

$$P_4(x)=x-\frac{1}{3}x^3$$

$$f(0.23)\approx 0.23-\frac{1}{3}(0.23)^3\approx 0.2259443333$$

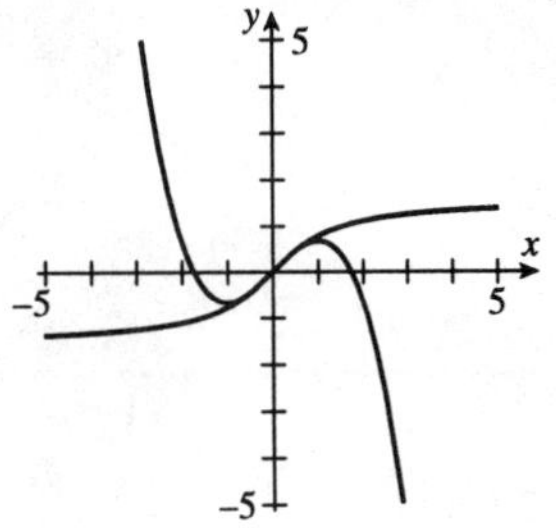

9. $f(2)=e^2$; $f'(2)=e^2$; $f''(2)=e^2$; $f'''(2)=e^2$

$$e^x\approx 7.3891+7.3891(x-2)+3.6945(x-2)^2+1.2315(x-2)^3$$

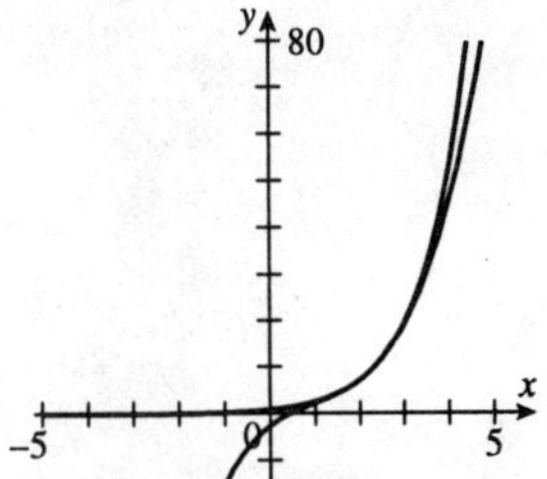

11. $f\left(\frac{\pi}{4}\right)=1$; $f'\left(\frac{\pi}{4}\right)=2$; $f''\left(\frac{\pi}{4}\right)=4$; $f^{(3)}\left(\frac{\pi}{4}\right)=16$

$$\tan x\approx 1+2\left(x-\frac{\pi}{4}\right)+2\left(x-\frac{\pi}{4}\right)^2+\frac{8}{3}\left(x-\frac{\pi}{4}\right)^3$$

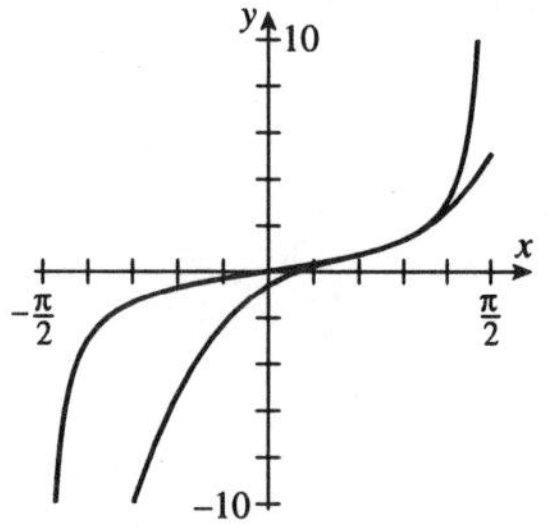

13. $f(1)=\frac{\pi}{4};\ f'(1)=\frac{1}{2};\ f''(1)=-\frac{1}{2};\ f^{(3)}(1)=\frac{1}{2}$

$\tan^{-1} x \approx \frac{\pi}{4}+\frac{1}{2}(x-1)-\frac{1}{4}(x-1)^2+\frac{1}{12}(x-1)^3$

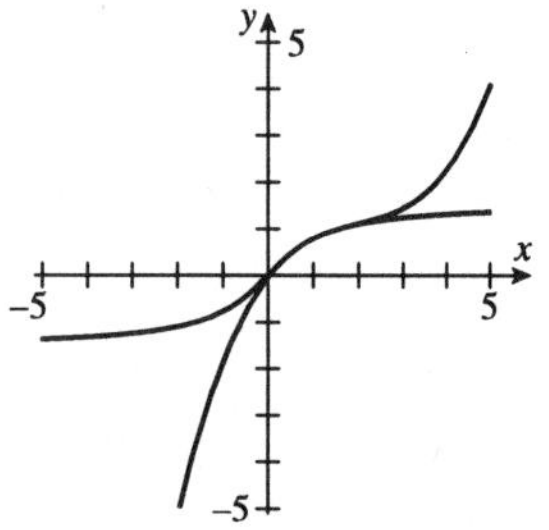

15. For $f(x)=e^x$, $R_n(x)=\frac{e^c}{(n+1)!}x^{n+1}$, so $R_n(1)=\frac{e^c}{(n+1)!}$, $0<c<1$.

$e^c<e^1<3$, so $|R_n(1)|<\frac{3}{(n+1)!}<0.001$ when $n\geq 6$. $e^1\approx 2.7181$

17. For $f(x)=\cos x$, $|R_n(x)|<\frac{1}{(n+1)!}x^{n+1}$, so $|R_n(0.5)|<\frac{1}{(n+1)!}<0.001$ when $n\geq 6$.

$\cos 0.5\approx 0.8776$

19. For $f(x)=\sin x$, $|R_n(x)|<\frac{1}{(n+1)!}x^{n+1}$, so $|R_n(1)|<\frac{1}{(n+1)!}<0.001$ when $n\geq 6$.

$\sin 1\approx 0.8417$

21. $f(1)=7;\ f'(1)=2;\ f''(1)=2;\ f^{(3)}(x)=6$

$f(x)\approx 7+2(x-1)+(x-1)^2+(x-1)^3$

$=5+3x-2x^2+x^3=f(x)$

23. $\sin x\approx x-\frac{x^3}{3!}+\frac{x^5}{5!}-\frac{x^7}{7!}+\cdots+\frac{(-1)^{(n-1)/2}x^n}{n!}$ for n odd.

Using $n=5$, $\sin x\approx x-\frac{x^3}{3!}+\frac{x^5}{5!}$ and $R_5(x)=\frac{\cos c}{720}x^6$; $|R_5(x)|<\frac{x^6}{720}$.

a. $\sin(0.1) \approx 0.099833;\ |R_5(0.1)| < 1.4 \times 10^{-9}$

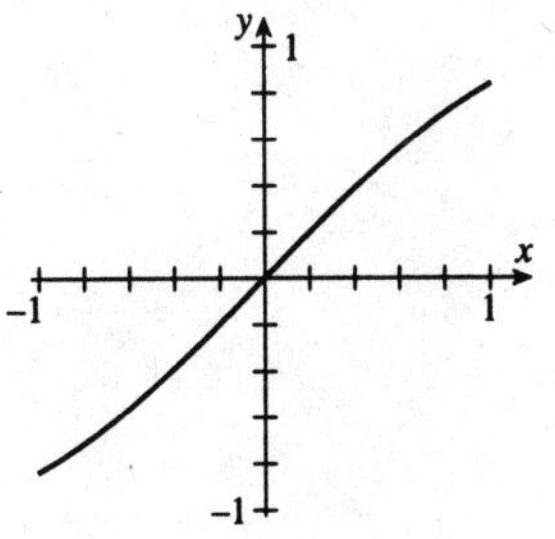

b. $\sin(0.5) \approx 0.47943;\ |R_5(0.5)| < 0.000022$

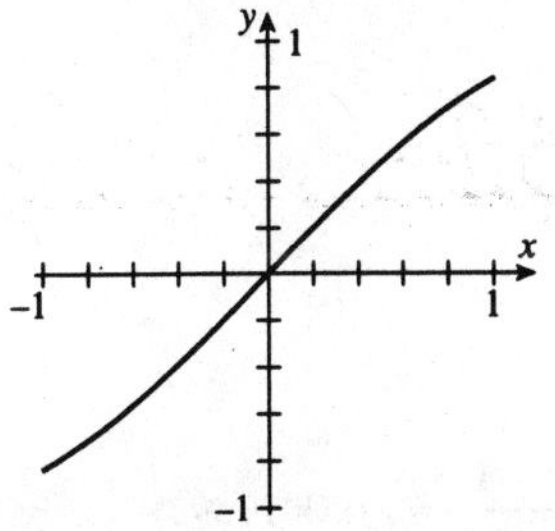

c. $\sin(1) \approx 0.84167;\ |R_5(1)| < 0.0014$

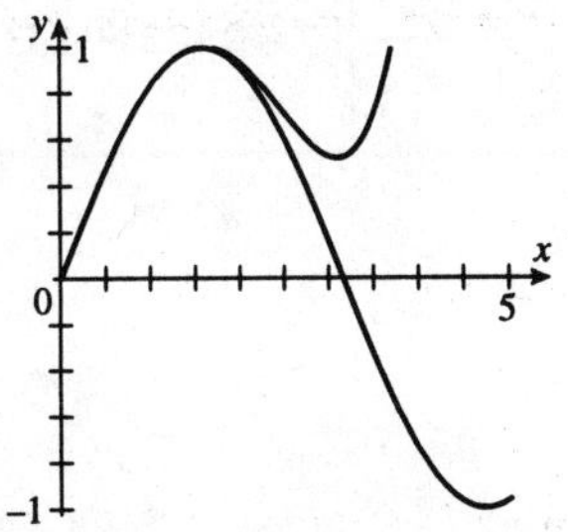

d. $\sin(10) \approx 677;\ |R_5(10)| < 1389$

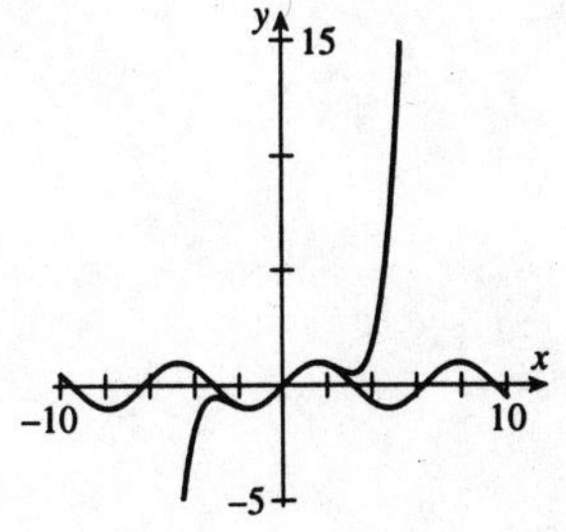

25. Let $m(v) = \dfrac{m_0}{\sqrt{1-\frac{v^2}{c^2}}}$. $m(0) = m_0;\ m'(0) = 0;\ m''(0) = \dfrac{m_0}{c^2}$.

The Maclaurin polynomial of order 2 is: $m(v) \approx m_0 + \dfrac{1}{2}\dfrac{m_0}{c^2}v^2 = m_0 + \dfrac{m_0}{2}\left(\dfrac{v}{c}\right)^2$.

27. The Maclaurin polynomial of order 2 for $1-e^{-(1+k)x}$ is $(1+k)x-\frac{(1+k)^2}{2}x^2$.

For $x=2k$, this is $2k-4k^3-2k^4 \approx 2k$ when k is very small.
$1+e^{-(1+0.01)(0.02)} \approx 0.019997 \approx 0.02$

29. a. $\sin x = x-\frac{x^3}{3!}+\frac{x^5}{5!}-\frac{x^7}{7!}+\cdots = x-\frac{x^3}{3!}+\frac{x^5}{5!}+\sum_{k=3}^{\infty}(-1)^k\frac{x^{2k+1}}{(2k+1)!}$,

so $\frac{\sin x - x+\frac{x^3}{6}}{x^5}=\frac{1}{5!}+\sum_{k=3}^{\infty}(-1)^k\frac{x^{2k-4}}{(2k+1)!}$.

$\lim_{x\to 0}\frac{\sin x - x+\frac{x^3}{6}}{x^5}=\lim_{x\to 0}\left(\frac{1}{5!}+\sum_{k=3}^{\infty}(-1)^k\frac{x^{2k-4}}{(2k+1)!}\right)=\frac{1}{5!}=\frac{1}{120}\approx 0.00833$

b. $\cos x = 1-\frac{x^2}{2!}+\frac{x^4}{4!}-\frac{x^6}{6!}+\frac{x^8}{8!}-\cdots = 1-\frac{x^2}{2!}+\frac{x^4}{4!}-\frac{x^6}{6!}+\sum_{k=4}^{\infty}(-1)^k\frac{x^{2k}}{(2k)!}$,

so $\frac{\cos x - 1+\frac{x^2}{2}-\frac{x^4}{24}}{x^6}=-\frac{1}{6!}+\sum_{k=4}^{\infty}(-1)^k\frac{x^{2k-6}}{(2k)!}$

and $\lim_{x\to 0}\frac{\cos x - 1+\frac{x^2}{2}-\frac{x^4}{24}}{x^6}=\lim_{x\to\infty}\left(-\frac{1}{6!}+\sum_{k=4}^{\infty}(-1)^k\frac{x^{2k-6}}{(2k)!}\right)=-\frac{1}{6!}=-\frac{1}{720}\approx -0.001389$.

31. $|R_4(x)|=\left|\frac{\cos(c)}{5!}x^5\right|\le\frac{x^5}{5!}<0.0002605$ if $0\le x\le 0.5$. Thus $\int_0^{0.5}\sin x\,dx \approx \left[\frac{x^2}{2}-\frac{x^4}{24}\right]_0^{0.5}\approx 0.122395833$.

A bound for the error is $[0.0002605x]_0^{0.5}=0.00013025$.

Section 9.5

Concepts Review

1. x_1; $f'(r)=0$

3. 1, 4, 2, 4, 2, ..., 4, 1

Problem Set 9.5

1.

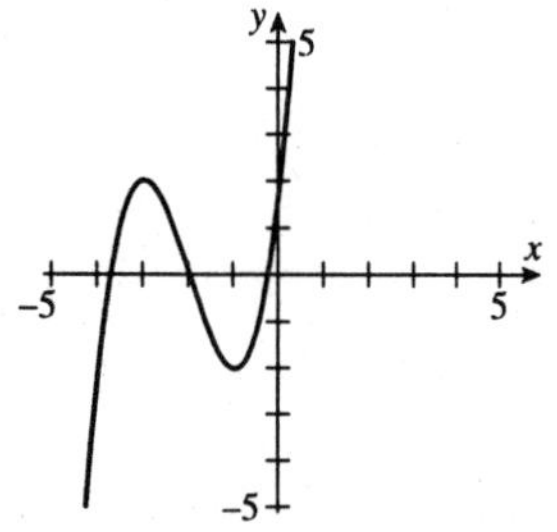

Use $x_1=0$ to start.

n	x_n
1	0
2	–0.22222222222
3	–0.26624338624
4	–0.26794667827
5	–0.26794919243
6	–0.26794919243

The root is at $x = -0.2679491924$. The rate of convergence depends on x_1.

3. 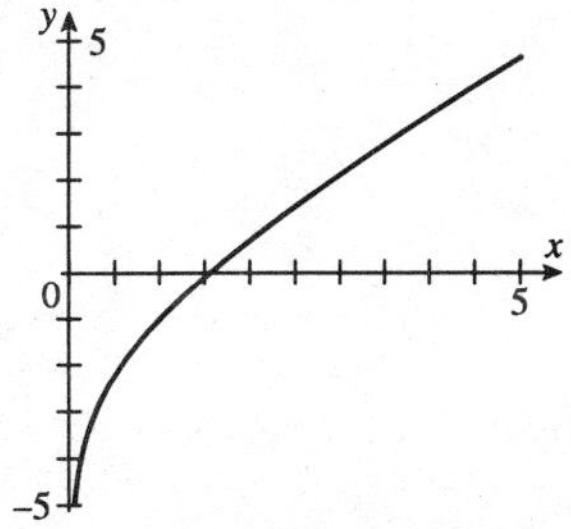

Use $x_1 = 1.5$ to start.

n	x_n
1	1.5
2	1.5567209351
3	1.5571455764
4	1.5571455990
5	1.5571455990

The root is at $x = 1.5571455990$. The number of digits of accuracy tripled at each step.

5. Let $f(x) = \cos x - x$.

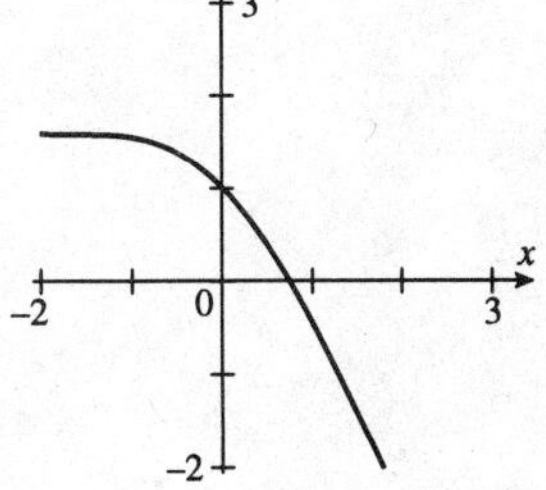

Use $x_1 = 0.5$ to start.

n	x_n
1	0.5
2	0.75522241711
3	0.73914166615
4	0.73908513392
5	0.73908513322
6	0.73908513322

The root is at $x = 0.7390851332$. The number of digits of accuracy tripled at each step.

7. 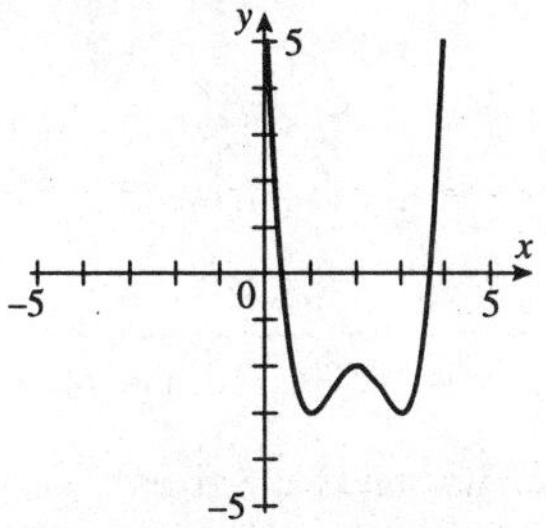

For the smaller root use $x_1 = 0$ to start.

n	x_n
1	0
2	0.25
3	0.33685064935
4	0.34697823828
5	0.34710832845
6	0.34710834972
7	0.34710834972

For the larger root use $x_1 = 3.5$ to start.

n	x_n
1	3.5
2	3.6916666667
3	3.6546710577
4	3.6528956184
5	3.6528916503
6	3.6528916503

The roots are $x = 0.3471083497$ and $x = 3.6258916503$.

9.

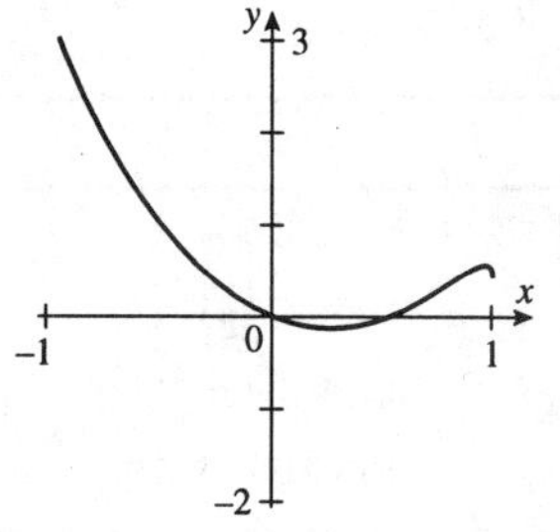

Use $x_1 = 0.5$ to start.

n	x_n
1	0.5
2	0.52791765128
3	0.52658258052
4	0.52657957179
5	0.52657957178

The root is at $x = 0.5265795718$. The number of digits of accuracy approximately doubled at each step.

11. With $n = 8$, the Parabolic Rule gives:

$$\int_1^2 \frac{1}{x^2}dx \approx \frac{0.125}{3}\left[\frac{1}{1^2}+\frac{4}{1.125^2}+\frac{2}{1.25^2}+\cdots+\frac{4}{1.875^2}+\frac{1}{2^2}\right] \approx 0.5000299.$$

With $n = 16$, the Parabolic Rule gives:

$$\int_1^2 \frac{1}{x^2}dx \approx \frac{0.0625}{3}\left[\frac{1}{1^2}+\frac{4}{1.0625^2}+\frac{2}{1.125^2}+\cdots+\frac{4}{1.9375^2}+\frac{1}{2^2}\right] \approx 0.5000019.$$

$$\int_1^2 \frac{1}{x^2}dx = \left[-\frac{1}{x}\right]_1^2 = \frac{1}{2}$$

Written response

13. With $n = 8$, the Parabolic Rule gives:

$$\int_0^4 \sqrt{x}\,dx \approx \frac{0.5}{3}\left[\sqrt{0}+4\sqrt{0.5}+2\sqrt{1}+\cdots+4\sqrt{3.5}+\sqrt{4}\right] \approx 5.3046342.$$

With $n = 16$, the Parabolic Rule gives

$$\int_0^4 \sqrt{x}\,dx \approx \frac{0.25}{3}\left[\sqrt{0}+4\sqrt{0.25}+2\sqrt{0.5}+\cdots+4\sqrt{3.75}+\sqrt{4}\right] \approx 5.3231855.$$

$$\int_0^4 \sqrt{x}\,dx = \left[\frac{2}{3}x^{3/2}\right]_0^4 = \frac{16}{3} \approx 5.3333333$$

Written response

15. For fixed x_{i-1} and x_i, $\bar{x}_i = \frac{x_i + x_{i-1}}{2}$ and $f'(\bar{x}_i)$ are constants. Thus,

$$\int_{x_{i-1}}^{x_i} f'(\bar{x}_i)(\bar{x}_i - x)dx = f'(\bar{x}_i)\int_{x_{i-1}}^{x_i} (\bar{x}_i - x)dx$$

$$= f'(\bar{x}_i)\left[\bar{x}_i x - \frac{x^2}{2}\right]_{x_{i-1}}^{x_i}$$

$$= f'(\bar{x}_i)\left(\bar{x}_i x_i - \frac{x_i^2}{2} - \bar{x}_i x_{i-1} + \frac{x_{i-1}^2}{2}\right)$$

$$= f'(\bar{x}_i)\left[\bar{x}_i(x_i - x_{i-1}) - \frac{1}{2}(x_i^2 - x_{i-1}^2)\right]$$
$$= f'(\bar{x}_i)\left[\frac{1}{2}(x_i + x_{i-1})(x_i - x_{i-1}) - \frac{1}{2}(x_i^2 - x_{i-1}^2)\right]$$
$$= f'(\bar{x}_i)[0] = 0.$$

17. Let $f(x) = \dfrac{1+\ln x}{x} = \dfrac{1}{x} + \dfrac{1}{x}\ln x$. Then $f'(x) = -\dfrac{1}{x^2}\ln x$ so

$$x_{n+1} = x_n - \frac{f(x_n)}{f'(x_n)} = x_n - \frac{\frac{1}{x_n}(1+\ln x_n)}{-\frac{1}{x_n^2}\ln x_n} = 2x_n + \frac{x_n}{\ln x_n}.$$

Using $x_1 = 1.2$, Newton's Method gives:

n	x_n
1	1.2
2	8.9818
3	22.055
4	51.2396

The values get larger without converging.
Using $x_1 = 0.5$, Newton's Method gives:

n	x_n
1	0.5
2	0.27865247956
3	0.3392311687
4	0.36467130311
5	0.36783767798
6	0.36787943406
7	0.36787944117
8	0.36787944117

The root is at $x = 0.3678794412$.

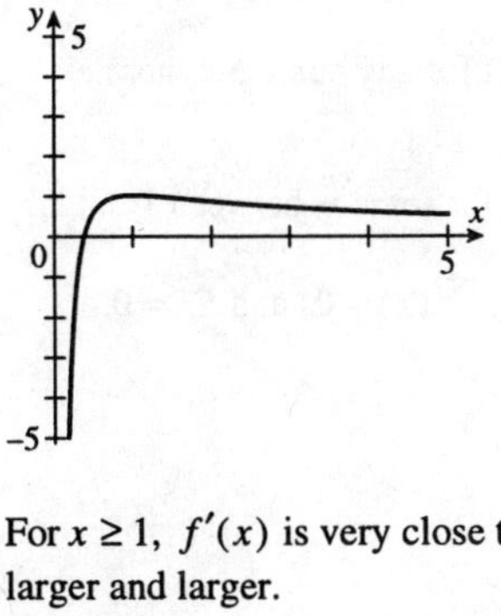

For $x \ge 1$, $f'(x)$ is very close to 0 and negative. Thus the approximations that Newton's Method yield for $x_1 \ge 1$ get larger and larger.

19. a. For Tom's car, $P = 2000$, $R = 100$, and $k = 24$, thus

$$2000 = \frac{100}{i}\left[1 - \frac{1}{(1+i)^{24}}\right] \text{ or } 20i = 1 - \frac{1}{(1+i)^{24}}, \text{ which is equivalent to } 20i(1+i)^{24} - (1+i)^{24} + 1 = 0.$$

b. Let $f(i) = 20i(1+i)^{24} - (1+i)^{24} + 1 = (1+i)^{24}(20i - 1) + 1$.
Then

$$f'(i) = 20(1+i)^{24} + 480i(1+i)^{23} - 24(1+i)^{23} = (1+i)^{23}(500i - 4), \text{ so}$$

$$i_{n+1} = i_n - \frac{f(i_n)}{f'(i_n)} = i_n - \frac{(1+i_n)^{24}(20i_n - 1) + 1}{(1+i_n)^{23}(500i_n - 4)} = i_n - \left[\frac{20i_n^2 + 19i_n - 1 + (1+i_n)^{-23}}{500i_n - 4}\right].$$

c.

n	i_n
1	0.012
2	0.01652971
3	0.0152651
4	0.0151323
5	0.0151308
6	0.0151308

$i = 0.0151308$
$r = 18.157\%$

21. a. To show that the Parabolic Rule is exact, examine it on the interval $[m - h, m + h]$.
Let $f(x) = ax^3 + bx^2 + cx + d$, then

$$\int_{m-h}^{m+h} f(x)dx$$

$$= \frac{a}{4}\left[(m+h)^4 - (m-h)^4\right] + \frac{b}{3}\left[(m+h)^3 - (m-h)^3\right] + \frac{c}{2}\left[(m+h)^2 - (m-h)^2\right] + d[(m+h) - (m-h)]$$

$$= \frac{a}{4}(8m^3h + 8h^3m) + \frac{b}{3}(6m^2h + 2h^3) + \frac{c}{2}(4mh) + d(2h).$$

The Parabolic Rule with $n = 2$ gives

$$\int_{m-h}^{m+h} f(x)dx = \frac{h}{3}[f(m-h) + 4f(m) + f(m+h)]$$

$$= 2am^3h + 2amh^3 + 2bm^2h + \frac{2}{3}bh^3 + 2chm + 2dh$$

$$= \frac{a}{4}(8m^3h + 8mh^3) + \frac{b}{3}(6m^2h + 2h^3) + \frac{c}{2}(4mh) + d(2h)$$

which agrees with the direct computation. Thus, the Parabolic Rule is exact for any cubic polynomial.

b. The error in using the Parabolic Rule is given by $E_n = -\frac{(l-k)^5}{180n^4} f^{(4)}(m)$ for some m between l and k.

However, $f'(x) = 3ax^2 + 2bx + c$, $f''(x) = 6ax + 2b$, $f^{(3)}(x) = 6a$, and $f^{(4)}(x) = 0$, and $E_n = 0$.

Section 9.6

Concepts Review

1. Power Series

3. (–1, 1)

Problem Set 9.6

1. $\displaystyle\sum_{n=1}^{\infty}\frac{(-1)^{n+1}x^n}{n(n+1)}$; $\displaystyle p=\lim_{n\to\infty}\left|\frac{x^{n+1}}{(n+1)(n+2)}\div\frac{x^n}{n(n+1)}\right|=\lim_{n\to\infty}|x|\left|\frac{n}{n+2}\right|=|x|$

The largest open interval of convergence is (–1, 1).

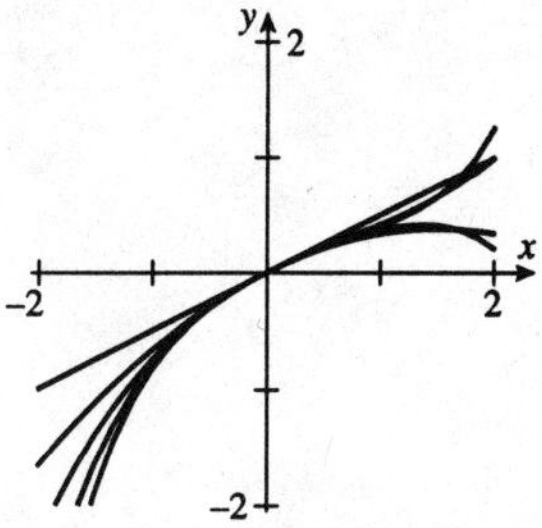

3. $\displaystyle\sum_{n=1}^{\infty}\frac{(-1)^{n+1}x^{2n-1}}{(2n-1)!}$; $\displaystyle p=\lim_{n\to\infty}\left|\frac{x^{2n+1}}{(2n+1)!}\div\frac{x^{2n-1}}{(2n-1)!}\right|=\lim_{n\to\infty}\left|x^2\right|\left|\frac{1}{2n(2n+1)}\right|=0$

The series converges for all x.

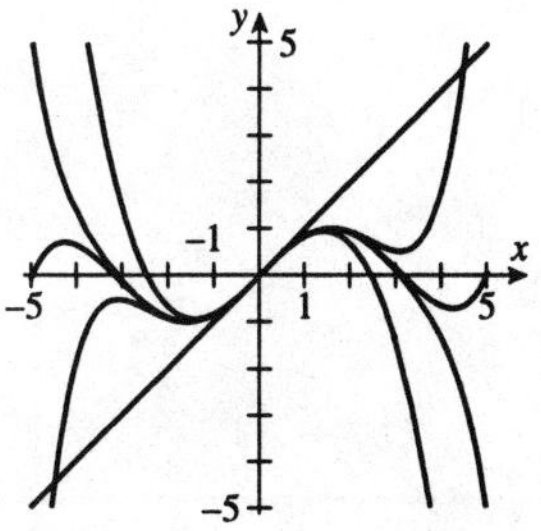

5. $\displaystyle\sum_{n=0}^{\infty}nx^n$; $\displaystyle p=\lim_{n\to\infty}\left|\frac{(n+1)x^{n+1}}{nx^n}\right|=\lim_{n\to\infty}|x|\left|\frac{n+1}{n}\right|=|x|$

The largest open interval of convergence is (–1, 1).

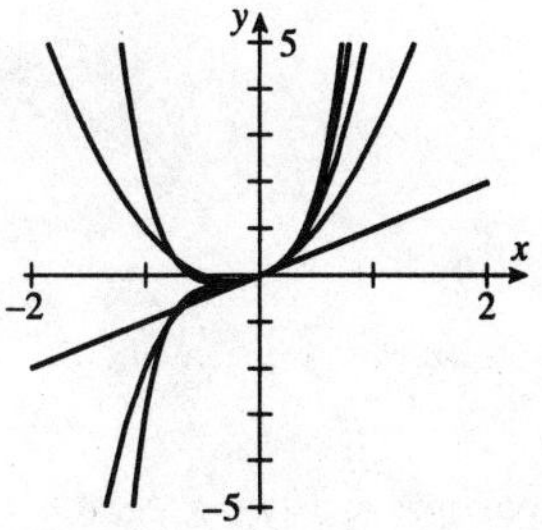

7. $\displaystyle 1+\sum_{n=1}^{\infty}\frac{(-1)^n x^n}{n}$; $\displaystyle p=\lim_{n\to\infty}\left|\frac{x^{n+1}}{n+1}\div\frac{x^n}{n}\right|=\lim_{n\to\infty}|x|\left|\frac{n}{n+1}\right|=|x|$

The largest open interval of convergence is (–1, 1).

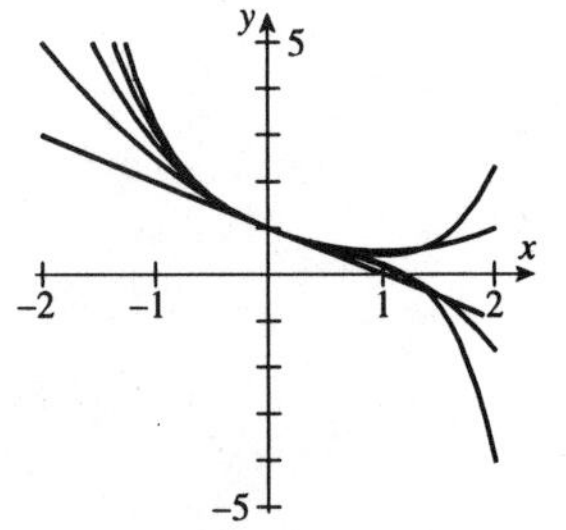

9. $1+\sum_{n=1}^{\infty}\frac{(-1)^n x^n}{n(n+2)}$; $p=\lim_{n\to\infty}\left|\frac{x^{n+1}}{(n+1)(n+3)}\div\frac{x^n}{n(n+2)}\right| = \lim_{n\to\infty}|x|\left|\frac{n^2+2n}{n^2+4n+3}\right| = |x|$

The largest open interval of convergence is (–1, 1).

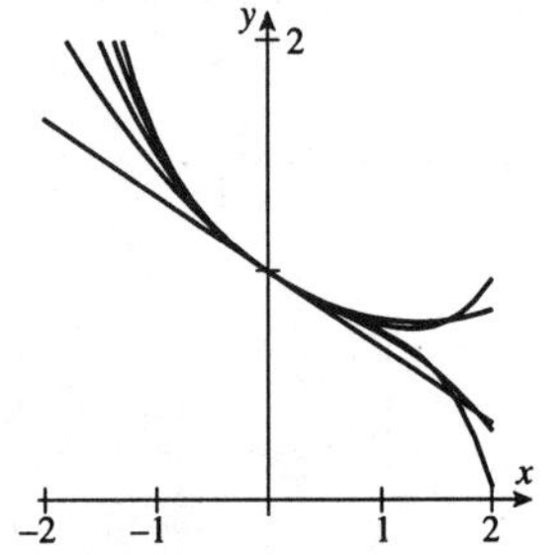

11. $\sum_{n=0}^{\infty}\frac{(-1)^n x^n}{2^n}$; $p=\lim_{n\to\infty}\left|\frac{x^{n+1}}{2^{n+1}}\div\frac{x^n}{2^n}\right| = \lim_{n\to\infty}\left|\frac{x}{2}\right| = \left|\frac{x}{2}\right|$

The largest open interval of convergence is (–2, 2).

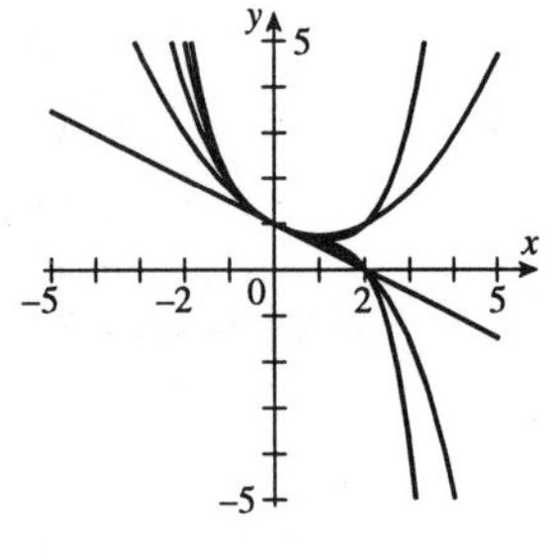

13. $\sum_{n=0}^{\infty}\frac{2^n x^n}{n!}$; $p=\lim_{n\to\infty}\left|\frac{2^{n+1}x^{n+1}}{(n+1)!}\div\frac{2^n x^n}{n!}\right| = \lim_{n\to\infty}|2x|\left|\frac{1}{n+1}\right| = 0$

The series converges for all x.

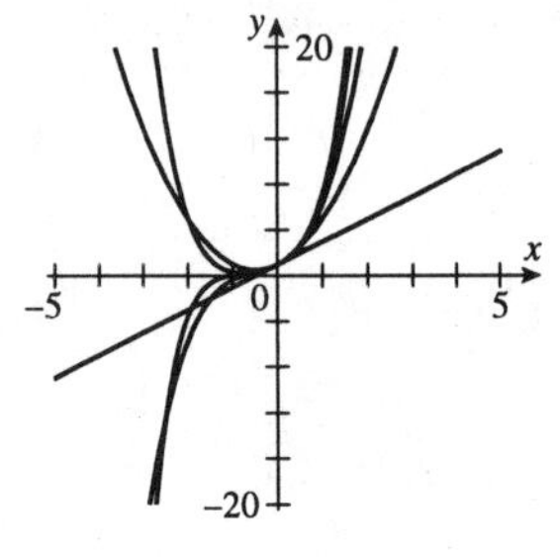

15. $\sum_{n=1}^{\infty} \frac{(x-1)^n}{n}$; $p = \lim_{n\to\infty} \left| \frac{(x-1)^{n+1}}{n+1} \div \frac{(x-1)^n}{n} \right| = \lim_{n\to\infty} |x-1| \left| \frac{n}{n+1} \right| = |x-1|$

The largest open interval of convergence is (0, 2).

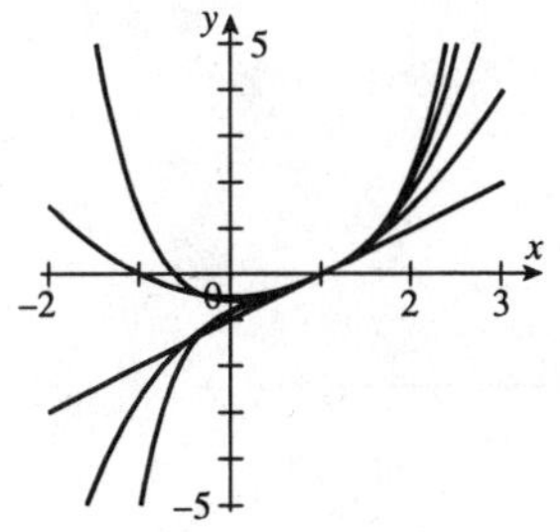

17. $\sum_{n=0}^{\infty} \frac{(x+1)^n}{2^n}$; $p = \lim_{n\to\infty} \left| \frac{(x+1)^{n+1}}{2^{n+1}} \div \frac{(x+1)^n}{2^n} \right| = \lim_{n\to\infty} \left| \frac{x+1}{2} \right| = \left| \frac{x+1}{2} \right|$

The largest open interval of convergence is (–3, 1).

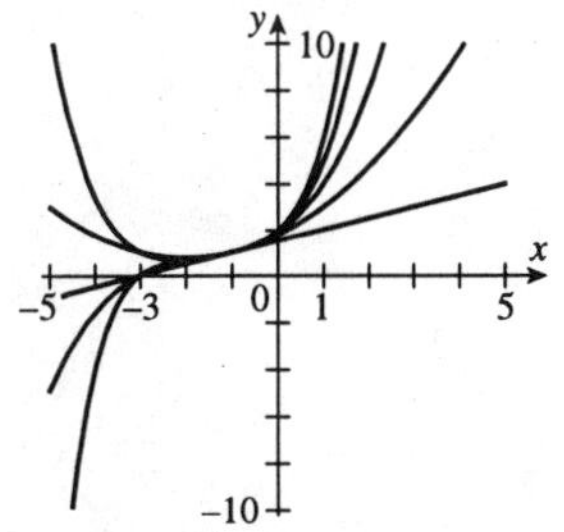

19. $\sum_{n=1}^{\infty} \frac{(x+5)^n}{n(n+1)}$; $p = \lim_{n\to\infty} \left| \frac{(x+5)^{n+1}}{(n+1)(n+2)} \div \frac{(x+5)^n}{n(n+1)} \right| = \lim_{n\to\infty} |x+5| \left| \frac{n}{n+2} \right| = |x+5|$

The largest open interval of convergence is (–6, –4).

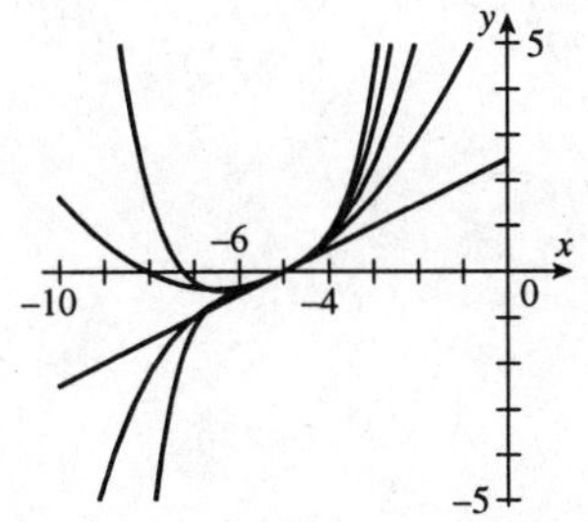

21. Let *f*(*x*) = sinh *x*. Then $f^{(n)}(x) = \sinh x$ if *n* is even and $f^{(n)}(x) = \cosh x$ if *n* is odd. Since sinh 0 = 0 and cosh 0 = 1,

$$\sinh x = x + \frac{x^3}{3!} + \frac{x^5}{5!} + \frac{x^7}{7!} + \cdots$$

$$= \sum_{n=1}^{\infty} \frac{x^{2n-1}}{(2n-1)!}$$

$$p = \lim_{n\to\infty} \left| \frac{x^{2n+1}}{(2n+1)!} \div \frac{x^{2n-1}}{(2n-1)!} \right| = \lim_{n\to\infty} \left| x^2 \right| \left| \frac{1}{2n(2n+1)} \right| = 0$$

The series converges for all *x*.

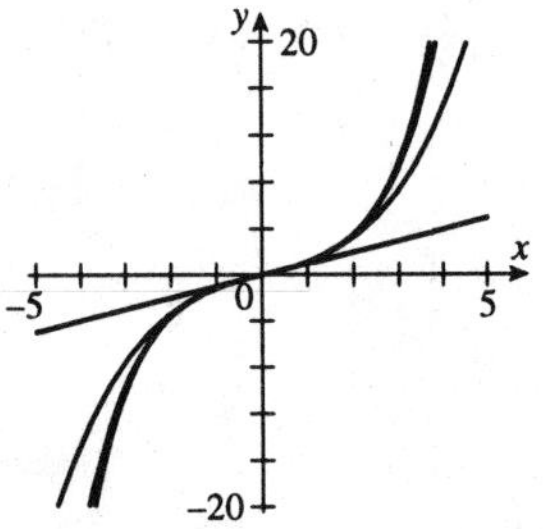

23. Let $f(x) = \ln x$. Then $f'(x) = \frac{1}{x}$, $f''(x) = -\frac{1}{x^2}$, $f'''(x) = \frac{2}{x^3}$, and $f^{(n)}(x) = \frac{(-1)^{n+1}(n-1)!}{x^n}$, so

$$\ln x \approx 1.609 + \frac{1}{5}(x-5) - \frac{1}{25}\cdot\frac{(x-5)^2}{2!} + \frac{2}{125}\cdot\frac{(x-5)^3}{3!} + \cdots$$

$$= 1.609 + \sum_{n=1}^{\infty} \frac{(-1)^{n+1}(x-5)^n}{n \cdot 5^n}$$

$$p = \lim_{n\to\infty}\left|\frac{(x-5)^{n+1}}{(n+1)5^{n+1}} \div \frac{(x-5)^n}{n\cdot 5^n}\right| = \lim_{n\to\infty}\left|\frac{x-5}{5}\right|\left|\frac{n}{n+1}\right| = \left|\frac{x-5}{5}\right|$$

The largest open interval of convergence is (0, 10).

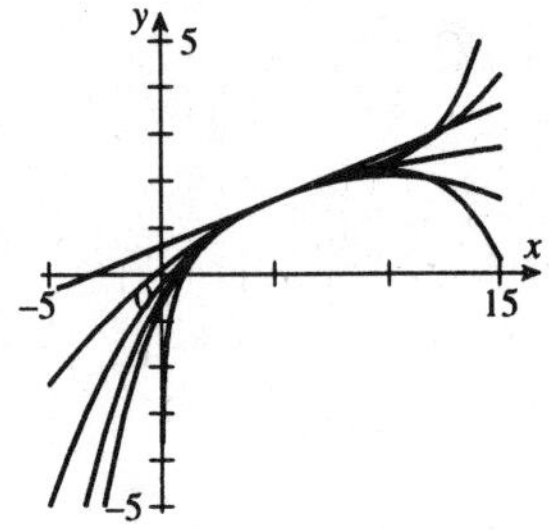

25. Let $f(x) = \frac{1}{1+x}$. Then $f'(x) = -\frac{1}{(1+x)^2}$, $f''(x) = \frac{2}{(1+x)^3}$, $f'''(x) = -\frac{6}{(1+x)^4}$, and $f^{(n)}(x) = \frac{(-1)^n n!}{(1+x)^{n+1}}$,

so $\frac{1}{1+x} = 1 - x + x^2 - x^3 + x^4 + \cdots$

$$= \sum_{n=0}^{\infty}(-1)^n x^n$$

$p = \lim_{n\to\infty}\left|\frac{x^{n+1}}{x^n}\right| = |x|$. The largest open interval of convergence is (–1, 1).

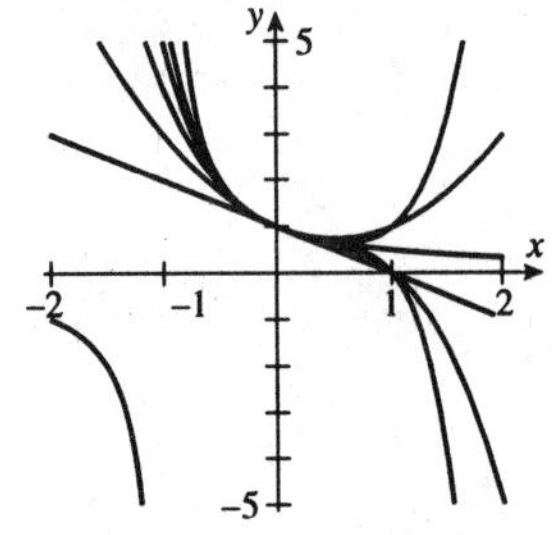

27. Let $f(x) = \sin x$. Then $f'(x) = \cos x,\ f''(x) = -\sin x,\ f'''(x) = -\cos x,\ f^{(4)}(x) = \sin x$, and so on. Thus

$$\sin x = \frac{1}{2} + \frac{\sqrt{3}}{2}\left(x - \frac{\pi}{6}\right) - \frac{1}{2}\cdot\frac{1}{2!}\left(x - \frac{\pi}{6}\right)^2 - \frac{\sqrt{3}}{2}\cdot\frac{1}{3!}\left(x - \frac{\pi}{6}\right)^3 + \cdots$$

$$= \frac{1}{2} - \frac{1}{2}\cdot\frac{1}{2!}\left(x - \frac{\pi}{6}\right)^2 + \frac{1}{2}\cdot\frac{1}{4!}\left(x - \frac{\pi}{6}\right)^4 - \cdots + \frac{\sqrt{3}}{2}\left(x - \frac{\pi}{6}\right) - \frac{\sqrt{3}}{2}\cdot\frac{1}{3!}\left(x - \frac{\pi}{6}\right)^3 + \frac{\sqrt{3}}{2}\cdot\frac{1}{5!}\left(x - \frac{\pi}{6}\right)^5 - \cdots$$

$$= \sum_{n=0}^{\infty}\frac{1}{2}\cdot\frac{(-1)^n}{(2n)!}\left(x - \frac{\pi}{6}\right)^{2n} + \sum_{n=1}^{\infty}\frac{\sqrt{3}}{2}\cdot\frac{(-1)^{n+1}}{(2n-1)!}\left(x - \frac{\pi}{6}\right)^{2n-1}.$$

For the first series,

$$p = \lim_{n\to\infty}\left|\frac{\left(x - \frac{\pi}{6}\right)^{2n+2}}{2\cdot(2n+2)!} \div \frac{\left(x - \frac{\pi}{6}\right)^{2n}}{2\cdot(2n)!}\right| = \lim_{n\to\infty}\left|\left(x - \frac{\pi}{6}\right)^2\right|\left|\frac{1}{(2n+2)(2n+1)}\right| = 0.$$

This series converges for all x.
For the second series,

$$p = \lim_{n\to\infty}\left|\frac{\sqrt{3}\left(x - \frac{\pi}{6}\right)^{2n+1}}{2(2n+1)!} \div \frac{\sqrt{3}\left(x - \frac{\pi}{6}\right)^{2n-1}}{2(2n-1)!}\right| = \lim_{n\to\infty}\left|\left(x - \frac{\pi}{6}\right)^2\right|\left|\frac{1}{2n(2n+1)}\right| = 0$$

This series also converges for all x.

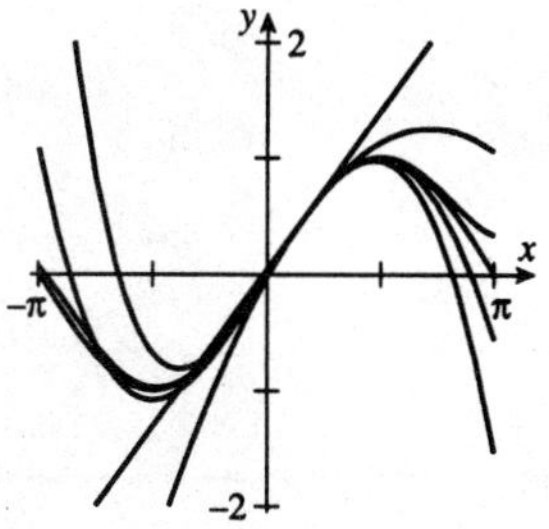

29. Let $f(x) = 1 + x^2 + x^3$. Then $f'(x) = 2x + 3x^2,\ f''(x) = 2 + 6x,\ f'''(x) = 6$, and $f^{(n)}(x) = 0$ for all $n \ge 4$. Thus,

$$1 + x^2 + x^3 = 3 + 5(x-1) + \frac{8}{2!}(x-1)^2 + \frac{6}{3!}(x-1)^3 = 3 + 5(x-1) + 4(x-1)^2 + (x-1)^3.$$

Since this series is a finite sum, it converges for all x.

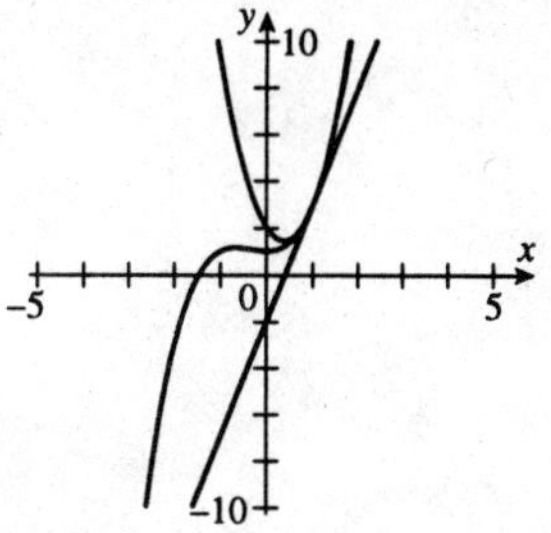

31. Using Mathematica, the graph looks like

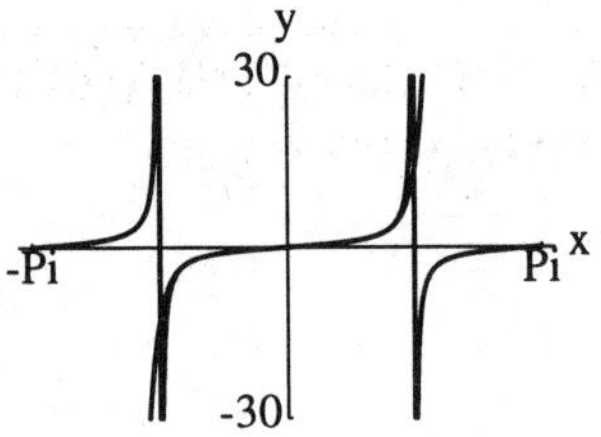

The largest open interval of convergence appears to be $\left(-\frac{\pi}{2}, \frac{\pi}{2}\right)$.

33. Using Mathematica, the graph looks like

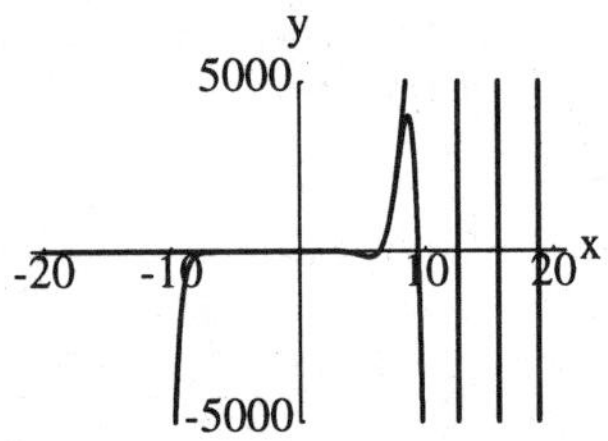

The largest open interval of convergence appears to be (–9, 9).

35. Using Mathematica, the graph looks like

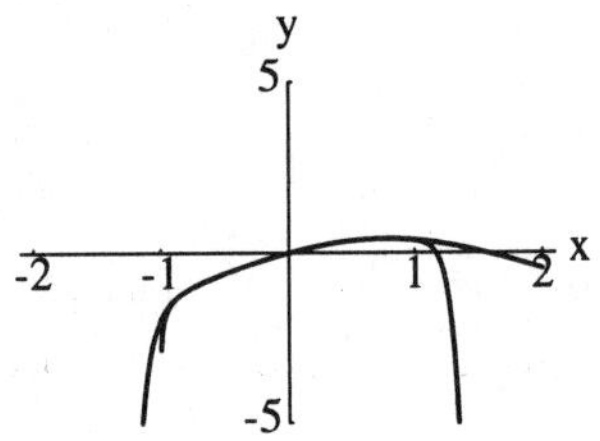

The largest open interval of convergence appears to be (–1, 1).

37. If for some x_0, $\lim_{n\to\infty} \frac{x_0^n}{n!} \neq 0$, then $\sum \frac{x_0^n}{n!}$ could not converge.

39. Using the Absolute Ratio Test, examine $\lim_{n\to\infty} \left| \frac{(pn+p)!}{((n+1)!)^p} x^{n+1} \div \frac{(pn)!}{(n!)^p} x^n \right|$

$$= \lim_{n\to\infty} |x| \left| \frac{(pn+p)(pn+p-1)\ldots(pn+p-(p-1))}{(n+1)^p} \right|$$

$$= \lim_{n\to\infty} |x| \left| p\left(p-\frac{1}{n+1}\right)\left(p-\frac{2}{n+1}\right)\cdots\left(p-\frac{p-1}{n+1}\right) \right|$$

$$= |x| p^p$$

The radius of convergence is p^{-p}.

41. $\sum_{n=0}^{\infty} a_n(x-3)^n$ is a Taylor series based at 3. If it converges at $x=-1$, then the radius of convergence must be at least $3-(-1)=4$. Since $|3-6|=3<4$, $x=6$ is within the radius of convergence so the series must converge at $x = 6$. If the radius of convergence is 4, then $x = 7$ is an endpoint, so the series may or may not converge at $x = 7$.

43. If $a_n = a_{n+p}$, then $a_0 = a_p = a_{2p} = a_{np}$, $a_1 = a_{p+1} = a_{2p+1} = a_{np+1}$, etc. Thus,

$$\sum_{n=0}^{\infty} a_n x^n = a_0 + a_1x + \cdots + a_{p-1}x^{p-1} + a_0x^p + a_1x^{p+1} + \cdots + a_{p-1}x^{2p-1} + \cdots$$

$$= (a_0 + a_1x + \cdots + a_{p-1}x^{p-1})(1 + x^p + x^{2p} + \cdots)$$

$$= (a_0 + a_1x + \cdots + a_{p-1}x^{p-1})\sum_{n=0}^{\infty} x^{np}$$

$a_0 + a_1x + \cdots + a_{p-1}x^{p-1}$ is a polynomial, which will converge for all x.

$\sum_{n=0}^{\infty} x^{np} = \sum_{n=0}^{\infty} (x^p)^n$ is a geometric series which converges for $\left|x^p\right| < 1$, or, equivalently, $|x| < 1$.

Since $\sum_{n=0}^{\infty} (x^p)^n = \frac{1}{1-x^p}$ for $|x|<1$, $S(x) = (a_0 + a_1x + \cdots + a_{p-1}x^{p-1})\left(\frac{1}{1-x^p}\right)$ for $|x|<1$.

Section 9.7

Concepts Review

1. Integrated; interior

3. $1 + x^2 + \frac{x^4}{2} + \frac{x^6}{6}$

Problem Set 9.7

1. From the geometric series for $\frac{1}{1-x}$ with x replaced by $-x$, we get $\frac{1}{1+x} = 1 - x + x^2 - x^3 + x^4 - x^5 + \cdots$ for $-1 < -x < 1$, or $-1 < x < 1$.

3. $\frac{d}{dx}\left(\frac{1}{1-x}\right) = \frac{1}{(1-x)^2}$; $\frac{d}{dx}\left(\frac{1}{(1-x)^2}\right) = \frac{2}{(1-x)^3}$, so $\frac{1}{(1-x)^3}$ is $\frac{1}{2}$ of the second derivative of $\frac{1}{1-x}$. Thus,

$$\frac{1}{(1-x)^3} = 1 + 3x + 6x^2 + 10x^3 + \cdots$$

$$= \sum_{n=0}^{\infty} \frac{n(n+1)}{2} x^n \text{ for } -1 < x < 1.$$

5. From the geometric series for $\frac{1}{1-x}$ with x replaced by $\frac{3}{2}x$, we get $\frac{1}{2-3x} = \frac{1}{2} + \frac{3}{4}x + \frac{9}{8}x^2 + \frac{27}{16}x^3 + \cdots$ for $-1 < \frac{3}{2}x < 1$, or $-\frac{2}{3} < x < \frac{2}{3}$.

7. From the geometric series for $\frac{1}{1-x}$ with x replaced by x^4, we get $\frac{x^2}{1-x^4} = x^2 + x^6 + x^{10} + x^{14} + \cdots$ for $-1 < x^4 < 1$, or $-1 < x < 1$.

9. From the geometric series for $\ln(1+x)$ with x replaced by t, we get $\int_0^x \ln(1+t)dt = \frac{1}{2}x^2 - \frac{1}{6}x^3 + \frac{1}{12}x^4 - \frac{1}{20}x^5 + \cdots$ for $-1 < x \le 1$.

11. Since $\tan x = \frac{\sin x}{\cos x}$,

$$\tan x = \frac{x - \frac{x^3}{3!} + \frac{x^5}{5!} - \frac{x^7}{7!} + \cdots}{1 - \frac{x^2}{2!} + \frac{x^4}{4!} - \frac{x^6}{6!} + \cdots}$$

The terms through x^5 are: $x + \frac{x^3}{3} + \frac{2x^5}{15}$.

13. $e^x \sin x = \left(1 + x + \frac{x^2}{2!} + \frac{x^3}{3!} + \frac{x^4}{4!} + \cdots\right)\left(x - \frac{x^3}{3!} + \frac{x^5}{5!} - \cdots\right)$

The terms through x^5 are: $x + x^2 + \frac{x^3}{3} - \frac{x^5}{30}$.

15. $\cos x \ln(1+x) = \left(1 - \frac{x^2}{2!} + \frac{x^4}{4!} - \cdots\right)\left(x - \frac{x^2}{2} + \frac{x^3}{3} - \frac{x^4}{4} + \frac{x^5}{5} - \cdots\right)$

The terms through x^5 are: $x - \frac{x^2}{2} - \frac{x^3}{6} + \frac{3x^5}{40}$.

17. Substitute $-x$ for x in the series for e^x to get:

$e^{-x} = 1 - x + \frac{x^2}{2!} - \frac{x^3}{3!} + \frac{x^4}{4!} - \frac{x^5}{5!} + \cdots$.

19. Add the result of Problem 17 to the series for e^x to get:

$e^x + e^{-x} = 2 + \frac{2x^2}{2!} + \frac{2x^4}{4!} + \frac{2x^6}{6!} + \cdots$.

21. $e^{-x} \cdot \frac{1}{1-x} = \left(1 - x + \frac{x^2}{2!} - \frac{x^3}{3!} + \cdots\right)(1 + x + x^2 + \cdots)$

$= 1 + \frac{x^2}{2} + \frac{x^3}{3} + \frac{3x^4}{8} + \frac{11x^5}{30} + \cdots$

23. $\frac{\tan^{-1} x}{e^x} = e^{-x} \tan^{-1} x$

$= \left(1 - x + \frac{x^2}{2!} + \frac{x^3}{3!} + \cdots\right)\left(x - \frac{x^3}{3} + \frac{x^5}{3} + \frac{x^7}{7} + \cdots\right)$

$= x - x^2 + \frac{x^3}{6} + \frac{x^4}{6} + \frac{3x^5}{40} + \cdots$

25. $(\tan^{-1} x)(1 + x^2 + x^4) = \left(x - \frac{x^3}{3} + \frac{x^5}{5} - \frac{x^7}{7} + \cdots\right)(1 + x^2 + x^4)$

$= x + \frac{2x^3}{3} + \frac{13x^5}{15} - \frac{29x^7}{105} + \cdots$

27. The series representation of $\dfrac{e^x}{1+x}$ is $1+\dfrac{x^2}{2}-\dfrac{x^3}{3}+\dfrac{3x^4}{8}-\dfrac{11x^5}{30}+\cdots$, thus

$$\int_0^x \frac{e^t}{1+t}\,dt = x+\frac{1}{6}x^3-\frac{1}{12}x^4+\frac{3}{40}x^5-\cdots.$$

29. $\tan^{-1}(e^x-1)=(e^x-1)-\dfrac{(e^x-1)^3}{3}+\dfrac{(e^x-1)^5}{5}\cdots$

$$=\left(x+\frac{x^2}{2!}+\frac{x^3}{3!}+\cdots\right)-\frac{1}{3}\left(x+\frac{x^2}{2!}+\cdots\right)^3+\cdots$$

$$=x+\frac{x^2}{2}-\frac{x^3}{6}-\cdots$$

31. From the Maclaurin series for $\sqrt{1+x}$ in Example 8, with x replaced by $-x^2$, we get

$$\sqrt{1-x^2}=1-\frac{1}{2}x^2-\frac{1}{8}x^4-\frac{1}{16}x^6-\frac{5}{128}x^8+\cdots.$$

Using division, we find that

$$\frac{1}{\sqrt{1-x^2}}=1+\frac{1}{2}x^2+\frac{3}{8}x^4+\frac{5}{16}x^6+\cdots, \text{ hence}$$

$$\sin^{-1}x=\int_0^x\frac{1}{\sqrt{1+t^2}}\,dt=x+\frac{1}{6}x^3+\frac{3}{40}x^5+\frac{5}{112}x^7+\cdots.$$

33. Since $\cos x=1-\dfrac{x^2}{2!}+\dfrac{x^4}{4!}-\dfrac{x^6}{6!}+\dfrac{x^8}{8!}-\cdots$

$$\cos(x^2)=1-\frac{x^4}{2!}+\frac{x^8}{4!}-\frac{x^{12}}{6!}+\frac{x^{16}}{8!}-\cdots$$

and $\int_0^1\cos(x^2)dx=\left[x-\dfrac{x^5}{5\cdot 2!}+\dfrac{x^9}{9\cdot 4!}-\dfrac{x^{13}}{13\cdot 6!}+\cdots\right]_0^1$

$$=1-\frac{1}{5\cdot 2!}+\frac{1}{9\cdot 4!}-\frac{1}{13\cdot 6!}+\frac{1}{17\cdot 8!}-\cdots$$

$\approx 0.9045.$

35. $f(t)$ cannot be represented by a Maclaurin series since all derivatives of $f'(t)$ are 0 at $t = 0$, yet the function is not identically 0. For the same reason, $g(t)$ cannot be represented by a Maclaurin series.

37. **a.** Since $\dfrac{1}{1-x}=1+x+x^2+x^3+x^4+\cdots,$

$$\frac{1}{1-x^2}=1+x^2+x^4+x^6+x^8+\cdots.$$

b. Again using $\dfrac{1}{1-x}=1+x+x^2+x^3+x^4+\cdots,$

$$\frac{1}{1-\cos x}-1=\cos x+\cos^2 x+\cos^3 x+\cdots.$$

c. $\ln(1-x) = -x - \frac{x^2}{2} - \frac{x^3}{3} - \frac{x^4}{4} - \cdots$, so

$\ln(1-x^2) = -x^2 - \frac{x^4}{2} - \frac{x^6}{3} - \frac{x^8}{4} - \cdots$, and

$-\frac{1}{2}\ln(1-x^2) = \ln\frac{1}{\sqrt{1-x^2}} = \frac{x^2}{2} + \frac{x^4}{4} + \frac{x^6}{6} + \frac{x^8}{8} + \cdots.$

39. $\frac{x}{x^2-3x+2} = \frac{x}{(x-2)(x-1)} = \frac{2}{x-2} - \frac{1}{x-1}$

$= -\frac{1}{1-\frac{x}{2}} + \frac{1}{1-x}$

$= -\left(1 + \frac{x}{2} + \frac{x^2}{4} + \frac{x^3}{8} + \cdots\right) + \left(1 + x + x^2 + x^3 + \cdots\right)$

$= \frac{x}{2} + \frac{3x^2}{4} + \frac{7x^3}{8} + \cdots = \sum_{n=1}^{\infty} \frac{(2^n-1)x^n}{2^n}$

41. $\pi \approx 16\left(\frac{1}{5} - \frac{1}{375} + \frac{1}{15{,}625} - \frac{1}{546{,}875} + \frac{1}{17{,}578{,}125}\right) - 4\left(\frac{1}{239}\right)$

≈ 3.14159

Section 9.8 Chapter Review

Chapter Test

1. False. The sequence does not converge.

3. True. The sequence can also be written as $\left\{\left(-\frac{1}{\sqrt{2}}\right)^n\right\}$, which converges since $\left|-\frac{1}{\sqrt{2}}\right| < 1$.

5. True. Since $a_{n+1} = a_n\left(1 - \frac{\sin|a_n|}{2}\right) < a_n$ and $a_n > 0$, the sequence converges by the Sequence Theorem.

7. True. If $\{a_n\}$ converges, then for some N, there are numbers m and M with $m \le a_n \le M$ for all $n \ge N$. Thus $\frac{m}{n} \le \frac{a_n}{n} \le \frac{M}{n}$ for all $n \ge N$. Since $\left\{\frac{m}{n}\right\}$ and $\left\{\frac{M}{n}\right\}$ both converge to 0, $\left\{\frac{a_n}{n}\right\}$ must also converge to 0.

9. False. This is true if $|f'(x)| < 1$, but not if $f'(x) < -1$ which is possible if $f'(x) < 1$.

11. True. $\left|\frac{e}{\pi}\right| < 1$ so the geometric series converges.

13. False. Many series cannot be summed exactly.

15. True. For this series $\lim_{n\to\infty} a_n = 1$, so the Ratio Test is inconclusive.

17. False. This series is $\sum_{n=1}^{\infty} \frac{1}{2n-1} = 1 + \frac{1}{3} + \frac{1}{5} + \cdots$ which diverges.

19. False. Consider the series with $a_n = \frac{(-1)^{n+1}}{n}$.

Then $(-1)^n a_n = \frac{(-1)^{2n+1}}{n} = \frac{-1}{n}$ so

$\sum_{n=1}^{\infty}(-1)^n a_n = -1 - \frac{1}{2} - \frac{1}{3} - \cdots$ which diverges.

21. False. The difference is less than or equal to 0.01.

23. True. The Maclaurin series exists since $f''(0)$ exists.

25. True. Any Maclaurin polynomial for $\cos x$ involves only even powers of x.

27. False. Newton's Method fails to converge because $f'(0)$ is undefined.

29. False. If the radius of convergence is 2, then the convergence at $x = 2$ is independent of the convergence at $x = -2$.

31. False. The convergence set of a power series may consist of a single point.

33. True. On $(-1, 1)$, $f(x) = \frac{1}{1-x}$.

$$f'(x) = \frac{1}{(1-x)^2} = [f(x)]^2$$

Sample Test Problems

1. $\lim_{n\to\infty} \frac{9n}{\sqrt{9n^2+1}} = \lim_{n\to\infty} \frac{9}{\sqrt{9+\frac{1}{n^2}}} = 3$

The sequence converges to 3.

3. $\lim_{n\to\infty}\left(1+\frac{4}{n}\right)^n = \lim_{n\to\infty}\left(\left(1+\frac{4}{n}\right)^{n/4}\right)^4 = e^4$

The sequence converges to e^4.

5. Note that $a_1 = 3.5(0.1)^2$, $a_2 = (3.5)^3(0.1)^4$, $a_3 = (3.5)^7(0.1)^8$, ..., so

$$a_n = (3.5)^{2^n-1}(0.1)^{2^n} = \frac{1}{3.5}(0.35)^{2^n}.$$

$$\lim_{n\to\infty} \frac{1}{3.5}(0.35)^{2^n} = 0$$

The sequence converges to 0.

7. $a_n \ge 0$; $\lim_{n\to\infty} \frac{\sin^2 n}{\sqrt{n}} \le \lim_{n\to\infty} \frac{1}{\sqrt{n}} = 0$

The sequence converges to 0.

9. $\sum_{k=1}^{\infty} \cos k\pi = -1+1-1+1-1+\cdots$

The series diverges since the partial sums oscillate between –1 and 0.

11. $\sum_{k=0}^{\infty} \frac{3}{2^k} = \sum_{k=0}^{\infty} 3\left(\frac{1}{2}\right)^k = \frac{3}{1-\frac{1}{2}} = 6$

$\sum_{k=0}^{\infty} \frac{4}{3^k} = \sum_{k=0}^{\infty} 4\left(\frac{1}{3}\right)^k = \frac{4}{1-\frac{1}{3}} = 6$

Since both series converge, the sum also converges.

$$\sum_{k=0}^{\infty}\left(\frac{3}{2^k}+\frac{4}{3^k}\right) = 12$$

13. $\sum_{k=1}^{\infty}\left(\frac{1}{\ln 2}\right)^k$ diverges since $\left|\frac{1}{\ln 2}\right| > 1.$

15. $e^x = 1+\frac{x}{1!}+\frac{x^2}{2!}+\frac{x^3}{3!}+\cdots$, so

$1-\frac{1}{1!}+\frac{1}{2!}-\frac{1}{3!}+\cdots = e^{-1} \approx 0.3679.$

17. Numerical methods indicate that the series converges to 10.112.

19. Since the series alternates, $\frac{1}{\sqrt[3]{n}} > \frac{1}{\sqrt[3]{n+1}} > 0$, and

$\lim_{n\to\infty} \frac{1}{\sqrt[3]{n}} = 0$, the series converges. Numerical methods indicate that the series converges to 0.57175.

21. $\sum_{n=1}^{\infty} \frac{2^n+3^n}{4^n} = \sum_{n=1}^{\infty}\left(\left(\frac{1}{2}\right)^n+\left(\frac{3}{4}\right)^n\right)$

$= \left(\frac{1}{1-\frac{1}{2}}-1\right)+\left(\frac{1}{1-\frac{3}{4}}-1\right) = 1+3 = 4$

The series converges to 4. The 1's must be subtracted since the index starts with $n = 1$.

23. $\lim_{n\to\infty} \frac{n+1}{10n+12} = \frac{1}{10} \ne 0$, so the series diverges.

25. $p = \lim_{n\to\infty}\left|\frac{(n+1)^2}{(n+1)!} \div \frac{n^2}{n!}\right| = \lim_{n\to\infty}\left|\frac{n+1}{n^2}\right| = 0$, so the series converges. Numerical methods indicate that the series converges to 5.4366.

27. $p = \lim_{n\to\infty}\left|\dfrac{(n+1)^2\left(\frac{2}{3}\right)^{n+1}}{n^2\left(\frac{2}{3}\right)^n}\right| = \lim_{n\to\infty}\left|\dfrac{2}{3}\right|\left|\dfrac{(n+1)^2}{n^2}\right| = \dfrac{2}{3} < 1$, so the series converges. Numerical methods indicate that the series converges to 30.

29. **a.** $xe^x \approx x + x^2 + \dfrac{x^3}{2!} + \dfrac{x^4}{3!}$

$0.1e^{0.1} \approx 0.11052$

b. $\cosh x \approx 1 + \dfrac{x^2}{2!} + \dfrac{x^4}{4!}$

$\cosh 0.1 \approx 1.005$

31. From Problem 30, we get that $g(x) = 3 + 9(x-2) + 4(x-2)^2 + (x-2)^3$. Thus,

$g(2.1) = 3 + 9(0.1) + 4(0.1)^2 + (0.1)^3 = 3.941.$

33. If $f(x) = \dfrac{1}{x+1}$, then $f^{(5)}(x) = -\dfrac{120}{(x+1)^6}$ and $R_4(x) = -\dfrac{(x-1)^5}{(c+1)^6}$. Thus, $R_4(1.2) = -\dfrac{0.00032}{(c+1)^6}$ for some c between 1 and 1.2.

$|R_4(1.2)| < 0.000005$

35. Using Problem 34,

$$\int_{0.8}^{1.2}\ln x\,dx \approx \int_{0.8}^{1.2}\left[(x-1) - \frac{1}{2}(x-1)^2 + \frac{1}{3}(x-1)^3 - \frac{1}{4}(x-1)^4\right]dx$$

$$= \left[\frac{1}{2}(x-1)^2 - \frac{1}{6}(x-1)^3 + \frac{1}{12}(x-1)^4 - \frac{1}{20}(x-1)^5\right]_{0.8}^{1.2}$$

$$= -\frac{1}{3}(0.2)^3 - \frac{1}{10}(0.2)^5 \approx -0.00269867.$$

$|R_4(x)| \le 0.00015625$ for $0.8 \le x \le 1.2$ so the error involved in the integral is not greater than

$\int_{0.8}^{1.2} 0.00015625dx = [0.00015625x]_{0.8}^{1.2} = 0.0000625.$

37. Use $x_1 = \dfrac{11\pi}{8}$ to start

n	x_n
1	$\dfrac{11\pi}{8}$
2	4.64661795
3	4.60091050
4	4.54662258
5	4.50658016
6	4.49422443
7	4.49341259
8	4.49340946

The root is at $x = 4.49341$.

39. The Absolute Ratio Test gives

$$p = \lim_{n\to\infty}\left|\frac{(-2)^{n+2}x^{n+1}}{2n+5} \div \frac{(-2)^{n+1}x^{n}}{2n+3}\right| = \lim_{n\to\infty}|2x|\left|\frac{2n+3}{2n+5}\right| = |2x|.$$

The largest open interval of convergence is $\left(-\frac{1}{2}, \frac{1}{2}\right)$.

41. The Absolute Ratio Test gives

$$p = \lim_{n\to\infty}\left|\frac{3^{n+1}x^{3n+3}}{(3n+3)!} \div \frac{3^{n}x^{3n}}{(3n)!}\right| = \lim_{n\to\infty}|3x^3|\left|\frac{1}{(3n+3)(3n+2)(3n+1)}\right| = 0.$$

The series converges for all x.

43. The Absolute Ratio Test gives

$$p = \lim_{n\to\infty}\left|\frac{(n+1)!(x+1)^{n+1}}{3^{n+1}} \div \frac{n!(x+1)^{n}}{3^{n}}\right| = \lim_{n\to\infty}\left|\frac{x+1}{3}\right||n+1| = \infty \text{ unless } x = -1.$$

There is no open interval of convergence.

45. $\frac{1}{1+x} = 1 - x + x^2 - x^3 + \cdots$ on the interval (–1, 1). If $f(x) = \frac{1}{1+x}$, then $f''(x) = \frac{2}{(1+x)^3}$. Differentiating the series for $\frac{1}{1+x}$ twice and dividing by $\frac{1}{2}$ gives

$$\frac{1}{(1+x)^3} = 1 - 3x + \frac{1}{2}(4\cdot 3)x^2 - \frac{1}{2}(5\cdot 4)x^3 + \cdots = \sum_{n=0}^{\infty}\frac{1}{2}(-1)^n(n+1)(n+2)x^n.$$

The interval of convergence is (–1, 1).

47. If $f(x) = e^x$, then $f^{(n)}(x) = e^x$. Thus,

$$e^x = e^2 + e^2(x-2) + \frac{e^2}{2!}(x-2)^2 + \frac{e^3}{3!}(x-2)^3 + \frac{e^4}{4!}(x-2)^4 + \cdots.$$

49. $\cos x^2 = 1 - \frac{x^4}{2!} + \frac{x^8}{4!} - \frac{x^{12}}{6!} + \frac{x^{16}}{8!} - \cdots$

Thus $\int_0^1 \cos x^2\,dx = \left[x - \frac{x^5}{5\cdot 2!} + \frac{x^9}{9\cdot 4!} - \frac{x^{13}}{13\cdot 6!} + \frac{x^{17}}{17\cdot 8!} - \cdots\right]_0^1 \approx 0.9045.$

Four terms are required to compute this value correct to four decimal places.

51. One million terms are needed to approximate the sum to within 0.001 since $\frac{1}{\sqrt{n+1}} < 0.001$ is equivalent to $999{,}999 < n$.

53. **a.** From the Maclaurin series for $\frac{1}{1-x}$, we have

$$\frac{1}{1-x^3} = 1 + x^3 + x^6 + x^9 + \cdots.$$

b. In Example 8 it is shown that $\sqrt{1+x} = 1 + \frac{1}{2}x - \frac{1}{8}x^2 + \frac{1}{16}x^3 - \frac{5}{128}x^4 + \cdots$ so

$$\sqrt{1+x^2} = 1 + \frac{1}{2}x^2 - \frac{1}{8}x^4 + \cdots.$$

c. $e^{-x} = 1 - x + \frac{x^2}{2!} - \frac{x^3}{3!} + \frac{x^4}{4!} - \frac{x^5}{5!} + \cdots$, so

$e^{-x} - 1 + x = \frac{x^2}{2!} - \frac{x^3}{3!} + \frac{x^4}{4!} - \cdots$.

d. Using division with the Maclaurin series for cos *x*, we get $\sec x = 1 + \frac{x^2}{2} + \frac{5x^4}{4!} + \frac{61x^6}{6!} + \cdots$.

Thus, $x \sec x = x + \frac{x^3}{2} + \frac{5x^5}{4!} + \cdots$.

e. $e^{-x} \sin x = \left(1 - x + \frac{x^2}{2!} - \frac{x^3}{3!} + \cdots\right)\left(x - \frac{x^3}{3!} + \frac{x^5}{5!} - \frac{x^7}{7!} + \cdots\right)$

$= x - x^2 + \frac{x^3}{3} - \cdots$

f. $1 + \sin x = 1 + x - \frac{x^3}{3!} + \frac{x^5}{5!} - \frac{x^7}{7!} + \cdots$

Using division, we find that

$\frac{1}{1 + \sin x} = 1 - x + x^2 - \cdots$.

55. $\int_0^1 e^{x^3}\, dx \approx 1.342190101$

This gives approximately 3 digits of accuracy.

Chapter 10

Section 10.1

Concepts Review

1. $\frac{x^2}{a^2}+\frac{y^2}{b^2}=1$

3. To the other focus

Problem Set 10.1

1. Vertical ellipse
$a=4,\ b=3,\ c=\sqrt{16-9}=\sqrt{7}\approx 2.65$
Foci are at (0, ±2.65) and vertices are at (0, ±4).

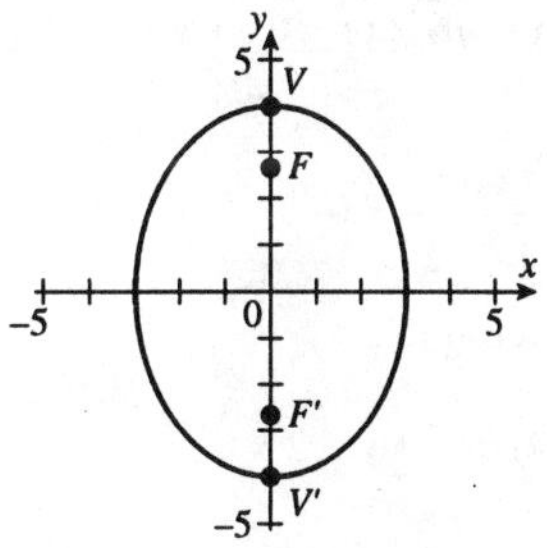

3. Vertical hyperbola; $a=3,\ b=4,\ c=\sqrt{16+9}=5$
Foci are at (0, ±5) and vertices are at (0, ±3).
Asymptotes are $y=\pm\frac{3}{4}x$.

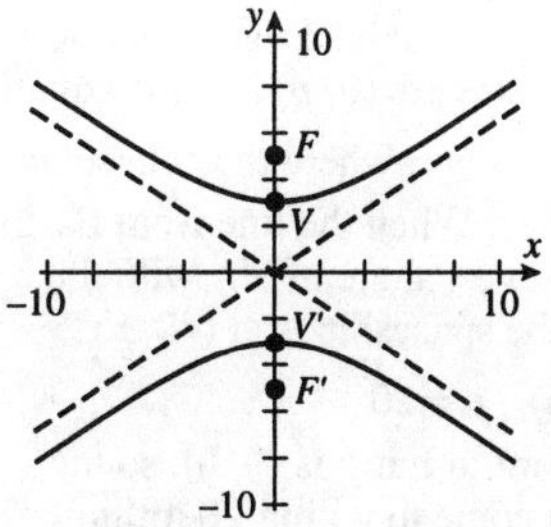

5. Horizontal parabola; $y^2=4(4)x,\ \ p=4$
Focus is at (0, 4) and directrix is at $x=-4$.

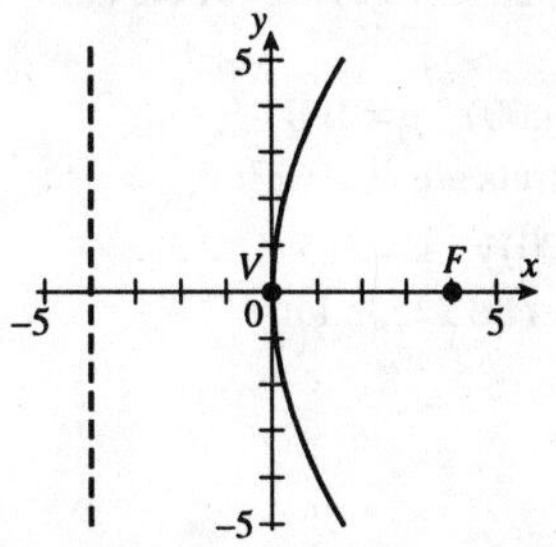

7. $\frac{x^2}{36}+\frac{y^2}{9}=1$; horizontal ellipse
$a=6,\ b=3,\ c=\sqrt{36-9}=\sqrt{27}\approx 5.20$
Foci are at (±5.20, 0) and vertices are at (±6, 0).

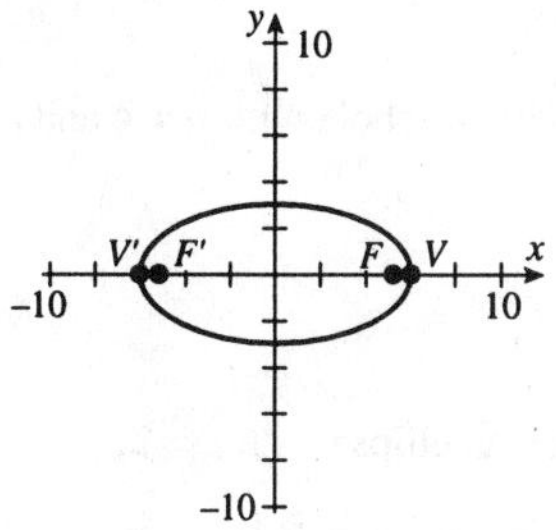

9. $x^2=6y$; vertical parabola; $x^2=4\left(\frac{3}{2}\right)y;\ p=\frac{3}{2}$
Focus is at $\left(0,\ \frac{3}{2}\right)$ and directrix is $y=-\frac{3}{2}$.

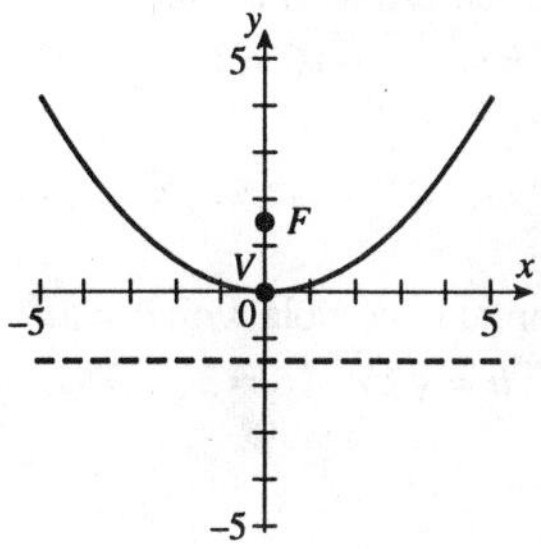

11. $\frac{x^2}{25}-\frac{y^2}{4}=1$; horizontal hyperbola
$a=5,\ b=2,\ c=\sqrt{25+4}=\sqrt{29}\approx 5.39$
Foci are at (±5.39, 0) and vertices are at (±5, 0).
Asymptotes are $y=\pm\frac{2}{5}x$.

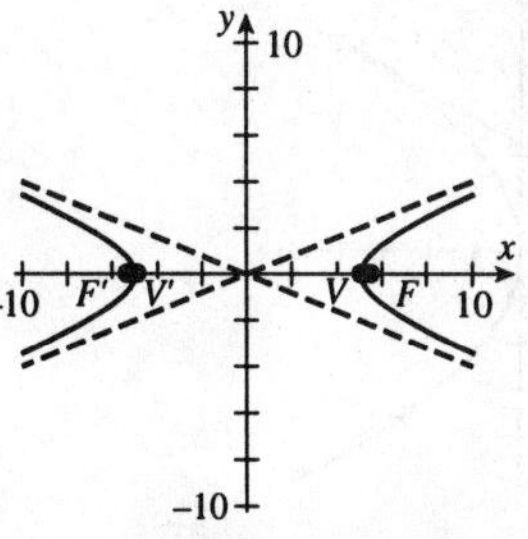

13. This is a horizontal parabola that opens to the right with $p=3$.
$y^2=12x$

15. This is a vertical parabola that opens upward with $p = 4$.

$x^2 = 16y$

17. This is a horizontal ellipse with $a = 6$ and $c = 3$.

$b = \sqrt{36-9} = \sqrt{27}$

$\frac{x^2}{36} + \frac{y^2}{27} = 1$

19. This is a vertical hyperbola with $a = 4$ and $c = 5$.

$b = \sqrt{25-16} = 3$

$\frac{y^2}{16} - \frac{x^2}{9} = 1$

21. This is a horizontal ellipse with $c = 2$.

$8 = \frac{a}{e},\ 8 = \frac{a}{\frac{c}{a}},$ so $a^2 = 8c = 16$.

$b = \sqrt{16-4} = \sqrt{12}$

$\frac{x^2}{16} + \frac{y^2}{12} = 1$

23. This is a vertical ellipse with $c = 4$.

$2a = 10,\ a = 5,\ b = \sqrt{25-16} = 3$

$\frac{x^2}{9} + \frac{y^2}{25} = 1$

25. This is a horizontal hyperbola with $c = 5$.

$2a = 8,\ a = 4,\ b = \sqrt{25-16} = 3$

$\frac{x^2}{16} - \frac{y^2}{9} = 1$

27. $y^2 = 5x;\ 2yy' = 5$

$2y\left(\frac{\sqrt{5}}{4}\right) = 5,\ y = 2\sqrt{5}$

$\left(2\sqrt{5}\right)^2 = 5x,\ x = 4$

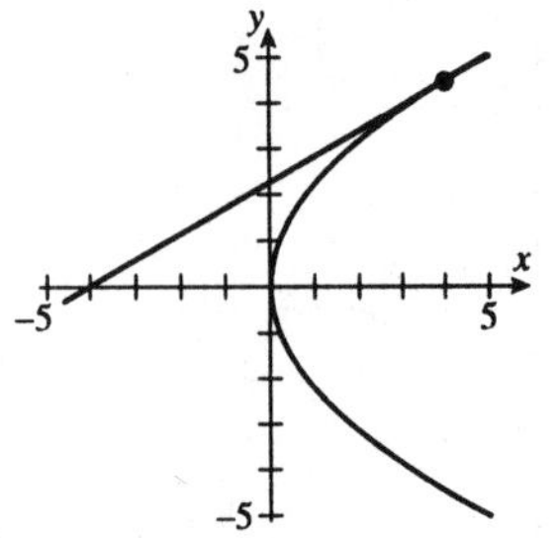

29. The slope of the line is $\frac{3}{2}$.

$y^2 = -18x;\ 2yy' = -18$

$2y\left(\frac{3}{2}\right) = -18;\ y = -6$

$(-6)^2 = -18x;\ x = -2$

The equation of the tangent line is

$y + 6 = \frac{3}{2}(x+2)$ or $3x - 2y - 6 = 0$.

31. $\frac{x^2}{9} + \frac{y^2}{27} = 1;\ \frac{2x}{9} + \frac{2yy'}{27} = 0$

$y' = -\frac{3x}{y};\ y' = -\sqrt{6}$ at $\left(\sqrt{6},\ 3\right)$.

$y - 3 = \left(-\sqrt{6}\left(x-\sqrt{6}\right)\right)$ or $\sqrt{6}x + y - 9 = 0$

33. $\frac{x^2}{8} - \frac{y^2}{4} = 1;\ \frac{x}{4} - \frac{yy'}{2} = 0$

$y' = \frac{x}{2y};\ y' = -1$ at $(-4, 2)$

$y - 2 = -1(x+4)$ or $x + y + 2 = 0$

35. $x^2 + y^2 = 25;\ 2x + 2yy' = 0$

$y' = -\frac{x}{y};\ y' = -\frac{3}{4}$ at $(3, 4)$

$y - 4 = -\frac{3}{4}(x-3)$ or $3x + 4y - 25 = 0$

37. Let the y-axis be the axis of the parabola, so Earth's coordinates are $(0, p)$ and the equation of the path is $x^2 = 4py$, where the coordinates are in millions of miles. When the line from Earth to the spaceship makes an angle of 90° with the axis of the parabola, the spaceship is at $(40, p)$.

$(40)^2 = 4p(p),\ p = 20$

The closest point to Earth is (0, 0), so the spaceship will come to within 20 million miles of Earth.

39. $x^2 = 4py$

The cables are attached to the towers at (±400, 400).

$(400)^2 = 4p(400),\ p = 100$

The vertical struts are at $x = \pm 300$.

$(300)^2 = 4(100)y,\ y = 225$

The struts must be 225 m long.

41. The slope of the line is $\frac{1}{\sqrt{2}}$.

$x^2+2y^2-2=0;\ 2x+4yy'=0$

$y'=-\frac{x}{2y};\ \frac{1}{\sqrt{2}}=-\frac{x}{2y};\ x=-\sqrt{2}y$

Substitute $x=-2\sqrt{y}$ into the equation of the ellipse.

$2y^2+2y^2-0;\ y=\pm\frac{1}{\sqrt{2}}$

The tangent lines are tangent at $\left(-1, \frac{1}{\sqrt{2}}\right)$ and $\left(1, -\frac{1}{\sqrt{2}}\right)$. The equations of the tangent lines are

$y-\frac{1}{\sqrt{2}}=\frac{1}{\sqrt{2}}(x+1)$ and $y+\frac{1}{\sqrt{2}}=\frac{1}{\sqrt{2}}(x-1)$ or

$x-\sqrt{2}y+2=0$ and $x-\sqrt{2}\,y-2=0$.

43. $x=\pm a\sqrt{1-\frac{y^2}{b^2}}$

$V=2\cdot\pi\int_0^b a^2\left(1-\frac{y^2}{b^2}\right)dy=2\pi a^2\left(y-\frac{y^3}{3b^2}\right)\Big]_0^b$

$=\frac{4\pi a^2 b}{3}$

45. $x=\pm a\sqrt{1-\frac{y^2}{b^2}}$

$A=4xy=4y\sqrt{1-\frac{y^2}{b^2}}=4\sqrt{y^2-\frac{y^4}{b^2}}$

$\frac{dA}{dy}=\frac{2\left(2y-\frac{4y^3}{b^2}\right)}{\sqrt{y^2-\frac{y^4}{b^2}}}=0$

$y-\frac{2y^3}{b^2}=0$

$y\left(1-\frac{2y^2}{b^2}\right)=0$

$y=0$ or $y=\pm\frac{b}{\sqrt{2}}$

The Second Derivative Test shows that $y=\frac{b}{\sqrt{2}}$ is a maximum.

$x=a\sqrt{1-\frac{\left(\frac{b}{\sqrt{2}}\right)^2}{b^2}}=\frac{a}{\sqrt{2}}$

Therefore, the rectangle is $a\sqrt{2}$ by $b\sqrt{2}$.

47. Let the y-axis run through the center of the arch and the x-axis lie on the floor. Thus $a=5$ and $b=4$ and the equation of the arch is $\frac{x^2}{25}+\frac{y^2}{16}=1$.

When $y=2$, $\frac{x^2}{25}+\frac{(2)^2}{16}=1$, so $x=\pm\frac{5\sqrt{3}}{2}$.

The width of the box can at most be $5\sqrt{3}\approx 8.66$ ft.

49. $a=18.09, b=4.56,$

$c=\sqrt{(18.09)^2-(4.56)^2}\approx 17.51$

The comet's closest approach is $18.09-17.51\approx 0.58$ AU

51. $a=4583,\ b=4132,$

$c=\sqrt{(4583)^2-(4132)^2}\approx 1982.54$

$e=\frac{c}{a}\approx 0.43$

53. Let (x, y) be the coordinates of P as the ladder slides. Using a property of similar triangles, $\frac{x}{a}=\frac{\sqrt{b^2-y^2}}{b}$.

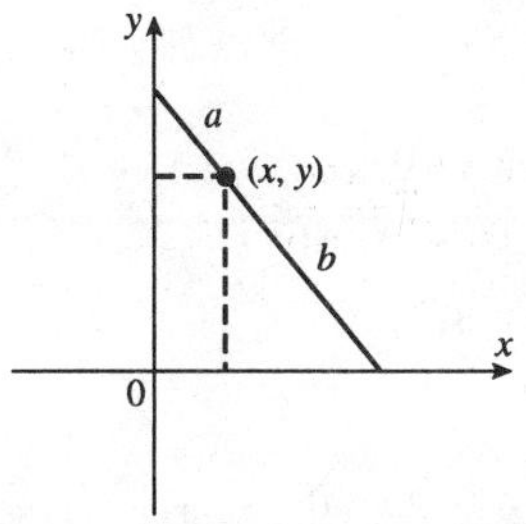

Square both sides to get

$\frac{x^2}{a^2}=\frac{b^2-y^2}{b^2}$ or $b^2x^2+a^2y^2=a^2b^2$.

55. If the original path is not along the major axis, the ultimate path will approach the major axis.

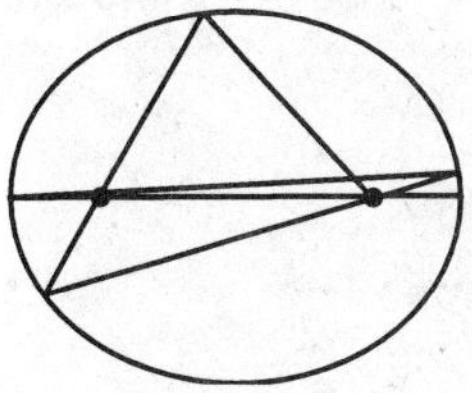

57. Possible answer: Attach one end of a string to F and attach one end of another string to F'. Place a spool at a vertex. Tightly wrap both strings in the same direction around the spool. Insert a pencil through the spool. Then trace out a branch of the hyperbola by unspooling the strings while keeping both strings taut.

Section 10.2

Concepts Review

1. $\frac{a^2}{4}$

3. A line, parallel lines, intersecting lines, a point, the empty set

Problem Set 10.2

1. $x^2 + y^2 - 2x + 4y + 4 = 0$
$(x^2 - 2x + 1) + (y^2 + 4y + 4) = -4 + 1 + 4$
$(x-1)^2 + (y+2)^2 = 1$
This is a circle.

3. $4x^2 + 9y^2 - 16x + 72y + 124 = 0$
$4(x^2 - 4x + 4) + 9(y^2 + 8y + 16) = -124 + 16 + 144$
$4(x-2)^2 + 9(y+4)^2 = 36$
This is an ellipse.

5. $4x^2 + 9y^2 - 16x + 72y + 160 = 0$
$4(x^2 - 4x + 4) + 9(y^2 + 8y + 16) = -160 + 16 + 144$
$4(x-2)^2 + 9(y+4)^2 = 0$
This is a point.

7. $y^2 - 10x - 8y - 14 = 0$
$y^2 - 8y + 16 = 10x + 14 + 16$
$(y-4)^2 = 10(x+3)$
This is a parabola.

9. $x^2 + y^2 - 2x + 4y + 20 = 0$
$(x^2 - 2x + 1) + (y^2 + 4y + 4) = -20 + 1 + 4$
$(x-1)^2 + (y+2)^2 = -15$
This is the empty set.

11. $4x^2 - 4y^2 + 16x - 20y - 9 = 0$
$4(x^2 + 4x + 4) - 4\left(y^2 + 5y + \frac{25}{4}\right) = 9 + 16 - 25$
$4(x+2)^2 - 4\left(y + \frac{5}{2}\right)^2 = 0$
Let $u = x + 2$ and $v = y + \frac{5}{2}$.
$4u^2 - 4v^2 = 0$
$(u - v)(u + v) = 0$
This is two intersecting lines.

13. $4x^2 - 16x + 15 = 0$
$(2x - 5)(2x - 3) = 0$
This is two parallel lines.

15. $25x^2 - 4y^2 + 150x - 8y + 129 = 0$
$25(x^2 + 6x + 9) - 4(y^2 + 2y + 1) = -129 + 225 - 4$
$25(x+3)^2 - 4(y+1)^2 = 92$
This is a hyperbola.

17. $\frac{(x+3)^2}{4} + \frac{(y+2)^2}{16} = 1$

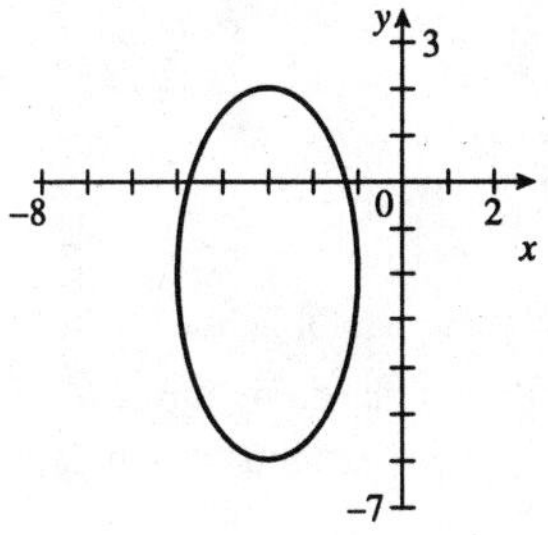

19. $\frac{(x+3)^2}{4} - \frac{(y+2)^2}{16} = 1$

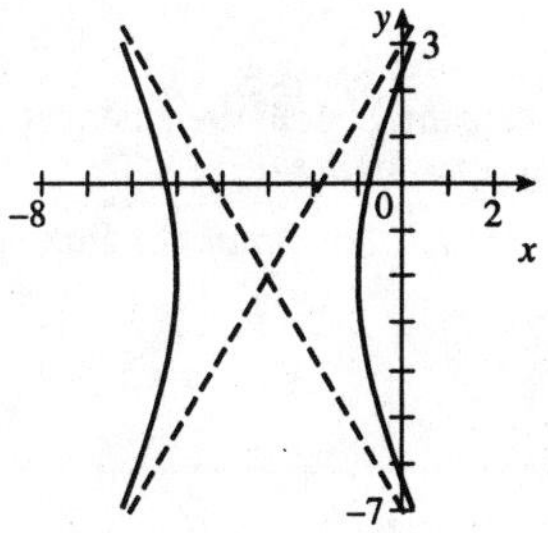

21. $(x+2)^2 = 8(y-1)$

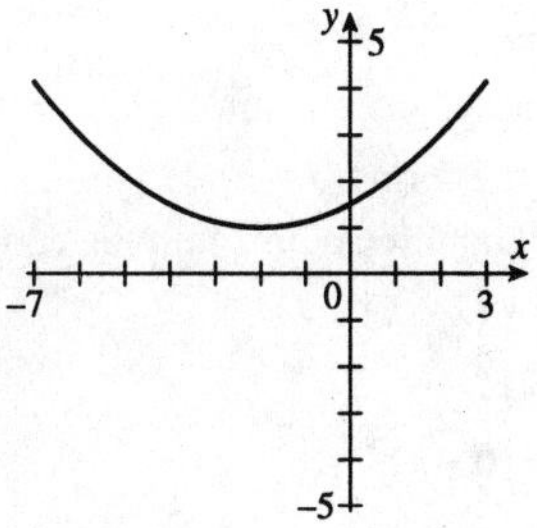

23. $(y-1)^2 = 16$
$y - 1 = \pm 4$
$y = 5,\ y = -3$

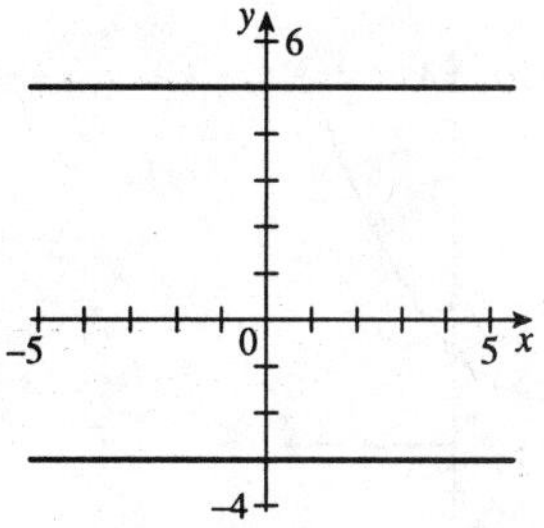

25. $x^2 + 4y^2 - 2x + 16y + 1 = 0$
$(x^2 - 2x + 1) + 4(y^2 + 4y + 4) = -1 + 1 + 16$
$(x-1)^2 + 4(y+2)^2 = 16$
$\frac{(x-1)^2}{16} + \frac{(y+2)^2}{4} = 1$

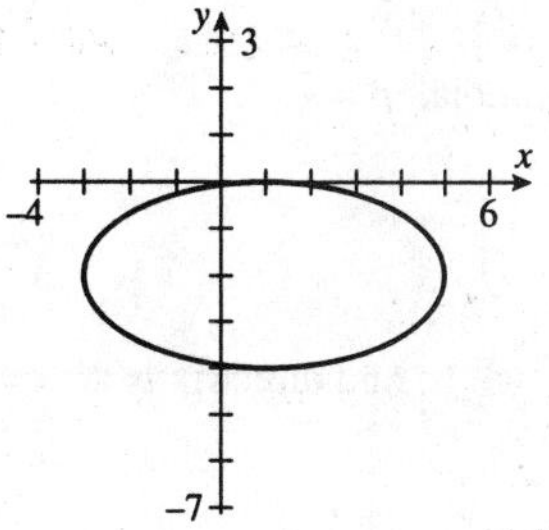

27. $9x^2 - 16y^2 + 54x + 64y - 127 = 0$
$9(x^2 + 6x + 9) - 16(y^2 - 4y + 4) = 127 + 81 - 64$
$9(x+3)^2 - 16(y-2)^2 = 144$
$\frac{(x+3)^2}{16} - \frac{(y-2)^2}{9} = 1$

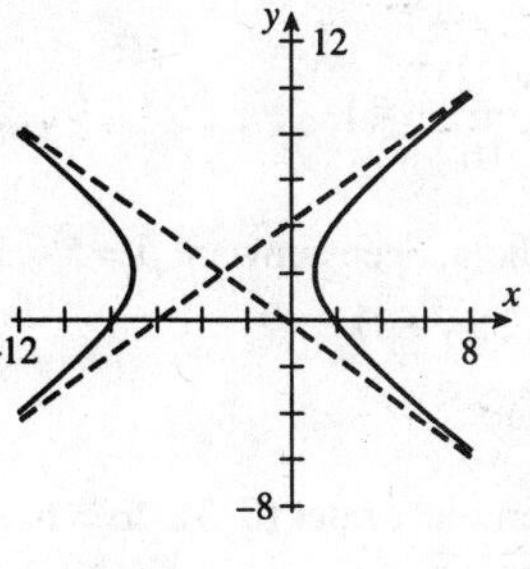

29. $4x^2 + 16x - 16y + 32 = 0$

$4(x^2 + 4x + 4) = 16y - 32 + 16$

$4(x+2)^2 = 16(y-1)$

$(x+2)^2 = 4(y-1)$

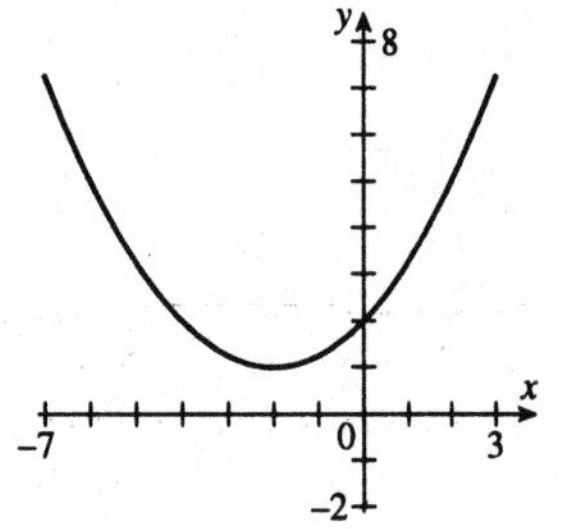

31. $2y^2 - 4y - 10x = 0$

$2(y^2 - 2y + 1) = 10x + 2$

$(y-1)^2 = 5\left(x + \frac{1}{5}\right)$

$(y-1)^2 = 4\left(\frac{5}{4}\right)\left(x + \frac{1}{5}\right)$

Horizontal parabola, $p = \frac{5}{4}$

Vertex $\left(-\frac{1}{5},\ 1\right)$

Focus is at $\left(\frac{21}{20},\ 1\right)$ and directrix is at $x = -\frac{29}{20}$.

33. $16(x-1)^2 + 25(y+2)^2 = 400$

$\frac{(x-1)^2}{25} + \frac{(y+2)^2}{16} = 1$

Horizontal ellipse, center (1, –2), $a = 5$, $b = 4$,
$c = \sqrt{25 - 16} = 3$
Foci are at (–2, –2) and (4, –2).

35. $a = 5$, $b = 4$

$\frac{(x-5)^2}{25} + \frac{(y-1)^2}{16} = 1$

37. Vertical parabola, opens upward $p = 5 - 3 = 2$

$(x-2)^2 = 4(2)(y-3)$

$(x-2)^2 = 8(y-3)$

39. Vertical hyperbola, center (0, 3), $2a = 6$, $a = 3$,
$c = 5$, $b = \sqrt{25 - 9} = 4$

$\frac{(y-3)^2}{9} - \frac{x^2}{16} = 1$

41. Horizontal parabola, opens to the left

Vertex (6, 5), $p = \frac{10-2}{2} = 4$

$(y-5)^2 = -4(4)(x-6)$

$(y-5)^2 = -16(x-6)$

43. Horizontal ellipse, center (0, 2), $c = 2$
Since it passes through the origin and center is at (0, 2), $b = 2$.

$a = \sqrt{2^2 + 2^2} = \sqrt{8}$

$\frac{x^2}{8} + \frac{(y-2)^2}{4} = 1$

45. a. If C is a vertical parabola, the equation for C can be written in the form $y = ax^2 + bx + c$.
Substitute the three points into the equation.
$2 = a - b + c$
$0 = c$
$6 = 9a + 3b + c$
Solve the system to get $a = 1$, $b = -1$, $c = 0$.

$y = x^2 - x$

b. If C is a horizontal parabola, an equation for C can be written in the form
$x = ay^2 + by + c$. Substitute the three points into the equation.
$-1 = 4a + 2b + c$
$0 = c$
$3 = 36a + 6b + c$

Solve the system to get $a = \frac{1}{4}$, $b = -1$, $c = 0$.

$x = \frac{1}{4}y^2 - y$

c. If C is a circle, an equation for C can be written in the form $(x-h)^2 + (y-k)^2 = r^2$.
Substitute the three points into the equation.

$(-1-h)^2 + (2-k)^2 = r^2$

$h^2 + k^2 = r^2$

$(3-h)^2 + (6-k)^2 = r^2$

Solve the system to get $h = \frac{5}{2}$, $k = \frac{5}{2}$, and $r^2 = \frac{25}{2}$.

$\left(x - \frac{5}{2}\right)^2 + \left(y - \frac{5}{2}\right)^2 = \frac{25}{2}$

47. Parabola: horizontal parabola, opens to the right,
$p = c - a$
$y^2 = 4(c-a)(x-a)$
Hyperbola: horizontal hyperbola, $b^2 = c^2 - a^2$
$\frac{x^2}{a^2} - \frac{y^2}{b^2} = 1$
$\frac{y^2}{b^2} = \frac{x^2}{a^2} - 1$
$y^2 = \frac{b^2}{a^2}(x^2 - a^2)$
Now show that y^2 (hyperbola) is greater than y^2 (parabola).
$\frac{b^2}{a^2}(x^2 - a^2) = \frac{c^2 - a^2}{a^2}(x^2 - a^2) = \frac{(c+a)(c-a)}{a^2}(x+a)(x-a)$
$= \frac{(c+a)(x+a)}{a^2}(c-a)(x-a) > \frac{(2a)(2a)}{a^2}(c-a)(x-a) = 4(c-a)(x-a)$
$c + a > 2a$ and $x + a > 2a$ since $c > a$ and $x > a$ except at the vertex.

Section 10.3

Concepts Review

1. Infinitely many

3. Circle; line

Problem Set 10.3

1.

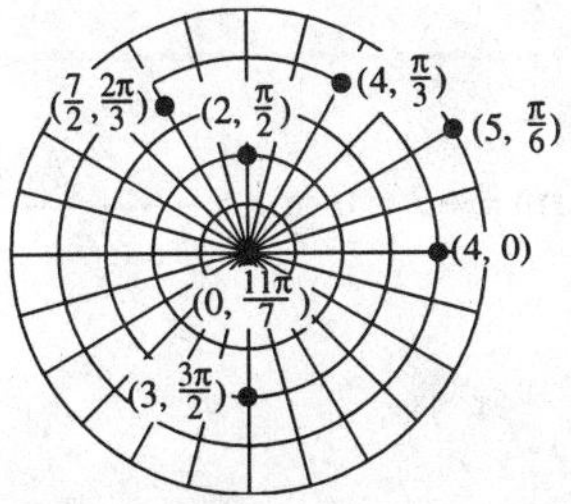

3.

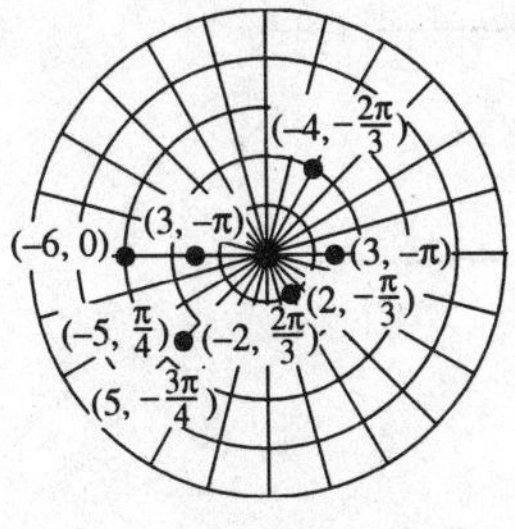

5.

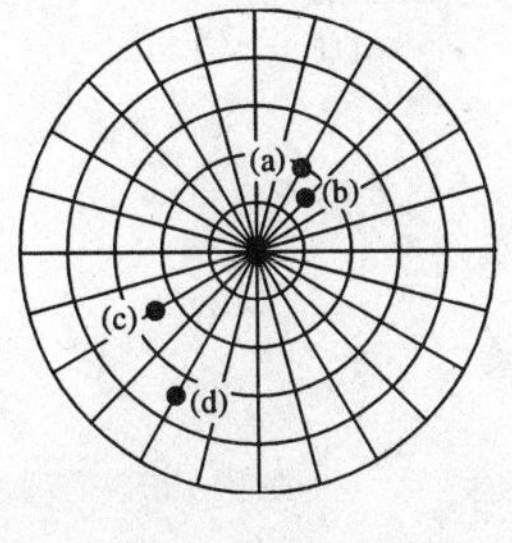

a. Possible answers: $\left(4, -\frac{5\pi}{3}\right), \left(4, \frac{7\pi}{3}\right), \left(-4, -\frac{2\pi}{3}\right), \left(-4, \frac{4\pi}{3}\right)$

b. Possible answers: $\left(3, \frac{\pi}{4}\right), \left(3, \frac{9\pi}{4}\right), \left(-3, -\frac{3\pi}{4}\right), \left(-3, \frac{13\pi}{4}\right)$

c. Possible answers: $\left(5, -\frac{5\pi}{6}\right), \left(5, \frac{7\pi}{6}\right), \left(-5, \frac{-11\pi}{6}\right), \left(-5, \frac{13\pi}{6}\right)$

d. Possible answers: $\left(7, \frac{4\pi}{3}\right), \left(7, \frac{10\pi}{3}\right), \left(-7, \frac{\pi}{3}\right), \left(-7, \frac{7\pi}{3}\right)$

7. a. $x = 4\cos\frac{\pi}{3} = 2,\ y = 4\sin\frac{\pi}{3} = 2\sqrt{3}$

$\left(2, 2\sqrt{3}\right)$

b. $x = -3\cos\frac{5\pi}{4} = \frac{3\sqrt{2}}{2};\ y = -3\sin\frac{5\pi}{4} = \frac{3\sqrt{2}}{2}$

$\left(\frac{3\sqrt{2}}{2}, \frac{3\sqrt{2}}{2}\right)$

c. $x = -5\cos\frac{\pi}{6} = \frac{-5\sqrt{3}}{2},\ y = -5\sin\frac{\pi}{6} = -\frac{5}{2}$

$\left(-\frac{5\sqrt{3}}{2}, -\frac{5}{2}\right)$

d. $x = 7\cos\left(-\frac{2}{3}\pi\right) = -\frac{7}{2}$

$y = 7\sin\left(-\frac{2}{3}\pi\right) = -\frac{7\sqrt{3}}{2}$

$\left(-\frac{7}{2}, -\frac{7\sqrt{3}}{2}\right)$

9. a. $r^2 = \left(-2\sqrt{3}\right)^2 + (-2)^2 = 16,\ \tan\theta = \frac{1}{\sqrt{3}}$

Possible answer: $\left(4, \frac{7\pi}{6}\right)$

b. $r^2 = (1)^2 + \left(\sqrt{3}\right)^2 = 4,\ \tan\theta = \frac{1}{\sqrt{3}}$

Possible answer: $\left(2, \frac{\pi}{6}\right)$

c. $r^2 = \left(\sqrt{2}\right)^2 + \left(\sqrt{2}\right)^2 = 4,\ \tan\theta = -1$

Possible answer: $\left(2, -\frac{\pi}{4}\right)$

d. Possible answer: (0, 0)

11. $x - 4y + 2 = 0$

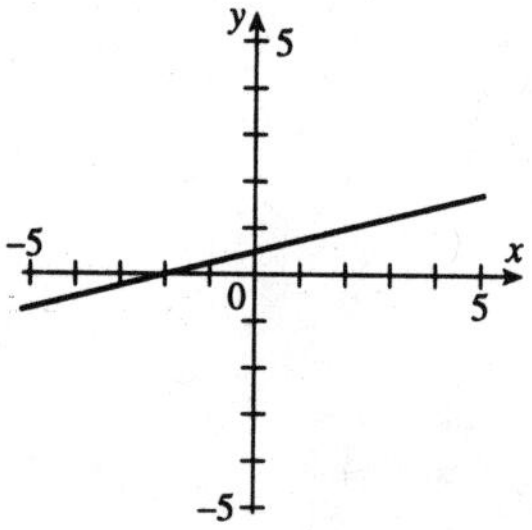

$r\cos\theta - 4r\sin\theta + 2 = 0$ or $r = \frac{2}{4\sin\theta - \cos\theta}$

13. $y = -5$

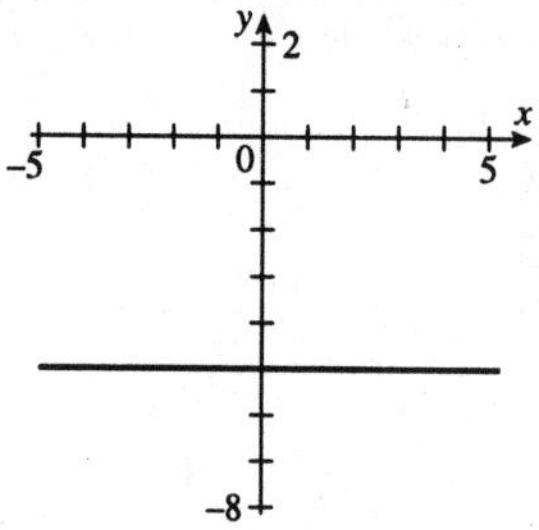

$r\sin\theta = -5$ or $r = -\frac{5}{\sin\theta}$

15. $x^2 + y^2 = 16$

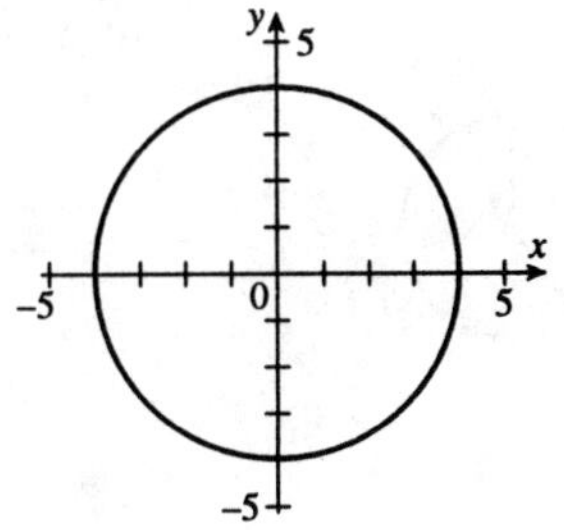

$r^2 = 16$ or $r = 4$

17. $\theta = \frac{\pi}{3}$

$\tan\frac{\pi}{3} = \frac{y}{x}$

$\sqrt{3} = \frac{y}{x}$

$y = \sqrt{3}x$

19. $r\cos\theta + 6 = 0$
$x + 6 = 0$
$x = -6$

21. $r\sin\theta - 4 = 0$
$y - 4 = 0$
$y = 4$

23. $r = 6$, circle

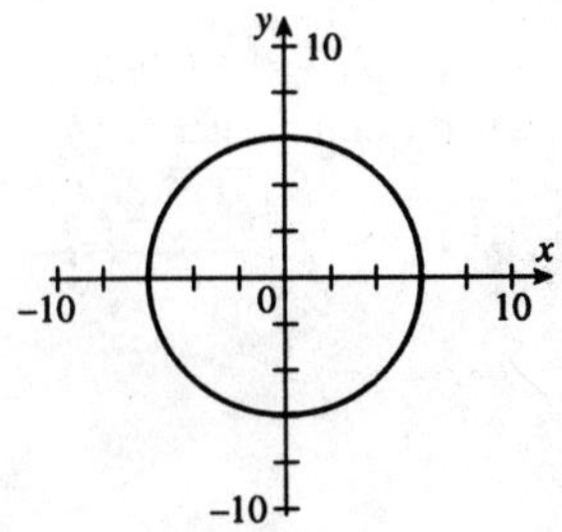

25. $r = \frac{3}{\sin\theta}$

$r = \frac{3}{\cos\left(\theta - \frac{\pi}{2}\right)}$, line

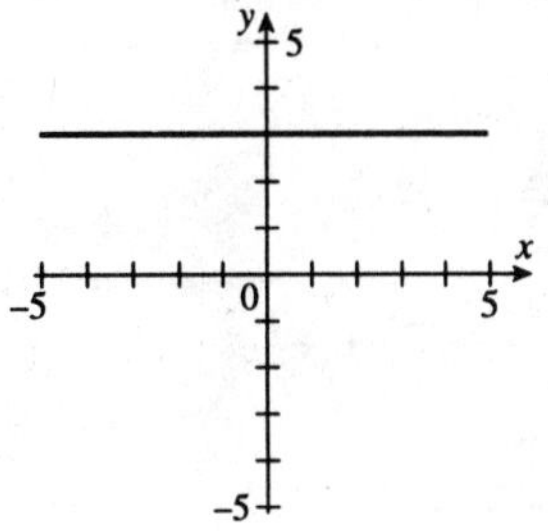

27. $r = 4\sin\theta$

$r = 2(2)\cos\left(\theta - \frac{\pi}{2}\right)$, circle

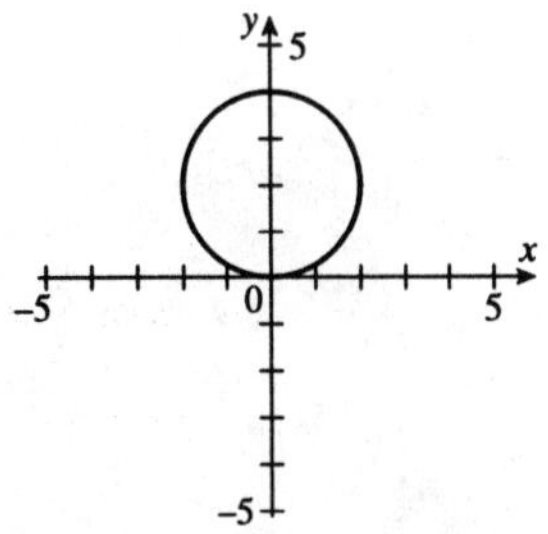

29. $r = \frac{4}{1 + \cos\theta}$

$r = \frac{(1)(4)}{1 + (1)\cos\theta}$, parabola

$e = 1$

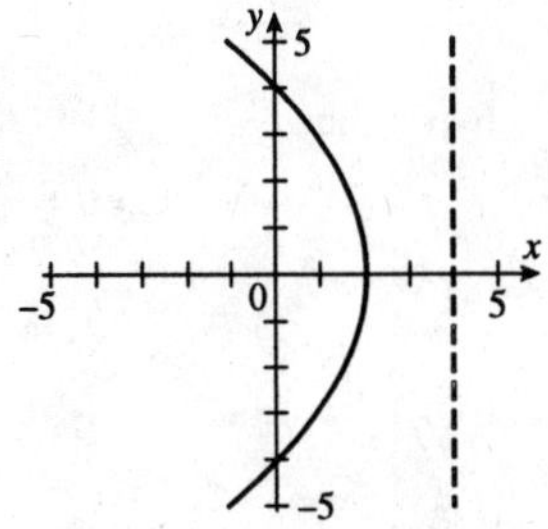

31. $r = \frac{6}{2 + \sin\theta}$

$r = \frac{\left(\frac{1}{2}\right)6}{1 + \left(\frac{1}{2}\right)\cos\left(\theta - \frac{\pi}{2}\right)}$, ellipse

$e = \frac{1}{2}$

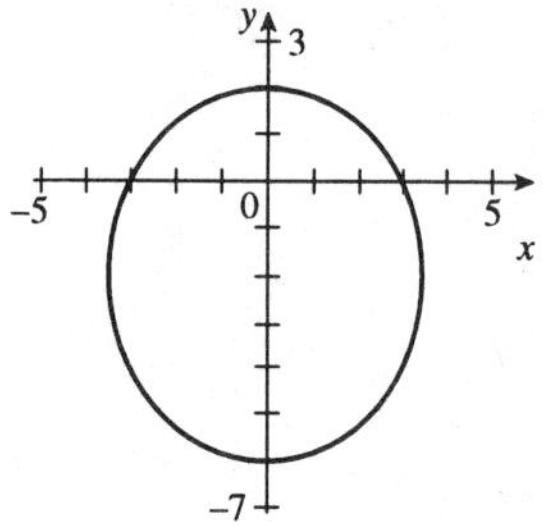

33. $r = \frac{4}{2 + 2\cos\theta}$

$r = \frac{(1)(2)}{1 + (1)\cos\theta}$, parabola

$e = 1$

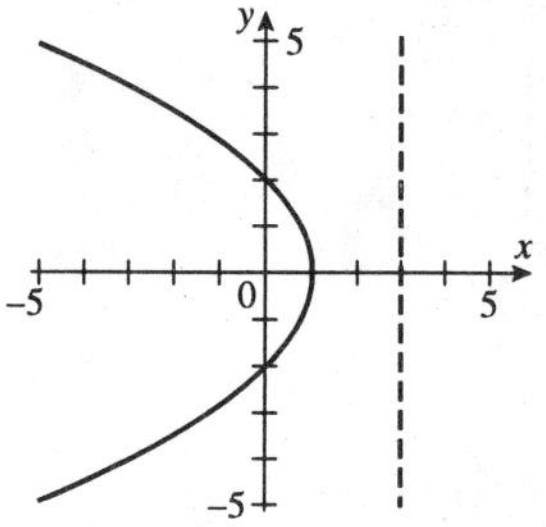

35. $r = \frac{4}{\frac{1}{2} + \cos(\theta - \pi)}$

$r = \frac{(2)(4)}{1 - 2\cos\theta}$, hyperbola

$e = 2$

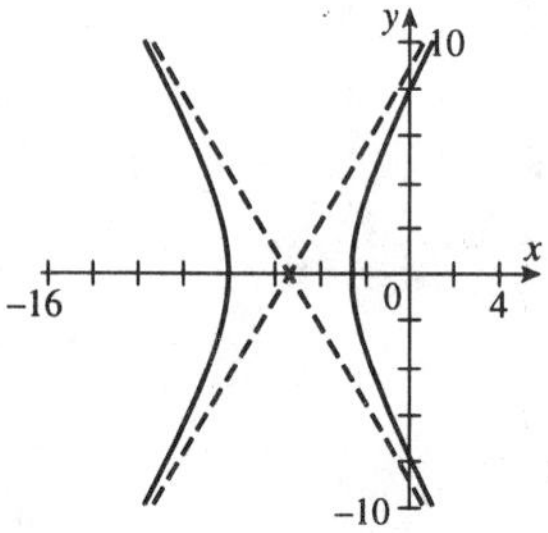

37. By the Law of Cosines,
$a^2 = r^2 + c^2 - 2rc\cos(\theta - \alpha)$ (see figure below).

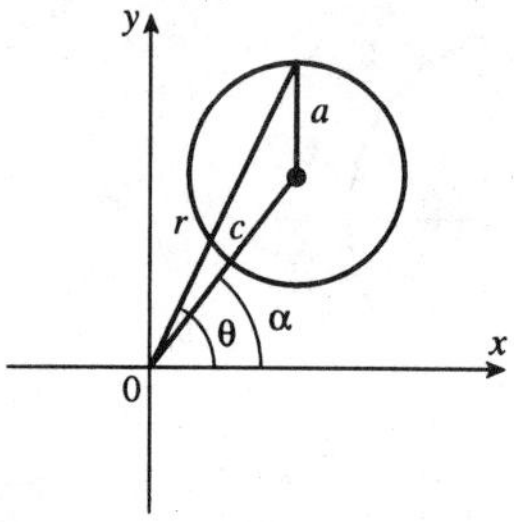

39. a. Since $\cos(\theta - \theta_0)$ is at most 1, the minimum distance is when $\theta = \theta_0$, so $r_1 = \frac{ed}{1+e}$.
Since $\cos(\theta - \theta_0)$ is at least –1, the minimum distance is when $\theta = \theta_0 + \pi$, so $r_2 = \frac{ed}{1-e}$.

b. The endpoints of the major axis are when $\theta = \theta_0$ and $\theta = \theta_0 + \pi$.
Therefore the major diameter
$= r_1 + r_2 = \frac{ed}{1+e} + \frac{ed}{1-e} = \frac{2ed}{1-e^2}$.
Since r_1 is the distance from a focus to the closest vertex and r_2 is the distance to the other vertex, $a = \frac{r_1 + r_2}{2} = \frac{ed}{1-e^2}$ and
$c = \frac{r_2 - r_1}{2} = \frac{e^2 d}{1-e^2}$.
Therefore the major diameter
$= 2b = 2\sqrt{a^2 - c^2} = 2\sqrt{\frac{e^2d^2 - e^4d^2}{(1-e^2)^2}}$
$= \frac{2ed}{\sqrt{1-e^2}}$.

41. $a = \frac{185.8}{2} = 92.9$,
$c = ea = (0.0167)\frac{185.8}{2} = 1.55143$
Perihelion $= a - c \approx 91.3$ million miles

43. a. $4 = \dfrac{d}{1+\cos\left(\frac{\pi}{2}-\theta_0\right)}$

$4+4\left(\cos\dfrac{\pi}{2}\cos\theta_0+\sin\dfrac{\pi}{2}\sin\theta_0\right)=d$

$d = 4+4\sin\theta_0$

$3 = \dfrac{d}{1+\cos\left(\frac{\pi}{4}-\theta_0\right)}$

$3+3\left(\cos\dfrac{\pi}{4}\cos\theta_0+\sin\dfrac{\pi}{4}\sin\theta_0\right)=d$

$d = 3+\dfrac{3\sqrt{2}}{2}\cos\theta_0+\dfrac{3\sqrt{2}}{2}\sin\theta_0$

$4+4\sin\theta_0 = 3+\dfrac{3\sqrt{2}}{2}\cos\theta_0+\dfrac{3\sqrt{2}}{2}\sin\theta_0$

$\dfrac{3\sqrt{2}}{2}\cos\theta_0+\left(\dfrac{3\sqrt{2}}{2}-4\right)\sin\theta_0-1=0$

$3\sqrt{2}\cos\theta_0+\left(3\sqrt{2}-8\right)\sin\theta_0-2=0$

$4.24\cos\theta_0-3.76\sin\theta_0-2=0$

b. A graph shows that a root lies near 0.5. Using Newton's Method, $\theta_0 \approx 0.485$.

c. $d = 4+4\sin\theta_0 \approx 5.86$

The closest the sun gets is $r = \dfrac{d}{1+\cos(\theta_0-\theta_0)} = \dfrac{d}{2} \approx 2.93$ AU

Section 10.4

Concepts Review

1. Limaçon

3. Rose

Problem Set 10.4

1. $\theta^2-1=0$
$\theta=-1,\ \theta=1$

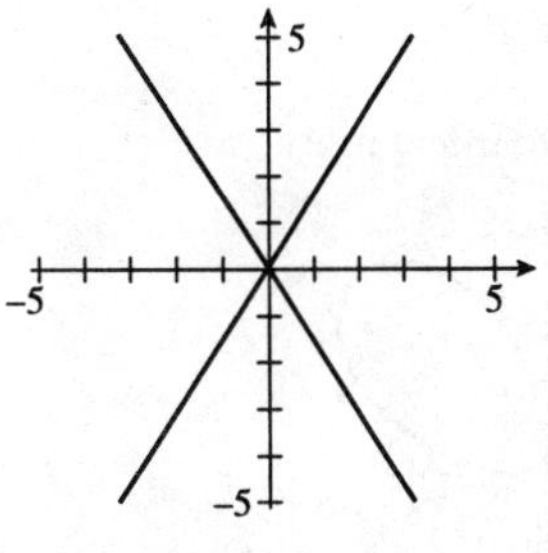

3. $r\sin\theta+6=0$
$y=-6$

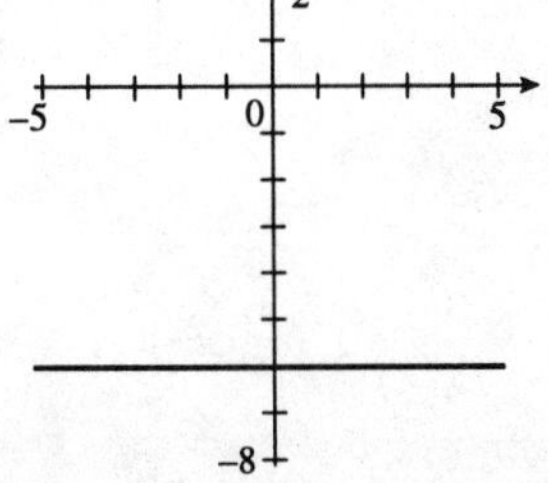

5. $r=6\sin\theta$

Circle, radius 3, center $\left(3, \dfrac{\pi}{2}\right)$

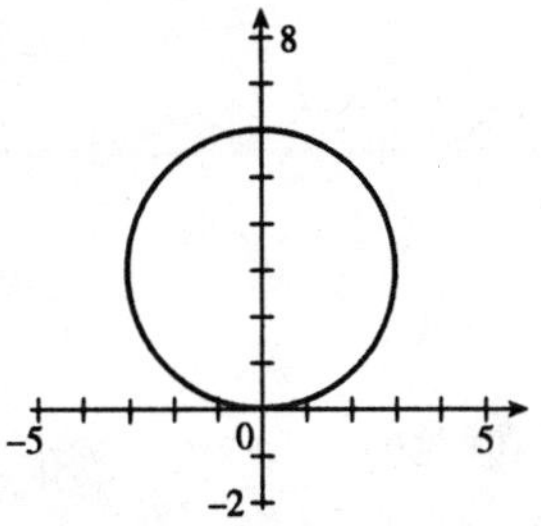

7. $r=\dfrac{4}{1-\cos\theta}$

Parabola

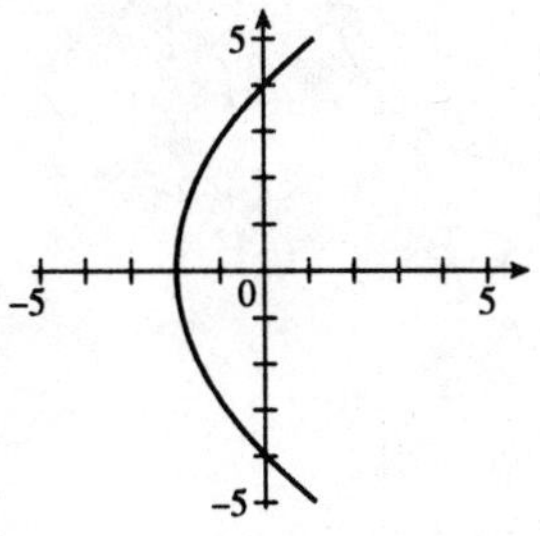

9. $r = 5 - 5\sin\theta$ (cardioid)

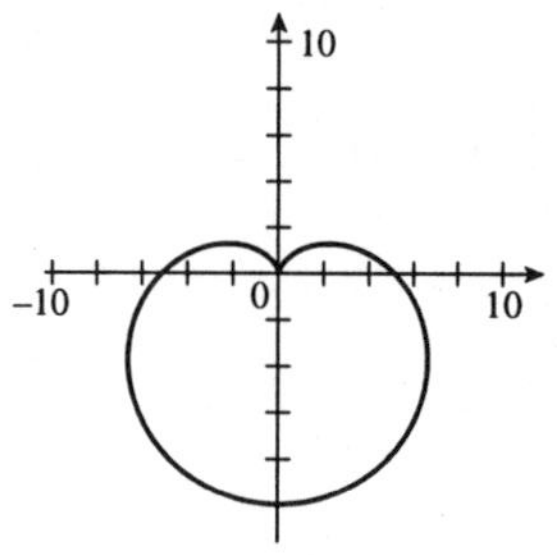

11. $r = 3 - 3\cos\theta$ (cardioid)

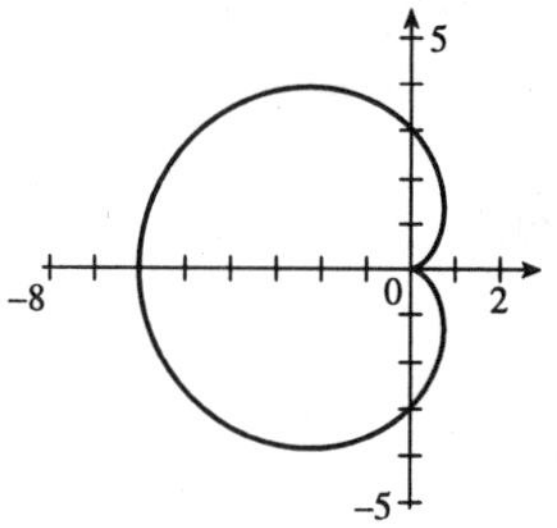

13. $r = 2 - 4\cos\theta$ (limaçon)

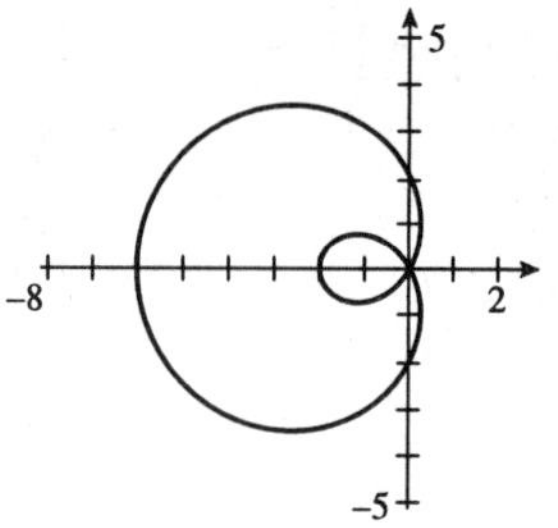

15. $r = 4 - 3\sin\theta$ (limaçon)

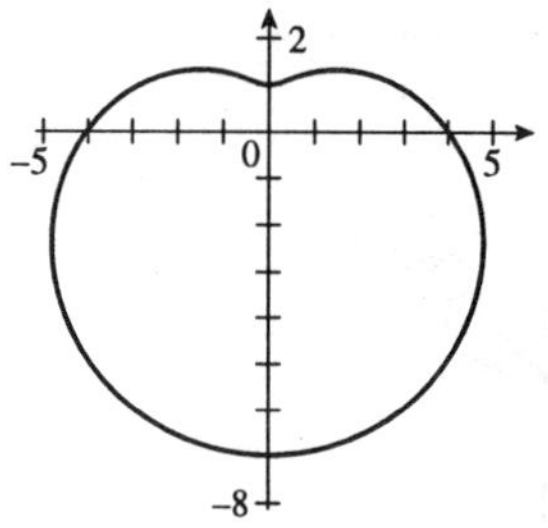

17. $r^2 = 9\sin 2\theta$ (lemniscate)
$r = \pm 3\sqrt{\sin(2\theta)}$

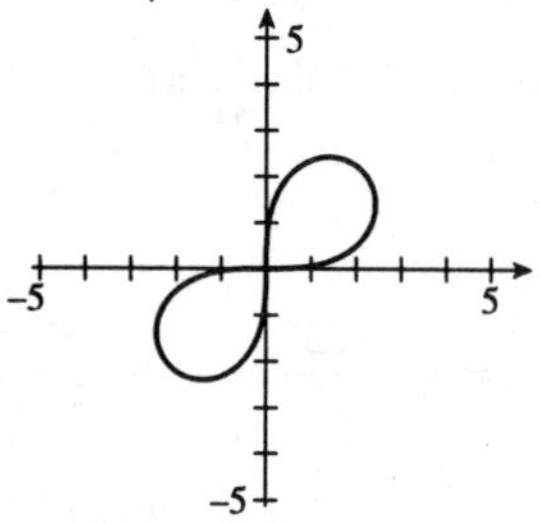

19. $r^2 = -16\cos 2\theta$ (lemniscate)
$r = \pm\sqrt{-16\cos 2\theta}$

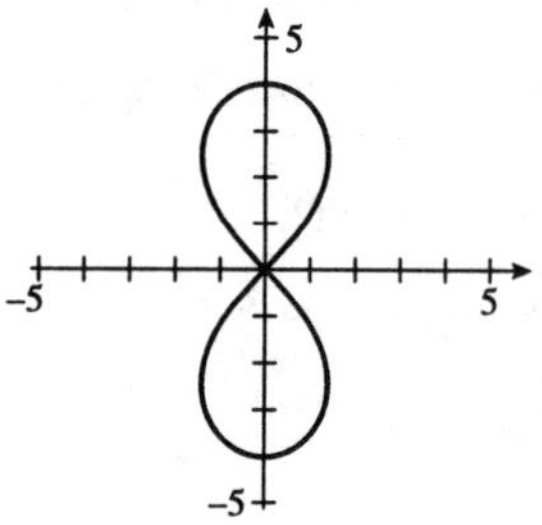

21. $r = 5\cos 3\theta$ (three-leaf rose)

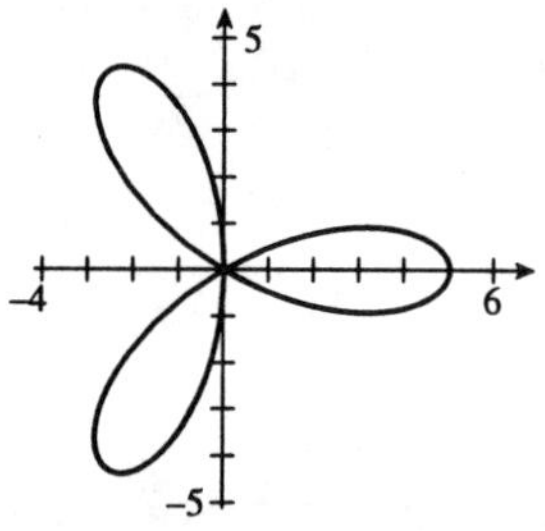

23. $r = 6\sin 2\theta$ (four-leaf rose)

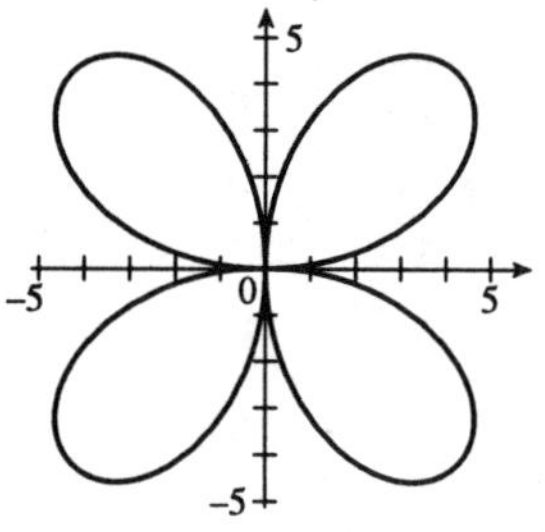

25. $r = 7\cos 5\theta$ (five-leaf rose)

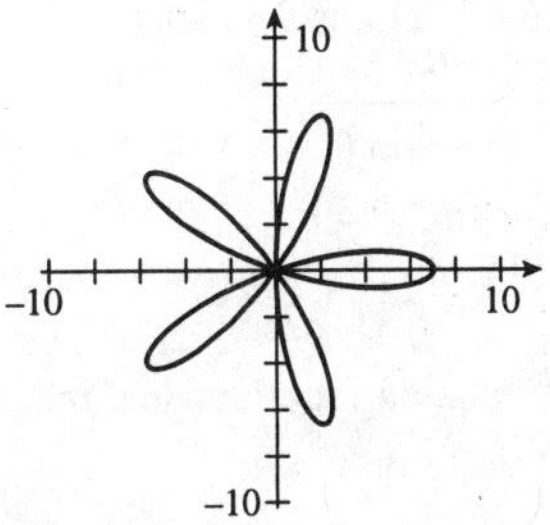

27. $r = \frac{1}{2}\theta,\ \theta \geq 0$ (spiral of Archimedes)

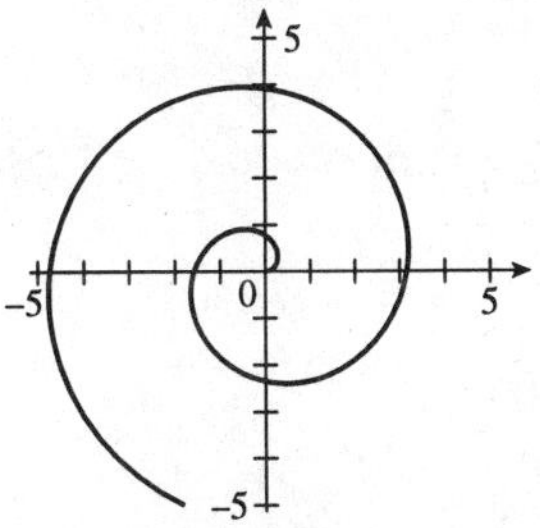

29. $r = e^{\theta},\ \theta \geq 0$ (logarithmic spiral)

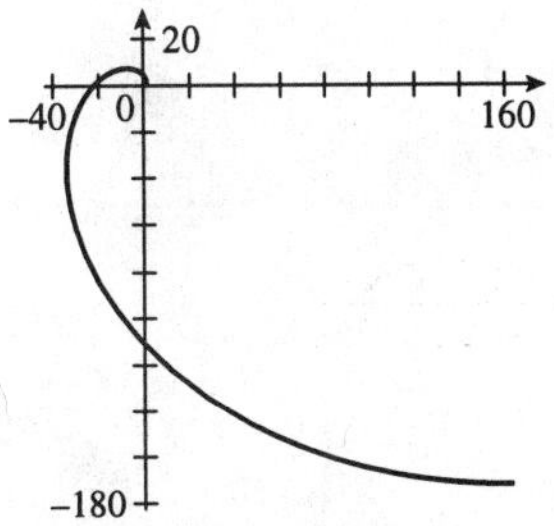

31. $r = \frac{2}{\theta},\ \theta > 0$ (reciprocal spiral)

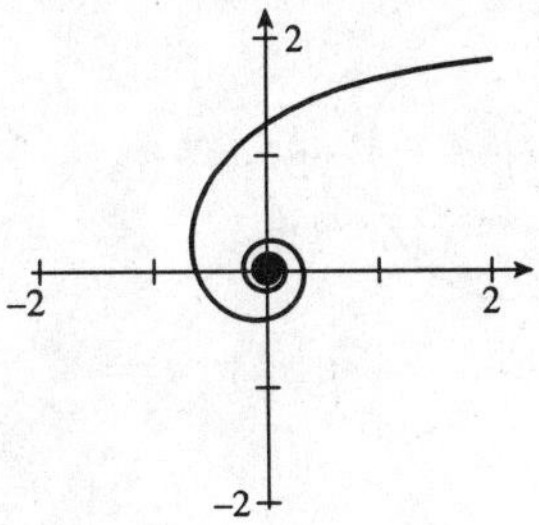

33. $r = 6,\ r = 4 + 4\cos\theta$

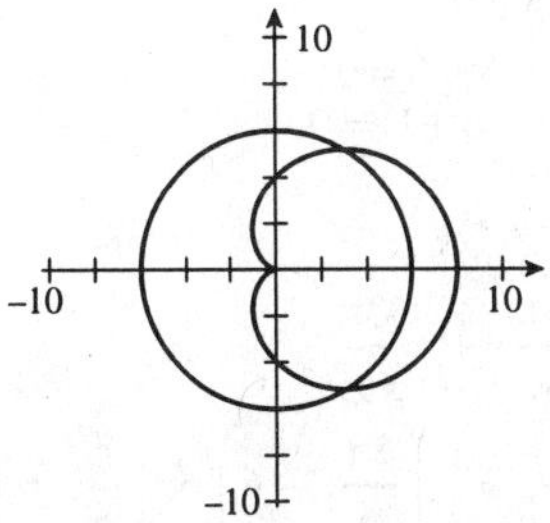

$6 = 4 + 4\cos\theta$

$\cos\theta = \frac{1}{2}$

$\theta = \frac{\pi}{3},\ \theta = \frac{5\pi}{3}$

$\left(6, \frac{\pi}{3}\right), \left(6, \frac{5\pi}{3}\right)$

35. $r = 3\sqrt{3}\cos\theta,\ r = 3\sin\theta$

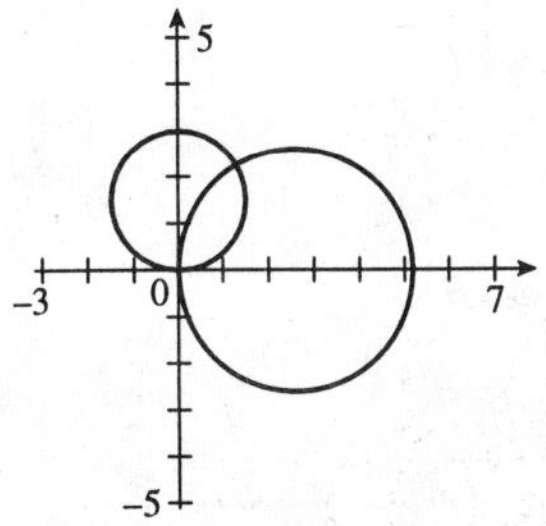

$3\sqrt{3}\cos\theta = 3\sin\theta$

$\tan\theta = \sqrt{3}$

$\theta = \frac{\pi}{3},\ \theta = \frac{4\pi}{3}$

$\left(\frac{3\sqrt{3}}{2}, \frac{\pi}{3}\right) = \left(-\frac{3\sqrt{3}}{2}, \frac{4\pi}{3}\right)$

(0, 0) is also a solution since both graphs include the pole.

37. $r = 6\sin\theta,\ r = \frac{6}{1 + 2\sin\theta}$

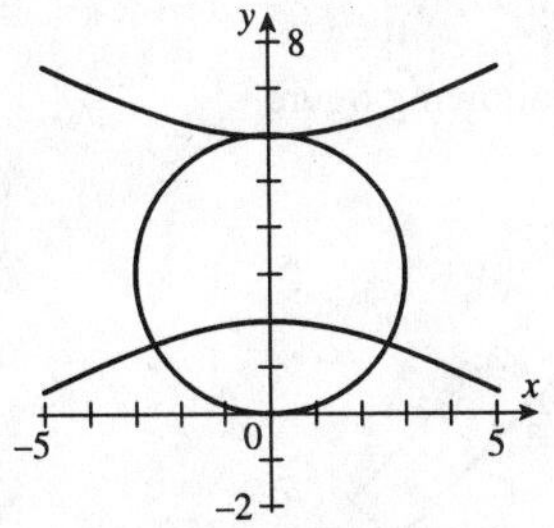

$$6\sin\theta = \frac{6}{1+2\sin\theta}$$

$$12\sin^2\theta + 6\sin\theta - 6 = 0$$

$$6(2\sin\theta - 1)(\sin\theta + 1) = 0$$

$$\sin\theta = \frac{1}{2},\ \sin\theta = -1$$

$$\theta = \frac{\pi}{6},\ \theta = \frac{5\pi}{6},\ \theta = \frac{3\pi}{2}$$

$$\left(\frac{\pi}{6},\ 3\right), \left(\frac{5\pi}{6},\ 3\right), \left(\frac{3\pi}{2},\ -6\right) \text{ or } \left(\frac{\pi}{2},\ 6\right)$$

39. Consider the following figure.

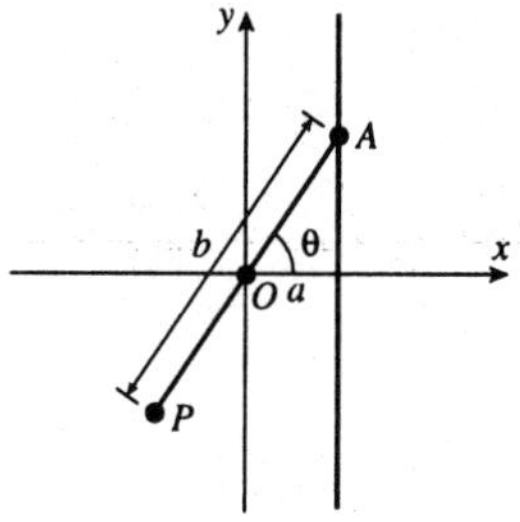

$$r = \frac{a}{\cos\theta} - b$$

$$r\cos\theta = a - b\cos\theta$$

$$x = a - b\cos\theta$$

$$xr = ar - br\cos\theta$$

$$(x-a)r = -bx$$

$$(x-a)^2 r^2 = b^2 x^2$$

$$(x-a)^2(x^2+y^2) = b^2x^2$$

$$y^2 = \frac{b^2x^2}{(x-a)^2} - x^2$$

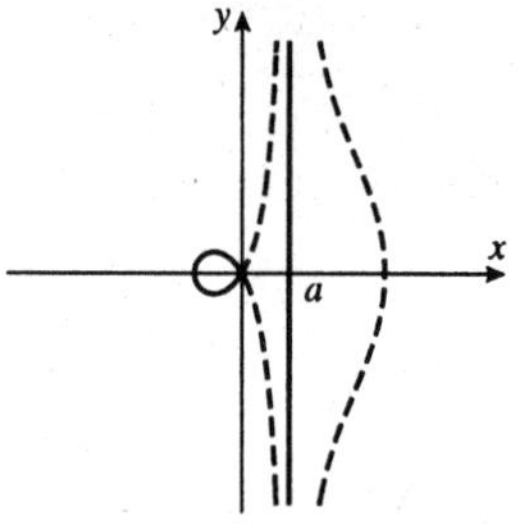

41. Consider the following figure.

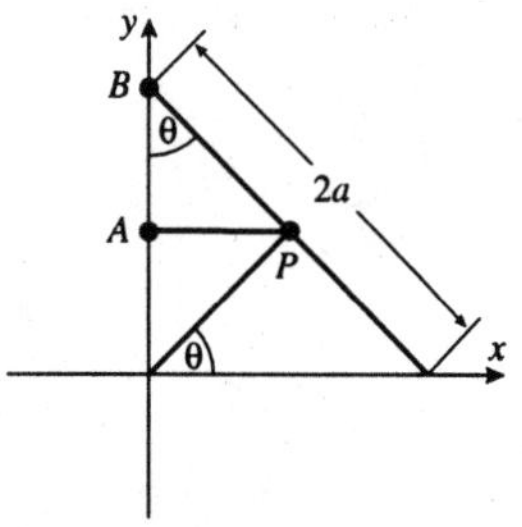

Then $\tan\theta = \frac{\overline{AP}}{\overline{BA}} = \frac{r\cos\theta}{2a\cos\theta - r\sin\theta}$

$$\frac{\sin\theta}{\cos\theta} = \frac{r\cos\theta}{2a\cos\theta - r\sin\theta}$$

$$2a\sin\theta\cos\theta - r\sin^2\theta = r\cos^2\theta$$

$$r\cos^2\theta + r\sin^2\theta = 2a\sin\theta\cos\theta$$

$$r = a\sin 2\theta$$

This is a polar equation for a four-leaf rose.

Section 10.5

Concepts Review

1. $\frac{1}{2}r^2\theta$

3. $\frac{1}{2}\int_0^{2\pi}(2+2\cos\theta)^2\,d\theta$

Problem Set 10.5

1. $r = a,\ a > 0$

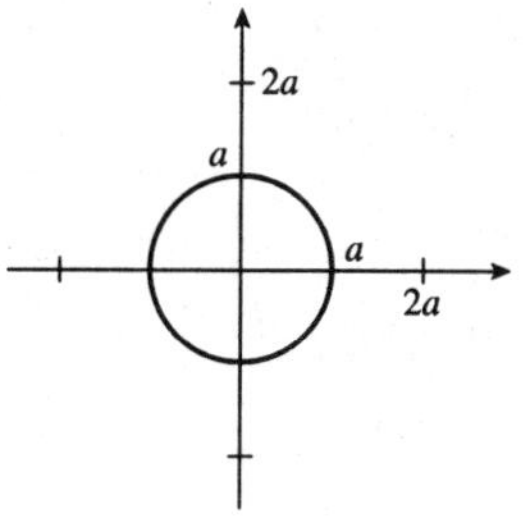

$$A = \frac{1}{2}\int_0^{2\pi} a^2\,d\theta = \pi a^2$$

3. $r = 3 + \cos\theta$

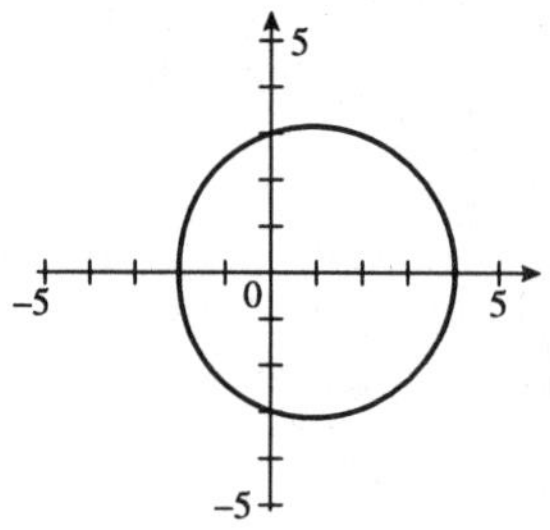

$$A = 2 \cdot \frac{1}{2} \int_0^{\pi} (3 + \cos\theta)^2 \, d\theta$$
$$= \int_0^{\pi} (9 + 6\cos\theta + \cos^2\theta) d\theta$$
$$= \int_0^{\pi} \left(\frac{19}{2} + 6\cos\theta + \frac{\cos 2\theta}{2} \right) d\theta$$
$$= \left[\frac{19}{2}\theta + 6\sin\theta + \frac{\sin 2\theta}{4} \right]_0^{\pi}$$
$$= \frac{19}{2}\pi$$

5. $r = 4 - 4\cos\theta$

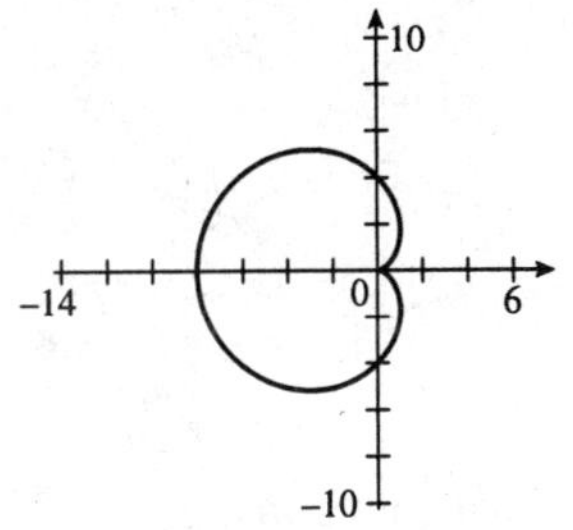

$$A = 2 \cdot \frac{1}{2} \int_0^{\pi} (4 - 4\cos\theta)^2 \, d\theta$$
$$= \int_0^{\pi} (16 - 32\cos\theta + 16\cos^2\theta) d\theta$$
$$= \int_0^{\pi} (24 - 32\cos\theta + 8\cos 2\theta) d\theta$$
$$= [24\theta - 32\sin\theta + 4\sin 2\theta]_0^{\pi}$$
$$= 24\pi$$

7. $r = a(1 + \sin\theta), \; a > 0$

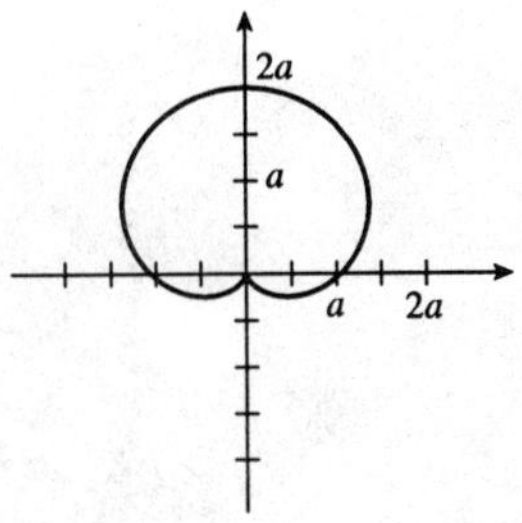

$$A = 2 \cdot \frac{1}{2} \int_{-\pi/2}^{\pi/2} a^2 (1 + \sin\theta)^2 \, d\theta$$
$$= a^2 \int_{-\pi/2}^{\pi/2} (1 + 2\sin\theta + \sin^2\theta) d\theta$$
$$= a^2 \int_{-\pi/2}^{\pi/2} \left(\frac{3}{2} + 2\sin\theta - \frac{\cos 2\theta}{2} \right) d\theta$$
$$= a^2 \left[\frac{3}{2}\theta - 2\cos\theta - \frac{\sin 2\theta}{4} \right]_{-\pi/2}^{\pi/2}$$
$$= \frac{3a^2\pi}{2}$$

9. $r^2 = 4\sin 2\theta$

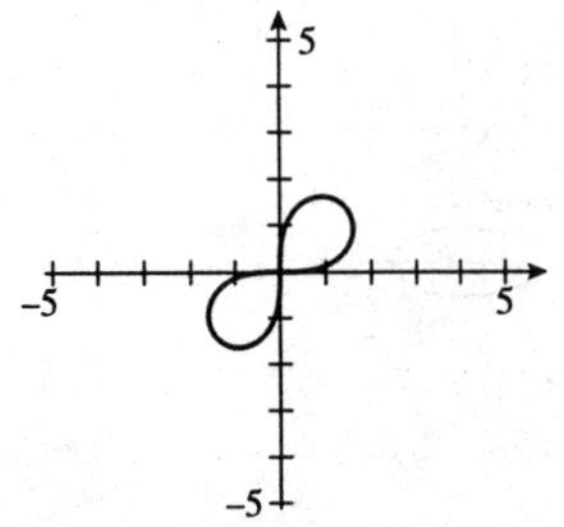

$$A = 2 \cdot \frac{1}{2} \int_0^{\pi/2} 4\sin 2\theta \, d\theta = [-2\cos 2\theta]_0^{\pi/2} = 4$$

11. $r = 2 - 4\cos\theta$

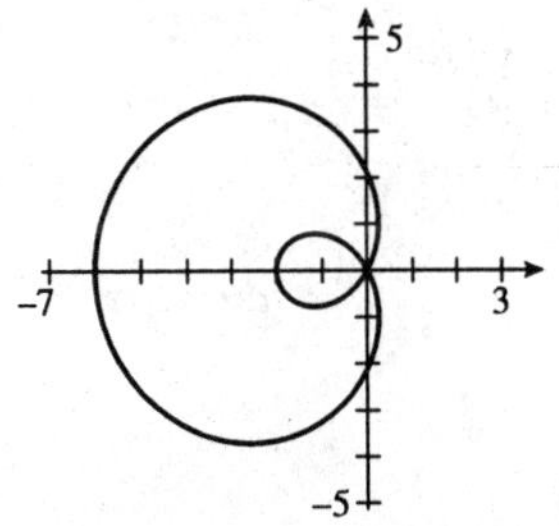

$$A = 2 \cdot \frac{1}{2} \int_0^{\pi/3} (2 - 4\cos\theta)^2 \, d\theta$$
$$= \int_0^{\pi/3} (4 - 16\cos\theta + 16\cos^2\theta) d\theta$$
$$= \int_0^{\pi/3} (12 - 16\cos\theta + 8\cos 2\theta) d\theta$$
$$= [12\theta - 16\sin\theta + 4\sin 2\theta]_0^{\pi/3}$$
$$= 4\pi - 6\sqrt{3}$$

13. $r = 2 - 4\sin\theta$

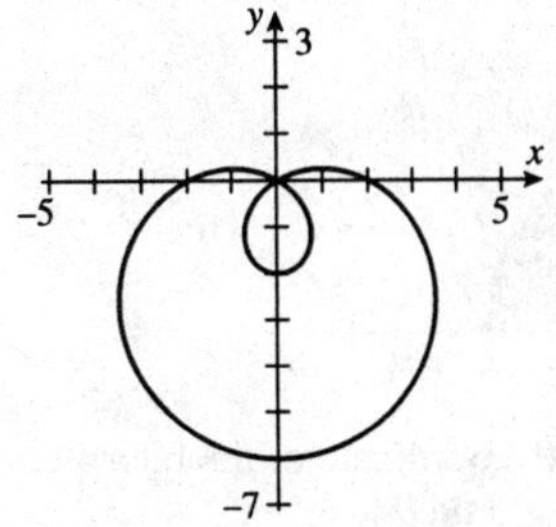

$$A = 2 \cdot \frac{1}{2} \int_{-\pi/2}^{\pi/6} (2 - 4\sin\theta)^2 \, d\theta$$
$$= \int_{-\pi/2}^{\pi/6} (4 - 16\sin\theta + 16\sin^2\theta) d\theta$$
$$= \int_{-\pi/2}^{\pi/6} (12 - 16\sin\theta - 8\cos 2\theta) d\theta$$
$$= [12\theta + 16\cos\theta - 4\sin 2\theta]_{-\pi/2}^{\pi/6}$$
$$= 8\pi + 6\sqrt{3}$$

15. $r = 4\cos 3\theta$

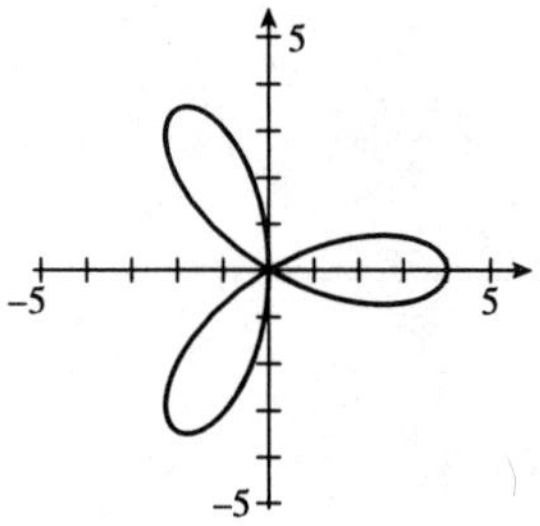

$$A = 6 \cdot \frac{1}{2}\int_0^{\pi/6} (4\cos 3\theta)^2\, d\theta$$

$$= 48\int_0^{\pi/6} \cos^2 3\theta\, d\theta = 24\int_0^{\pi/6}(1+\cos 6\theta)d\theta$$

$$= 24\left[\theta + \frac{1}{6}\sin 6\theta\right]_0^{\pi/6} = 4\pi$$

17. $A = \frac{1}{2}\int_0^{2\pi} 100 d\theta - \frac{1}{2}\int_0^{2\pi} 49 d\theta = 51\pi$

19. $r = 2,\ r^2 = 8\cos 2\theta$

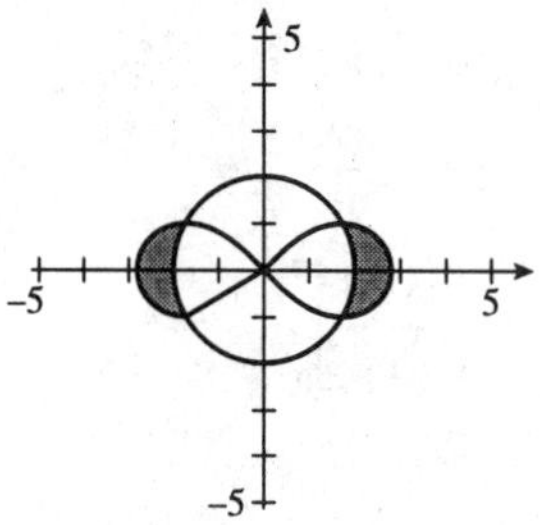

Solve for the θ-coordinate of the first intersection point.

$4 = 8\cos 2\theta$

$\cos 2\theta = \frac{1}{2}$

$2\theta = \frac{\pi}{3}$

$\theta = \frac{\pi}{6}$

$$A = 4 \cdot \frac{1}{2}\int_0^{\pi/6} (8\cos 2\theta - 4) d\theta$$

$$= 2[4\sin 2\theta - 4\theta]_0^{\pi/6}$$

$$= 4\sqrt{3} - \frac{4\pi}{3}$$

21. $r = 3 + 3\cos\theta,\ r = 3 + 3\sin\theta$

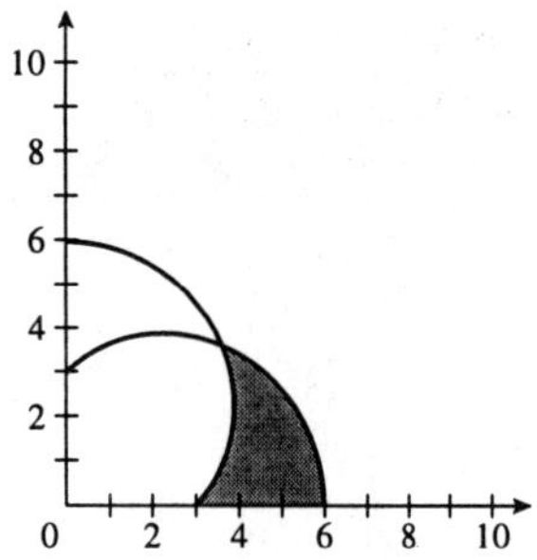

Solve for the θ-coordinate of the intersection point.

$3 + 3\cos\theta = 3 + 3\sin\theta$

$\tan\theta = 1$

$\theta = \frac{\pi}{4}$

$$A = \frac{1}{2}\int_0^{\pi/4}\left[(3+3\cos\theta)^2 - (3+3\sin\theta)^2\right]d\theta = \frac{1}{2}\int_0^{\pi/4}\left(18\cos\theta + 9\cos^2\theta - 18\sin\theta - 9\sin^2\theta\right)d\theta$$

$$= \frac{1}{2}\int_0^{\pi/4}(18\cos\theta - 18\sin\theta + 9\cos 2\theta)d\theta = \frac{1}{2}\left[18\sin\theta + 18\cos\theta + \frac{9}{2}\sin 2\theta\right]_0^{\pi/4}$$

$$= 9\sqrt{2} - \frac{27}{4}$$

23. **a.** $f(\theta)=2\cos\theta,\ f'(\theta)=-2\sin\theta$

$$m=\frac{(2\cos\theta)\cos\theta+(-2\sin\theta)\sin\theta}{-(2\cos\theta)\sin\theta+(-2\sin\theta)\cos\theta}=\frac{2\cos^2\theta-2\sin^2\theta}{-4\cos\theta\sin\theta}=\frac{\cos 2\theta}{-\sin 2\theta}$$

At $\theta=\frac{\pi}{3},\ m=\frac{-\frac{1}{2}}{-\frac{\sqrt{3}}{2}}=\frac{1}{\sqrt{3}}.$

b. $f(\theta)=1+\sin\theta,\ f'(\theta)=\cos\theta$

$$m=\frac{(1+\sin\theta)\cos\theta+(\cos\theta)\sin\theta}{-(1+\sin\theta)\sin\theta+(\cos\theta)\cos\theta}=\frac{\cos\theta+2\sin\theta\cos\theta}{\cos^2\theta-\sin^2\theta-\sin\theta}=\frac{\cos\theta+\sin 2\theta}{\cos 2\theta-\sin\theta}$$

At $\theta=\frac{\pi}{3},\ m=\frac{\frac{1}{2}+\frac{\sqrt{3}}{2}}{-\frac{1}{2}-\frac{\sqrt{3}}{2}}=-1.$

c. $f(\theta)=\sin 2\theta,\ f'(\theta)=2\cos 2\theta$

$$m=\frac{(\sin 2\theta)\cos\theta+(2\cos 2\theta)\sin\theta}{-(\sin 2\theta)\sin\theta+(2\cos 2\theta)\cos\theta}$$

At $\theta=\frac{\pi}{3},\ m=\frac{\left(\frac{\sqrt{3}}{2}\right)\left(\frac{1}{2}\right)+(-1)\left(\frac{\sqrt{3}}{2}\right)}{-\left(\frac{\sqrt{3}}{2}\right)\left(\frac{\sqrt{3}}{2}\right)+(-1)\left(\frac{1}{2}\right)}=\frac{-\frac{\sqrt{3}}{4}}{-\frac{5}{4}}=\frac{\sqrt{3}}{5}.$

d. $f(\theta)=4-3\cos\theta,\ f'(\theta)=3\sin\theta$

$$m=\frac{(4-3\cos\theta)\cos\theta+(3\sin\theta)\sin\theta}{-(4-3\cos\theta)\sin\theta+(3\sin\theta)\cos\theta}=\frac{4\cos\theta-3\cos^2\theta+3\sin^2\theta}{-4\sin\theta+6\sin\theta\cos\theta}=\frac{4\cos\theta-3\cos 2\theta}{-4\sin\theta+3\sin 2\theta}$$

At $\theta=\frac{\pi}{3},\ m=\frac{4\left(\frac{1}{2}\right)-3\left(-\frac{1}{2}\right)}{-4\left(\frac{\sqrt{3}}{2}\right)+3\left(\frac{\sqrt{3}}{2}\right)}=\frac{\frac{7}{2}}{-\frac{\sqrt{3}}{2}}=-\frac{7}{\sqrt{3}}.$

25. $f(\theta)=1-2\sin\theta,\ f'(\theta)=-2\cos\theta$

$$m=\frac{(1-2\sin\theta)\cos\theta+(-2\cos\theta)\sin\theta}{-(1-2\sin\theta)\sin\theta+(-2\cos\theta)\cos\theta}=\frac{\cos\theta-4\sin\theta\cos\theta}{-\sin\theta+2\sin^2\theta-2\cos^2\theta}=\frac{\cos\theta(1-4\sin\theta)}{-\sin\theta+2\sin^2\theta-2\cos^2\theta}$$

Solve for θ such that $m=0$.

$$\frac{\cos\theta(1-4\sin\theta)}{-\sin\theta+2\sin^2\theta-2\sin^2\theta}=0$$

$\cos\theta=0,\ 1-4\sin\theta=0$

$\theta=\frac{\pi}{2},\ \theta=\frac{3\pi}{2},\ \theta=\sin^{-1}\left(\frac{1}{4}\right)\approx 0.25,\ \theta=\pi-\sin^{-1}\left(\frac{1}{4}\right)\approx 2.89$

$f\left(\frac{\pi}{2}\right)=-1,\ f\left(\frac{3\pi}{2}\right)=3,\ f\left(\sin^{-1}\left(\frac{1}{4}\right)\right)=\frac{1}{2},\ f\left(\pi-\sin^{-1}\left(\frac{1}{4}\right)\right)=\frac{1}{2}$

$\left(-1,\ \frac{\pi}{2}\right),\left(3,\ \frac{3\pi}{2}\right),\left(\frac{1}{2},\ 0.25\right),\left(\frac{1}{2},\ 2.89\right)$

27. $f(\theta)=a(1+\cos\theta),\ f'(\theta)=-a\sin\theta$

$$L=\int_0^{2\pi}\sqrt{[a(1+\cos\theta)]^2+[-a\sin\theta]^2}\,d\theta=a\int_0^{2\pi}\sqrt{2+2\cos\theta}\,d\theta$$

29. $f(t) = 2 + \cos t,\ f'(t) = -\sin t$

$L = 2\int_0^{\pi} \sqrt{[2+\cos t]^2 + [-\sin t]^2}\, dt = 2\int_0^{\pi} \sqrt{5 + 4\cos t}\, dt$

Using a numerical method to approximate the length, $L \approx 13.36$.

$f(t) = 2 + 4\cos t,\ f'(t) = -4\sin t$

$L = 2\int_0^{\pi} \sqrt{[2+4\cos t]^2 + [-4\sin t]^2}\, dt = 4\int_0^{\pi} \sqrt{5 + 4\cos t}\, dt$

Using a numerical method to approximate the length, $L \approx 26.73$.

31. $A = 4 \cdot \frac{1}{2}\int_0^{\pi/4} 8\cos 2t\, dt = 2[4\sin 2t]_0^{\pi/4} = 8$

$f(t) = \sqrt{8\cos 2t},\ f'(t) = \frac{-8\sin 2t}{\sqrt{8\cos 2t}}$

$L = 4\int_0^{\pi/4} \sqrt{8\cos 2t + \frac{8\sin^2 2t}{\cos 2t}}\, dt = \int_0^{\pi/4} \sqrt{\frac{8}{\cos 2t}}\, dt$

$= 8\sqrt{2}\int_0^{\pi/4} \frac{1}{\sqrt{\cos 2t}}\, dt$

Using a numerical method to approximate the length, $L \approx 9.4$.

33. If n is even, there are $2n$ leaves.

$A = 2n\frac{1}{2}\int_{-\pi/2n}^{\pi/2n} (a\cos n\theta)^2\, d\theta = na^2\int_{-\pi/2n}^{\pi/2n} \cos^2 n\theta\, d\theta = na^2\int_{-\pi/2n}^{\pi/2n} \frac{1+\cos 2n\theta}{2}\, d\theta$

$= na^2\left[\frac{1}{2}\theta + \frac{\sin 2n\theta}{4n}\right]_{-\pi/2n}^{\pi/2n} = \frac{1}{2}a^2\pi$

If n is odd, there are n leaves.

$A = n \cdot \frac{1}{2}\int_{-\pi/2n}^{\pi/2n} (a\cos n\theta)^2\, d\theta = \frac{na^2}{2}\int_{-\pi/2n}^{\pi/2n} \cos^2 n\theta\, d\theta = \frac{1}{4}a^2\pi$

35. a. Sketch the graph.

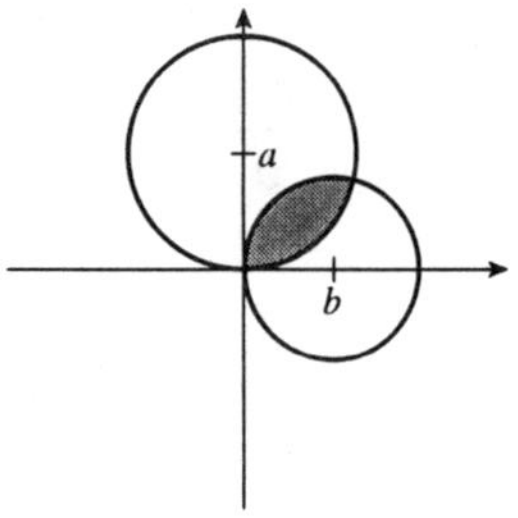

Solve for the θ-coordinate of the intersection.

$2a\sin\theta = 2b\cos\theta$

$\tan\theta = \frac{b}{a}$

$\theta = \tan^{-1}\left(\frac{b}{a}\right)$

Let $\theta_0 = \tan^{-1}\left(\frac{b}{a}\right)$.

$$A = \frac{1}{2}\int_0^{\theta_0} (2a\sin\theta)^2\, d\theta + \frac{1}{2}\int_{\theta_0}^{\pi/2} (2b\cos\theta)^2\, d\theta$$

$$= 2a^2\int_0^{\theta_0} \sin^2\theta\, d\theta + 2b^2\int_{\theta_0}^{\pi/2} \cos^2\theta\, d\theta = a^2\int_0^{\theta_0}(1-\cos 2\theta)d\theta + b^2\int_{\theta_0}^{\pi/2}(1+\cos 2\theta)d\theta$$

$$= a^2\left[\theta - \frac{\sin 2\theta}{2}\right]_0^{\theta_0} + b^2\left[\theta + \frac{\sin 2\theta}{2}\right]_{\theta_0}^{\pi/2} = a^2\theta_0 + b^2\left(\frac{\pi}{2} - \theta_0\right) - \frac{a^2+b^2}{2}\sin 2\theta_0$$

$$= a^2\theta_0 + b^2\left(\frac{\pi}{2} - \theta_0\right) - (a^2+b^2)\sin\theta_0\cos\theta_0 = a^2\tan^{-1}\left(\frac{b}{a}\right) + b^2\left(\frac{\pi}{2} - \tan^{-1}\left(\frac{b}{a}\right)\right) - ab.$$

Note that since $\tan\theta = \frac{b}{a}$, $\cos\theta = \frac{a}{\sqrt{a^2+b^2}}$ and $\sin\theta = \frac{b}{\sqrt{a^2+b^2}}$.

b. Let m_1 be the slope of $r = 2a\sin\theta$.

$$m_1 = \frac{2a\sin\theta\cos\theta + 2a\cos\theta\sin\theta}{-2a\sin\theta\sin\theta + 2a\cos\theta\cos\theta} = \frac{2\sin\theta\cos\theta}{\cos^2\theta - \sin^2\theta}$$

At $\theta = \tan^{-1}\left(\frac{b}{a}\right)$, $m_1 = \frac{2ab}{a^2-b^2}$.

At $\theta = 0$ (the pole), $m_1 = 0$.

Let m_2 be the slope of $r = 2b\cos\theta$.

$$m_2 = \frac{2b\cos\theta\cos\theta - 2b\sin\theta\sin\theta}{-2b\cos\theta\sin\theta - 2b\sin\theta\cos\theta} = \frac{\cos^2\theta - \sin^2\theta}{-2\sin\theta\cos\theta}$$

At $\theta = \tan^{-1}\left(\frac{b}{a}\right)$, $m_2 = -\frac{a^2-b^2}{2ab}$.

At $\theta = \frac{\pi}{2}$ (the pole), m_2 is undefined.

Therefore the two circles intersect at right angles.

37. The edge of the pond is described by the equation $r = 2a\cos\theta$.

Solve for intersection points of the circles $r = ak$ and $r = 2a\cos\theta$.

$ak = 2a\cos\theta$

$\cos\theta = \frac{k}{2}$, $\theta = \cos^{-1}\left(\frac{k}{2}\right)$

Let A be the grazing area.

$$A = \frac{1}{2}\pi(ka)^2 + 2\cdot\frac{1}{2}\int_{\cos^{-1}\left(\frac{k}{2}\right)}^{\pi/2}[(ka)^2 - (2a\cos\theta)^2]d\theta = \frac{1}{2}k^2a^2\pi + a^2\int_{\cos^{-1}\left(\frac{k}{2}\right)}^{\pi/2}(k^2 - 4\cos^2\theta)d\theta$$

$$= \frac{1}{2}k^2a^2\pi + a^2\int_{\cos^{-1}\left(\frac{k}{2}\right)}^{\pi/2}((k^2-2) - 2\cos 2\theta)d\theta = \frac{1}{2}k^2a^2\pi + a^2\left[(k^2-2)\theta - \sin 2\theta\right]_{\cos^{-1}\left(\frac{k}{2}\right)}^{\pi/2}$$

$$= \frac{1}{2}k^2a^2\pi + a^2\left[k^2\theta - 2\theta - 2\sin\theta\cos\theta\right]_{\cos^{-1}\left(\frac{k}{2}\right)}^{\pi/2}$$

$$= \frac{1}{2}k^2a^2\pi + a^2\left[\frac{k^2\pi}{2} - \pi - k^2\cos^{-1}\left(\frac{k}{2}\right) + 2\cos^{-1}\left(\frac{k}{2}\right) + \frac{k\sqrt{4-k^2}}{2}\right]$$

$$= a^2\left[(k^2-1)\pi + (2-k^2)\cos^{-1}\left(\frac{k}{2}\right) + \frac{k\sqrt{4-k^2}}{2}\right]$$

39. The untethered goat has a grazing area of πa^2. From Problem 38, the tethered goat has a grazing area of $a^2\left(\frac{\pi k^2}{2}+\frac{k^3}{3}\right)$.

$\pi a^2 = a^2\left(\frac{\pi k^2}{2}+\frac{k^3}{3}\right)$

$\pi = \frac{\pi k^2}{2}+\frac{k^3}{3}$

$2k^3 + 3\pi k^2 - 6\pi = 0$

Using a numerical method or graphing calculator, $k \approx 1.26$.

Section 10.6

Concepts Review

1. Sample; closed; simple

3. Cycloid

Problem Set 10.6

1. a.

t	x	y
–2	–4	–6
–1	–2	–3
0	0	0
1	2	3
2	4	6

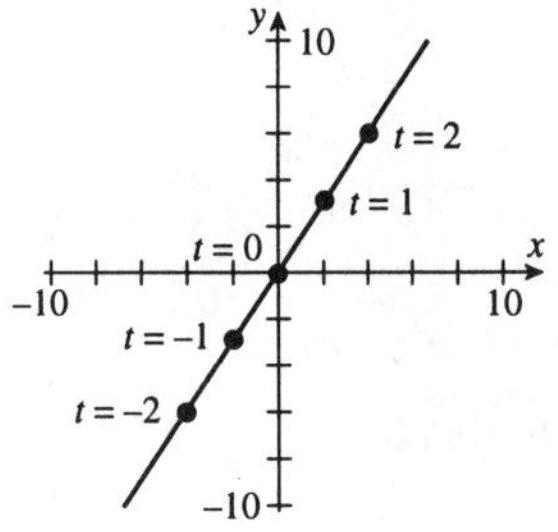

b. Simple

c. $t = \frac{x}{2}$

$y = \frac{3}{2}x$

3. a.

t	x	y
0	–4	0
1	–3	1
2	–2	$\sqrt{2}$
3	–1	$\sqrt{3}$
4	0	2

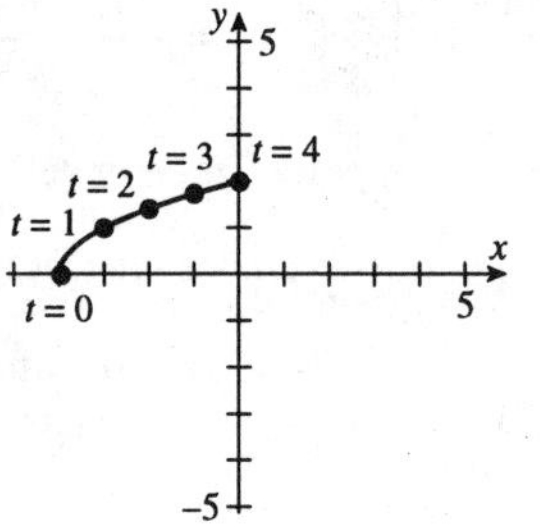

b. Simple

c. $t = x + 4$

$y = \sqrt{x+4}$

5. a.

t	x	y
–1	1	–1
0	0	0
1	1	1
2	4	8

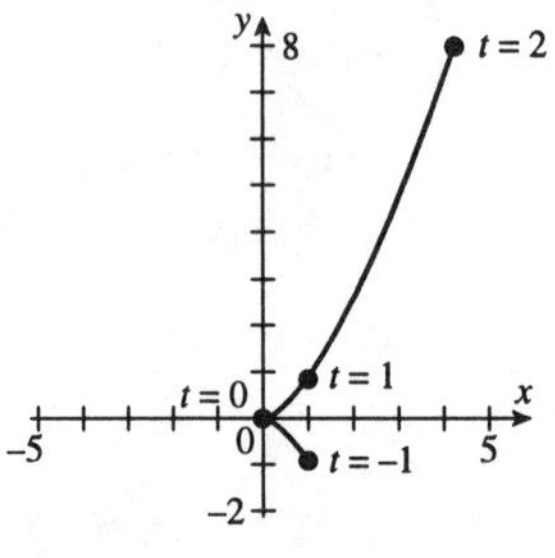

b. Simple

c. $t = \pm\sqrt{x}$

$y = \pm x^{3/2}$ or $y^2 = x^3$

7. a.

t	x	y
–3	–15	5
–2	0	0
–1	3	–3
0	0	–4
1	–3	–3
2	0	0
3	15	5

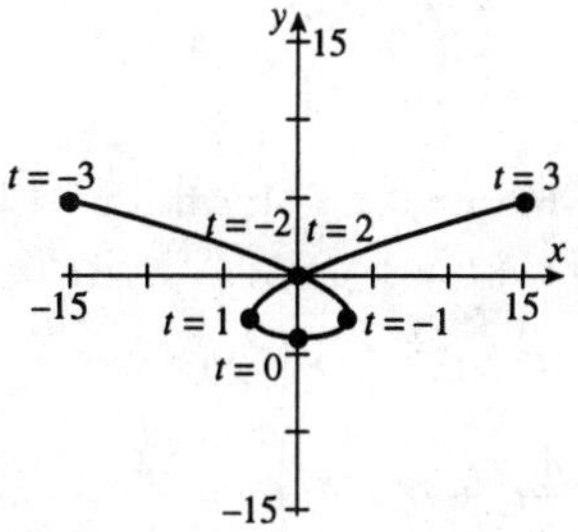

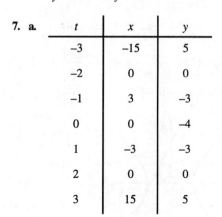

b. Neither simple nor closed

c. $t = \pm\sqrt{y+4}$

$x = t(t^2 - 4) = \left(\pm\sqrt{y+4}\right)(y)$ or

$x^2 = (y+4)y^2$

9. a.

t	x	y
0	0	5
$\frac{\pi}{2}$	3	0
π	0	–5
$\frac{3\pi}{2}$	–3	0
2π	0	5

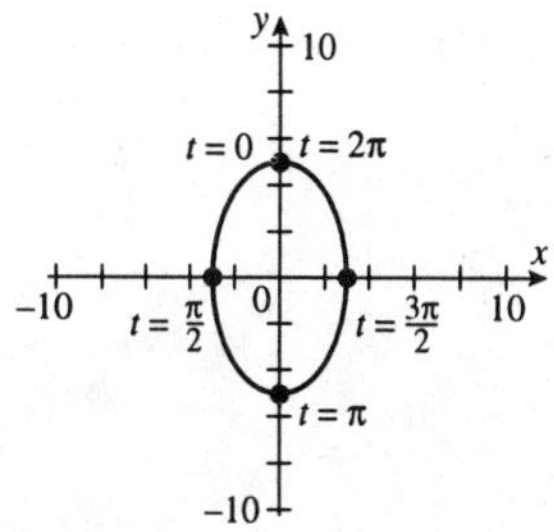

b. Simple and closed

c. $\frac{x}{3} = \sin t,\ \frac{y}{5} = \cos t$

$\frac{x^2}{9} + \frac{y^2}{25} = 1$

11. a.

t	x	y
0	0	4
$\frac{\pi}{6}$	$\frac{1}{4}$	$\frac{9}{4}$
$\frac{\pi}{4}$	1	1
$\frac{\pi}{3}$	$\frac{9}{4}$	$\frac{1}{4}$
$\frac{\pi}{2}$	4	0

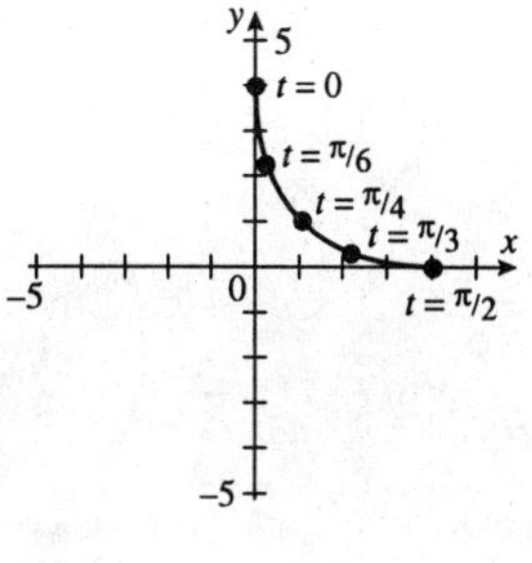

b. Simple

c. $\frac{\sqrt{x}}{2} = \sin^2 t,\ \frac{\sqrt{y}}{2} = \cos^2 t$

$\frac{\sqrt{x}}{2} + \frac{\sqrt{y}}{2} = 1$

13. $\frac{dx}{dt} = 6t,\ \frac{dy}{dt} = 6t^2$

$\frac{dy}{dx} = \frac{6t^2}{6t} = t$

$\frac{dy'}{dt} = 1$

$\frac{d^2y}{dx^2} = \frac{1}{6t}$

15. $\frac{dx}{dt} = 2 + \frac{3}{t^2},\ \frac{dy}{dt} = 2 - \frac{3}{t^2}$

$\frac{dy}{dx} = \frac{2 - \frac{3}{t^2}}{2 + \frac{3}{t^2}} = \frac{2t^2 - 3}{2t^2 + 3}$

$\frac{dy'}{dt} = \frac{4t(2t^2+3) - 4t(2t^2-3)}{(2t^2+3)^2} = \frac{24t}{(2t^2+3)^2}$

$\frac{d^2y}{dx^2} = \frac{\frac{24t}{(2t^2+3)^2}}{2 + \frac{3}{t^2}} = \frac{24t^3}{(2t^2+3)^3}$

17. $\frac{dx}{dt} = 3\sec^2 t,\ \frac{dy}{dt} = 5\sec t \tan t$

$\frac{dy}{dx} = \frac{5\sec t \tan t}{3\sec^2 t} = \frac{5\tan t}{3\sec t} = \frac{5}{3}\sin t$

$\frac{dy'}{dt} = \frac{5}{3}\cos t$

$\frac{d^2y}{dx^2} = \frac{\frac{5}{3}\cos t}{3\sec^2 t} = \frac{5}{9}\cos^3 t$

19. $\frac{dx}{dt} = 2t,\ \frac{dy}{dt} = 3t^2$

$\frac{dy}{dx} = \frac{3t^2}{2t} = \frac{3}{2}t$

At $t = 2$, $x = 4$, $y = 8$, and $\frac{dy}{dx} = 3$.

Tangent line: $y - 8 = 3(x - 4)$ or $3x - y - 4 = 0$

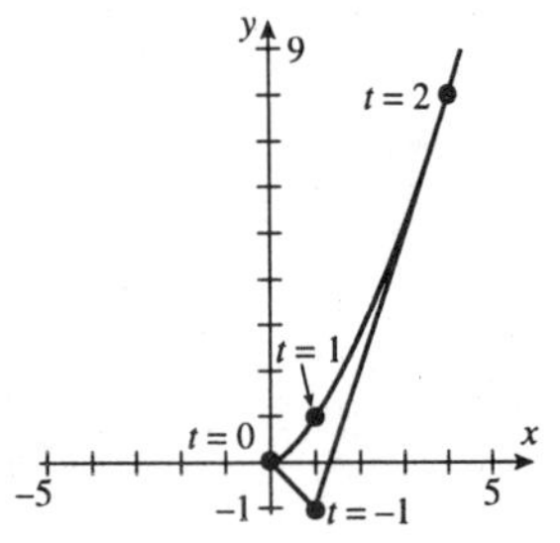

21. $\frac{dx}{dt} = 2\sec t \tan t,\ \frac{dy}{dt} = 2\sec^2 t$

$\frac{dy}{dx} = \frac{2\sec^2 t}{2\sec t \tan t} = \csc t$

At $t = -\frac{\pi}{6}$, $x = \frac{4}{\sqrt{3}}$, $y = -\frac{2}{\sqrt{3}}$, and $\frac{dy}{dx} = -2$.

Tangent line: $y + \frac{2}{\sqrt{3}} = -2\left(x - \frac{4}{\sqrt{3}}\right)$ or

$2\sqrt{3}x + \sqrt{3}y - 6 = 0$

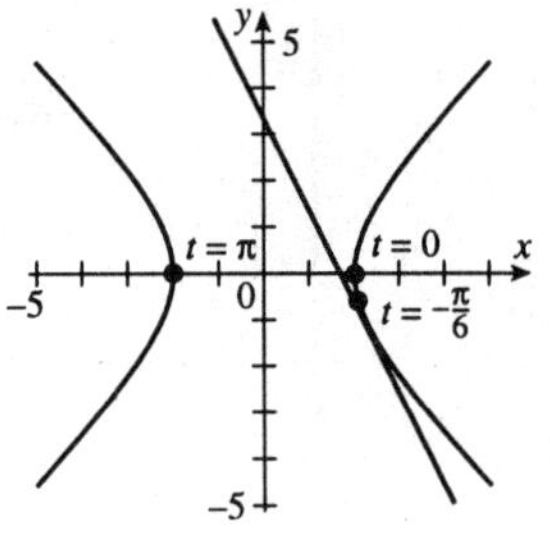

23. $dx = dt$; when $x = 0$, $t = -1$; when $x = 1$, $t = 0$.

$\int_0^1 (x^2 - 4y)dx = \int_{-1}^0 [(t+1)^2 - 4(t^3 - 4)]dt$

$= \int_{-1}^0 (-4t^3 + t^2 + 2t + 17)dt$

$= \left[-t^4 + \frac{1}{3}t^3 + t^2 + 17t\right]_{-1}^0 = \frac{52}{3}$

25. $dx = 2e^{2t}dt$

$A = \int_1^{25} y\,dx = \int_0^{\ln 5} 2e^t dt = \left[2e^t\right]_0^{\ln 5} = 8$

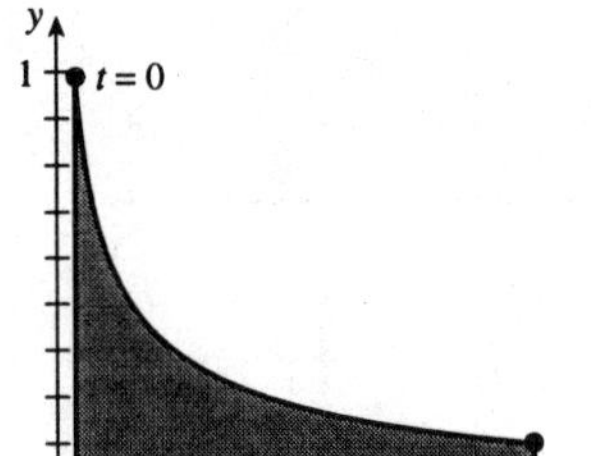

27.

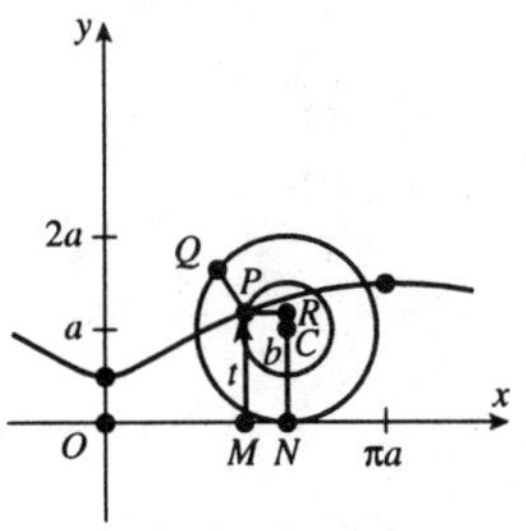

Let the wheel roll along the x-axis with P initially at $(0, a - b)$.

$|ON| = \text{arc } NQ = at$

$x = |OM| = |ON| - |MN| = at - b\sin t$

$y = |MP| = |RN| = |NC| + |CR| = a - b\cos t$

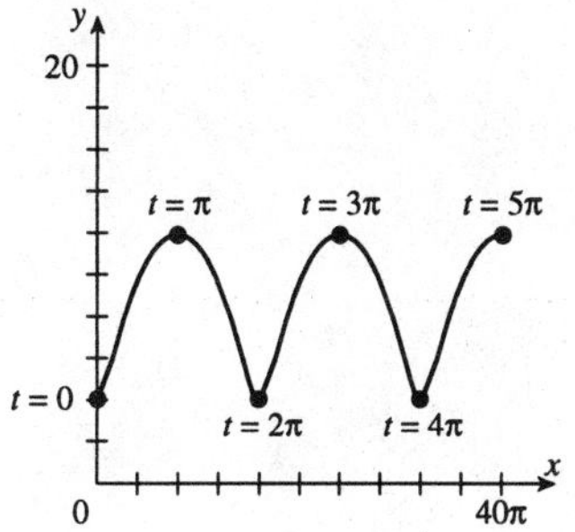

29. a. $t = \dfrac{x}{v_0 \cos\alpha}$

$y = -\dfrac{16x^2}{v_0^2 \cos^2\alpha} + \tan\alpha x$

This is an equation for a parabola.

b. Solve for t when $y = 0$.

$-16t^2 + (v_0 \sin\alpha)t = 0$

$t(-16t + v_0 \sin\alpha) = 0$

$t = 0, \ \dfrac{v_0 \sin\alpha}{16}$

The time of flight is $\dfrac{v_0 \sin\alpha}{16}$ seconds.

c. At $t = \dfrac{v_0 \sin\alpha}{16}$, $x = (v_0 \cos\alpha)\left(\dfrac{v_0 \sin\alpha}{16}\right)$

$= \dfrac{v_0^2 \sin\alpha\cos\alpha}{16} = \dfrac{v_0^2 \sin 2\alpha}{32}$.

d. Let R be the range as a function of α.

$R = \dfrac{v_0^2 \sin 2\alpha}{32}$

$\dfrac{dR}{d\alpha} = \dfrac{v_0^2 \cos 2\alpha}{16}$

$\dfrac{v_0^2 \cos 2\alpha}{16} = 0, \ \cos 2\alpha = 0, \ \alpha = \dfrac{\pi}{4}$

$\dfrac{d^2R}{d\alpha^2} = -\dfrac{v_0^2 \sin 2\alpha}{8}; \ \dfrac{d^2R}{d\alpha^2} < 0$ at $\alpha = \dfrac{\pi}{4}$.

The range is the largest possible when $\alpha = \dfrac{\pi}{4}$.

31. The x- and y-coordinates of the center of the circle of radius b are $(a - b)\cos t$ and $(a - b)\sin t$, respectively. The angle measure (in a clockwise direction) of arc BP is $\dfrac{a}{b}t$. The horizontal change from the center of the circle of radius b to P is

$b\cos\left(-\left(\dfrac{a}{b}t - t\right)\right) = b\cos\left(\dfrac{a-b}{b}t\right)$ and the vertical change is

$b\sin\left(-\left(\dfrac{a}{b}t - t\right)\right) = -b\sin\left(\dfrac{a-b}{b}t\right)$. Therefore,

$x = (a-b)\cos t + b\cos\left(\dfrac{a-b}{b}t\right)$ and

$y = (a-b)\sin t - b\sin\left(\dfrac{a-b}{b}t\right)$.

33. Consider the following figure similar to the one in the text.

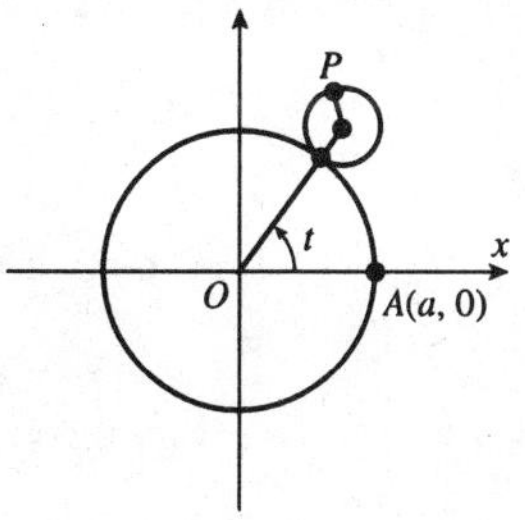

The x- and y-coordinates of the center of the circle of radius b are $(a + b)\cos t$ and $(a + b)\sin t$ respectively. The angle measure (in a counter-clockwise direction) of arc PB is $\dfrac{a}{b}t$. The horizontal change from the center of the circle of radius b to P is

$b\cos\left(\dfrac{a}{b}t + t + \pi\right) = -b\cos\left(\dfrac{a+b}{b}t\right)$ and the vertical change is

$b\sin\left(\dfrac{a}{b}t + t + \pi\right) = -b\sin\left(\dfrac{a+b}{b}t\right)$. Therefore,

$x = (a+b)\cos t - b\cos\left(\dfrac{a+b}{b}t\right)$ and

$y = (a+b)\sin t - b\sin\left(\dfrac{a+b}{b}t\right)$.

35. a. From Figure 14 in the text, $\tan\theta = \dfrac{2a}{x}$.

Recall from studying polar coordinates that $\overline{OB} = 2a\sin\theta$. Therefore, $x = \dfrac{2a}{\tan\theta}$ and

$y = \overline{OB}\sin\theta = 2a\sin^2\theta$.

b. $x^2 = \frac{4a^2}{\tan^2\theta} = 4a^2\cot^2\theta$

$\sin^2\theta = \frac{y}{2a}$, $\cos^2\theta = 1 - \frac{y}{2a} = \frac{2a-y}{2a}$

$\cot^2\theta = \frac{2a-y}{y}$

$x^2 = 4a^2\left(\frac{2a-y}{y}\right)$

37. $x = (a+b)\cos t - b\cos\left(\frac{a+b}{b}t\right)$, $y = (a+b)\sin t - b\sin\left(\frac{a+b}{b}t\right)$

$\frac{dx}{dt} = -(a+b)\sin t + (a+b)\sin\left(\frac{a+b}{b}t\right)$, $\frac{dy}{dt} = (a+b)\cos t - (a+b)\cos\left(\frac{a+b}{b}t\right)$

Note that $\cos\left(\frac{a+b}{b}t\right) = \cos\left(\frac{a}{b}t + t\right) = \cos\left(\frac{a}{b}t\right)\cos t - \sin\left(\frac{a}{b}t\right)\sin t$

and $\sin\left(\frac{a+b}{b}t\right) = \sin\left(\frac{a}{b}t + t\right) = \sin\left(\frac{a}{b}t\right)\cos t + \cos\left(\frac{a}{b}t\right)\sin t$.

Therefore, $\left(\frac{dx}{dt}\right)^2 = (a+b)^2\left[\sin^2 t - 2\sin t\left(\sin\left(\frac{a}{b}t\right)\cos t + \cos\left(\frac{a}{b}t\right)\sin t\right) + \sin^2\left(\frac{a+b}{b}t\right)\right]$

and $\left(\frac{dy}{dt}\right)^2 = (a+b)^2\left[\cos^2 t - 2\cos t\left(\cos\left(\frac{a}{b}t\right)\cos t - \sin\left(\frac{a}{b}t\right)\sin t\right) + \cos^2\left(\frac{a+b}{b}t\right)\right]$.

$\left(\frac{dx}{dt}\right)^2 + \left(\frac{dy}{dt}\right)^2 = (a+b)^2\left[2 - 2\cos\left(\frac{a}{b}t\right)\right] = 4(a+b)^2\left[\frac{1-\cos\left(\frac{a}{b}t\right)}{2}\right] = 4(a+b)^2\sin^2\left(\frac{a}{2b}t\right)$

$L = \int_0^{2\pi}\sqrt{\left(\frac{dx}{dt}\right)^2 + \left(\frac{dy}{dt}\right)^2}\,dt = \int_0^{2\pi} 2(a+b)\left|\sin\left(\frac{a}{2b}t\right)\right|dt = \frac{2a}{b}(a+b)\int_0^{2b\pi/a}\sin\left(\frac{a}{2b}t\right)dt$

$= \frac{2a(a+b)}{b}\left[-\frac{2b}{a}\cos\left(\frac{a}{2b}t\right)\right]_0^{2b\pi/a}$

$= -4(a+b)[-1-1] = 8(a+b)$

39. a. $x = 2(t - \sin t)$, $y = 2(1 - \cos t)$

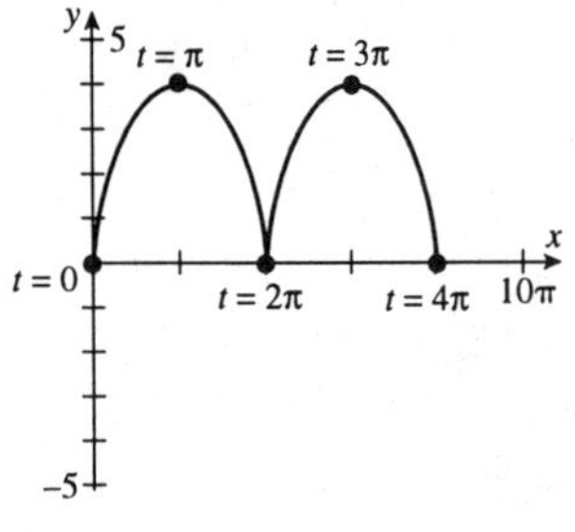

b. $x = 2t - \sin t, y = 2 - \cos t$

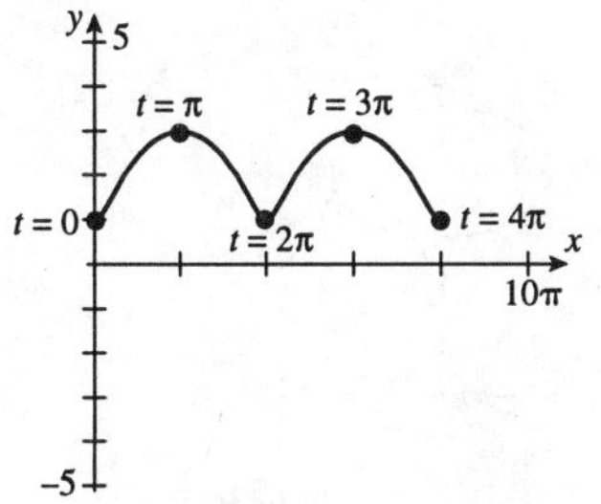

c. $x = t - 2\sin t, y = 1 - 2\cos t$

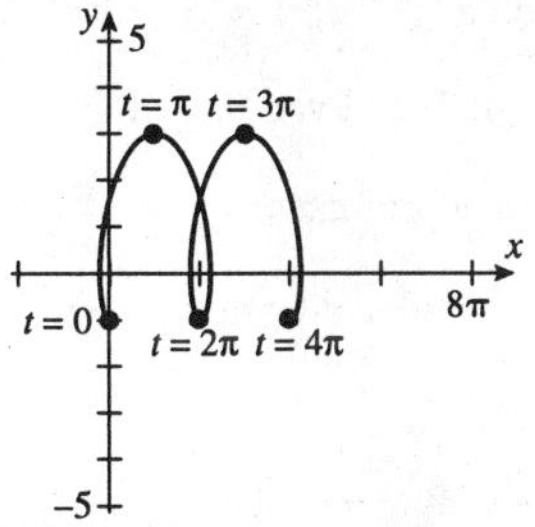

41. Drawn and written responses

Section 10.7 Chapter Review

Concepts Test

1. False. If $a = 0$, the graph is a line.

3. False. The defining condition of an ellipse is $|PF| = e|PL|$ where $0 < e < 1$. Hence the distance from the vertex to a directrix is a greater than the distance to a focus.

5. True. The asymptotes for both hyperbolas is $y = \pm\frac{b}{a}x$.

7. True. As e approaches 0, the ellipse becomes more circular.

9. False. The equation $x^2 + y^2 = 0$ represents the point (0, 0).

11. True. If $k > 0$, the equation is a horizontal hyperbola; if $k < 0$, the equation is a vertical hyperbola.

13. False. If $b > a$, then the distance is $2\sqrt{b^2 - a^2}$.

15. True. Since light from one focus reflects to the other focus, light away from the first focus will reflect beyond the other focus.

17. False. For example, if $C = D = F = 0$, then the graph is the point (0, 0).

19. True. If the point is (r_0, θ_0), then the other coordinates are $(r_0, \theta_0 + 2\pi k)$ and $(-r_0, \theta_0 + \pi k)$ where k is an integer.

21. True. The graph has 3 leaves and the area is exactly one quarter of the circle $r = 4$. (See Problem Set 10.5 #33.)

23. True. The point $\left(-\sqrt{\cos 2\theta}, \theta\right)$ of $r^2 = \cos 2\theta$ is represented by $\left(\sqrt{\cos 2(\theta + \pi)}, \theta + \pi\right) = \left(\sqrt{\cos 2\theta}, \theta + \pi\right)$ of $r = \sqrt{\cos 2\theta}$.

25. False. For example, $x = 0, y = t$, and $x = 0, y = -t$ both represent the line $x = 0$.

27. False. For example, the graph of $x = t^2$, $y = t$ does not represent y as a function of x. $y = \pm\sqrt{x}$, but $h(x) = \pm\sqrt{x}$ is not a function.

29. False. For example, if $x = t^3$, $y = t^3$ then $y = x$ so $\frac{d^2y}{dx^2} = 0$, but $\frac{g''(t)}{f''(t)} = 1$.

Sample Test Problems

1. a. $x^2 - 4y^2 = 0;\ y = \pm\frac{x}{2}$

(5) Two intersecting lines

b. $x^2 - 4y^2 = 0.001;\ \frac{x^2}{1000} - \frac{y^2}{250} = 1$

(9) A hyperbola

c. $x^2 - 4 = 0;\ x = \pm 2$

(4) Two parallel lines

d. $x^2 - 4x + 4 = 0;\ x = 2$

(3) A single line

e. $x^2 + 4y^2 = 0;\ (0, 0)$

(2) A single point

f. $x^2 + 4y^2 = x;\ x^2 - x + \frac{1}{4} + 4y^2 = \frac{1}{4};$

$$\frac{\left(x - \frac{1}{2}\right)^2}{\frac{1}{4}} + \frac{y^2}{\frac{1}{16}} = 1$$

(8) An ellipse

g. $x^2+4y^2=-x;\ x^2+x+\frac{1}{4}+4y^2=\frac{1}{4};$

$$\frac{\left(x+\frac{1}{2}\right)^2}{\frac{1}{4}}+\frac{y^2}{\frac{1}{16}}=1$$

(8) An ellipse

h. $x^2+4y^2=-1$

(1) No graph

i. $(x^2+4y-1)^2=0,\ x^2+4y-1=0$

(7) A parabola

j. $3x^2+4y^2=-x^2+1;\ x^2+y^2=\frac{1}{4}$

(6) A circle

3. $9x^2+4y^2-36=0;\ \frac{x^2}{4}+\frac{y^2}{9}=1$

Vertical ellipse; $a=3,\ b=2,\ c=\sqrt{5}$

Foci are at $\left(0,\ \pm\sqrt{5}\right)$ and vertices are at $(0, \pm 3)$.

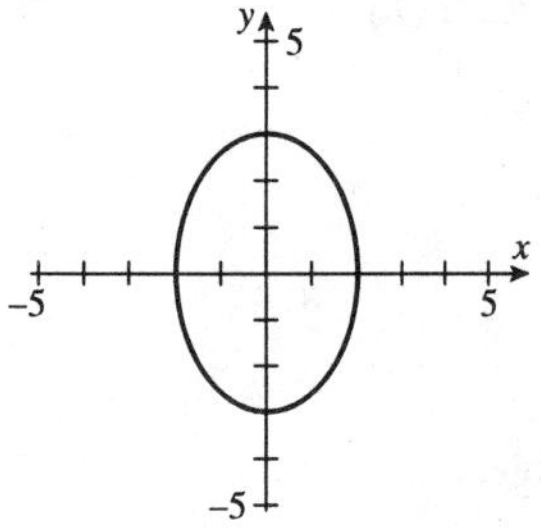

5. $x^2-9y=0;\ x^2=9y;\ x^2=4\left(\frac{9}{4}\right)y$

Vertical parabola; opens upward; $p=\frac{9}{4}$

Focus at $\left(0,\ \frac{9}{4}\right)$ and vertex at (0, 0).

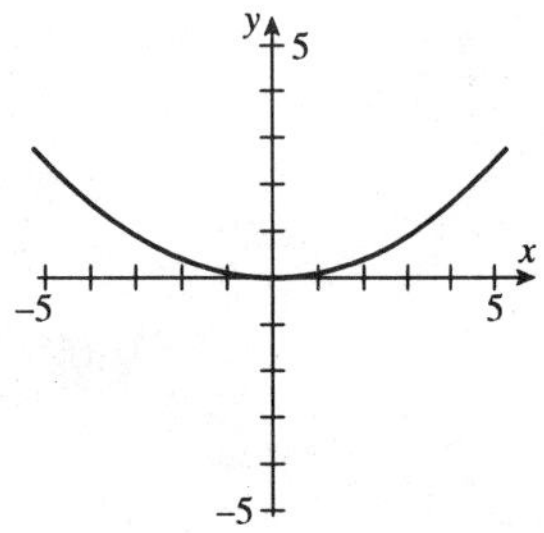

7. $9x^2+25y^2-225=0;\ \frac{x^2}{25}+\frac{y^2}{9}=1$

Horizontal ellipse, $a=5,\ b=3,\ c=4$

Foci are at (±2, 0) and vertices are at (±5, 0).

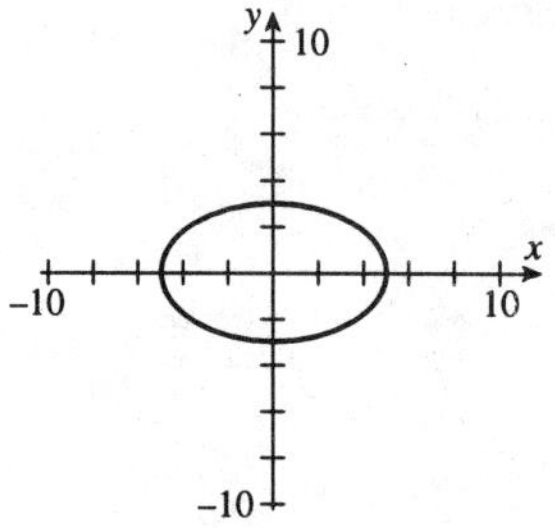

9. $r=\frac{5}{2+2\sin\theta}=\frac{\left(\frac{5}{2}\right)(1)}{1+(1)\cos\left(\theta-\frac{\pi}{2}\right)}$

$e=1$; parabola

Focus is at (0, 0) and vertex is at $\left(0,\ \frac{5}{4}\right)$

(in Cartesian coordinates).

11. Horizontal ellipse; center at (0, 0), $a=4$,

$e=\frac{c}{a}=\frac{1}{2},\ c=2,\ b=\sqrt{16-4}=2\sqrt{3}$

$$\frac{x^2}{16}+\frac{y^2}{12}=1$$

13. Horizontal parabola;

$y^2=ax,\ (3)^2=a(-1),\ a=-9$

$y^2=-9x$

15. Horizontal hyperbola, $a=2$,

$x=\pm 2y,\ \frac{a}{b}=2,\ b=1$

$$\frac{x^2}{4}-\frac{y^2}{1}=1$$

17. Horizontal ellipse; $2a=10,\ a=5,\ c=4-1=3,$

$b=\sqrt{25-9}=4$

$$\frac{(x-1)^2}{25}+\frac{(y-2)^2}{16}=1$$

19. $4x^2+4y^2-24x+36y+81=0$

$4(x^2-6x+9)+4\left(y^2+9y+\frac{81}{4}\right)=-81+36+81$

$4(x-3)^2+4\left(y+\frac{9}{2}\right)^2=36$

$(x-3)^2+\left(y+\frac{9}{2}\right)^2=9$, circle

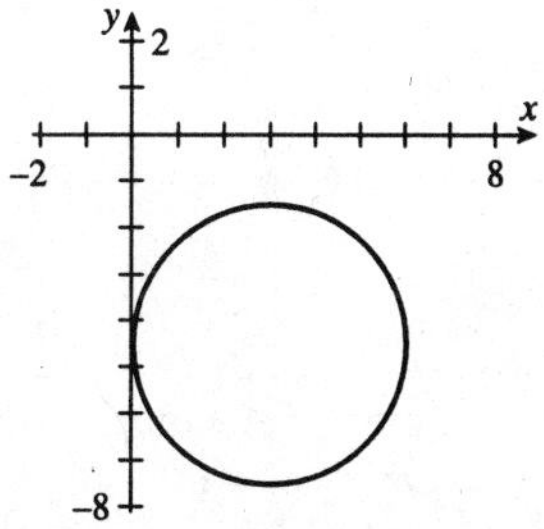

21. $x^2 - 8x + 6y + 28 = 0$

$x^2 - 8x + 16 = -6y - 28 + 16$

$(x-4)^2 = -6(y+2)$, parabola

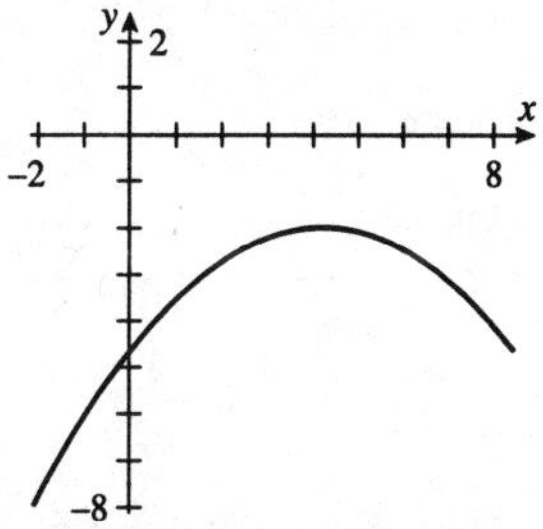

23. $r = 6\cos\theta$

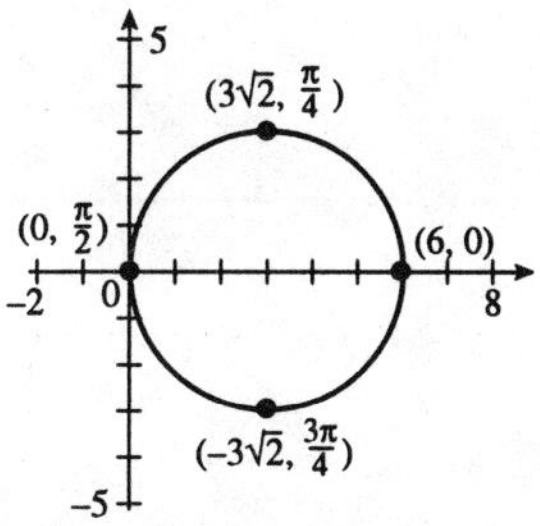

25. $r = \cos 2\theta$

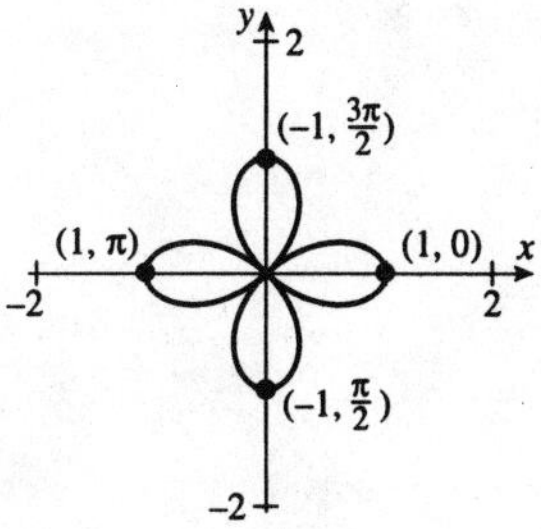

27. $r = 4$

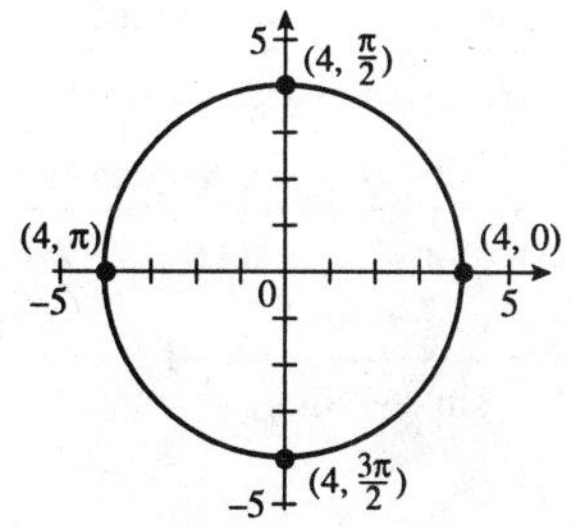

29. $r = 4 - 3\cos\theta$

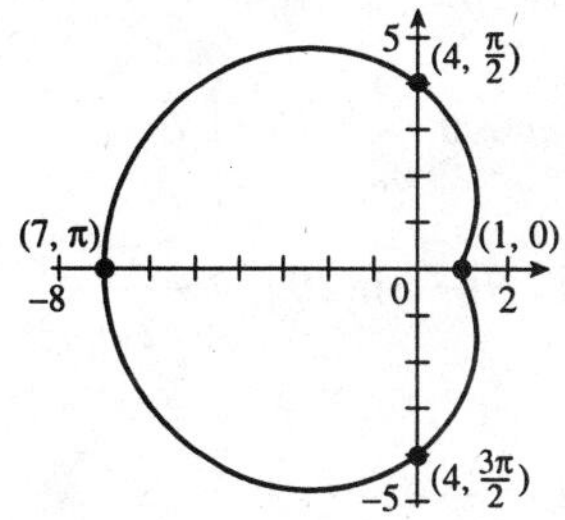

31. $\theta = \dfrac{2}{3}\pi$

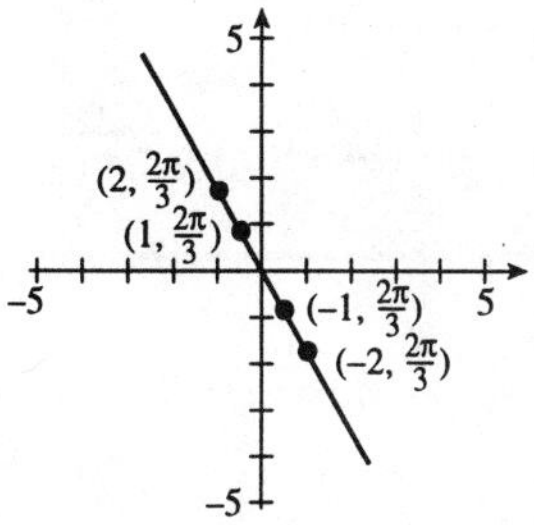

33. $r^2 = 16\sin 2\theta$

$r = \pm 4\sqrt{\sin 2\theta}$

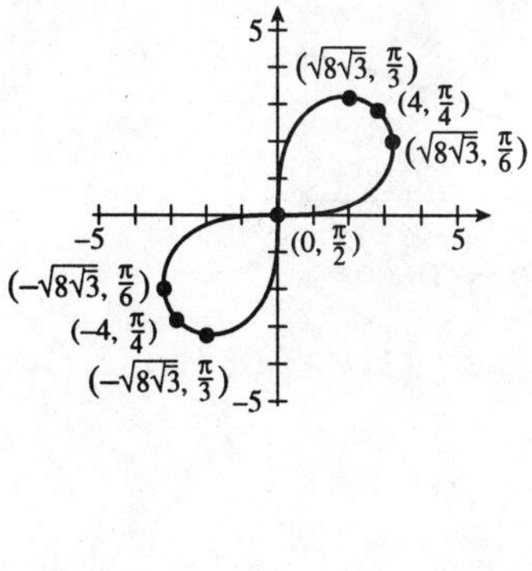

35. $f(\theta)=3+3\cos\theta,\ f'(\theta)=-3\sin\theta$

$$m=\frac{(3+3\cos\theta)\cos\theta+(-3\sin\theta)\sin\theta}{-(3+3\cos\theta)\sin\theta+(-3\sin\theta)\cos\theta}$$

$$=\frac{\cos\theta+\cos^2\theta-\sin^2\theta}{-\sin\theta-2\cos\theta\sin\theta}=\frac{\cos\theta+\cos 2\theta}{-\sin\theta-\sin 2\theta}$$

At $\theta=\frac{\pi}{6},\ m=\frac{\cos\frac{\pi}{6}+\cos\frac{\pi}{3}}{-\sin\frac{\pi}{6}-\sin\frac{\pi}{3}}=-1.$

37. $A=2\cdot\frac{1}{2}\int_0^{\pi}(5-5\cos\theta)^2\,d\theta$

$$=25\int_0^{\pi}(1-2\cos\theta+\cos^2\theta)d\theta$$

$$=25\int_0^{\pi}\left(\frac{3}{2}-2\cos\theta+\frac{1}{2}\cos 2\theta\right)d\theta$$

$$=25\left[\frac{3}{2}\theta-2\sin\theta+\frac{1}{4}\sin 2\theta\right]_0^{\pi}=\frac{75\pi}{2}$$

39. $\frac{x^2}{400}+\frac{y^2}{100}=1;\ \frac{x}{200}+\frac{yy'}{50}=0$

$y'=-\frac{x}{4y};\ y'=-\frac{2}{3}$ at (16, 6)

Tangent line: $y-6=-\frac{2}{3}(x-16)$

When $x=14,\ y=-\frac{2}{3}(14-16)+6=\frac{22}{3}.$

$k=\frac{22}{3}$

41. $t=\frac{y}{4}$

$x=\frac{y^2}{4}$ or $y^2=4x$

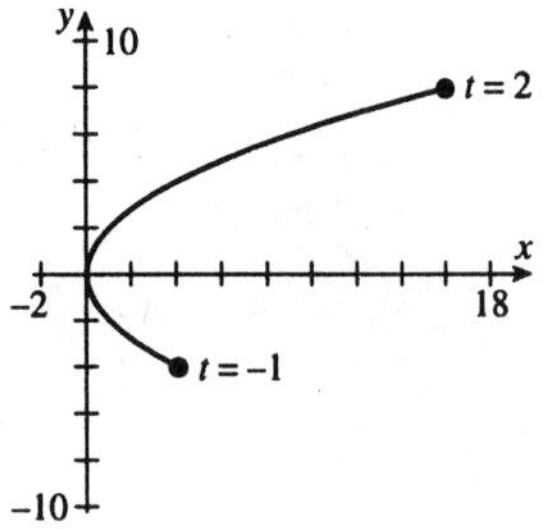

43. $\frac{x^2}{4}=\sec^2 t,\ y^2=\tan^2 t$

$\frac{x^2}{4}-y^2=1$

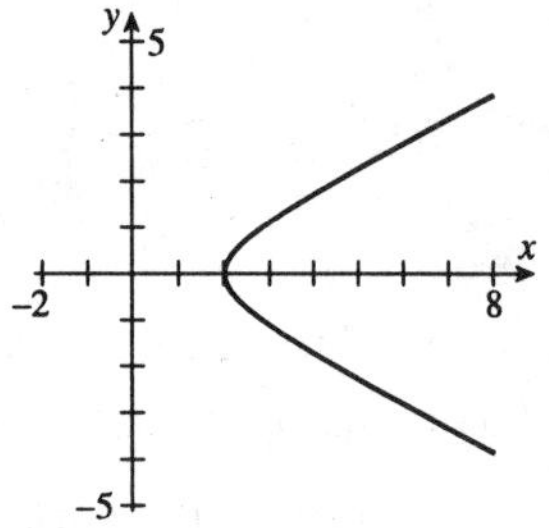

45. $\frac{dx}{dt}=-3e^{-t},\ \frac{dy}{dt}=\frac{1}{2}e^t$

$\frac{dy}{dx}=\frac{\frac{1}{2}e^t}{-3e^{-t}}=-\frac{1}{6}e^{2t}$

At $t=0,\ x=3,\ y=\frac{1}{2}$, and $\frac{dy}{dx}=-\frac{1}{6}.$

Tangent line: $y-\frac{1}{2}=-\frac{1}{6}(x-3)$ or $x+6y-6=0$

Technology Pages

These pages provide specific instructions and keystrokes for calculator and computer material directly related to the student text. Specific page references to the text are noted by a large bullet (●) in the left margin, followed by a page number. If the material is related to a specific example, the example is referenced beside the page number.

Section 2.1

● **(p.92) Function graphing**

Below we show how to graph the function $\frac{10}{x-2}+e^x$.

Graphing calculators

First store the function.

Casio 7700:
1 0÷(X–2)+e(X)
[SHIFT] [F]MEM STO 1

TI-81, TI-82:
[Y=]
10/(X–2)+e^X

TI-85:
[GRAPH] Y =
10 / (x – 2) + e^x
[EXIT]

Then graph the function $\frac{10}{x-2}+e^x$ stored above. Don't forget to set the window (range) first.

Casio 7700:
[Graph] [SHIFT] [F]MEM RCL 1 [EXE]

TI-81, TI-82:
[GRAPH]

TI-85:
GRAPH (the menu option, not the GRAPH key)

The TI-85 calculator can generate a default y-range using ZOOM ZFIT after the graph is generated; the window is then set to include all of the points in the internal table of values.

- **(p.93)** *Computer algebra systems*

To graph the function $\frac{10}{x-2}+e^x$ using the window $-10 \le x \le 10$, $-10 \le y \le 10$:

Mathematica:

```
Plot[10/(x-2)+E^x,{x,–10,10},PlotRange->{–10,10}]
```

Maple:

```
plot(10/(x–2)+exp(x),x=–10..10,y=–10..10);
```

Derive:

1) Author the expression $10/(x-2)+\hat{e}$^x.
2) Choose Plot to move to the plot window.
3) Choose Scale to set the x-scale to 5 and the y-scale to 5 (since $(10-(-10))/4=5$).
4) Choose Move to move the cross to (0, 0).
5) Choose Center to center the graph at (0, 0).
6) Choose Plot to generate the graph.

NOTE: If this is the first graph of your Derive session, you don't need steps 4) and 5), since the graph is already centered at (0, 0).

To graph the function $\frac{10}{x-2}+e^x$ using the default y-range the command would be:

Mathematica:

```
Plot[10/(x-2)+E^x,{x,–10,10}]
```

Maple:

```
plot(10/(x–2)+exp(x),x=–10..10);
```

Derive does not have a default y-range capability.

Section 2.2

- **(p.99) Generating tables for calculators without a table feature**

Graph and Trace after running the following program. The y-range is not set by the program since we are focusing on the table rather than the graph. NOTE: For the Casio 7700 the y-range is set very large so that all of the graph is likely to appear; the Trace will not work where the graph does not appear.

TI-85:

```
PROGRAM: TABLE
:Prompt X
:Prompt D
:X-D*63→xMin
:X-D*63→xMax
:10*D→xScl
```

Casio 7700

```
'TABLE
"X0"?→X
"DX"?→D
Range X,X+94D,94
D÷10, –1E50, 1E50,
1E49
```

TI-81:

```
PROGRAM: TABLE
:Disp "X0"
:Input X
:Disp "DX"
:Input D
:X-D*47→xMin
:X-D*47→xMax
:10*D→xScl
```

- **(p.99) Generating function values**

Casio 7700

```
'EVALUATE
"X"? → X
f1
```

TI-81

```
PROGRAM:EVALUATE
:Disp "X"
:Input X
:Disp Y1
```

TI-85 users should use the EVAL command under the GRAPH menu. TI-82 users should use the **value** command under the CALC menu.

- **(p.100, Example 2) Creating a graph and table for the function**

$$f(t) = -5e^{-10t}(\cos(10t) + \sin(10t))$$

Graphing calculators

Graphing calculators in function mode normally use x rather than t.
To store the function $f(x) = -5e^{-10x}(\cos(10x) + \sin(10x))$:

TI-81, TI-82:

$Y_1 = -5e\wedge(-10X)*(\cos(10X) + \sin(10X))$

TI-85:

$y_1 = -5e\wedge(-10x)*(\cos(10x) + \sin(10x))$

Casio 7700:

$f_1{:}-5e(-10X)*(\cos(10X) + \sin(10X))$

To create a table of values for the function stored above:

Use the program named Table shown above with X0=0 and DX=0.1. After running the program you must graph the function and then Trace to get the numbers in the table. Use the built-in table feature if one is available instead of a program.

Computer algebra systems

To define and graph the function $f(x) = -5e^{-10x}(\cos(10x) + \sin(10x))$ using the graph window $0 \le t \le 1$, $-7 \le y \le 7$:

Mathematica:

```
f[t_]:=-5*E^(-10*y)*(Cos[10*t]+Sin[10*t])
Plot[f[t],{t,0,1},PlotRange->{-7,7}]
```

Maple:

```
f(t):=-5*exp(-10*y)*(cos(10*t)+sin(10*t));
plot(f(t), t=0..1, y=-7..7);
```

Derive:

1) Author the expression
    f(t):=-5*ê^(-10*t)*(cos(10*t)+sin(10*t))
2) Choose Plot.
3) Choose Scale and set the x-scale to 0.25 and the y-scale to 3.5.
4) Choose Move and move the cross to (0.5, 0).
5) Choose Center.
6) Choose Plot.

To produce a table of values for the function $f(x) = -5e^{-10x}(\cos(10x) + \sin(10x))$ defined above:

Mathematica:

```
Table[{t,f[t]}, {t,0,1,0.1}]//TableForm//N
```

Derive:

1) Author the expression
    Vector([t, f(t)], t, 0, 1, 0.1)
2) Choose approXimate.

- **(p.103, Example 3) Graphing two functions together on the same set of axes**

Graphing Calculators

When two or more functions are graphed together, each function on the screen can be traced. Use the up and down arrow keys to toggle between graphs.

To graph the functions $50\sin(2\pi x)+5x+80$ and 150 together:

TI-81, TI-82:
Save the function 50sin(2πX)+5X+80 in Y_1 and save 150 in Y_2, and then Graph.

TI-85:
Save the cost function 50sin(2πx)+5x+80 in y_1 and save 150 in y_2, and then Graph.

Casio 7700:
1) Save the cost function 50sin(2πX)+5X+80 in function memory f_1 and save 150 in function memory f_2.
2) Give the command to graph function f_1, then type SHIFT EXE, then give the command to graph function f_2 and finally type EXE.

Computer Algebra Systems

To graph the functions $50\sin(2\pi x)+5x+80$ and 150 together:

Mathematica:

```
Plot[{50*Sin[2*Pi*t]+5*t+80, 150},{t,0,5}]
```

Maple:

```
plot({50*sin(2*Pi*t)+5*t+80, 150}, t=0..5);
```

Derive:
1) Author 50*sin(2*Pi*t)+5*t+80 and then Plot.
2) Author 150 and then Plot.

Section 2.3

- **(p.106-7) Graphing the function $100e^{kx}$ for several values of k on the screen at once using Procedure 1 from the text**

Calculator suggestions
TI calculators: Store the function as 100e^(KX) in function memory Y_1 and turn off function memory Y_1. Now store values in K (such as 0.2→K) and use DrawF Y_1 to create the graphs. A program with the single statement DrawF Y_1 reduces the number of keystrokes needed.

Casio 7700: Store the function as 100e(KX) in function memory f_1, store values in K (such $0.2 \rightarrow K$), and graph with the command Graph $Y=f_1$.

Computer suggestions

Derive: Author the expression 100ê^(k*t), then use Manage-Substitute to replace k with various numerical values (leaving t alone), and then Plot each time.

Mathematica: Define a function of both *t* and *k* and then plot for various *k* (either separately or together). To plot the graphs together the commands would be

```
f[t_,k_]:=100*E^(k*t);
Plot[{f[t,0.05], f[t,0.2], f[t,0.5]}, {t,0,5}]
```

- **(p.108) Plotting data and equation graphs together using the time-temperature data from the text and the function $70+28e^{-kx}$**

Calculator suggestions

TI-82:

1) Type [STAT] [1] to begin editing the data. Clear any existing data in list L1 by putting the cursor on the L1 symbol and typing [CLEAR] [ENTER]. Clear list L2 similarly. Now enter the data, with the time in list L1 and the temperature in list L2.

2) Type [2nd] [Y=] (STATPLT) [1] to set the statistical plotting options for Plot 1. Set the options on the screen as shown below.

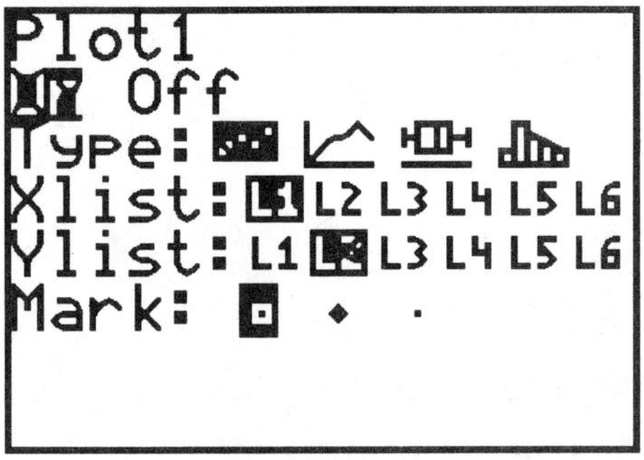

Set the graph window to $0 \leq x \leq 25$, $70 \leq y \leq 100$ with the [WINDOW] key. Press the [GRAPH] to produce the scatter plot of the data.

3) Store the function as 70+28e^(-KX) in function memory Y_1. Store various numerical values in memory K (such as $0.2 \rightarrow K$) and graph the function for each value using the [GRAPH] key. Keep adjusting K until a good fit is achieved.

TI-85:

1) Press STAT F2 (EDIT) to begin editing the statistical data. Press ENTER twice to choose the lists xStat and yStat as the lists for the time and temperature respectively (if xStat and yStat are not the currently chosen names they can be chosen from the menu). Choose F5 (CLRxy) to clear any previous data. Enter the first data pair by typing 0 ENTER 98.3 ENTER, then enter the rest of the data similarly.
2) Set the graph window to $0 \le x \le 25$, $65 \le y \le 100$ with GRAPH F2 (RANGE). Plot the data with STAT F3 (DRAW) F2 (SCAT).
3) Store the function as 70+28 e^(-K x) in function memory *y*1. Store various numerical values in memory K (such as $0.2 \rightarrow K$) and graph the function for each value using GRAPH F5. The data must be plotted again as in 2) each time. Keep adjusting K until a good fit is achieved.

Casio 7700:

1) Use the Range key to set the graph window to $0 \le x \le 25$, $70 \le y \le 100$. Type MODE ÷ to put the calculator into regression (two variable statistics) mode. Type MODE SHIFT 1 and then MODE SHIFT 3 to put the calculator in STORE and DRAW modes. Choose EDIT from the menu, then choose ERS and YES to erase any previous data. Type 0 SHIFT , 98.3 DT to put the first data point in. Enter the rest of the data similarly. The data points are graphed as they are put in.
2) Store the function in function memory f_1 as 70+28e(-KX), store various values in memory K (such as $0.2 \rightarrow K$) and graph using Graph Y=f_1. Adjust K until a good fit is achieved.

Computer suggestions

Derive :

1) Author the data as a vector of ordered pairs:

   [[0,98.3],[4,89.5],[8,81.3],[12,77.6],[16,76.5],[20,72.1],[24,71.3]]

2) Choose Plot to move to the graph window and then Plot again to plot the data.
3) Choose Algebra to move to the algebra window and then Author the function as:

   70+28ê^(-k*t)

4) Use Manage-Substitute to replace k with various numerical values. Plot for each new k value until a good fit is achieved.

Mathematica: Define the data as a list and define the function as a function of t and k. Choose various numerical values for k, plot the data and function separately and then combine using Show. Adjust k until a good fit is achieved. Below k is chosen to be 0.2.

```
data={{0,98.3},{4,89.5},{8,81.3},{12,77.6},{16,76.5},
     {20,72.1},{24,71.3}};
f[t_,k_]:=70+28*E^(-k*t);
p1=ListPlot[data];
p2=Plot[f[t,0.2],{t,0,24}];
Show[p1,p2]
```

Section 2.4

- **(p.116, Example 1) Zooming with Graphing Calculators**

One more zoom and Trace shows us that the root lies in the interval $1.5320854 \le x \le 1.5320915$, so $x = 1.53209$, accurate to five decimal places.

- **(p.117) Zooming with Derive**

Because the zoom occurs at the center of the screen, you need to move the cursor to the part of the curve where you want the zoom to occur, then use the Center option, and finally use the Zoom option (the F9 and F10 keys can also be used to zoom in or out respectively). Three consecutive zooms change the x- and y-scales by a factor of 10 (you can also zoom in the x- or y-directions separately). Hint: You don't have to wait for the entire graph to appear before zooming again.

Section 2.5

- **(p.126-7, Example 2) Finding roots of the function** $x^2 - 2^x$

Graphing Calculators

TI-85:

1) Store the function $f(x)$ in function memory $y1$ as

   y1 = x2 – 2^x

2) Set the graph window and graph the function. A good choice for the graph window can be obtained using ZOOM [MORE] ZDECM which sets the window to $-6.3 \le x \le 6.3$, $-3.1 \le x \le 3.1$ and gives "nice" decimal values for x when tracing.
3) Choose MATH from the menu (after pressing the [MORE] key), then choose ROOT from the next menu.
4) Use the left and right arrow keys to Trace to a point near the root you are looking for; this will be the initial guess of the root-finder. Our suggestion was to use $x = -1$, but by tracing you can get closer if you want.
5) Press [ENTER] and the calculator should give you the root you are looking for. If your initial guess was not close enough to the root, it might find a different root, or no root at all.

NOTE: The TI-85 can also solve equations using the SOLVER command. With this option, you can type in a numerical starting value, rather than getting one by tracing the graph.

TI-82:

1) Store the function $f(x)$ in function memory Y1 as

   Y1 = X^2 – 2^X

2) Set the graph window and graph the function. A good choice for the graph window can be obtained by pressing [ZOOM] and choosing 4:ZDecimal which sets the window to $-4.7 \le x \le 4.7$, $-3.1 \le x \le 3.1$ and gives "nice" decimal values for x when tracing.
3) Press [2nd] CALC and then choose 2:root from the menu.
4) Use the left and right arrow keys to Trace to a point to the left of the root and press [ENTER], then Trace to a point to the right of the root and press [ENTER]. This establishes a lower and upper bound for a starting interval. Finally, Trace to a point near the root itself for an initial guess. The TI-82 requires both an initial interval and an initial guess.
5) Press [ENTER] and the calculator should give you the root you are looking for.

NOTE: The TI-82 can also find roots using the solve command by pressing [MATH] [0]. With this option, you can type in a numerical starting value, rather than getting one by tracing the graph.

Casio 7700:

1) Enter the following program (the bisection method described in the text) into any empty program memory. We assume that the program is entered into program memory 1 for the rest of the instructions below.

```
'BISECT
"A"?→A
"B"?→B
A→X
f1→D
Lbl 2
(A+B)÷2→C
C→X◢
f1→E
DE<0⇒Goto 1
C→A
Goto 2
Lbl 1
C→B
Goto 2
```

2) Store the function in memory f_1 as

   $X^2 - 2x^yX$

3) Run the program from 1) by typing [SHIFT] PRGM Prg 1 [EXE] (if the program was stored in, say, program memory 2 you would run Prg 2 instead of Prg 1).
4) Type –2.5 [EXE] and 0 [EXE] for A and B when prompted for them. These are the endpoints of the starting interval.
5) The midpoint of the current interval is displayed. Each time you press [EXE] a new interval is calculated and the new midpoint is displayed. Keep pressing [EXE] until the same value appears three times to the desired number of decimal places.
6) Press [AC] to stop the program, and press [EXE] to start it again. Repeat 4) and 5) with the other starting intervals.

Computer Algebra Systems

Mathematica:

NOTE: Mathematica uses a numerical root-finder which requires an initial guess (as with Newton's method).

```
FindRoot[x^2==2^x, {x, -1}]
FindRoot[x^2==2^x, {x, 2}]
FindRoot[x^2==2^x, {x, 4}]
```

Derive:

NOTE: Derive uses a numerical root-finder which requires an initial interval (as with the bisection method).

1) Author the equation

    x^2=2^x

2) Put Derive into Approximate mode by choosing Options-Precision-Approximate.
3) Choose Solve.
4) Type in –2.5 for Lower and 0 for Upper.
5) Highlight the equation from 1) again, and choose Solve. Use 0 for lower and 2.5 for Upper this time. Repeat using 2.5 for Lower and 5 for Upper.
6) Put Derive back into exact mode by choosing Options-Precision-Approximate (failure to do this step can result in confusing results later).

- **(p.127, Example 3) Finding exact and approximate solutions to the equation** $2sin^2x + 3sinx = 2$

Mathematica:

```
sol=Solve[2*Sin[x]^2+3*Sin[x]==2, x]
        (solves the equation)
N[sol]
        (turns the exact result into a
        numerical approximation)
```

Derive:

1) Author the equation

    2 *sin(x)^2+3*sin(x)=2

2) Choose Options-Precision-Exact to put Derive in exact mode, if it is not already.
2) Choose Solve.
3) Choose approXimate to turn the exact solutions (if there are any) into numerical approximations.

Section 2.6

- **(p.137, Example 3) Fitting a linear equation to data and plotting the data and equation together**

Below we use the year versus temperature data from the text.

Graphing calculators

Casio 7700:

1) Use the [Range] key to set the graph window to $1970 \le x \le 1990$, $46.9 \le y \le 47.9$. Type [MODE] [÷] to put the calculator into regression mode and type [MODE] [4] to fit a linear model. Type [MODE] [SHIFT] [1] and then [MODE] [SHIFT] [3] to put the calculator in STORE and DRAW modes. Choose EDIT from the menu, then choose ERS and YES to erase any previous data. Type 1970 [SHIFT] , 46.97 DT to put the first data point in. Enter the rest of the data similarly. The data points are graphed as they are put in.
2) Type [G↔T] to move to the text screen and then choose REG from the menu. The menu now shows A (y-intercept) and B (slope). Type [F1] [EXE] to get A and type [F2] [EXE] to get B.
3) The data has already been graphed.
4) Type [Graph] [SHIFT] Line [1] to graph the regression line.

NOTE: If you make a mistake typing the data in, type [SHIFT] Cls [EXE] to clear the graph screen, then choose EDIT from the menu. After editing the data choose PRE to move to the previous menu, choose CAL from the menu, and then repeat steps 2) through 4).

TI-82:

1) Type [STAT] [1] to begin editing the data. Clear any existing data in list L_1 by putting the cursor on the L_1 symbol and typing [CLEAR] [ENTER]. Clear list L_2 similarly. Now enter the data, with the year in list L_1 and the temperature in list L_2.
2) Type [STAT] [▸] [9] [ENTER] to get the results of a linear regression.
3) Type [2nd] [Y=] (STATPLT) [1] to set the statistical plotting options for Plot 1. Set the options on the screen as shown below.

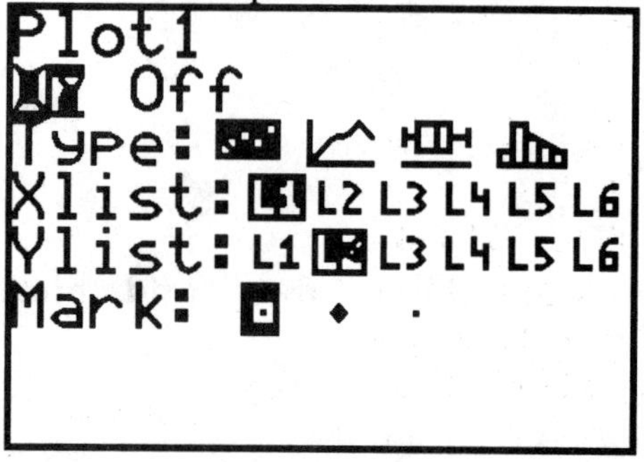

Set the graph window to $1970 \le x \le 1990$, $46.9 \le y \le 47.9$ with the [WINDOW] key. Press the [GRAPH] to produce the scatter plot of the data.

4) Press [Y=] and [CLEAR] to clear the function stored in Y1, or move to an empty storage location. Type [VARS] [5] [▸] [▸] [7] to bring the regression equation into the function memory. Type [GRAPH] to produce the graph of the equation along with the data.

TI-85:

1) Press [STAT] [F2] (EDIT) to begin editing the statistical data. Press [ENTER] twice to choose the lists xStat and yStat as the lists for year and temperature respectively (if xStat and yStat are not the currently chosen names they can be chosen from the menu). Choose [F5] (CLRxy) to clear any previous data. Enter the first data pair by typing 1970 [ENTER] 46.97 [ENTER], then enter the rest of the data similarly.
2) Press [STAT] [F1] (CALC) [ENTER] [ENTER] and then [F2] to choose a linear model. The results of a linear regression are given.
3) Set the graph window to $1970 \le x \le 1990$, $46.9 \le y \le 47.9$ with [GRAPH] [F2] (RANGE). Plot the data with [STAT] [F3] (DRAW) [F2] (SCAT).
4) Choose [F4] (DRREG) from the current menu (you should still have the DRAW menu from step 3) to draw the regression line.

Computer Algebra Systems

Mathematica:

1) data={{1970, 49.97}, {1971, 50.7}, ... , {1990, 50.75}}
   (You would, of course, actually enter all of the data).
2) f1=Fit[data, {1, x}, x]
3) p1=ListPlot[data,PlotStyle->PointSize[.02]]
   p2=Plot[f1, {x, 1970, 1990}]
   Show[p1,p2]

Derive:

1) Author the statement

   data:=[[1970, 49.97], [1971,50.7], ... , [1990, 50.75]]

   (You would, of course, actually enter all of the data).
2) Choose Transfer Merge Derive and then type the word "regress" in response to the "file:" prompt. Author the statement

   regr(data, 21)

and choose approXimate to get the data for the regression equation. The number 21 is the number of data points.

3) Author the regression equation in the form $a+b*x$ where a and b are given in part 2. Plot the regression equation, then go back to the Algebra screen, highlight the data statement (from part 1) and Plot again. The x- and y-scales should be set to (1990–1970)/4=5 and (47.9–46.9)/4=0.25 and the x- and y-coordinates of the center of the screen should be (1990+1970)/2=1980 and (47.9+46.9)/2=47.4.

Section 2.7

- **(p.143) Fitting an Exponential Model with Graphing Calculators**

The procedure for fitting and graphing a nonlinear model is basically the same as for fitting and graphing a linear model, except that you must choose the appropriate nonlinear model instead of a linear model. Therefore, you can follow the instructions for linear models from Section 2.6 of this manual with the following adjustments. You should get a graph of both the data and the fitted curve. Of course, an appropriate graph window must be chosen, based on the data values.

<u>Casio 7700</u>:
In step 1), choose [MODE] [6] to fit an exponential model instead of [MODE] [4] for a linear model.

<u>TI-82</u>:

In step 2), choose [STAT] [▸] [ALPHA] A [ENTER] to choose an exponential model instead of [STAT] [▸] [9] [ENTER] for a linear model.

<u>TI-85</u>:
In step 2), choose [STAT] [F1] [ENTER] [ENTER] [F4] for an exponential model instead of [STAT] [F1] [ENTER] [ENTER] [F2] for a linear model.

- **(p.143) Fitting an Exponential Model with Computer Algebra Systems**

In Section 2.6 of this manual we give instructions for fitting a linear model to data, and for plotting the data and the fitted curve together. Since we need both the original data and the natural logarithms of the y-

coordinates, we show below one way to use lists to produce the logarithms of the data points without typing them in separately. We use the data for the U.S. population from 1790 to 1860 as in the text.

Mathematica:

```
xdata={0,10,20,30,40,50,60,70}
ydata={3.929, 5.308, 7.24, 9.639, 12.861, 17.063, 23.192, 31.443}
data=Table[{xdata[[i]], ydata[[i]]}, {i, 1, Length[xdata]}]
lndata=Table[{xdata[[i]], Log[ ydata[[i]] ]},
        {i, 1, Length[xdata]}]
ListPlot[lndata, PlotStyle->PointSize[.02] ]
```

(The result of this command is a plot of x_i versus $\ln y_i$)

```
Fit[lndata, {1, x}, x]
```

(The result of this command is the linear regression equation 1.37528+0.029514 x)

```
p1=ListPlot[data,PlotStyle->PointSize[.02]]
p2=Plot[E^(1.37528)*E^(0.029514*x), {x, 0, 70}]
Show[p1,p2]
```

Derive:

1) Author the statements

```
1:   xdata:=[0,10,20,30,40,50,60,70]
2:   ydata:=[3.929, 5.308, 7.24, 9.639, 12.861, 17.063, 23.192,
          31.443]
3:   vector([ element(xdata, i), log( element(ydata, i) ) ], i,
          1, 8)
```

and then choose approXimate (return). Statement #4 will then be a list of ordered pairs of the form $[x_i, \ln y_i]$. Plot this statement to get a plot of x_i versus $\ln y_i$ (after choosing the scales and center appropriately).

2) Choose Transfer Merge Derive and then type "regress" in response to the "file:" prompt. Then Author the statement

```
regr(#4, 8)
```

and choose Simplify (return) and then approXimate (return) to get the data for the regression equation. The #4 in the statement above refers to the number of the statement which has the ordered pairs $[x_i, \ln y_i]$ and may be different for your Derive session. The result is a slope of b = 0.029513996 and a y-intercept of a = 1.375276903.

3) Author the regression equation in the form

```
ê^1.37528*ê^(0.029514*x)
```

Plot the regression equation, go back to the Algebra screen and Author the statement

vector([element(xdata, i), element(ydata, i)], i, 1, 8)

then approXimate and Plot the result. The x- and y-scales and center need to be reset after the plot from 1) (look at the x and y ranges of the data itself).

- **(p.144) Fitting a Power Model with Graphing Calculators**

As explained above for fitting exponential models, you can follow the instructions for linear models Section 2.6 with the following adjustments.

Casio 7700:
In step 1), choose [MODE] [7] to fit a power model instead of [MODE] [4] for a linear model.

TI-82:

In step 2), choose [STAT] [▸] [ALPHA] B [ENTER] to choose a power model instead of [STAT] [▸] [9] [ENTER] for a linear model.

TI-85:
In step 2), choose [STAT] [F1] [ENTER] [ENTER] [F5] for a power model instead of [STAT] [F1] [ENTER] [ENTER] [F2] for a linear model.

- **(p.146, Example 2) Fitting a Power Model with Computer Algebra Systems**

Below we use the data for distance from the sun vs. period for the planets.

Mathematica:

```
xdata={0.387, 0.732, 1, 1.524, 5.203, 9.555, 19.218, 30.11, 39.81}
ydata={0.24, 0.62, 1, 1.88, 11.86, 29.46, 84.01, 164.79, 247.68}
data=Table[{xdata[[i]], ydata[[i]]}, {i, 1, Length[xdata]}]
lndata=Table[{Log[ xdata[[i]] ], Log[ ydata[[i]] ]},
      {i, 1, Length[xdata]}]
ListPlot[lndata, PlotStyle->PointSize[.02] ]
      (The result of this command is a log-log plot)
Fit[lndata, {1, x}, x]
(The result of this command is the linear regression
```

```
equation -0.00305081 + 1.4993 x)
p1=ListPlot[data, PlotStyle->PointSize[.02]]
p2=Plot[E^(-0.00305081)*x^1.4993, {x, 0, 70}]
Show[p1,p2]
```

Derive:

1) Author the statements

   1: xdata:=[0.387, 0.732, 1, 1.524, 5.203, 9.555, 19.218, 30.11, 39.81]

   2: ydata:=[0.24, 0.62, 1, 1.88, 11.86, 29.46, 84.01, 164.79, 247.68]

   3: vector([log(element(xdata, i)), log(element(ydata, i))], i, 1, 9)

   and then choose approXimate (return). Statement #4 will then be a list of ordered pairs of the form $[\ln x_i, \ln y_i]$. Plot this statement to get a plot of $\ln x_i$ versus $\ln y_i$ as in Figure 11.

2) Choose Transfer Merge Derive and then type "regress" in response to the "file:" prompt. Then Author the statement

   regr(#4, 9)

   and choose Simplify (return) and then approXimate (return) to get the data for the regression equation. The #4 in the statement above refers to the number of the statement which has the ordered pairs $[\ln x_i, \ln y_i]$ and may be different for your Derive session. The result is a slope of b = 1.499297790 and a y-intercept of a = –0.003050806619.

3) Author the regression equation in the form

   ê^(–0.0030508)* x^1.4993

   Plot the regression equation, go back to the Algebra screen and Author the statement

   vector([element(xdata, i), element(ydata, i)], i, 1, 8)

   then approXimate and Plot the result. The x- and y-scales and the center need to be reset after the log-log plot from 1).

Section 3.1

- **(p.155, Example 6) Finding Derivatives with Computer Algebra**

Below we find $f'(x)$ for $f(x) = x^3$ using first the Limit command and then the Derivative command.

The Limit Command

We need to find $\lim_{h\to 0} \frac{(x+h)^3 - x^3}{h}$

Mathematica:
Limit[((x+h)^3–x^3)/h, h->0]

Derive:
1) Author
   ((x+h)^3–x^3)/h
2) Choose Calculus-Limit from the menu. Enter the correct expression number (the default should be correct after step 1), type h for the variable (the default is x), and enter 0 for the limit point (the default is 0).
3) Simplify the result from step 2.

The Derivative Command

Now we simply ask for the derivative of x^3 directly.

Mathematica:
   D[x^3, x]

Derive:
1) Author
   x^3
2) Choose Calculus-Derivative from the menu. Enter the correct expression number (the default should be correct after step 1), type x for the variable (the default is x), and enter 1 for the order (which is the default).
3) Simplify the result from step 2.

Section 3.6

- **(p.197) Making Circles Look Like Circles**

In order to get the "true" shape of certain figures such as circles, one unit should be the same length on the x-axis as on the y-axis. This can be accomplished on various computing devices as follows.

Graphing Calculators:

TI-81, TI-82, Casio 7700:
The width of the graph window in the x direction should be 3/2 times the height of the graph window in the y direction. The graph in Figure 2 in the text uses the window $-3 \le x \le 3$, $-2 \le x \le 2$ since $6 = \frac{3}{2} \cdot 4$.

TI-85:
The width of the graph window in the x direction should be 1.7 times the height of the graph window in the y direction. A graph similar to Figure 2

in the text can be produced with the window $-3.4 \le x \le 3.4$, $-2 \le x \le 2$ because $6.8 = (1.7) \cdot (4)$.

Computer Algebra Systems

Derive:
In the graph window, set the x-scale and the y-scale to be equal. A graph similar to Figure 2 in the text can be produced with both scales set to 1.

Mathematica:
Include "AspectRatio->Automatic" in the Plot statement. A graph similar to Figure 2 in the text can be generated with the statement

```
Plot[{Sqrt(4–x^2), –Sqrt(4–x^2)}, {x, –2, 2}, AspectRatio->Automatic]
```

- **(p.198) Graphing Implicit Equations Directly**

Mathematica:
The commands to generate the plot in Figure 3 in the text would be:

```
Needs["Graphics`ImplicitPlot`"]
ImplicitPlot[x^2+y^2==4, {x, -2,2},{y,-2,2}, PlotPoints->50]
```

NOTES: 1) The higher the value you use for "PlotPoints" the more accurate the graph will be. 2) With versions of Mathematica which do not have the ImplicitPlot package, you can get implicit plots using the ContourPlot command.

Section 4.1

- **(p.224) Numerical Max/Min Finding**

To find a maximum or minimum value of a function on an interval with a graphing calculator, one first graphs the function and then gives the command to find an extreme (maximum or minimum) value. The user is then prompted for an initial guess, which is given by tracing the curve until the cursor is near the extreme value. An interval may also be required, also determined by tracing. Details are given below.

TI-82: After graphing the function, press [2nd] CALC to access the calculate menu. Then press [3] to look for a minimum. You must now trace to determine an interval; move the cursor to the left of the minimum, press [ENTER], then move the cursor to the right of the minimum and press [ENTER] again. Finally, an initial guess is required; move the cursor near the minimum value and press [ENTER]. The cursor jumps to the minimum point on the curve and the coordinates are displayed.

TI-85: After graphing the function, press GRAPH MORE F1 (MATH) to access the MATH menu. Then press MORE F1 (FMIN) to look for a minimum. An initial guess is required by tracing; move the cursor near the minimum value and press ENTER. The cursor jumps to the minimum point on the curve and the coordinates are displayed.

Section 4.2

- **(p.238, Example 6) Numerical Inflection Point Finding**

TI-85: Graph the original function (not the second derivative), then press GRAPH MORE F1 (MATH) to access the MATH menu. Then press MORE F3 (INFLC) to look for an inflection point. An initial guess is required by tracing; move the cursor near the minimum value and press ENTER. The cursor jumps to the inflection point on the curve and the coordinates are displayed.

Section 5.1

- **(p.287, Example 6) Antiderivatives by Computer Algebra**

Below we show the specific command(s) you would give to various computer algebra systems to find $\int xe^{-x^2}\,dx$.

Mathematica:
Integrate[x*E^(-x^2), x]

Derive:
Author the statement
x*ê ^(-x^2)

and then choose Calculus-Integrate. Accept the defaults for expression number (the highlighted expression) and variable (x) and then press Return or Enter (do not enter anything for lower limit or upper limit; these are for definite integration to be explained later in this chapter). Simplify the result.

Section 5.3

- **(p.306, Example 3) Finding Riemann Sums**

We show how to estimate the area under the function $f(x) = e^{-x^2}$ between $a = 0$ and $b = 1$ using $n = 10$ rectangles with both left and right endpoint Riemann sums.

Computer Algebra Systems

1) First we define a function called *Riemann* which will compute Riemann sums for a function f. The arguments (inputs) are denoted below by a,b,n,r in that order. The left and right vertical boundaries are a and b, and the number of rectangles used is n. The remaining argument, r, determines the choice of the point $\bar{x}_i$ by the formula $\bar{x}_i = a+(i-1+r)\Delta x$. Thus $r=0$ corresponds to the choice $\bar{x}_i = a+(i-1)\Delta x = x_{i-1}$ (left endpoint Riemann sums) and $r=1$ corresponds to the choice $\bar{x}_i = a+i\cdot\Delta x = x_i$ (right endpoint Riemann sums).

Derive:
Author the statements

```
F(x):=
Riemann(a, b, n, r):=sum(F(a+(i–1+r)*(b–a)/n)*(b–a)/n, i, 1, n)
```

Mathematica:

```
Riemann[a_ , b_ , n_ , r_ ]:=Sum [f [a+(i–1+r)*(b–a)/n]*(b–a)/n, {i, 1, n}]
```

2) Next we define the function $f(x)=e^{-x^2}$ and estimate the area under f between $a=0$ and $b=1$ using $n=10$ rectangles with both left and right endpoint Riemann sums.

Derive: Author the statements

```
f(x):=ê^(–x^2)
Riemann(0, 1, 10, 0)        (left endpoint sums)
Riemann(0, 1, 10, 1)        (right endpoint sums)
```

and approXimate the second two statements.

Mathematica:

```
f[x_]:=E^(–x^2)
Riemann[0, 1, 10, 0]//N        (left endpoint sums)
Riemann[0, 1, 10, 1]//N        (right endpoint sums)
```

NOTE: The //N is needed to get a numerical rather than a symbolic result.

The results are 0.777818 for left endpoint sums and 0.714605 for right endpoint sums.

Graphing Calculators

1) First we write a program called *Riemann* which will compute Riemann sums for a function f stored in the function memory area. The inputs are denoted below by a,b,n,r in that order. The left and right vertical

boundaries are a and b, and the number of rectangles used is n. The remaining argument, r, determines the choice of the first point $\bar{x}_1$ by the formula $\bar{x}_1 = a + r\Delta x$. Thus $r = 0$ corresponds to the choice $\bar{x}_1 = a + (0)\Delta x = a = x_0$ (left endpoint Riemann sums) and $r = 1$ corresponds to the choice $\bar{x}_1 = a + 1 \cdot \Delta x = x_1$ (right endpoint Riemann sums).

Casio 7700:	TI-82:	TI-85:
'RIEMANN	PROGRAM: RIEMANN	PROGRAM: RIEMANN
"A"?→A	Prompt A	Prompt A
"B"?→B	Prompt B	Prompt B
"N"?→N	Prompt N	Prompt N
"R"?→R	Prompt R	Prompt R
0→S	0→S	0→S
(B-A) ÷N→H	(B-A)/N→H	(B-A)/N→H
A+RH→X	A+RH→X	A+R*H→x
0→I	For(I,1,N)	For(I,1,N)
Lbl 1	Y_1H+S→S	y1*H+S→S
f_1H+S→S	X+H→X	x+H→x
X+H→X	End	End
I+1→I	Disp S	Disp S
I<N⇒Goto 1		
S		

2) Next we store the function $f(x) = e^{-x^2}$ in the appropriate function memory as shown below.

Casio 7700:
Store the function as e(−X²) in function memory f_1.

TI-82:
Store the function as e^(−X%o) in function memory Y_1.

TI-85:
Store the function as e^(−x%o) in function memory y1.

Then run the program *Riemann* to estimate the area under f between $a = 0$ and $b = 1$ using $n = 10$ rectangles with both left and right endpoint Riemann sums. For left endpoint sums use 0 for r, and for right endpoint sums use 1 for r. The results are 0.777818 (left endpoint sums) and 0.714605 (right endpoint sums).

Section 5.4

- **(p.315, Example 2) Numerical Integration**

We show how to estimate the definite integral $\int_0^2 \sin(x^2)dx$ numerically.

Computer Algebra

Mathematica:

```
NIntegrate[ Sin[x^2], {x, 0, 2}]
```

Derive:
Author the statement

```
sin(x^2)
```

and choose Calculus-Integrate. Accept the defaults for expression number (the highlighted expression) and variable (x), and type in the limits of 0 and 2 using the Tab key to move from the lower to the upper limit. Then approXimate the result.

WARNING: It may be that not all of the digits reported by a computer algebra system are accurate when doing a numerical integration. It is wise to remain somewhat skeptical about the last digit or two reported.

Graphing Calculators

TI-85:
Give the command

```
fnInt(sin x^2,x,0,2)
```

The `fnInt` command is F5 on the CALC menu.

TI-82:
Give the command

```
fnInt(sin X^2,X,0,2)
```

The `fnInt` command is option 9: on the MATH menu.

Casio 7700:
Give the command

```
∫(sin Xxʸ2,0,2)
```

Begin the integration command by typing SHIFT ∫dx.

Section 5.5

- **(p.326, Ex. 9) Exact Definite Integration by Computer Algebra**

We attempt to find the definite integrals $\int_0^1 \sin(\sqrt{x})dx$ and $\int_0^1 \sqrt{\sin x}dx$ exactly.

Mathematica:

```
Integrate[ Sin[Sqrt[x]], {x, 0, 1}]
Integrate[ Sqrt[Sin[x]], {x, 0, 1}]
```

Maple:

```
integrate( sin(sqrt(x)), x=0..1);
integrate( sqrt(sin(x)), x=0..1);
```

Derive:
Author the expressions

```
sin(sqrt(x))
sqrt(sin(x))
```

Highlight the first expression and choose Calculus-Integrate from the menu. Choose the defaults for expression number and variable (x); then type 0 TAB 1 to enter the limits of integration. Finally Simplify the result. Repeat for the second expression.

- **(p.327) Finding a Numerical Approximation to a Symbolic Expression with Computer Algebra**

Any symbolic expression, such as 2π or sin(1), that results as the output of a computation can be approximated numerically by issuing the appropriate command, shown below, directly after the symbolic expression.

Mathematica:

```
N[%]
```

Derive:
Choose approXimate from the menu.

Maple:

```
evalf(");
```

NOTE: In Mathematica the symbol % refers the the previous output; in Maple the symbol " refers to the previous output.

Section 6.3

- **(p.368, Example 2) Parametric graphing**

We show how to plot the parametric equations $x = 2t + 1$ and $y = t^2 - 1$ for $0 \le t \le 3$.

TI-85:

1) To put the calculator in parametric mode type [2nd] [MORE] (MODE), then use the arrow keys to highlight Param in the fifth line of the mode screen and press [ENTER].
2) To enter the parametric equations, press the [GRAPH] key and choose [E(t)=] from the menu. Enter the equations for x and y in terms of t (use the [F1] key to enter the t variable).
3) Press [EXIT] and choose [RANGE] from the menu. Set tMin and tMax to 0 and 3 respectively, and set tStep to 0.15. Set the x- and y-ranges as indicated in the text.
4) Choose [GRAPH] from then menu.

Mathematica: Give the command

ParametricPlot[{2 t+1, t^2–1}, {t, 0, 3}]

Derive:

1) Author the statement

   [2 t+1, t^2–1]
2) Choose Plot from the algebra menu, and then choose Scale from the plot menu. Set both the x- and y-scales to 4 for a reasonable graph window.
3) Choose Plot from the plot menu. Choose Min to be 0 and Max to be 3 (the minimum and maximum t-values).

Section 7.5

- **(p.430) Creating a Loop to Calculate Points Based on Equations**

Graphing Calculators

The TI-82 and TI-85 programs for generating numerical and graphical solutions to first order differential equations $y' = f(y,t)$, $y(0) = y_0$ using Euler's method are given below. To use these programs, first store the right-hand-side of the differential equation in Y_1 (TI-82) or y1 (TI-85). For Example 2 of section 7.5 you would store 0.05Y(500–Y)/500. Then set the graph window appropriately. For Example 2 of section 7.5 set the window to $0 \le x \le 200$, $0 \le y \le 600$. Now run the program; T1 and T2 represent the beginning and ending t-values for the solution, H is the step size, and Y0 is the initial y-value. Thus for Example 2 of section 7.5 you would use 0 for T1, 200 for T2, 10, 1, or 0.1 for H, and 20 for Y0.

The graph of the numerical solution is displayed, and the last y-value is displayed. Thus for example 2 of section 7.5 we would get the graph shown in Figure 3 in the text and the values shown in Figure 4. In order to get the values corresponding to $y(100)$ rather than $y(200)$, use 100 for T2.

TI-82	TI-85
Disp "T1"	Disp "T1"
Input T	Input T
Disp "T2"	Disp "T2"
Input S	Input S
Disp "H"	Disp "H"
Input H	Input H
Disp "Y0"	Disp "Y0"
Input Y	Input Y
Lbl A	Lbl A
Y1*H+Y→V	y1*H+Y→V
Line(T, Y, T+H,V)	Line(T, Y, T+H,V)
V→Y	V→Y
T+H→T	T+H→T
If T<S—0.5H	If T<S—0.5H
Go to A	Go to A
Disp Y	Disp Y

Computer Algebra Systems

<u>Mathematica</u>:
First define the function euler by

```
euler[f_,t0_, y0_,h_, n_]:= Block[{g},g[{t_,y_}]:=
        {t+h,N[h f[t,y]+y]}; NestList[g,{t0,y0},n]]
```

where f is a function which represents the right-hand-side of the differential equation $y' = f(y,t)$, t0 is the initial t-value, y0 is the initial y-value, h is the step size, and n is the number of points generated. Thus t0+nh would be the final t-value generated in the solution.

Next, define the function f, which represents the right-hand-side of the differential equation $y' = f(y,t)$. For Example 2 of Section 7.5 this would be

```
f[t_,y_]:=0.05y(500–y)/500
```

Now generate the points which represent the solution. For Example 2 of Section 7.5 the inputs would be 0 for t0, 20 for y0, 10 for h and 20 for n (so that the final t-value is 0+(10)(20)=200). The command would be

```
points=euler[f,0,20,10,20]
```

Finally we plot the points to generate the graph as in the text with

```
p1=ListPlot[points, PlotJoined->True]
```

You could then generate another solution for a smaller h value and call that plot p2, and use the command Show[p1,p2] to combine the two plots.

Section 9.1

- **(p.525) Iterating Functions**

For Lab 17 following Section 9.1, you will need to be able to iterate a function and graph the results as a function of iteration number. This technique may also be helpful for some of the problems from Section 9.1.

TI-85 calculator:

To use the program below, store the function to be iterated in y1 using x as the iteration variable. You are prompted for the initial value of x, called x0, and for the number of iterations N. Thus to iterate the function $f(x)=x+\frac{k}{L}x(L-x)$ for 100 iterations using k=0.1, L=140 and using the initial x value of x0=30 as in Lab 17 you would store x+0.1/140*x(140–x) in y1, run the program, and input 30 for x0 and 100 for N.

The results of the iteration are stored in the lists xStat and yStat, with iteration number in xStat and the values for x in yStat. These values can then be inspected using STAT EDIT or can be plotted using STAT DRAW xyLINE (connected points) or STAT DRAW SCAT (point not connected).

```
Prompt x0
Prompt N
0→T
0→xStat(1)
x0→yStat(1)
x0→x
For(I,2,N+1)
y1→x
x→yStat(I)
T+1→T
T→xStat(I)
END
```

TI-82 calculator:

The TI-82 has built in function iteration and graphing. Press the MODE key and choose Seq from the fourth line of the screen. The iteration variable is referred to as U_n. To iterate the function $f(x) = x + \frac{k}{L}x(L-x)$ from Lab 17 using k=0.1 and L=140, press the Y= key and type in $U_{n-1} + 0.1/140 * U_{n-1}(140 - U_{n-1})$ for U_n. To use the initial value x0=30 and to iterate 100 times as in the problem from Lab 17, set the window by pressing the WINDOW key and using the values

U_nStart=30
V_nStart=1
nStart=0
nMin=0
nMax=100
Xmin=0
Xmax=100
Xscl=10
Ymin=0
Ymax=200
Yscl=20

To see the graph, press the GRAPH key. To see numerical values, either use the TRACE or press 2nd TABLE. Note that the value of V_nStart is not actually used in the problem in Lab 17 (it is used for iterating systems of equations).

Mathematica:

To iterate the function $f(x) = x + \frac{k}{L}x(L-x)$ from Lab 17 using k=0.1 and L=140, we first define a function $h(n,x) = (n+1, f(x))$ which when iterated gives us the values of both the iteration number n and the iteration variable x.

```
Clear[h];
h[{n_,x_}]:={n+1,x+0.1/140*x(140-x)}//N
```

We next iterate this function 100 times using initial values of n=0 and x0=30.

```
points=NestList[h,{0,30},100]
```

We can now inspect the values in the list or plot the points using

ListPlot[points,PlotRange->All]

or

ListPlot[points,PlotRange->All,PlotJoined->True]

depending on whether we want to connect the points or not.

Derive:

To iterate the function $f(x)=x+\frac{k}{L}x(L-x)$ from Lab 17 using k=0.1 and L=140, we first define the function $F(x)$ as above, and then define a function $H(v)$ which when iterated gives us the values of both the iteration number n and the iteration variable x (v will be the ordered pair (n, x)).

```
F(x) := x+0.1/140* x (140 – x)
G(v) := [ELEMENT(v, 1) + 1, F(ELEMENT(v, 2))]
```

Next we iterate G(v) 100 times using the ITERATES command with initial ordered pair (0, 30) for (n, x).

```
ITERATES(G(v), v, [0, 30], 100)
```

Now give the approXimate command and the results can be inspected using right arrow key, or plotted using the Plot command. The graph window can be set with center at (50, 100) (use Move to move the cross to (50, 100) and then use Center) and x- and y- scales set to 25 and 50 respectively.

Section 9.2

- **(p.532) Summing Series Numerically**

TI-82 and TI-85:
The sum and seq commands can be combined to numerically sum a finite series. On the TI-82, press 2nd LIST MATH 5 followed by 2nd LIST 5 to get the sum and seq commands. On the TI-85 press 2nd LIST OPS MORE F1 followed by 2nd LIST OPS MORE F3 to get the sum and seq commands.

In order to find the approximation $\sum_{k=0}^{10}\frac{3}{2^k}$ to the infinite series $\sum_{k=0}^{\infty}\frac{3}{2^k}$ as in part 1 of Example 7 from Section 9.2, you would input

```
sum seq(3/2^K,K,0,10,1)
```

which results in the value 5.9970703125 (TI-85) or 5.997070313 (TI-82) as in Figure 3 of Section 9.2 in the text.

Mathematica:

In order to find the approximation $\sum_{k=0}^{10} \frac{3}{2^k}$ to the infinite series $\sum_{k=0}^{\infty} \frac{3}{2^k}$ as in part 1 of Example 7 from Section 9.2, you would give the command

```
Sum[3/2^k,{k,0,10}]//N
```

for which the result is 5.99707 as in Figure 3 of Section 9.2 in the text.

Derive:

In order to find the approximation $\sum_{k=0}^{10} \frac{3}{2^k}$ to the infinite series $\sum_{k=0}^{\infty} \frac{3}{2^k}$ as in part 1 of Example 7 from Section 9.2, you would Author

```
sum(3/2^k,k,0,10)
```

and then approXimate the result to get the value 5.99707 as in Figure 3 of Section 9.2 in the text.